# 46 Springer Series in Chemical Physics
Edited by Fritz Peter Schäfer

# Springer Series in Chemical Physics
Editors: Vitalii I. Goldanskii   Fritz P. Schäfer   J. Peter Toennies

Volume 40 **High-Resolution Spectroscopy of Transient Molecules**
Editor: E. Hirota

Volume 41 **High Resolution Spectral Atlas of Nitrogen Dioxide 559–597 nm**
By K. Uehara and H. Sasada

Volume 42 **Antennas and Reaction Centers of Photosynthetic Bacteria**
Structure, Interactions, and Dynamics
Editor: M. E. Michel-Beyerle

Volume 43 **The Atom-Atom Potential Method for Organic Molecular Solids**
By A. J. Pertsin and A. I. Kitaigorodsky

Volume 44 **Secondary Ion Mass Spectrometry SIMS V**
Editors: A. Benninghoven, R. J. Colton, D. S. Simons, and H. W. Werner

Volume 45 **Thermotropic Liquid Crystals, Fundamentals**
By G. Vertogen and W. H. de Yeu

Volume 46 **Ultrafast Phenomena V**
Editors: G. R. Fleming and A. E. Siegman

Volumes 1–39 are listed on the back inside cover

# Ultrafast Phenomena V

Proceedings of the Fifth OSA Topical Meeting
Snowmass, Colorado, June 16–19, 1986

Editors: G. R. Fleming and A. E. Siegman

With 427 Figures

Springer-Verlag Berlin Heidelberg New York
London Paris Tokyo

Professor Graham R. Fleming
Department of Chemistry, The University of Chicago,
Chicago, IL 60637, USA

Professor Anthony E. Siegman
Electrical Engineering, E.L. Gintzton Laboratory,
Stanford University, Stanford, CA 94305, USA

*Series Editors*

Professor Dr. Fritz Peter Schäfer
Max-Planck-Institut für
Biophysikalische Chemie
D-3400 Göttingen-Nikolausberg
Fed. Rep. of Germany

Professor Vitalii I. Goldanskii
Institute of Chemical Physics
Academy of Sciences
Kosygin Street 4
Moscow V-334, USSR

Professor Dr. J. Peter Toennies
Max-Planck-Institut für Strömungsforschung
Böttingerstraße 6–8
D-3400 Göttingen
Fed. Rep. of Germany

ISBN 3-540-17077-4 Springer-Verlag Berlin Heidelberg New York
ISBN 0-387-17077-4 Springer-Verlag New York Berlin Heidelberg

Library of Congress Cataloging-in-Publication Data. OSA Topical Meeting (5th : 1986 : Snowmass, Colo.) Ultrafast phenomena V. (Springer series in chemical physics ; v. 46). Includes index. 1. Laser pulses, Ultrashort–Congresses. 2. Picosecond pulses–Congresses. I. Fleming, Graham R. II. Siegman, A.E. III. Title. IV. Series. QC689.5.L37073 1986 535.5'8 86-24871

This work is subject to copyright. All rights are reserved, whether the whole or part of the material is concerned, specifically those of translation, reprinting, reuse of illustrations, broadcasting, reproduction by photocopying machine or similar means, and storage in data banks. Under § 54 of the German Copyright Law where copies are made for other than private use, a fee is payable to "Verwertungsgesellschaft Wort", Munich.

© Springer-Verlag Berlin Heidelberg 1986
Printed in Germany

The use of registered names, trademarks, etc. in this publication does not imply, even in the absence of a specific statement, that such names are exempt from the relevant protective laws and regulations and therefore free for general use.

Offsetprinting: Druckhaus Beltz, 6944 Hemsbach/Bergstr. Bookbinding: J. Schäffer OHG, 6718 Grünstadt
2153/3150-543210

# Preface

The first Optical Society of America (OSA) Topical Meeting on Picosecond Phenomena, held at Hilton Head, South Carolina, in 1978, brought together in a congenial setting an interdisciplinary group of laser engineers and physicists who were exploring the emerging technologies for generating and applying picosecond optical pulses, together with scientists from the fields of chemistry, physics, biology, and electronics who saw in those pulses capabilities for studying atomic and molecular phenomena on time scales previously unrealizable.

The technology in this field has since developed even more rapidly and remarkably than foreseen eight years ago, and the applications to science and technology, in physics, chemistry, biology, electronics, and communications, have proven to be equally extraordinary. Optical pulses with pulse widths shorter than 10 femtosecond – only a few optical cycles in duration – along with monocycle infrared pulses, complex nonlinear optical solitons, electrooptic techniques with subpicosecond time resolutions, and a full toolkit of measurement and detection techniques have now emerged, including new methods for making ultrafast measurements in some cases even without ultrafast optical pulses. These tools are now being widely applied to study the internal motions of complex molecules and atomic lattices, the relaxation times of superheated electrons in solids, the ultrafast dynamics of chemical reactions, the excited-state lifetimes of photosynthetic and visual pigments, and the response times of the fastest electronic circuits yet developed.

This volume records the technical program presented at the fifth in this series of meetings – now renamed the Topical Meetings on Ultrafast Phenomena – which was held in Snowmass, Colorado, June 16–19, 1986, under the sponsorship of the Optical Society of America. The papers in this volume will document the high quality of the technical material presented as well as the remarkable advances made both in the ultrafast optical technologies themselves and in the fundamental science accomplished using these technologies. What the printed pages can, unfortunately, only indirectly convey is the remarkable stimulation and exchange of ideas that took place among the participants at the meeting.

Much appreciation is owed to all those who contributed to the success of this conference – most notably Joan Carlisle and her colleagues from the Optical Society of America for their irreplaceable assistance in organizing and managing the meeting, and all the past chairmen and members of the program committee for their advice and counsel. The scientific interchange so well demonstrated in this volume would not have been possible without the primary financial support for this meeting provided by a grant jointly from the Air Force Office of Scientific Research and the National Science Foundation. Additional generous support from industrial sources, including Amoco Corporation, Coherent Inc., E.I. du Pont de Nemours & Company, Exxon Research and Engineering, Hamamatsu Photonic Systems, Hughes Research Laboratories, Lambda Physik, Marco Scientific, Newport Corporation, Quantronix, Spectra-Physics, and Standard Oil Company (Ohio), contributed greatly to the success of the meeting.

Chicago, Stanford, USA  G.R. Fleming
July 1986  A.E. Siegman

# Contents

| Part I | Mode Locking and Ultrashort Pulse Generation |
|---|---|

Passive and Hybrid Femtosecond Operation of a Linear
Astigmatism Compensated Dye Laser
By J.-C. Diels, N. Jamasbi, and L. Sarger (With 1 Figure) ......... 2

Generation of 55-fs Pulses and Variable Spectral Windowing in a
Linear-Cavity Synchronously Pumped cw Dye Laser
By M.D. Dawson, T.F. Boggess, D.W. Garvey, and A.L. Smirl
(With 4 Figures) ..................................................... 5

Cavity-Mirror Dispersion Dependence of Pulse Duration Generated
from a Simple CPM Laser: An Experimental Study
By M. Yamashita, K. Torizuka, T. Sato, and M. Ishikawa
(With 2 Figures) ..................................................... 8

Femtosecond Pulse Generation from Passively Mode Locked
Continuous Wave Dye Lasers 550–700 nm
By P.M.W. French and J.R. Taylor (With 3 Figures) ............... 11

Stabilisation of a CPM Dye Laser Synchronously Pumped by a
Frequency Doubled ML YAG Laser
By J. Chesnoy and L. Fini (With 3 Figures) ........................ 14

Fluctuations and Chirp in Colliding-Pulse Mode-Locked
Dye Lasers. By D. Kühlke, T. Bonkhofer, U. Herpers,
and D. von der Linde (With 2 Figures) ............................. 17

Experimental Observation of High Order Solitons in a Colliding
Pulse Mode-Locked Laser
By F. Salin, P. Grangier, G. Roger, and A. Brun (With 4 Figures)  20

Advances in the Theory of Mode-Locking by Synchronous
Pumping. By G.H.C. New and J.M. Catherall (With 3 Figures) ...  24

Collective Modes - An Analytical Model for Active Mode Locking
in the Transient Case
By P. Aechtner, P. Heinz, and A. Laubereau (With 2 Figures) .....  27

Generation of Picosecond Pulses from a Continuous Wave
Neodymium:Phosphate Glass Laser
By L. Yan, J.D. Ling, P.-T. Ho, and C.H. Lee (With 3 Figures) ....    30

## Part II    Ultrafast Optical Generation and Measurement Techniques

Fourier Transform Picosecond Pulse Shaping and Spectral Phase
Measurement in a Grating Pulse-Compressor. By J.P. Heritage,
A.M. Weiner, and R.N. Thurston (With 4 Figures) ................    34

Picosecond Pulse Amplification Using Pulse Compression
Techniques. By D. Strickland, P. Maine, M. Bouvier,
S. Williamson, and G. Mourou (With 4 Figures) ....................    38

New Optical Design for a Jet Amplifier. By O. Seddiki, A. Goddi,
R. Mounet, J.-F. Morhange, and C. Hirlimann (With 2 Figures) ...    43

Fiber Raman Amplification Soliton Laser (FRASL)
By M.N. Islam, L.F. Mollenauer, and R.H. Stolen (With 4 Figures)    46

Dispersion Compensated Fiber Raman Oscillator
By J.D. Kafka, D.F. Head, and T. Baer (With 2 Figures) ..........    51

80-fs Soliton-like Pulses from an Optical Nonlinear Fiber
Resonator. By B. Zysset, P. Beaud, W. Hodel, and H.P. Weber
(With 4 Figures) ......................................................    54

The Stabilized Soliton Laser
By F.M. Mitschke and L.F. Mollenauer (With 4 Figures) ..........    58

The Soliton Self Frequency Shift. By F.M. Mitschke,
L.F. Mollenauer, and J.P. Gordon (With 2 Figures) ................    62

Solitons at the Zero Dispersion Wavelength of Single-Mode Fibers
By P.K.A. Wai, C.R. Menyuk, H.H. Chen, and Y.C. Lee
(With 1 Figure) ......................................................    65

Active Mode-Locking of an InGaAsP Optical-Fiber Ring Laser
By G. Eisenstein, R.M. Jopson, M.S. Whalen, K.L. Hall,
and G. Raybon (With 4 Figures) ......................................    68

Femtosecond Resolved Fluorescence
By W. Rudolph and J.-C. Diels (With 2 Figures) ....................    71

Parametric Amplification Sampling Spectroscopy (PASS): A
New Technique for Resolving Near-Infrared Luminescence on a
Subpicosecond Time Scale. By D. Hulin, A. Migus, A. Antonetti,
I. Ledoux, J. Badan, and J. Zyss (With 3 Figures) ................    75

Measurement of Optical Phase with Subpicosecond Resolution
by Time Domain Interferometry. By J.E. Rothenberg
(With 6 Figures) ...................................................... 78

Real Time Picosecond Optical Oscilloscope
By J.A. Valdmanis (With 8 Figures) ................................ 82

Beam Overlap for Long Delay Lines Using Active Feedback
By C. Doland, W.B. Jackson, and A. Andersson (With 3 Figures) . 86

Ultrashort Dye Laser Pulses Using the Sweeping Oscillator
Method. By Y.H. Meyer, M.M. Martin, E. Bréhéret,
and O. Benoist d'Azy (With 4 Figures) ............................ 89

An Investigation on Ultrashort Light Pulse Generation by
Travelling-Wave Amplified Spontaneous Emission
By W. Lee, C. Ning, Z. Huang, and W. Wang (With 5 Figures) ... 92

## Part III  Electrooptic Sampling Techniques

Electrooptic Sampling of Gallium Arsenide Integrated Circuits
By K.J. Weingarten, M.J.W. Rodwell, J.L. Freeman,
S.K. Diamond, and D.M. Bloom (With 6 Figures) ................... 98

Picosecond Characterization of Ultrafast Phenomena: New
Devices and New Techniques. By D.R. Dykaar, R. Sobolewski,
J.F. Whitaker, T.Y. Hsiang, G.A. Mourou, M.A. Hollis,
B.J. Clifton, K.B. Nichols, C.O. Bozler, and R.A. Murphy
(With 4 Figures) ...................................................... 103

Precise Measurement of Signal Propagation Characteristics in
GaAs Integrated Circuits by Picosecond Electro-Optic Sampling
By R.K. Jain, X.-C. Zhang, M.G. Ressl, and T.J. Pier
(With 2 Figures) ...................................................... 107

Propagation of Ultrashort Electrical Pulses on Superconducting
Transmission Lines
By I.N. Duling III, C.-C. Chi, W.J. Gallagher, D. Grischkowsky,
N.J. Halas, M.B. Ketchen, and A.W. Kleinsasser (With 4 Figures)  110

High Repetition Rate Electro-Optic Sampling with an Injection
Laser. By A.J. Taylor, R.S. Tucker, J.M. Wiesenfeld, G. Eisenstein,
and C.A. Burrus (With 5 Figures) ................................... 114

Picosecond Optoelectronic Sampling of Electrical Waveforms
Produced by an Optically Excited Field Effect Transistor
By D.E. Cooper and S.C. Moss (With 1 Figure) .................... 117

Picosecond Electrical Pulses in Microelectronics. By P.G. May,
G.P. Li, J.-M. Halbout, M.B. Ketchen, C.-C. Chi, M. Scheuermann,
I.N. Duling III, D. Grischkowsky, and M. Smyth (With 3 Figures) . 120

High Speed Circuit Measurements Using Photoemission Sampling
By J. Bokor, A.M. Johnson, R.H. Storz, and W.M. Simpson
(With 2 Figures) ...................................................... 123

Photoemissive Sampling of Picosecond Electrical Waveforms
By A.M. Weiner, R.B. Marcus, P.S.D. Lin, and J.H. Abeles
(With 5 Figures) ...................................................... 127

Nonlinear Responses of Picosecond Photodetectors to
Photogenerated Carriers
By T.F. Carruthers and J.F. Weller (With 3 Figures) ............... 131

Direct Generation of Picosecond to Subpicosecond Optical Pulses
Using Electrooptic Modulation Methods
By T. Kobayashi, A. Morimoto, T. Fujita, K. Amano, T. Uemura,
and T. Sueta (With 6 Figures) ........................................ 134

Elimination of Dynamic Flash in a Picosecond Streak Image Tube
By Huanwen Zhang (With 1 Figure) ................................. 137

## Part IV  Nonlinear Optics and Continuum Generation

Parametric Chirp Reversal and Enhancement: Application
in Femtosecond Optics. By A. Piskarskas, D. Podenas,
A. Stabinis, A. Umbrasas, A. Varanavichius, A. Yankauskas,
and G. Yonushauskas (With 7 Figures) ............................. 142

Supercontinuum Generation in Gases: A High Order
Nonlinear Optics Phenomenon. By P.B. Corkum, C. Rolland,
and T. Srinivasan-Rao (With 3 Figures) ............................ 149

New Excitation and Probe Continuum Sources for Subpicosecond
Absorption Spectroscopy
By J.H. Glownia, J. Misewich, and P.P. Sorokin (With 2 Figures) . 153

Induced Phase Modulation and Spectral Broadening of a Weak
530-nm Picosecond Pulse by an Intense 1060-nm Picosecond Pulse
in Glass. By R.R. Alfano, Q.X. Li, T. Jimbo, J.T. Manassah,
and P.P. Ho (With 2 Figures) ........................................ 157

The Observation of Chirped Stimulated Raman Scattered Light
in Fibers. By A.M. Johnson, R.H. Stolen, and W.M. Simpson
(With 3 Figures) ...................................................... 160

Observation of 7.2-THz Beats Between the D-Lines of Atomic Rb
By J.E. Golub and T.W. Mossberg (With 1 Figure) ................. 164

Coherent Multiphoton Resonant Interaction and Harmonic
Generation. By A. Mukherjee, N. Mukherjee, J.-C. Diels,
and G. Arzumanyan (With 4 Figures) .............................. 166

Ultrafast Chaos from Semiconductor Lasers. By Y. Cho,
T. Umeda, I. Jun Cha, M. Koishi, and M. Miwa (With 5 Figures) . 169

---

| Part V | Applications to Semiconductors, Quantum Wells, and Solid State Physics |
|---|---|

Thermodynamics and Kinetics of Melting, Evaporation and
Crystallization, Induced by Picosecond Pulsed Laser Irradiation
By F. Spaepen (With 1 Figure) ....................................... 174

Investigation of Nonthermal Population Distributions with 10-fs
Optical Pulses. By C.V. Shank, R.L. Fork, C.H. Brito Cruz,
and W. Knox (With 3 Figures) ........................................ 179

Superheating During Ultrafast Laser Heating of Semiconductors
By D. von der Linde, N. Fabricius, B. Danielzik, and P. Hermes
(With 4 Figures) .................................................... 182

Non-equilibrium Carriers in GaAs: Secondary Emission During the
First Two Picoseconds
By J.A. Kash and J.C. Tsang (With 5 Figures) ...................... 188

Ultrafast Carrier Dynamics in GaAs and $Al_xGa_{1-x}As$
By W.Z. Lin, J.G. Fujimoto, E.P. Ippen, and R.A. Logan
(With 4 Figures) .................................................... 193

Subpicosecond Optical Non-linearities in GaAs Multiple-Quantum-
Well Structures. By D. Hulin, A. Antonetti, A. Migus,
A. Mysyrowicz, H.M. Gibbs, N. Peyghambarian, W.T. Masselink,
and H. Morkoç (With 5 Figures) ...................................... 197

Picosecond Relaxation of Nonthermal Wannier Excitons in GaAs
By L. Schultheis, J. Kuhl, A. Honold, and C.W. Tu
(With 1 Figure) ..................................................... 201

Picosecond Observation of the Photorefractive Effect in GaAs
By A.L. Smirl, G.C. Valley, M.B. Klein, K. Bohnert,
and T.F. Boggess (With 2 Figures) ................................... 203

Time-Resolved Photoluminescence Measurements in $Al_xGa_{1-x}As$
Under Intense Picosecond Excitation
By K. Bohnert, H. Kalt, D.P. Norwood, T.F. Boggess, A.L. Smirl,
and R.Y. Loo (With 2 Figures) ........................................ 207

Picosecond Excite-Probe and Transient Grating Studies of
$Ga_xIn_{1-x}As_yP_{1-y}$. By R.J. Manning, A. Miller, A.M. Fox,
and J.H. Marsh (With 2 Figures) ...................................... 210

Ultrafast Dynamics in GaAlAs Diode Laser Amplifiers
By M.S. Stix, M.P. Kesler, and E.P. Ippen (With 8 Figures) ....... 213

Electronic Energy Relaxation and Localization in Two II-VI
Compound Semiconductor Quantum Well Structures
By Y. Hefetz, W.C. Goltsos, D. Lee, and A.V. Nurmikko
(With 6 Figures) ........................................................ 218

Transient Raman Scattering in Multiple Quantum Well Structures
By D.Y. Oberli, D.R. Wake, M.V. Klein, J. Klem, and H. Morkoç
(With 2 Figures) ........................................................ 223

Fast Energy Relaxation of Hot Electrons in Bulk GaAs and Multi-
Quantum Wells. By C.H. Yang and S.A. Lyon (With 2 Figures) ... 227

Picosecond Photoluminescence and Energy-Loss Rates in GaAs
Quantum Wells Under High-Density Excitation
By T. Kobayashi, H. Uchiki, Y. Arakawa, and H. Sakaki
(With 4 Figures) ........................................................ 231

Broad Tuning of the Photoluminescence Energy and Lifetime
by the Quantum-Confined Stark Effect. By H.-J. Polland,
L. Schultheis, J. Kuhl, E.O. Göbel, and C.W. Tu (With 2 Figures)   234

Auger Heating of Silicon-on-Sapphire by Femtosecond Optical
Pulses. By M.C. Downer and C.V. Shank (With 2 Figures) ........ 238

The Origin of Picosecond Photoinduced Absorption Decays in
Hydrogenated Amorphous Silicon
By W.B. Jackson, C. Doland, and C.C. Tsai (With 2 Figures) ..... 242

Picosecond Decay of Photoinduced Absorption in Hydrogenated
Amorphous Silicon
By D.M. Roberts and T.L. Gustafson (With 2 Figures) ............. 245

Femtosecond Spectroscopy of Hot Carriers in Germanium
By P.M. Fauchet, D. Hulin, G. Hamoniaux, A. Orszag, J. Kolodzey,
and S. Wagner (With 3 Figures) ...................................... 248

Spin Dephasing Kinetics of Free Carriers in Alloy Semimagnetic
Semiconductors $Cd_{1-x}Mn_xSe$ by One and Two Photon Excitation
By M.R. Junnarkar and R.R. Alfano (With 1 Figure) ............... 251

Detection of Higher Order Fourier Components of Index Gratings
in Picosecond Transient Grating Experiments
By E.O. Göbel and H. Saito (With 3 Figures) ...................... 254

Transient Thermoreflectance Studies of Thermal Transport
in Compositionally Modulated Metal Films. By G.L. Eesley,
C.A. Paddock, and B.M. Clemens (With 2 Figures) ................. 257

Femtosecond Studies of Nonequilibrium Electronic Processes
in Metals. By R.W. Schoenlein, W.Z. Lin, J.G. Fujimoto,
and G.L. Eesley (With 4 Figures) ................................. 260

Time-Resolved Observation of Electron-Phonon Relaxation
During Femtosecond Laser Heating of Copper. By H. Elsayed-Ali,
M. Pessot, T. Norris, and G. Mourou (With 2 Figures) ............. 264

Femtosecond Carrier Relaxation in Semiconductor-Doped Glasses
By M.C. Nuss, W. Zinth, and W. Kaiser (With 2 Figures) .......... 267

Femtosecond Dynamics of Electron-Hole Plasma in Semiconductor
Microcrystallite Doped Glass. By G.R. Olbright, B.D. Fluegel,
S.W. Koch, and N. Peyghambarian (With 2 Figures) ............... 270

High-Contrast Ultrafast Phase Conjugation in Semiconductor-
Doped Glass. By D. Cotter (With 3 Figures) ...................... 274

Femtosecond Vibrational Relaxation of the $F_2^+$ Center in LiF
By W.H. Knox, L.F. Mollenauer, and R.L. Fork (With 2 Figures) . 277

Determination of the Rapid Quenching Rates of Excited State
F-Centers by $OH^-$ Defects in KCl. By Du-Jeon Jang,
T.C. Corcoran, M.A. El-Sayed, L. Gomes, and F. Luty
(With 4 Figures) ................................................. 280

Propagation of Coherent Phonon Polaritons in $LiTaO_3$ Measured
by FIR-Cherenkov-Pulses
By M.C. Nuss and D.H. Auston (With 3 Figures) .................... 284

---

## Part VI  Chemical Reaction Dynamics

Cages, Crossings and Correlations – Theoretical Perspectives on
Solution Reaction Dynamics. By J.T. Hynes ....................... 288

Polarity Dependent Barriers and the Photoisomerization Dynamics of Polar Molecules in Solution. By J.M. Hicks, M.T. Vandersall, E.V. Sitzmann, and K.B. Eisenthal (With 3 Figures) .............. 293

Dynamic Solvent Effects on Small Barrier Isomerizations
By P.F. Barbara and V. Nagarajan (With 3 Figures) .............. 299

Solvation Dynamics in Polar Liquids: Experiment and Simulation
By M. Maroncelli, E.W. Castner, Jr., S.P. Webb, and G.R. Fleming
(With 4 Figures) ...................................................... 303

Femtosecond Study of Electron Localization and Solvation in Pure Water. By Y. Gauduel, J.L. Martin, A. Migus, N. Yamada, and A. Antonetti (With 2 Figures) .................................... 308

Time-Dependent Fluorescence Shift in Alcoholic Solvents:
A Non-Debye Behaviour Related to Hydrogen Bonds
By C. Rullière, A. Declémy, and Ph. Kottis (With 4 Figures) ...... 312

Picosecond Dynamics of Proton-Anion Ion Pair Geminate Recombination. By D. Huppert and E. Pines (With 1 Figure) ..... 315

Excited State Proton Transfer in Matrix Isolated Water and Methanol Complexes of 2-Hydroxy-4,5-benzotropone and 3-Hydroxyflavone
By D.F. Kelley and G.A. Brucker (With 2 Figures) ................ 319

Detection of the Inverted Region in Photo-induced Intramolecular Electron Transfer. By R.J. Harrison, G.S. Beddard, J.A. Cowan, and J.K.M. Sanders (With 2 Figures) ............................... 322

Ultrafast Studies Designed to Test the Fundamental Statistical Assumptions Underlying Chemical Reactivity in Liquids
By C.B. Harris, J.K. Brown, M.E. Paige, D.E. Smith, and D.J. Russell (With 4 Figures) ..................................... 326

Geminate Recombination and Relaxation of Condensed Phase Molecular Halogens
By D.F. Kelley and N.A. Abul Haj (With 1 Figure) ................ 330

Fast Photochemical Processes of Aromatic Nitro Compounds in Solution. By B.B. Craig, S.K. Chattopadhyay, and J.C. Mialocq (With 4 Figures) ...................................................... 334

Cage Recombination and Unimolecular $\beta$-Scission Reactions of Sulfur Centered Free Radicals
By T.W. Scott and S.N. Liu (With 3 Figures) ...................... 338

The Influence of Friction and Deuteration on Stilbene
Isomerization. By S.H. Courtney, M.W. Balk, S. Canonica,
S.K. Kim, and G.R. Fleming (With 3 Figures) ....................... 341

Kramers-Hubbard Approach to the Solvent Dependence of
Isomerization. By M. Lee and R.M. Hochstrasser (With 3 Figures) 344

Photoisomerization Studies of Substituted Stilbenes: 4,4'-
Dihydroxystilbene and 4,4'-Dimethoxystilbene
By D.M. Zeglinski and D.H. Waldeck (With 2 Figures) ............. 347

Picosecond Studies of Barrierless Torsional Diffusion
By D. Ben-Amotz and C.B. Harris (With 2 Figures) ................ 350

Time-Resolved Fluorescence Spectra of Ethidium Bromide
By J.H. Sommer, T.M. Nordlund, M. McGuire, and G. McLendon
(With 3 Figures) ...................................................... 353

Picosecond and Femtosecond Molecular Beam Chemistry:
Coherence and Fragment Recoil Dynamics
By A.H. Zewail (With 4 Figures) ..................................... 356

Picosecond Laser Study of the Collisionless UV Photodissociation
of Energetic Materials
By J.-C. Mialocq and J.C. Stephenson (With 3 Figures) ............ 362

Experimental Study of Harmonic Generation with Picosecond
248 nm Radiation. By T.S. Luk, A. McPherson, H. Jara,
U. Johann, I.A. McIntyre, A.P. Schwarzenbach, K. Boyer,
and C.K. Rhodes (With 2 Figures) ................................... 366

Time-Resolved Measurement of Laser-Induced Desorption
of a Molecular Monolayer. By G. Arjavalingam, T.F. Heinz,
and J.H. Glownia (With 1 Figure) ................................... 370

## Part VII   Dynamics of Biological Processes

Picosecond Electron Transfer and Stimulated Emission in
Reaction Centers of *Rhodobacter sphaeroides* and *Chloroflexus
aurantiacus*. By M. Becker, D. Middendorf, N.W. Woodbury,
W.W. Parson, and R.E. Blankenship (With 4 Figures) ............. 374

Femtosecond Spectroscopy of the Primary Events of Bacterial
Photosynthesis
By W. Zinth, J. Dobler, and W. Kaiser (With 3 Figures) .......... 379

An Accumulated Photon Echo Study of Sub-picosecond Processes
in Photosynthetic Reaction Centers
By S.R. Meech, A.J. Hoff, and D.A. Wiersma (With 2 Figures) .... 384

Ultrafast Electron and Energy Transfer in Reaction Center and
Antenna Proteins from Photosynthetic Bacteria
By M.R. Wasielewski, D.M. Tiede, and H.A. Frank
(With 5 Figures) .................................................... 388

Femtosecond Spectroscopy of Excitation Energy Transfer
and Initial Charge Separation in the Reaction Center of the
Photosynthetic Bacterium *Rhodopseudomonas sphaeroides*
By J. Breton, J.-L. Martin, A. Migus, A. Antonetti, and A. Orszag
(With 4 Figures) .................................................... 393

Picosecond Transient Absorption Spectroscopy of Green Plant
Photosystem I Reaction Centres
By B.L. Gore, L.B. Giorgi, and G. Porter (With 1 Figure) ......... 398

Femtosecond-Pulse Spectroscopy of Primary Photoprocesses in
Reaction Centers of *Rhodopseudomonas sphaeroides* R-26
By S.V. Chekalin, Yu.A. Matveets, and A.P. Yartsev
(With 3 Figures) .................................................... 402

Detergent Effects upon the Picosecond Dynamics of Higher Plant
Light Harvesting Chlorophyll Complex (LHC). By J.P. Ide,
D.R. Klug, W. Kuhlbrandt, G. Porter, and J. Barber
(With 1 Figure) ..................................................... 406

Picosecond Conformational Intermediates in the Bacteriorhodopsin
Photocycle. By G.H. Atkinson, T.L. Brack, D. Blanchard,
G. Rumbles, and L. Siemankowski (With 3 Figures) ................. 409

Electron Transfer and Rapid Restricted Motion in Homologous
Azurins. By J.W. Petrich, J.W. Longworth, and G.R. Fleming
(With 3 Figures) .................................................... 413

Primary Process of Vision: Hypsorhodopsin
By T. Kobayashi, H. Ohtani, and M. Tsuda (With 3 Figures) ...... 416

Reactivity and Dynamics of Hemeproteins in the Femtosecond and
Picosecond Time Domains. By D. Houde, J.W. Petrich, O.L Rojas,
C. Poyart, A. Antonetti, and J.L. Martin (With 3 Figures) ......... 419

Picosecond Raman Hole Burning as a Probe of Conformational
Heterogeneity: Applications to Oxyhemoglobin
By B.F. Campbell and J.M. Friedman (With 4 Figures) ............. 423

Ultrafast Studies of Nitrosylmyoglobin. By K.A. Jongeward,
J.C. Marsters, and D. Magde (With 4 Figures) ..................... 427

Molecular Dynamics Study of Vibrational Cooling in
Optically Excited Hemeproteins. By E.R. Henry, W.A. Eaton,
and R.M. Hochstrasser (With 1 Figure) ............................. 430

Chemical Reaction in a Glassy Matrix: Dynamics of Ligand
Binding to Protoheme in Glycerol:Water. By J.R. Hill, M.J. Cote,
D.D. Dlott, J.F. Kauffman, J.D. McDonald, P.J. Steinbach,
J.R. Berendzen, and H. Frauenfelder (With 3 Figures) .............. 433

## Part VIII     Energy Transfer and Relaxation

Energy and Electron Transfer of Adsorbed Dyes on Molecular
Single Crystals and Other Substrates. By K. Kemnitz,
N. Nakashima, and K. Yoshihara (With 5 Figures) .................. 438

Optical Pump-Probe Spectroscopy of Dyes on Surfaces: Ground-
State Recovery of Rhodamine 640 on ZnO and Fused Silica
By P.A. Anfinrud, T.P. Causgrove, and W.S. Struve
(With 1 Figure) ...................................................... 442

Picosecond Fluorescence Spectroscopy on Molecular Association in
Langmuir-Blodgett Films
By I. Yamazaki, N. Tamai, and T. Yamazaki (With 3 Figures) ..... 444

Fluorescence Concentration Depolarization of DODCI in Glycerol:
A Photon-Counting Test of Three-Dimensional Excitation
Transport Theory
By D.E. Hart, P.A. Anfinrud, and W.S. Struve (With 1 Figure) ... 447

Fractal Behaviors in Two-Dimensional Excitation Energy Transfer
on Vesicle Surfaces. By N. Tamai, T. Yamazaki, I. Yamazaki,
and N. Mataga (With 3 Figures) ..................................... 449

Transient Vibrational Heating of Molecules After Internal
Conversion. By A. Seilmeier, U. Sukowski, W. Kaiser,
and S.F. Fischer (With 2 Figures) ................................... 454

Nonlinear Absorption Spectroscopy of Liquids with Ultrashort IR
Pulses
By H. Graener, R. Dohlus, and A. Laubereau (With 2 Figures) .... 458

Femtosecond Relaxation Dynamics of Large Organic Molecules
By M.J. Rosker, F.W. Wise, C.L. Tang, and A.J. Taylor
(With 4 Figures) ..................................................... 461

Population Lifetimes of $OH(v=1)$ and $OD(v=1)$ Vibrations
in Alcohols, Silanols and Crystalline Micas. By E.J. Heilweil,
M.P. Casassa, R.R. Cavanagh, and J.C. Stephenson ................ 465

$S_0$-$S_n$ Two-Photon Absorption Dynamics of Rhodamine Dyes
By P. Sperber, M. Weidner, and A. Penzkofer (With 3 Figures) .... 469

Nonlinear Optical Response of One-Dimensional Excitons in
Polydiacetylene. By B.I. Greene, J. Orenstein, R.R. Millard,
and L.R. Williams (With 2 Figures) ................................. 472

Picosecond Photoconductivity and Nonlinear Optical Phenomena
in *trans*-Polyacetylene
By D. Moses, M. Sinclair, and A.J. Heeger (With 1 Figure) ........ 475

Singlet Exciton Fusion in Molecular Solids
By R.R. Millard and B.I. Greene (With 2 Figures) ................... 478

Matrix Effect on Vibrational Relaxation in Molecular Crystals
By J.R. Hill, E.L. Chronister, J.C. Postlewaite, and D.D. Dlott
(With 1 Figure) ....................................................... 482

Optical Damage in Molecular Crystals: A Solid State Explosion
By D.D. Dlott, T.J. Kosic, and J.R. Hill (With 4 Figures) .......... 485

Rotational Relaxation of Free and Solvated Rotors
By A.J. Bain, C. Han, P.L. Holt, P.J. McCarthy, A.B. Myers,
M.A. Pereira, and R.M. Hochstrasser (With 5 Figures) ............. 489

Ultrafast Dynamics at the Interface: Probing the Transition from
Solution to Surface Interactions in Charged Micelles
By E.F. Templeton, K. Brinker, S. Paone, and G.A. Kenney-
Wallace (With 1 Figure) .............................................. 495

Shock Moderated Photophysics and Photochemistry at Multi-
kilobar Pressures
By B.L. Justus, A.L. Huston, and A.J. Campillo (With 5 Figures) . 499

## Part IX  Coherent Spectroscopic Techniques

Phase Grating Approach to Susceptibility Tensors: Determination
in Isotropic Media. By J. Etchepare, G. Grillon, I. Thomazeau,
G. Hamoniaux, and A. Orszag (With 4 Figures) .................... 504

Nonlinear Response Function for Four-Wave Mixing: Application
to Coherent Raman Lineshapes in Polyatomics and to the Optical
Anderson Transition
By S. Mukamel, Z. Deng, and R.F. Loring (With 2 Figures) ........ 510

Picosecond Laser Pulse Shaping and Phase Shifting for Molecular
Spectroscopy
By M. Haner, F. Spano, and W.S. Warren (With 2 Figures) ....... 514

Third-Order Nonlinear Optical Interactions in Thin Films of
Organic Polymers Investigated by Picosecond and Subpicosecond
Four-Wave Mixing. By P.N. Prasad, D. Narayana Rao,
J. Swiatkiewicz, P. Chopra, and S.K. Ghoshal ..................... 518

Picosecond Raman-Induced Phase Conjugation Spectroscopy
By R. Dorsinville, P. Delfyett, and R.R. Alfano (With 2 Figures) .. 521

Polarization Dependence of Time-Resolved CARS in Liquids
By N. Kohles and A. Laubereau (With 3 Figures) .................... 524

Direct Measurement of Wave-Vector-Dependent Polariton Energy Velocity and Dephasing in $NH_4Cl$
By G.M. Gale, F. Vallée, and C. Flytzanis (With 4 Figures) ....... 528

Impulsive Stimulated Rayleigh, Brillouin, and Raman Scattering: Experiments and Theory of Light Scattering Spectroscopy in the Time Domain. By M.R. Farrar, L.R. Williams, Yong-Xin Yan, Lap-Tak Cheng, and K.A. Nelson (With 4 Figures) ................. 532

Ultrafast Transient Spectroscopy with Broadband Non-Transform-Limited Light Sources
By T. Yajima and N. Morita (With 4 Figures) ...................... 536

Picosecond Dephasing Time Measurement by CSRS Using Temporally Incoherent Nanosecond Laser with Short Correlation Time
By T. Kobayashi, T. Hattori, and A. Terasaki (With 3 Figures) ... 541

Anomalous Pulse Duration Dependence of the Quasicontinuum Absorption Spectrum
By P. Mukherjee and H.S. Kwok (With 3 Figures) ................... 544

**Index of Contributors** .............................................. 549

# Part I

# Mode Locking and Ultrashort Pulse Generation

# Passive and Hybrid Femtosecond Operation of a Linear Astigmatism Compensated Dye Laser

*J.-C. Diels*[1], *N. Jamasbi*[1], *and L. Sarger*[2]

[1] Center for Applied Quantum Electronics, P.O. Box 5368,
North Texas State University, Denton, TX 76203, USA
[2] Centre de Physique Moléculaire Optique et Herzienne,
Université de Bordeaux I, F-33405 Talence, France

## 1. Introduction

An antiresonant (AR) ring has been proposed by Siegman [1] to obtain standing wave saturation in a linear mode-locked dye laser. This laser configuration is equivalent to that of a mode-locked ring laser if the saturable absorber is located at equal optical distance from the beam splitter (Fig. 1). Two technical problems seem to make this configuration impractical:
a) The requirement of tight focusing in the absorber [2, 3] implies a large beam size through the beam splitter, hence a large angle of the AR triangle, and a large astigmatism impeding tight focusing.
b) The "colliding pulse" condition requires that the two arms of the AR ring be made equal with an accuracy of 10 microns.

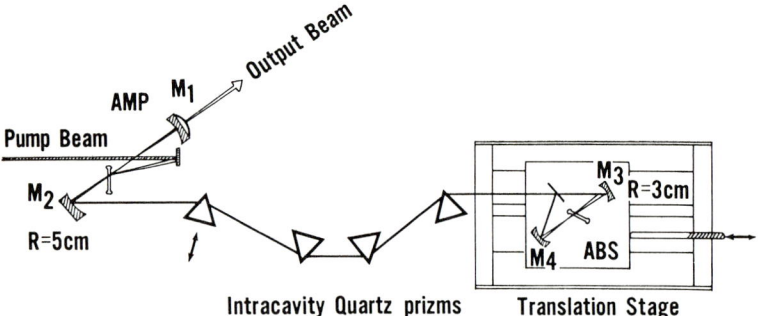

Figure 1: Sketch of the laser cavity.

We demonstrate that the astigmatism of the AR ring (even for angles as large as 16°) can be compensated by the geometry of the cavity. Furthermore, this cavity configuration enables us to test for the first time experimentally the "colliding pulse" condition, independently of any other parameter. It is shown that the accuracy requirement for the equality of the two arms of the AR ring is much less stringent than anticipated. The operation of the laser is not affected by a departure from the central position of less than 400 $\mu$m!

## 2. Cavity Configuration

The laser cavity is simple enough to make a complete analytical ABCD matrix analysis possible [4] and map the stability regions of the laser. We have determined cavity configurations for which a tight focusing (round spot of a few micron diameter) is achieved in the absorber jet. In the optimized cavity, a large angle at the amplifier end ($34°$) introduces an astigmatism compensating that at the absorber jet. All the parameters (angles, mirror distances, positions of the images of the fluorescence spots) of the laser having been predetermined, the laser alignment follows a systematic and straightforward procedure. Using a 0.7 mm thick beam splitter, a very small amount of positive dispersion is needed to compensate the small phase modulation in the strongly saturated absorber dye, resulting in a stable mode-locked train of pulses in a 50 fsec to 70 fsec range.

## 3. Colliding Pulse Effect

The "colliding pulse" effect can be investigated by changing the relative length of the two arms of the antiresonant ring. As the beam splitter is moved away from the position corresponding to the symmetric ring, the threshold pump power for short pulse operation remains constant in a 800 micron wide region. It is only for larger departures from the "equal arms" condition that this threshold power is seen to increase from 1.7 to 3.5 W. If the pump power is maintained at 3.5 W, none of the following parameters of the laser is affected by departing from the symmetric ring: the pulse output power, the intensity autocorrelation, and the losses of the antiresonant ring. Below the "threshold power for short pulse operation", the laser output consists in a train of long (psec) bursts of noise. It can be concluded that the "colliding pulse operation" is important in reducing the threshold for stable mode-locked operation, and in the formation stage of the steady state pulse from noise.

## 4. Determination of the Pulse Shape In Amplitude and Phase

We have demonstrated previously a method to determine the pulse shape in amplitude and phase through iterative fittings of the pulse spectrum, intensity and interferometric autocorrelations [5]. We developed an alternate method [6] in which a complete pulse description is extracted from a single measurement. As opposed to a technique proposed recently by ROTHENBERG [7], our methods do not require the availability of a **shorter** pulse than the one to be measured. The beam from the laser is sent through a standard autocorrelator in which a block of glass has been inserted (typically 5 cm of BK7 or SF5 glass). As the delay is being scanned, the second harmonic detection coupled to a minicomputer records the interferometric second-order crosscorrelation between the pulse broadened by dispersion through the glass and the (shorter) original laser pulse. Numerical averaging of this function yields the intensity crosscorrelation, which can be easily deconvoluted (given some coarse and noncritical assumptions on the short pulse $E_1$) to give the temporal profile - in amplitude only - of the pulse after glass $E_2$. Both $E_2$ and the rough estimate of $E_1$ are used to deconvolute the complete interferometric crosscorrelation, yielding the phase $\varphi_2$ of the pulse dispersed through glass. Since the dispersive properties of the glass are accurately documented, the complete characterization ($E_1$, $\varphi_1$) of the laser pulse can be calculated by computing the "reverse propagation" of ($E_2$, $\varphi_2$) through the glass. The method has been checked against the iterative fitting procedure [5]. Another simple verification is to compare the calculated and measured pulse spectrum.

The pulse from the passively mode-locked linear laser is found to have the same asymmetric pulse shape as that from the ring laser [5]. However, in contrast to the ring laser which could be adjusted to generate bandwidth-limited pulses, or pulses with a slight downchirp, even the shortest pulses from the antiresonant laser are found to have a significant residual phase modulation, as confirmed by an asymmetric spectrum. The frequency modulation is found to be a combination of a saturation induced chirp [2] at the leading edge of the pulse, followed by a Kerr-effect type modulation, superimposed on a linear chirp [6].

5. Hybrid Mode Locking

The cavity length of the laser can be adjusted continuously over a range of several cm, hereby providing a possibility of continuous tuning of the pulse repetition rate without affecting otherwise the mode-locked operation. It can therefore be very easily synchronously pumped. Hybrid operation of this laser exhibits the same pulse duration and stability as passive operation. An average output power of 100 mW is obtained, for a train of pulses of 60 fsec duration. The intensity autocorrelation of the pulses have the steep wings characteristic of the passively mode-locked lasers. The interferometric autocorrelation and pulse spectra are however drastically different, indicating a more symmetric shape and a reduced residual chirp [4].

6. Conclusion

In conclusion, we have successfully designed a laser cavity for which a straightforward, simple alignment procedure replaces the skills and artful expertise normally required for fsec lasers.

This work was supported by the National Science Foundation, under grant Nb ECS-8406985.

**REFERENCES.**

1. A. E. Siegman, IEEE J. Quantum Electron. **QE-9**, 247-250 (1973).
2. O. E. Martinez, R. L. Fork and J. P. Gordon,, Optics Letters **9**, 156-158 (1984); J.-C. Diels, W. Dietel, J. J. Fontaine, W. Rudolph and B. WIlhelmi, J. Opt. Soc. Am. B, **2**, 680-686 (1985).
3. J.-C. Diels, W. Dietel, J. J. Fontaine, W. Rudolph. and B. Wilhelmi, Journal of the Opt. Soc. B, **2**, 680-686 (1985).
4. J.-C. Diels, N. Jamasbi and L. Sarger, submitted to Rev. of Sci. Instr. (1986).
5. J.-C. Diels, J. J. Fontaine, I. C. McMichael, and F. Simoni, Applied Optics **24**, 1270-1282 (1985).
6. J.-C. Diels, N. Jamasbi and L. Sarger, to be published (1986).
7. J. Rothenberg, Topical Meeting on Ultrafast Phenomena V, paper WC4, Aspen, Co (June 1986).

# Generation of 55-fs Pulses and Variable Spectral Windowing in a Linear-Cavity Synchronously Pumped cw Dye Laser

*M.D. Dawson[1], T.F. Boggess[1], D.W. Garvey[1], and A.L. Smirl[2]*

[1]Center for Applied Quantum Electronics, North Texas State University, Denton, TX 76203, USA
[2]Hughes Research Laboratories, 3011 Malibu Canyon Road, Malibu, CA 90265, USA

Direct generation of optical pulses with duration below 100 fs from cw dye lasers has almost exclusively been limited to the spectral region 605-630 nm.[1-4] Typically, conventional ring[1] or antiresonant ring[2,3] cavity arrangements designed to utilize the colliding pulse mode-locking (CPM) effect,[1] where counterpropagating pulses meet in a saturable absorber jet-stream, are also required.

We describe the generation of pulses as short as 55 fs at 675 nm and 69 fs at 583 nm from a linear-cavity synchronously pumped, hybridly mode-locked cw dye laser incorporating Brewster prisms, which does not require the CPM technique. In addition, we have demonstrated variable spectral windowing of the spatially dispersed frequency components occurring within the intracavity prism sequence in this laser.

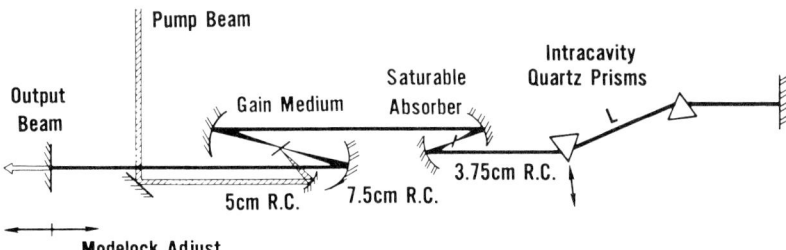

Fig. 1 Laser cavity configuration

The laser oscillator is shown in Fig. 1. A related cavity has produced 33-fs pulses by purely passive mode-locking.[4] The laser was pumped by 650 mW average power, 70-ps pulses from a frequency-doubled cw mode-locked Nd:YAG laser. The dual-jet linear resonator was terminated by a two-Brewster-prism and plane mirror arrangement used for adjustment of intracavity dispersion.[4] High reflectivity cavity mirrors with single stack coatings, having their reflectivity centered at 632.8 nm were used throughout, except for the output coupler which was 95% reflecting at 632.8 nm. Pulsewidth measurements were performed by both collinear type I and non-collinear background-free autocorrelation techniques.

Two different dye combinations yielded femtosecond pulses from this laser, provided the cavity length was carefully matched to that of the pump. Using Sulforhodamine 101 as gain medium ($1.8 \times 10^{-3}$M in ethylene glycol) and a $5 \times 10^{-4}$M solution of DQTCI[5,6] as saturable absorber, stable

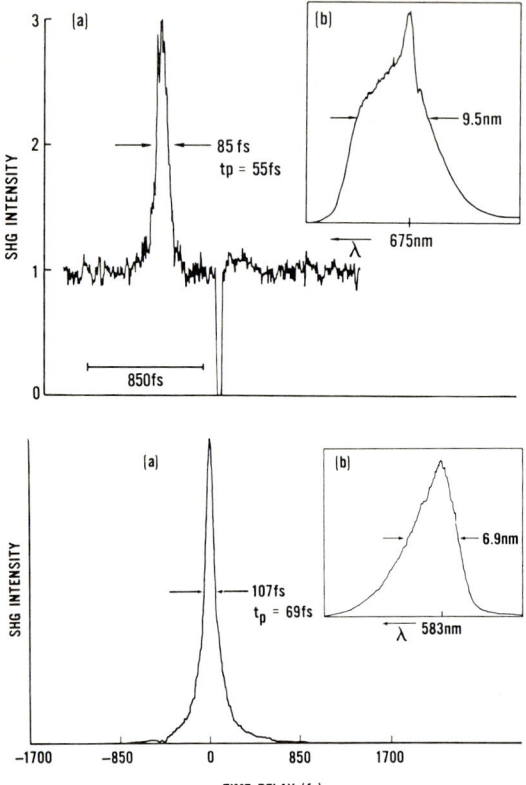

Fig. 2 Collinear type I autocorrelation at 675 nm

Fig. 3 Background-free autocorrelation at 583 nm

pulses as short as 55 fs were obtained (Fig. 2, sech$^2$ pulse shape assumed) with time-averaged output powers ~35 mW. The pulses were spectrally centered near 675 nm with a bandwidth of 9.5 nm, giving a time-bandwidth product $\Delta\nu\Delta t = 0.34$. A $2 \times 10^{-3}$M solution of Rhodamine 6G and a mixed $1.8 \times 10^{-5}$M DQOCI/$1.2 \times 10^{-5}$M DODCI saturable absorber solution yielded pulses as short as 69 fs (Fig. 3) with output power of 55 mW. These pulses were spectrally centered near 583 nm with a bandwidth of 6.9 nm, which also implies operation close to the transform limit.

Inclusion of a simple variable aperture between the second prism and the end mirror, a region of the cavity in which the frequency components are spatially dispersed, allowed adjustment of the <u>intracavity</u> laser spectral width (in direct analogy to the <u>extracavity spectral windowing</u> techniques developed by Thurston <u>et al.</u>[7]). This simple expedient allowed the output pulsewidth to be <u>directly</u> controlled in the range up to 500 fs (Fig. 4) and wavelength tuning of the laser could be achieved by translating the aperture transversely.

In conclusion, we have demonstrated a linear-cavity, hybridly modelocked cw dye laser capable of producing pulses well below 100 fs duration outside the "common" wavelength range for femtosecond generation and in which the laser bandwidth can be directly controlled. Its advantages over other designs are principally its simplicity, ease of alignment and non-

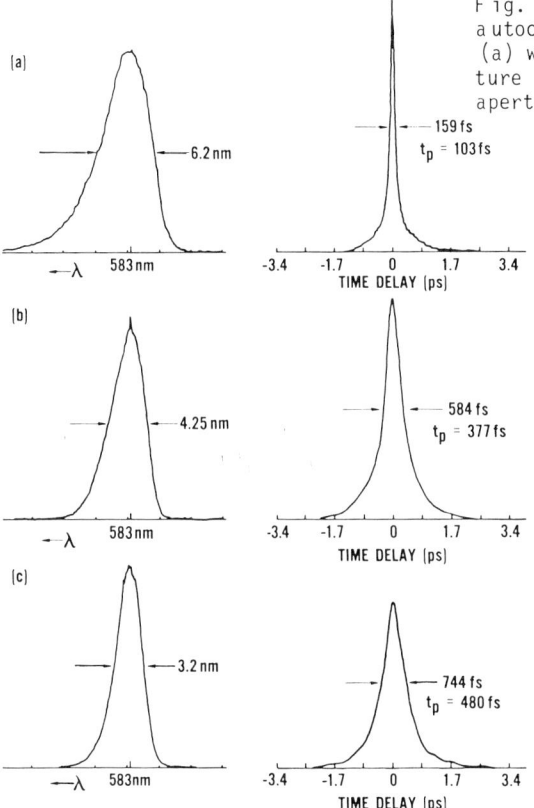

Fig. 4 Spectrum and corresponding autocorrelation for hybrid dye laser (a) without aperture (b) with aperture width 1.1 mm and (c) with aperture width 0.95 mm

critical positioning of the saturable absorber-jet, combined with its high efficiency and stability.

Acknowledgements

This research was supported by the U. S. Office of Naval Research, The Robert A. Welch Foundation and the North Texas State University Faculty Research Fund.

1. R. L. Fork, B. I. Greene and C. V. Shank: Appl. Phys. Lett. 38, 671 (1981).
2. H. Vanherzeele, R. Torti and J.-C. Diels: Appl. Opt. 23, 4182 (1984).
3. T. Norris, T. Sizer, II and G. Mourou: J. Opt. Soc. Am. B 2, 613 (1985).
4. J. A. Valdmanis, R. L. Fork and J. P. Gordon: Opt. Lett. 10, 131 1985).
5. J. R. Taylor: Opt. Commun. 57, 117 (1986).
6. P. M. W. French and J. R. Taylor: IEEE J. Quantum Electron. (to be published).
7. R. N. Thurston, J. P. Heritage, A. M. Weiner and W. J. Tomlinson: IEEE J. Quantum Electron. QE-22, 682 (1986).

# Cavity-Mirror Dispersion Dependence of Pulse Duration Generated from a Simple CPM Laser: An Experimental Study

M. Yamashita[1], K. Torizuka[1], T. Sato[1], and M. Ishikawa[2]

[1]Laser Research Section, Electrotechnical Laboratory, 1-1-4 Umezono, Sakura, Niihari, Ibaraki 305, Japan
[2]Tsukuba Research Laboratory, Hamamatsu Photonics K.K., 5-9-2 Tokodai, Toyosato, Tsukuba, Ibaraki, 300-26, Japan

Since the advent of light pulses shorter than 100 fsec from a CPM laser [1], many studies to directly generate shorter pulses from a laser oscillator have been carried out [2]. These studies have shown that, for the generation of shorter pulses, the most important factor is compensation of chirp arising from dispersion and phase modulation inherent to intracavity optical elements [3-7]. For example, DIETEL et al.[5] produced pulses shorter than 60 fsec by the adjustment of the optical path of a positive dispersion prism-glass in the cavity for compensation of down-chirp. On the other hand, VALDMANIS et al.[3], have produced pulses as short as 27 fsec by the adjustment of the insertion of a four-prism sequence which results in negative cavity dispersion to compensate for up-chirp. These two results insist on the completely opposite conclusion for the sign of the dominant chirp in a CPM laser. In addition, the insertion of prisms for chirp compensation complicated the alignment of the cavity configuration. In this paper, it is shown quantitatively that chirp in a simple CPM laser is compensated by the use of cavity mirror dispersion only. Consequently, this enables us to clarify the criterion of the best mirror coating for the generation of the short pulses from a CPM laser without additional intracavity elements, and to generate 50-fsec pulses.

We investigated pulse durations as a function of the cavity mirror dispersion by using different coating mirrors with different incident angles. Our CPM laser [8] simply consists of a amplifying medium (rhodamine 6G, 233 μm thick), a saturable absorber (DODCI, 39 μm thick) and seven multilayer dielectric mirrors. The pulse duration was measured by a conventional background-free SHG autocorrelator (0.2 mm thick KDP). As a pulse intensity profile, $sech^2$ was assumed.

In general, the effect of the multilayer dielectric mirror on the amplitude and phase of an incident EM wave $E_i(\omega)$ depends on its angular frequency $\omega$ and hence its wavelength. The reflected complex amplitude of the field $E_r(w)$ is written $E_r(\omega) = r(\omega) \cdot \exp[i\phi(\omega)]E_i(\omega)$, where $R(\omega) = |r(\omega)|^2$ is the intensity reflectance and $\phi(\omega)$ is the phase shift. The reflectance and phase shift are numerically calculated by using a matrix formulation for multilayer filters [9]. The effective quantity for chirp compensation is the second derivatives of the phase shift $\ddot{\phi}(\omega) = \partial^2\phi(\omega)/\partial\omega^2$ [4]. The quantity $\ddot{\phi}(\omega)$ is related to group-velocity dispersion $\partial^2 k(\omega)/\partial\omega^2$ of a dispersive material with an effective length $\ell$, by an equation $\ddot{\phi}(\omega) = -\ell \, \partial^2 k(\omega)/\partial\omega^2$. Therefore, the sign of $\ddot{\phi}(\omega)$ is opposite to that of the group-velocity dispersion. For all the cavity mirrors, the values of $R(w)$ and $\ddot{\phi}(\omega)$ in the vicinity of the lasing wavelengths were calculated as a function of the wavelengths taking the incident angle and p-component of polarization into consideration. Absorption and dispersion of the layer materials of $TiO_2$ ($n_H$=2.25) and $SiO_2$ ($n_L$=1.46) were neglected. All the mirrors were carefully made from uniform multilayers with thickness variation smaller than 6%.

Intracavity dispersion was changed over a wide range from positive to negative values by using various flat mirrors. At each mirror configuration, the generated pulse duration was measured by careful adjustment of the operating conditions such as pump power and cavity alignment, while monitoring the fast scanned autocorrelation trace on an oscilloscope.

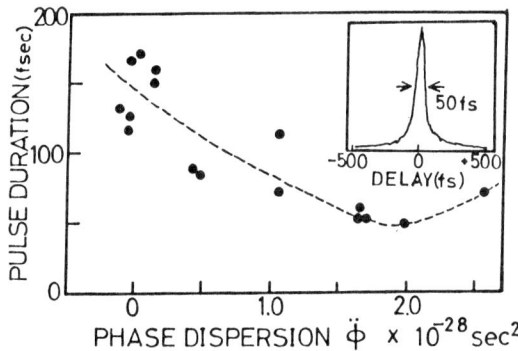

Fig.1 The dependence of the pulse duration on the the total dispersion of the seven cavity mirrors.

The shortest pulse duration of 50 fsec with average output power of 28 mW at the center wavelength of 636 nm (with the spectral width $\Delta\lambda$ = 7.8 nm of the nearly transform limited value) was generated around $\ddot{\phi}(\omega)$=+1.8 × $10^{-28} sec^2$ and not around $\ddot{\phi}(\omega)$=0. This result implies that in our simple CPM laser, up-chirp is dominant and is compensated by negative group velocity dispersion (corresponding to amount of $\ddot{\phi}(\omega)$=+1.8 × $10^{-28} sec^2$ ) from only one cavity mirror. The mirror was made from a doubly coated stack of 13-$\lambda_{o.up}$/4-layers ($\lambda_{o.up}$=535 nm) and 10-$\lambda_{o.low}$/4-layers ($\lambda_{o.low}$=650 nm) for normal incidence. The dominant up-chirp is caused by positive self-phase modulation which is probably due to fast response time-dependent nonlinear refractive indices from the electronic hyperpolarizability of ethylene glycol, rhodamine 6G, and DODCI [3],[10]. This effect occurs more remarkably in our CPM laser, because the cavity configuration is free from negative cavity dispersion from prism sequences, and the pulse beam to the absorber jet is tightly focused (1.25 cm of focal length).

In order to broaden the pulse spectrum and shorten the pulse duration, we mixed an organic compound having a higher nonlinear refractive index ($n_2$) than that of ethylene glycol ($n_2$=7.1 × $10^{-14}$ esu) to the DODCI solution. MNA (2-methyl-4- nitroaniline) was chosen because of its high solubility in ethylene glycol, high $n_2$ (1.9 × $10^{-13}$ esu in 10 mM solution), and no distinct absorption

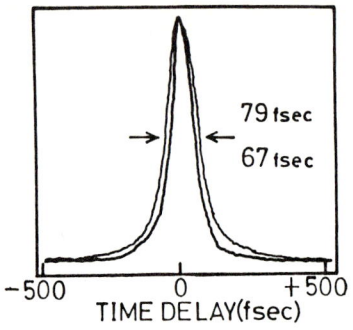

Fig.2
Autocorrelation functions without (79 fsec) and with (67 fsec) addition of MNA (10 mM) under the same operation conditions except for the pumping power.

around the lasing wavelengths. A high nonlinear refractive index of MNA is due to electronic hyperpolarizability of the $\pi$-electron cloud, and hence responds to ultrafast optical pulses. The effect of a nonlinear compound on the reduction of the pulse duration was proved experimentally for the first time. The further experiment is now in progress.

References

1.  R.L.Fork, B.I.Greene and C.V.Shank, Appl.Phys. Lett. 38, 671 (1981).
2.  For example Ultrafast Phenomena IV, Ed. by D.H.Auston & K.B.Eisenthal (Springer, Berlin, 1984).
3.  J.A.Valdmanis, R.L.Fork and J.P.Gordon, Opt.Lett. 10, 131 (1985).
4.  S.D.Silvestri, P.Laporta and O.Svelto, IEEE J.Quantum Electron., QE-20, 533 (1984).
5.  W.Dietel, J.J.Fontaine and J.-C. Diels, Opt.Lett. 8, 4 (1983).
6.  R.L.Fork, O.E.Maltinez and J.P.Gordon, Opt.Lett. 9, 150 (1984).
7.  W.Dietel, E.Dopel, K.Hehl, W.Rudolph and E.Schmidt, Opt.Commun., 50, 179 (1984).
8.  M.Yamashita, S.Aoshima and T.Sato, Rev. Laser Engineering, 12, 576 (1984) in Japanese.
9.  H.M.Liddell, Computer-Aided Techniques for the Design of Multilayer Filters, (Adam Hilger Ltd., Bristol). (1981).
10. Y.Ishida, K.Naganuma and T.Yajima. IEEE J.Quantum Electron., QE-21, 69 (1985).

# Femtosecond Pulse Generation from Passively Mode Locked Continuous Wave Dye Lasers 550–700 nm

*P.M.W. French and J.R. Taylor*

Laser Optics Group, Physics Department, Imperial College,
Prince Consort Road, London SW7 2BZ, UK

The passively mode locked cw dye laser has provided the source of the temporally shortest pulses since its inception in 1972 (1). As well as being capable of producing pulses as short as 27fs (2) in an optimised and dispersion controlled cavity, it exhibits very high amplitude stability and considerably lower interpulse jitter than its synchronously or hybridly mode locked counterparts. Requiring no mode locked pump source or pulse compressor, it is the cheapest and simplest of femtosecond laser systems. However, until very recently all such lasers have been restricted to the spectral region 610 - 640 nm, mainly using the standard active/passive combination of Rhodamine 6G and DODCI. Supplementary saturable absorbers for c.w. operation such as Malachite Green (3) and DQOCI (4) have been demonstrated but essentially only over the same spectral region.

In hybrid systems femtosecond pulses have been generated from 535 - 590 nm (5) using DODCI and Rhodamine B as the saturable absorbers with Rhodamine 110 and disodium fluorescein; and from 815 - 850nm using the dye 1R - 140 with Styryl 9 (6). None of these combinations have been shown to work in purely passively mode locked c.w. dye lasers. The first subpicosecond passively mode locked c.w. dye at new wavelengths was the Krypton ion pumped Rhodamine 700 laser system in the spectral ranges 727-740nm and 762-778nm (7). We now present results on seven new active / passive dye combinations which have yielded subpicosecond pulses continuously from 550 nm to 700 nm in a simple linear cavity with no optimisation of the dispersion.

The basic five mirror linear cavity arrangement designed to comply with New's criteria (8) for stable pulse evolution, is shown in figure 1. The active gain media were circulated in a 100μm thick free flowing vertical jet stream located between 100mm radius of curvature mirrors, while the passive folded section was a simple retroreflector with a focussing mirror of 50 mm radius of curvature. For these initial measurements, broadband, 100% reflecting dielectric coatings were used for all mirrors. The neglect of dispersion compensation probably precluded the generation of pulses shorter than 100 fs. In addition, the intracavity dielectric coated tuning wedge used for wavelength selection, and

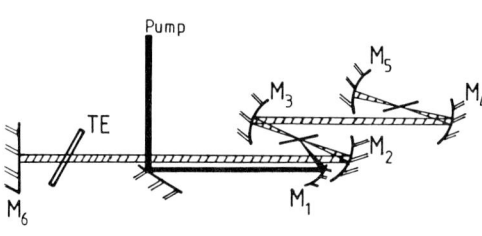

Fig.1

also to provide two outputs via its surface reflections, also limited the minimum pulse width achievable by restricting the bandwidth. The pump source was a c.w. Spectra Physics 2020 Argon ion laser with either 7W maximum power in 'all lines' operation or 3W maximum in single line operation at 514 nm. In many of the passive systems described here femtosecond pulse generation was possible with only 2 - 3W continuous pump power. Throughout this work, pulse durations were measured using a standard collinear autocorrelation technique, using $LiIO_3$ or KDP as the frequency doubling crystals. $Sech^2$ pulse profiles have been assumed in all instances. Initially DASBTI was used as the saturable absorber for a Rhodamine 6G laser operating in the spectral region 570 - 600 nm where average output powers of 15mW and pulse durations of 500fs were obtained. The same saturable absorber was used to mode lock Rhodamine 110 producing pulses as short as 150fs over the range 553-570nm. Another saturable absorber, HICI, gave optimum performance in the region 550 - 570 nm where it was possible to pump the dye laser 2W above threshold threshold giving peak output powers of 3kW with high stability, while maintaining single pulse operation with pulse widths $\sim$ 250fs.

Rhodamine B was mode locked over the range 616 - 658 nm using DQTCI, a benzthiazole derivative - giving pulses of 200fs duration. Depending on the cavity round trip time and the pump power, the laser would support one to three pulses in the cavity with average output powers of up to $\sim$100mW. The same saturable absorber was used to mode lock an energy transfer laser in which Sulphur Rhodamine 101 was pumped by a transfer from Rhodamine 6G. Mode locking was achieved from 652 - 694nm with pulses as short as 120fs - the shortest ever obtained from a simple linear passive system. A typical autocorrelation trace is shown in Figure 2. This laser was pumped 1.5W above threshold to give average powers of up to 24mW. A further saturable absorber, DCCI, was also used with the Sulphur Rhodamine 101 energy transfer laser over the same tuning range. Pulses of typically 200 fs duration were obtained with up to 36 mW of average output power. With both saturable absorbers, the laser would support either one or two pulses in the cavity depending on the cavity round trip time and the pump power. The dye DCM was also passively mode locked over the range 655 - 673 nm using DQTCI as the saturable absorber, producing pulses as short as 680fs. This is the first time that dye not belonging to the Xanthene family has been passively mode locked in a c.w. laser. All the above systems exhibited strong bistability and hysteresis - the laser switching between on and off, and also between states which supported different numbers of pulses in the cavity. In contrast to other reported passively mode locked systems, most of the lasers described here could be pumped well above threshold while still maintaining stable, short pulse operation. All of the active/passive combinations could be made to support only one pulse in the cavity but this did not always correspond to the genera- tion of the shortest pulse. The long term stability of all the systems was excellent - as would be expected from passive mode locking.

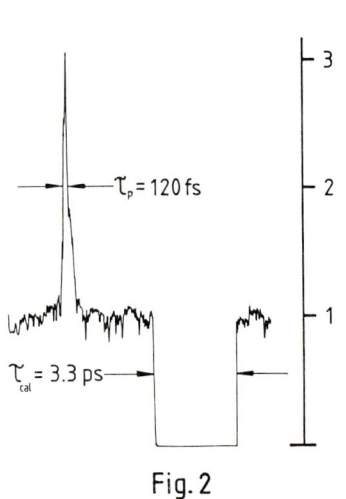

Fig. 2

Spectral investigations revealed that none of the pulses were transform limited. This observed chirp may have been due to the broadband "double-stack" dielectric mirror coatings (9,10).

In conclusion we have extended the technique of passive mode locking from 500 nm to 700 nm in a simple linear cavity with no optimisation of the intracavity dispersion. Pulses < 500 fs have been generated over the complete range, the shortest pulses being of 120fs duration We are confident that with higher pump powers than those available for this work all these systems would generate pulses of 200fs duration. Cavities with optimised dispersion characteristics should yield pulses of less than 100fs from 500 nm to 700 nm with average output powers sufficient to permit fibre optic compression directly with no decrease in the repetition rate. Figure 3 is a schematic of our reported achieved tuning ranges with the appropriate active and passive dye combinations. Operation outside the range indicated above is possible and we shall report presently on the extension to the range from 480nm - 900nm As well as providing the simplest and cheapest source of femtosecond pulses over the visible and near infra red spectrum, passive mode locking is likely to be the only source of tunable femtosecond pulses below 540nm where there is no readily available synchronous pump source.

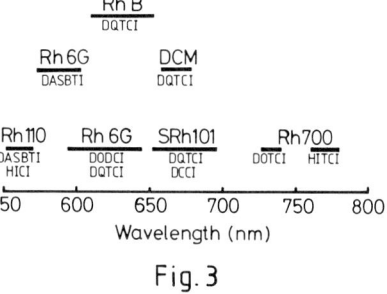

Fig. 3

References

1.  E.P.Ippen, C.V. Shank and A. Dienes, Appl.Phys.Lett. 21, 348 (1972)
2.  J.A. Valdmanis, R.L. Fork and J.P. Gordon, Opt. Lett. 9, 552 (1985).
3.  E.P. Ippen and C.V. Shank, Appl. Phys. Lett. 27, 488 (1975).
4.  R.S. Adrain, E.G. Arthurs, D.J. Bradley, A.G. Roddie and J.R. Taylor, Optics Commun. 12 140 (1974).
5.  Y. Ishida, K. Naganuma and T. Yajima, Jap. J. Appl. Phys. 21, 312 (1982).
6.  K. Smith, W. Sibbett and J.R. Taylor, Opt. Commun. 49 359 (1984).
7.  K. Smith, N. Langford, W. Sibbett and J.R. Taylor, Opt. Lett. 10, 559 (1985).
8.  G.H.C. New, IEEE J. Quantum Electron QE101, 115 (1974).
9.  S. De Silvestri, P. Laporta and O. Svelto, Opt. Lett. 9 335 (1984).
10. W. Dietel, E. Dopel, K. Hehl, W. Rudolph and E. Schmidt, Opt. Commun. 50, 179 (1984).

# Stabilisation of a CPM Dye Laser Synchronously Pumped by a Frequency Doubled ML YAG Laser

*J. Chesnoy and L. Fini*

Laboratoire d'Optique Quantique du CNRS, Ecole Polytechnique, F-91128 Palaiseau Cedex, France

We present a simple improvement of a synchronously pumped femtosecond dye laser which can be used to achieve active stabilisation of the cavity length. The frequency doubled mode locked Nd : YAG laser employed as the pump also gives us the possibility of operating our femtosecond system at two frequencies without any amplification. As we show here, a frequency shifted femtosecond pulse is readily obtained by frequency mixing the dye laser pulse with the infrared YAG pulse train.

Although the original pumping scheme of a femtosecond laser was based on a continuous Argon laser [1] some unique advantages were recognised for synchronous pumping of such systems. Recent development demonstrated operation starting either from a frequency doubled mode locked YAG laser [2] or from a mode locked Argon laser [3] as the pumping source. Following a similar approach, we developed a femtosecond dye laser source based on a continuously mode locked YAG laser in order to exploit the new possibility of dual wavelength operation without amplification.

Our dye laser system (Fig. 1) consists of an astigmatically compensated linear cavity containing a central Rhodamine 6G amplifier jet and terminated by a Sagnac interferometer in which a saturable absorber jet (20 µm DODCI) is inserted to preserve the colliding pulse mode locking operation (CPM). An intracavity prism pair is used to compensate for group velocity dispersion [4]. With a pumping level of 700 mW average power of $\lambda = 0.532$ µm doubled YAG at 100 MHz repetition rate, the dye laser delivers 35 mW average power around 0.62 µm in a 65 fs (femtosecond) pulse at 50 MHz repetition rate (the dye laser is twice as long as the pump laser). Besides the usual observation that a correct modelocking of the pump source is needed, we noticed that the dye laser cavity length has to be maintained fixed with a high precision in agreement with a recent report [3].

Figure 2 shows the measured variation of the laser pulse width as a function of cavity length (notice the 1 µm full scale of the coordinate). When

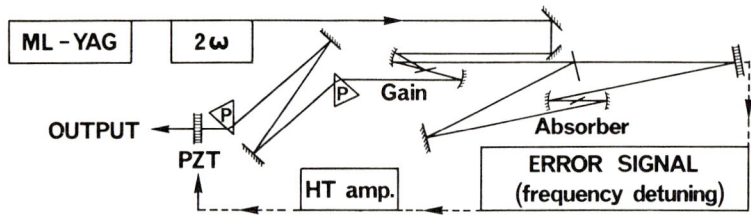

Fig. 1. Dye laser setup. The cavity length stabilisation uses the variation of the central wavelength as the error signal

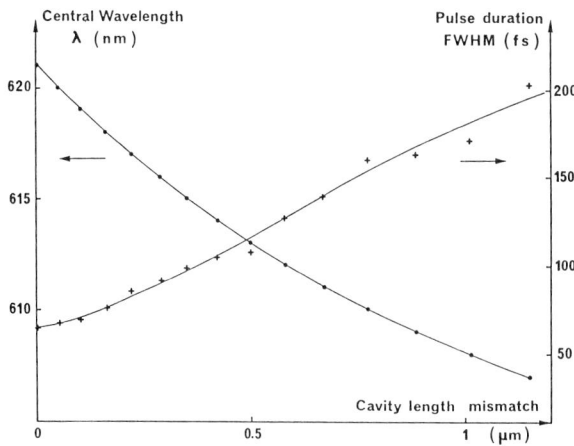

Fig. 2. Variation of the femtosecond laser pulse-width (+) versus cavity length mismatch (arbitrary zero). Variation of the central laser wavelength (.) used as the error signal for feedback to cavity length

the laser is operated, the duration of the generated pulse fluctuates accordingly following the cavity length variations. As various sources give drift of the cavity length (i.e. temperature or moisture of the air, acoustical noises...), an active stabilisation of the optical path length is highly desirable to maintain it in a tolerable range not exceeding 0.1 µm.

To achieve active stabilisation of the dye laser cavity length, one needs an error signal which monotonically depends on cavity mismatch. On fig. 2, we also show the variation of the dye laser central wavelength with the cavity length. A variation of $\Delta\lambda$ exceeding 10 nm is readily observed for a 1-µm cavity detuning. With this observation, the feedback loop setup is straight-forward : a small fraction of the dye laser beam is spectrally analysed by a grating and sent to a pair of photodiodes which provide a differential signal proportional to the wavelength drift referenced with regard to the desired wavelength. The loop is closed by applying the amplified output to a piezoelectric stack supporting the output mirror. This active control of the laser cavity length ensures the stabilisation of the pulse width, the laser central frequency (achieved within 0.5 nm) and pulse train amplitude (achieved within 5 percent despite the pump fluctuations exceeding 10 percent).

After the pumping of the femtosecond dye laser, one is left with a strong infrared YAG pulse train at 1.064 µm hardly altered by the 10 percent harmonic conversion. This 85-ps pulse is synchronised with the femtosecond laser and can readily be used to obtain a new femtosecond pulse train by frequency mixing. Compared to the case of a standard synchronously pumped picosecond laser [5], the conversion efficiency is favoured in our conditions by the powerful femtosecond pulse. Nevertheless, the length of the mixing nonlinear crystal must remain short enough to avoid broadening by group velocity mismatch. We performed sum frequency generation of the $\lambda = 1.06$ µm pulse with the femtosecond pulse $\lambda \approx 0.62$ µm and obtained a shifted blue femtosecond pulse around 0.39 µm. With a $LiIO_3$ crystal length of 140 µm cut at 42° from the birefringence axis, the pulse broadening is negligible (Fig. 3). The average power is around 100 µW (30 W peak power). Difference frequency generation is possible as well, but the power expected for the femtosecond pulse generated around 1.5 µm is somewhat lower due to frequency factors.

In conclusion, we have demonstrated that the synchronous pumping of a femtosecond dye laser by a frequency doubled ML YAG laser permits stable

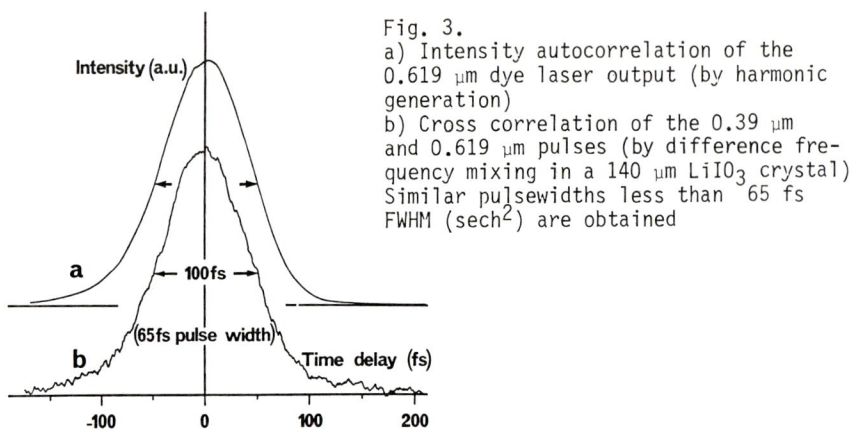

Fig. 3.
a) Intensity autocorrelation of the 0.619 μm dye laser output (by harmonic generation)
b) Cross correlation of the 0.39 μm and 0.619 μm pulses (by difference frequency mixing in a 140 μm $LiIO_3$ crystal) Similar pulsewidths less than 65 fs FWHM ($sech^2$) are obtained

dual wavelength operation without amplification. This is a simple way to eliminate the coherent artefact attached to single frequency pump-probe experiments.

This work was supported in part by "Direction des Recherches - Etudes et Techniques" under grant n° 85 050

References

1. R.L. Fork, B.I. Greene and C.V. Shank : Appl. Phys. Letters 38, 671 (1981)
2. T. Norris, T. Sizer and G. Mourou : J.O.S.A.B. 2, 613 (1985)
3. M.C. Nuss, R. Leonhardt and W. Zinth : Optics Letters 10, 16 (1985)
4. R.L. Fork, O.E. Martinez and J.P. Gordon : Optics Letters 9, 150 (1984)
5. D. Cotter and K.I. White : Optics Communications 49, 205 (1984)

# Fluctuations and Chirp
# in Colliding-Pulse Mode-Locked Dye Lasers

*D. Kühlke, T. Bonkhofer, U. Herpers, and D. von der Linde*

Fachbereich Physik, Universität Essen,
D-4300 Essen 1, Fed. Rep. of Germany

To ensure the shortest possible pulse duration in passively mode-locked dye lasers, intracavity self-phase modulation (SPM) and group velocity dispersion (GVD) must be properly balanced /1,2/. SPM is caused mainly by time-dependent saturation of the bleachable absorber /3/, and also by the Kerr-type nonlinearity of the index of refraction of the dye solvent. Two methods have been employed for introducing adjustable GVD. Valdmanis et al. /2/ have used a four-prism sequence /4/ in which the amount of GVD is determined by the geometrical distance between the prisms and the optical path in the prism material. Kuhl et al. /5/ have employed a Gires-Tournois interferometer (GTI), which provides GVD that can be varied by changing the angle of incidence.

Very little attention has so far been paid to the interrelation between SPM and GVD on the one hand, and fluctuations of the laser pulses on the other hand. Pulse noise is clearly of considerable importance for applications of the pulses. The noise behavior also provides interesting insight into pulse formation mechanisms. The purpose of this paper is to present data on fluctuations of pulse energy and pulse duration of a colliding pulse passively mode-locked dye (CPM) laser. In our experiments GVD is controlled either with the help of a four-prism sequence or else with a GTI, or a combination of both. We show that the different methods of controlling GVD have a strong effect on the fluctuations of the pulses.

The CPM laser is operated with rhodamine 6G as gain medium and DODCI as saturable absorber and pumped with 4 - 5 W at 515 nm from a cw argon ion laser. The separation of the prisms of the four-prism sequence is adjusted to 39.6 cm. The GTI is a thin-film interferometer of a thickness of 5 half-waves coated on a highly reflecting layer sequence. The noise of the pulses is measured with a fast photodiode in conjunction with an electronic spectrum analyzer /6/. The pulse duration is determined from measurements of the autocorrelation function using background-free second harmonic generation.

In a first series of experiments we use the four-prism sequence for controlling the intracavity GVD. In Fig. 1a we plot the measured pulse duration and the r.m.s. value of the energy fluctuations as a function of the total glass length travelled by the pulses in the four-prism sequence. The lower and the upper horizontal scale show the glass length and the corresponding change of the GVD, respectively. It is seen that the pulse duration decreases rapidly as the amount of negative GVD is decreased. The shortest pulses correspond to about 50 fs. The important point to be noticed, however, is the dramatic increase of the energy fluctuations that accompanies the decrease of the pulse width. The dashed vertical line marks the limit where the laser pulses become completely unstable. The fluctuations of the pulse width exhibit a similar behavior, increasing from 3% to as much as 16% at the boundary of the stability range.

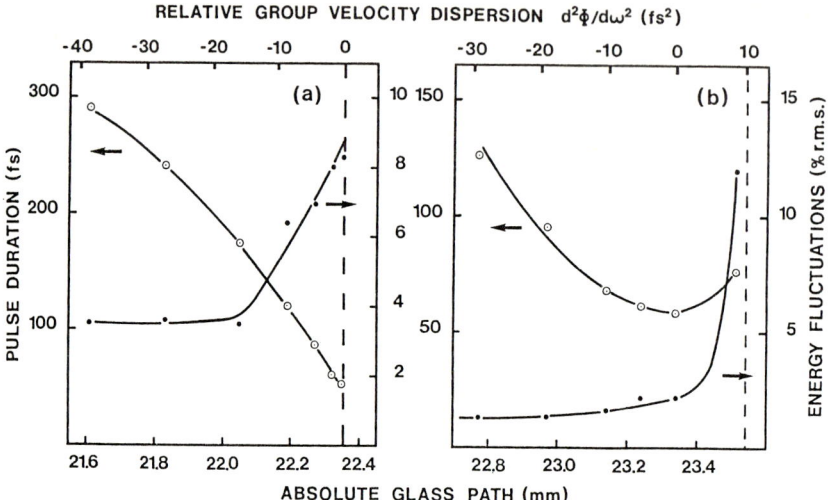

Fig. 1 Pulsewidth (open circles) and energy fluctuations (full circles) versus absolute glass path (lower scale) and corresponding change of the group velocity dispersion (upper scale) for the four-prism sequence (a) and for the combination prism sequence and GTI at an angle of incidence of 24° (b). The vertical dashed line marks the boundary of the stability range.

When the four-prism sequence is replaced by the GTI we obtained pulses of 85 fs with energy fluctuations as low as 2.5%.

In a further series we used a combination GTI (angle of incidence 24°) and prism sequence. The measured pulse duration and energy fluctuations for this configuration are plotted as a function of the glass length in Fig. 1b. Comparing with the results of Fig. 1a we notice that quite a different behavior is observed with the combination GTI and prisms. The shortest pulses do not occur at the boundary of the stability regime where the energy fluctuations are maximal. Rather, the minimum of the pulse duration (about 55 fs) now occurs in the low fluctuation regime with energy fluctuations of only about 2%.

From the comparison of the results in Fig. 1 the question arises whether there is an optimum angle of incidence for which the energy fluctuations are minimal, while the shortest pulse duration is maintained. We have therefore varied the angle of incidence. For each angle the glass length of the prism sequence was adjusted to give the shortest pulse duration. In Fig. 2 we plot the measured minimum values of the pulse duration. It is seen that a distinct minimum of the fluctuations of only 1.5% is measured at an angle of 26.5°. The fluctuations of the pulse duration exhibit a very similar behavior decreasing from 18% to less than 3% at 26.5°.

We believe that the explanation of the observed correlations between pulse fluctuations and GVD adjustment is the fact that the GVD changes substantially over a frequency range corresponding to the bandwidth of the pulses. These changes have opposite signs for the prism sequence and the GTI. The variation of the GVD for a four-prism sequence can be estimated by means of the results given in /4/. For a pulse spectrum with a bandwidth of 10 nm we obtain a variation of about 160 fs$^2$, i. e. the GVD increases with increasing wavelength. This estimate reveals that the four-prism sequence exhibits a

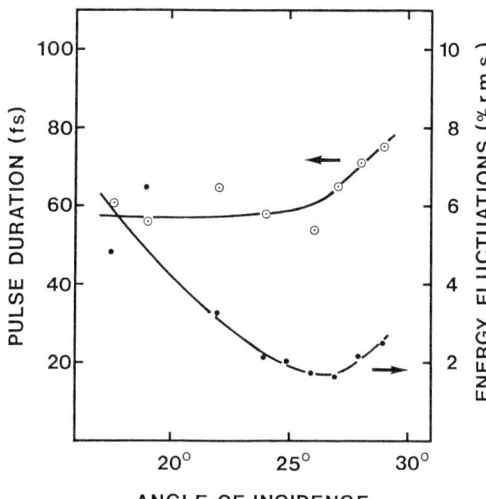

Fig. 2 Pulse width (open circles) and energy fluctuations (full circles) versus angle of incidence on the Gires-Tournois interferometer.

considerable variation of the GVD over the spectrum of pulses with a duration shorter than 100 fs. The change of GVD over the pulse bandwidth in the GTI depends essentially on the angle of incidence and the wavelength. The calculation of the change of GVD in our GTI for a wavelength of 625 nm yields a variation of $-160$ fs$^2$ at an angle of incidence of about $26°$. At this angle the resulting variation of GVD in the combination four-prism sequence and GTI is zero. Fig. 2 reveals that this cancelation angle agrees well with the angle corresponding to minimal fluctuations.

These observations suggest that the pulse fluctuations are significantly influenced by the variation in GVD over the bandwidth of the pulses. Pulses with minimum energy fluctuations are obtained for an intracavity GVD which is nearly constant over the pulse bandwidth. In our experiment this situation is achieved by setting the GTI at an angle of incidence for which the variation in GVD of the four-prism sequence is compensated for by that of the GTI.

References

/1/ W. Dietel, J. J. Fontaine, J. C. Diels, Opt. Lett. 8, 4 (1983)
/2/ J. A. Valdmanis, R. L. Fork, J. P. Gordon, Opt. Lett. 10, 131 (1985)
/3/ D. Kühlke, W. Rudolph, B. Wilhelmi, IEEE J. Quant. Electron. QE-19, 526 (1983)
/4/ R. L. Fork, O. E. Martinez, J. Gordon, Opt. Lett. 9, 150 (1984)
/5/ J. Heppner, J. Kuhl, Appl. Phys. Lett. 47, 453 (1985)
/6/ D. von der Linde, Appl. Phys B 39, 201 (1986)

# Experimental Observation of High Order Solitons in a Colliding Pulse Mode-Locked Laser

F. Salin, P. Grangier, G. Roger, and A. Brun

Institut d'Optique Théorique et Appliquée, Bât. 503,
Centre Universitaire d'Orsay, B.P. 43, F-91406 Orsay Cedex, France

In recent years, generation of femtosecond light pulses from colliding pulse mode-locked (CPM) ring dye lasers [1] has led to pulses shorter than 0.1 ps. To explain the formation of such ultra-short pulses, recent theories [2] introduce both self phase modulation (SPM) and group velocity dispersion (GVD) in the equations describing the pulse evolution during a single cavity round trip. The mathematical form of these equations has led to the "soliton-type shaping mechanism" but, to our knowledge, the soliton character of the pulses produced by a CPM laser was not clearly evidenced. We present here an observation of an evolution of the shape of pulses produced by a CPM laser, which is consistent with an N = 3 soliton behaviour.

We have built a CPM laser pumped by a cw Argon ion laser. The cavity includes a sequence of four prisms allowing precise adjustment of the intracavity amount of GVD. As was previously theoretically predicted and experimentally observed [3] by other authors, the pulse duration decreases from 300 fs to less than 40 fs when we translate one prism (introducing more glass). If we go on introducing more glass, we observe an unstable regime characterized by a noisy autocorrelation trace and pulse-train envelope. However, this apparently unstable regime may be stabilized by using a pump power just above the laser threshold-power. The autocorrelation function presents then a triple humped structure (Fig. 1 a) while a regular modulation, with two characteristic frequencies at 35 kHz and 70 kHz appears on the pulse-train envelope (Fig. 1 b). We think that it is possible to interpret these phenomena as an N = 3 solitonlike behaviour [4].

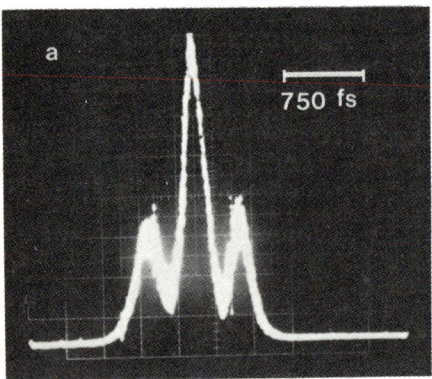 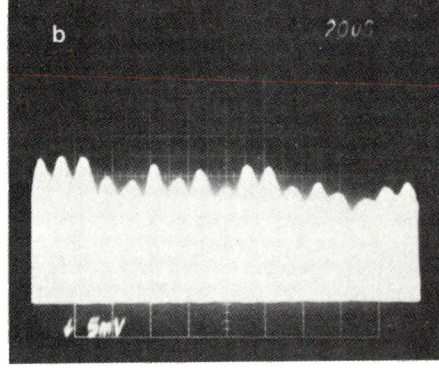

Fig.1 : Autocorrelation trace (a) and pulse-train envelope (b) with an excess of positive GVD

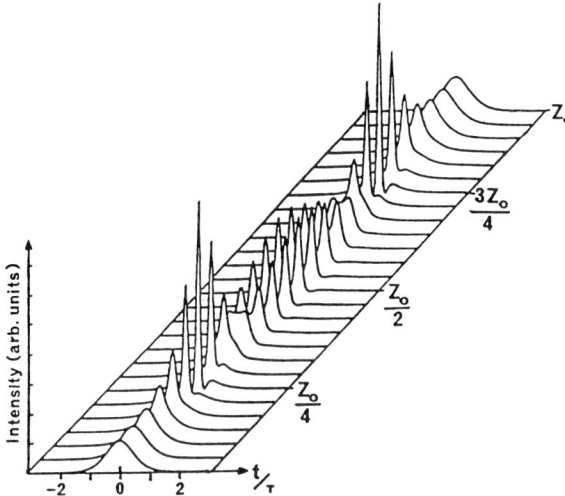

Fig.2 : Perspective plot of an N = 3 soliton temporal shape at different points of the soliton period $Z_o$. $\tau$ is the FWHM of the initial pulse intensity

Soliton bound states (N-order soliton) are the solutions of a non-linear Schrödinger equation [5]. The evolution of the shape of an N-soliton is characterized by N-1 frequencies. For instance, Fig.2 shows a perspective plot of the theoretical evolution of a N = 3 soliton. The parameter $Z_0$ corresponds to the length of non-linear medium at which the pulse shape is restored. $Z_0$, which is actually independent of the soliton order, is given by

$$Z_0 = 0.322 \; \Pi^2 \; \tau^2 \; c/\lambda^2 \; D \; , \qquad (1)$$

where D is the GVD of the medium (in fs/nm. km), $\lambda$ is the vacuum wavelength. c is the vacuum velocity of light and $\tau$ the usual experimental pulse width (intensity FWHM). Using the usual cavity dispersion $\phi''$ (in fs$^2$) given by

$$\phi = (\lambda^2/2\Pi c) \; D\ell \; , \qquad (2)$$

where $\ell$ is the length of non-linear medium in the cavity, we find that the pulse shape is restored after $N_o$ cavity round trips, with

$$N_o = Z_o/\ell = 0.322 \; \Pi \; \tau^2/2|\phi''| \; . \qquad (3)$$

Since our cavity round trip time is T = 12 ns the corresponding soliton frequency is

$$f_o = 1/N_o T = 2|\phi''|/0.322 \; \Pi \; \tau^2 \; T \; . \qquad (4)$$

For fixed $\phi''$ and $\tau$ parameters, $f_o$ is always the lowest of the N-1 characteristic frequencies of an N-soliton. In our case we have observed two frequencies in the pulse train envelope. According to our interpretation the lowest (f = 35 kHz) is the fundamental soliton frequency corresponding to $N_o$ = 2 380 cavity round trips.

Direct evidence of the above assertions is obtained by our recording the evolution of the pulse autocorrelation function during the soliton period. This was achieved by analysis of the output signal of the optical autocorrelator using a sampler averager triggered synchronously with the

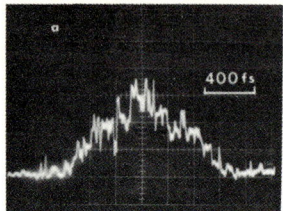 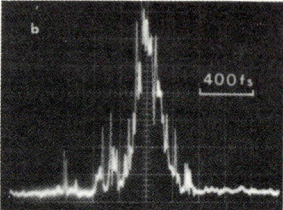 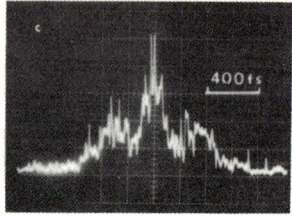

Fig.3 : Experimental autocorrelation function taken at $Z = 0$ (a), $Z = Z_o/4$ (b) and $Z = Z_o/2$ (c).

35 kHz modulation. The sample window could be shifted relative to the 35 kHz modulation phase. This experimental procedure is equivalent to the recording of the autocorrelation function of pulses selected at a given time in the soliton period. Figure 3 shows three autocorrelation functions obtained for different values of the phase shift. These phases shifts correspond to $Z = 0$, $Z = Z_o/4$ and $Z = Z_o/2$ on Fig. 2. These results are clearly consistent with the theoretical $N = 3$ soliton evolution sketched in this figure.

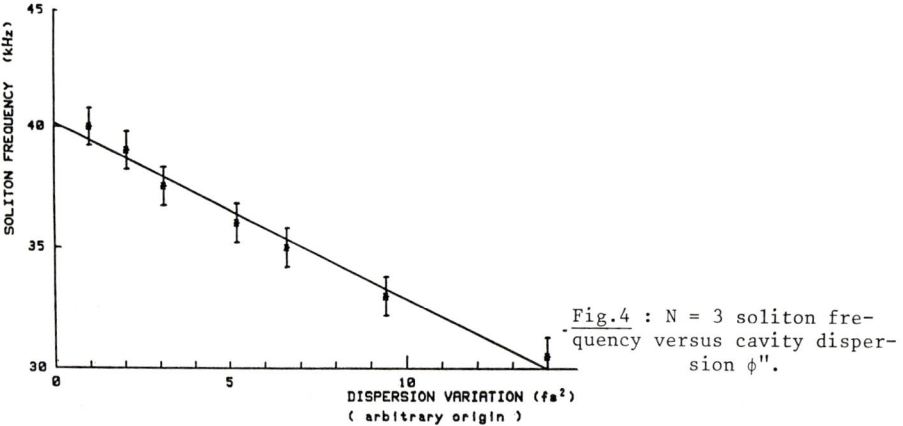

Fig.4 : $N = 3$ soliton frequency versus cavity dispersion $\phi''$.

We have also checked the relation (4) derived from the non-linear Schrödinger equation. Figure 4 shows the dependance of the soliton frequency $f_o$ upon the total amount of cavity dispersion $\phi''$. It exhibits a nearly linear relationship between $\Delta f_o$ and $\Delta \phi''$ which indicates that $\tau$ is roughly constant. The slope is $- 0.725 \pm 0.05$ kHz/fs$^2$ which yields a $\tau$ value of 475 fs (4). This value remains consistent with values measured on experimental autocorrelation traces (Fig.3.a).

From Eq. (4), we can obtain an estimation of the intracavity GVD. For our laser we obtain $\phi'' = - 55$ fs$^2$ (for prism positions giving $f_o = 40$ kHz). Using the fact that it only needs a very small prism translation (about 100 µm) to change from this $N = 3$ soliton (with a 40 kHz frequency) to the usual 50 fs pulses, we estimate that, for these latter pulses, the intracavity GVD is about $- 60$ fs$^2$. Note that this negative GVD must be accompanied by a positive SPM to obtain soliton propagation. Valdmanis et al. have already suggested that negative GVD and positive SPM (arising from gain saturation and dye solvent non-linear index) could introduce solitonlike shaping in a

CPM laser, leading to pulses as short as 27 fs [3]. The sign of $\phi"$ that we deduce from N = 3 soliton observations, seems to be in accordance with such a soliton shaping involving positive SPM. From our GVD estimation we can obtain an order of magnitude for the equivalent non-linear index of our laser. We obtain $n_2 \simeq 5.10^{-16}$ cm$^2$/W. This value is very close to the non-linear index of the solvent (Ethylene glycol = $n_2$ = $3.10^{-16}$ cm$^2$/W).

In conclusion, we have observed, for the first time to our knowledge, N = 3 solitons produced by a CPM laser.

These observations could be interpreted using a theoretical model in which the pulse shape suffers only a slight change during a single cavity round trip, but is restored after a very large number of them. Such a theoretical model could lead to further progress in the production of higher order solitons, which offers new possibilities of intracavity pulse compression by extraction of the pulse at the narrower point of its periodical evolution.

This work was supported in part by Direction des Recherches Etudes et Techniques (Division Optique).

1. R.L. Fork, B.I. Greene and C.V. Shank, Appl. Phys. Lett.38, 671 (1981).
2. O.E. Martinez, R.L. Fork and J.P. Gordon, J. Opt. Soc. Am. B2, 753 (1985).
3. J.A. Valdmanis, R.L. Fork and J.P. Gordon, Opt. Lett. 10, 131 (1985)
4. F. Salin, P. Grangier, G. Roger and A. Brun, Phys. Rev. Lett. 56, 1132 (1986).
5. V.E. Zakharov and A.B. Shabat, Zh. Eksp. Teor. Fiz., 61, 118 (1971) (Sov. Phys. JETP 34, 62 (1972)).

# Advances in the Theory of Mode-Locking by Synchronous Pumping

## G.H.C. New and J.M. Catherall

Department of Physics, Imperial College of Science and Technology, Prince Consort Road, London SW7 2BZ, UK

In a previous paper on the theory of mode-locking by synchronous pumping (MLSP) [1], we reported a discrepancy between the pulse profiles predicted by the standard "Self-Consistent Profile" (SCP) technique and those obtained using a simple difference equation model. Here, we confirm that the SCP method fails near optimum cavity tuning and we show why this occurs. We also demonstrate that fluctuations originating in the spontaneous emission background can cause severe perturbations in the mode-locked pulse profiles, even under nominally steady-state conditions [2-3].

As in earlier work, we adopt a simple ring cavity model in which the round-trip gain and loss are represented by a single time-dependent factor $g(t)$, and bandwidth limitation is provided by a generalised filter of characteristic time $t_f$. If $t_m$ is the temporal mismatch between the cavity transit period and the pumping period, the self-consistent field profile $V(t)$ in a synchronously mode-locked laser is given by the equation

$$(t_m t_f + t_m^2/2) d^2 V/dt^2 + (t_m + t_f) dV/dt = (g - 1)V + S(t), \qquad (1)$$

where $S(t)$ is a source term representing spontaneous emission. Optimum laser operation might be expected to occur when $t_m = -t_f$, since the pumping period is then exactly matched to the effective cavity transit period, allowing for the group time delay introduced by the filter. While this inference is correct for standard active mode-locking (AML), it does not hold for MLSP where the step-like delivery of energy to the laser by the pump leads to strong asymmetry of the laser characteristics with respect to cavity mismatch; in this case optimum performance is realised near $t_m = 0$.

For $t_m = -t_f$, (1) reduces essentially to the mode-locking equation of Haus [4] which, for AML by loss modulation for instance, has Gaussian pulse solutions. For $t_m = 0$ on the other hand, (1) becomes a first-order differential equation and, as we have shown previously [5], this and the rate equation controlling the laser gain are the familiar Wagner-Lengyel equations of giant-pulse laser theory. The Wagner-Lengyel equations are known to be highly resistant to approximate analytical solutions, and this is the reason for the failure of the SCP method in the case of MLSP. Most SCP analyses for example assume a symmetrical pulse profile at the outset; however, detailed examination of numerical solutions shows that terms involving the 3rd derivative of the field at the peak of the profile can play a dominant role.

Spontaneous emission plays a crucial role in MLSP especially for $t_m \geqslant t_f$ in which case the pulse evolves directly from it. While to obtain truly steady-state solutions the stochastic nature of spontaneous emission has naturally to be neglected and $S(t)$ in (1) represented by a constant, fluctuations in $S(t)$ can induce severe fluctuations in the mode-locked pulses. The stochastic effects are less important for negative values of $t_m$ and are extremely slight near $t_m = -t_f$.

The perturbative effect of the noise background is illustrated in Fig. 1, which shows the evolution of MLSP for $t_m = 0$ and typical values of other parameters. For $t_m = -t_f$, the perturbations are less evident but, as shown in Fig. 2, short pulses are not formed. For $t_m = t_f$ on the other hand (not shown), the satellite pulse in Fig. 1 is eliminated but the fluctuations are somewhat more severe.

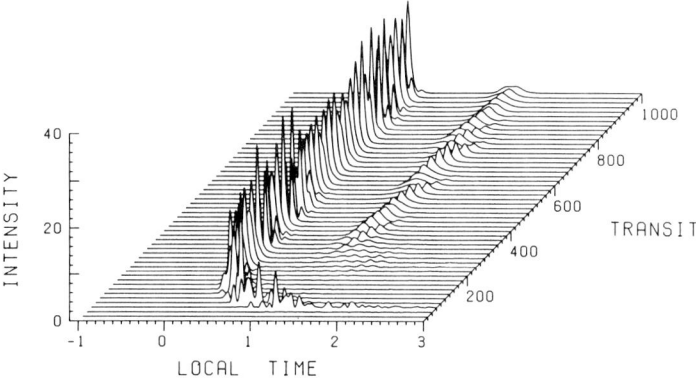

Fig. 1  Numerical simulation of pulse evolution in a synchronously mode-locked laser for $t_m = 0$

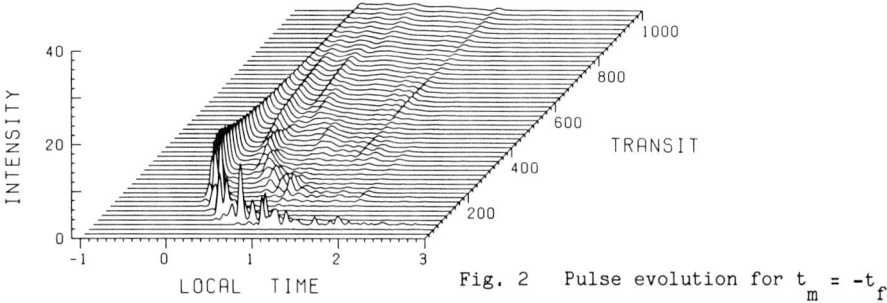

Fig. 2  Pulse evolution for $t_m = -t_f$

The autocorrelation function for $t_m = +t_f$, time-averaged over 90 profiles, is presented in Fig. 3. This shows that the effect of taking time-averages over the fluctuating pulses is to produce cusp-shaped records of the type commonly observed in experimental work on MLSP.

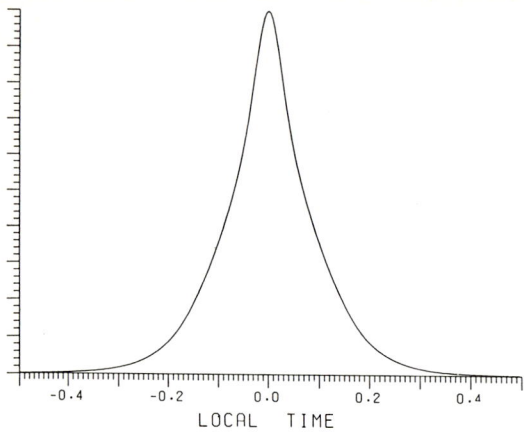

Fig. 3 Autocorrelation function of pulses in a synchronously pumped mode-locked laser time averaged over 90 transits

We conclude (i) that numerical solution is the only safe way to study the dynamics of MLSP and (ii) that there is a fundamental optimisation problem for MLSP since the conditions for eliminating the large-scale fluctuations of Fig. 1 and for obtaining short pulses are mutually exclusive.

1. G.H.C. New and J.M. Catherall, Opt. Commun., 50, 111 (1984).
2. G.H.C. New and J.M. Catherall, in <u>Laser Instabilities</u> (Cambridge University Press 1986), 97.
3. J.M. Catherall, and G.H.C. New, IEEE J. Quantum Electron., in press.
4. H.A. Haus, J. Appl. Phys., 46, 3049 (1975).
5. J.M. Catherall, G.H.C. New and P.M. Radmore, Opt. Lett., 7, 319 (1982).

# Collective Modes - An Analytical Model for Active Mode Locking in the Transient Case

P. Aechtner, P. Heinz, and A. Laubereau

Physikalisches Institut, Universität Bayreuth,
D-8580 Bayreuth, Fed. Rep. of Germany

The demand for short and reproducible pulses has renewed the interest in the theoretical understanding and description of active mode locking of pulsed laser systems. While the theory of stationary mode locking is well established /1/, the transient situation of pulsed lasers is still subject to discussion. Recently an analytical model has been proposed for the transient case /2/; a special initial condition starting with a single axial mode was assumed. The initially statistical character of mode locked radiation starting from spontaneous emission is not contained in this approach. On the other hand, numerical calculations of transient mode locking necessitate a huge computational effort because of the large number of $10^3$ to $10^5$ axial modes involved.

We have developed an alternative analytical approach which accounts for the complicated temporal and spectral structure of the initial lasing condition /3/. Our model considers a parabolic gain profile with constant net amplification $K(\omega)$ per cavity round trip switched on at $t=0$. Gain depletion and additional non-linear effects are omitted; i.e. the treatment applies for the linear regime of pulse generation. The novel concept of collective (temporal) modes instead of the axial cavity (frequency) modes is introduced by the help of Fourier transformation; in this way analytical solutions are derived in closed form. Salient features of our analytical results have been shown to agree with experimental data on pulsed solid state lasers.

The electromagnetic field $E(t)$ representing the laser emission is given by the superposition of collective modes $\tilde{E}_\ell$:

$$E(t) = \sum_\ell^N \tilde{E}_\ell(t). \qquad (1)$$

$\tilde{E}_\ell$ is evaluated by Fourier series from N conventional cavity modes exceeding threshold. We arrive at

$$\tilde{E}_\ell(t) = \tilde{E}_{\ell o} \times V_\ell(t_M) \left[\frac{\pi}{(\alpha_1+\beta_1)t_M}\right]^{1/2} \exp\left[-(t-t_M-\tau_\ell)^2/t_c^2 + i\omega_o t\right], \qquad (2)$$

where $\tilde{E}_{\ell o}$ represents the (random) initial amplitude. $V_\ell$ and $t_c$ respectively denote the amplification and duration of collective mode $\ell$. $t_M$ states the build-up time of the lasing process; $\tau_\ell$ indicates the temporal position of mode $\ell$ with respect to the

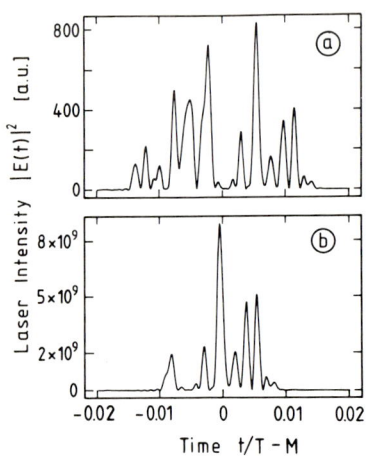

Fig. 1:

Intensity of laser emission vs time for N=1000 axial cavity modes and rectangular gain profile; K=1.01; $\eta_{min}$=0.3; $\Delta\omega = T^{-1} - \Omega' = 0$:
a) after M=500 round trips,
b) after M=1300 round trips.
The intensity spikes represent the collective modes; the strong selection of collective modes due to the mode locker should be noted

phase of the active mode locker. $\alpha_1$ and $\beta_1$ are well-known model parameters determined by the net amplification $K(\omega)$ and transmission $\eta = \eta_0 + \eta_1 \cos\Omega' t$ of the mode locker (modulator frequency $\Omega'$). The explicit result of $V_\ell$, $t_c$ is discussed in Ref. /3/.

The formation of a mode locked pulse originates from a selection process of collective modes. Fig. 1 illustrates this point. Assuming $10^3$ frequency modes of the cavity to be above threshold, the laser starts equivalently with $10^3$ collective modes. The initial spontaneous emission corresponds to a statistical distribution of amplitudes $\hat{E}_{\ell_0}$. After 500 round trips the radiation is shown in Fig. 1 a), where only $\sim$ 28 collective modes have survived in a small interval $|\tau_\ell| \leq 1.5 \times 10^{-2}$ T (cavity round trip time T). After 1300 round trips the selection mechanism has continued and only $\sim$ 17 spikes survive in the laser cavity. The different ordinate scales of Figs. a) and b) due to the amplification process should be noted.

Perfect mode locking corresponds in this picture to the survival of a single collective mode at $\tau_\ell = 0$. The build-up time $t_{ml}$ which is required to come to this point is evaluated in our model. $t_{ml}$ is a measure of the necessary time to reach stationary conditions. For the Nd:glass laser we estimate $t_{ml} \simeq 0.5$ ms; i.e. stationary conditions are not expected for common pulsed operation. For a cw Nd-YAG laser one arrives at $t_{ml} \simeq 30$ μs in agreement with published data /2,4/. Our result for the pulse duration in the stationary limit agrees with steady state theory /1/.

For the transient region $t_M < t_{ml}$ analytic results are also obtained, e.g. laser pulse duration, spectral width and tolerable frequency detuning of the active mode locker. Most important, a substructured pulse (= sequence of collective modes) is predicted which is not Fourier transform limited. The bandwidth limitation can be reached only in the stationary regime.

Some features of active mode locking under transient conditions have been demonstrated experimentally for a pulsed Nd: glass laser /3/. For example, a large product $t_p \times \delta\nu_L \simeq 70$ has

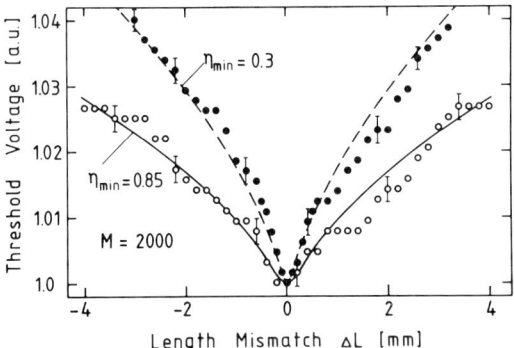

Fig. 2: Threshold voltage vs cavity length mismatch $\Delta L = -\Delta\omega \times L^2/2\pi c$ for a pulsed Nd:glass laser (L = 3 m); two mode locker efficiencies ($\eta_{min}$ = 0.85 and 0.3; $\eta_{max} \simeq 1$); experimental points, theoretical curves for M = 2000

been measured in agreement with our theoretical expectations. Experimental data on the frequency matching of the cavity round trip frequency to the active mode locker are shown in Fig. 2. The theoretical curves in the Figure agree well with the experimental points. The only fitting parameter in the calculation is the effective number of round trips $M = t_M/T$ while the other parameters are known from independent measurements. The observed minimum of the threshold curve has experimental significance to adjust frequency resonance $\Delta\omega=0$.

In conclusion we point out that a novel analytical model has been developed for active mode locking, which predicts substructured pulses in the transient case. Due to the strong selection process of collective modes only a few modes have to be retained in the calculation in contrast to the conventional frequency mode picture where this selection mechanism is missing so that large mode numbers have to be considered (reduction of $10^2$ to $10^3$). Use of our theoretical approach is highly recommendable for numerical calculations, e.g. computations of the nonlinear regime with passive mode locking or gain depletion of the laser material.

1. D.J. Kuizenga and A.E. Siegman, IEEE J. Quant. Electron. QE-6, 694 (1970); 709 (1970); Opto-Electron. 6, 43 (1974)
2. D.J. Kuizenga et al, Opt. Commun. 9, 221 (1973)
3. P. Aechtner and A. Laubereau, Appl. Phys. B 40, (1986)
4. G.F. Albrecht, M.T. Gruneisen, D. Smith, IEEE J. Quant. Electron. QE-21, 1189 (1985)

# Generation of Picosecond Pulses from a Continuous Wave Neodymium:Phosphate Glass Laser

L. Yan, J.D. Ling, P.-T. Ho, and Chi H. Lee

Department of Electrical Engineering, University of Maryland, College Park, MD 20742, USA

Up to now, the only way to generate very short (a few picosecond) pulses from neodymium:glass lasers has been Q-switching at repetition rates of a few hertz [1-3]. For some experiments and applications, a solid state laser able to generate picosecond pulses at high repetition rates is preferred. Here, we report the generation of 7 ps pulses at 1.054 μm from a CW actively mode-locked Nd:Phosphate glass laser. These pulses are about 14 times shorter than those from a CW mode-locked Nd:YAG laser and 10 times shorter than those previously obtained by our group [4].

The laser oscillator (Fig. 1) consists of a slab of Nd:Phosphate glass [5] longitudinally pumped by the 514 nm output of an argon ion laser, and mode-locked by an acousto-optic modulator. Several improvements have been made over the previous system [4]. First, a standing wave, instead of a travelling wave [4], modulator is used for active mode-locking, which is driven at about 50 MHz, half the round trip frequency of the ring cavity. A standing wave modulator is believed to give better amplitude modulation of light. Second, use of a standing wave modulator allows us to eliminate the two intracavity lenses used previously [4]; this results in a reduction of the intracavity loss and optical distortion. Third, the water-cooling system of the glass has been redesigned to dissipate faster the heat caused by pumping. Thus, we can increase the pump power over 3 watts. In addition, the etalon used previously [4] is also removed to relieve the spectral bandwidth constraint.

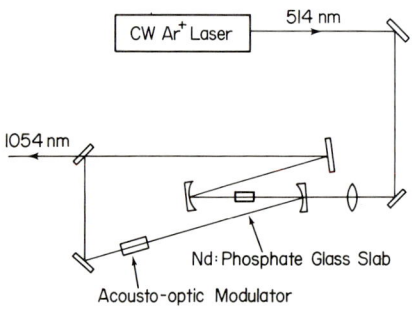

Figure 1. Schematic of the Nd: Phosphate glass ring laser. The transmittance of the output mirror is 3%.

We have found experimentally that the Nd:glass laser becomes more stable, and pulses become shorter and coherent, as the pump power increases. Good laser performance occurs at pumping levels well above the threshold of ∿1.1 watts. Below 1.7 watts, long and noisy pulses are observed. As the pump power increases to 1.8 to 2 watts, partially coherent pulses under 100 ps are obtained. At pump level above 2 watts, even shorter and coherent pulses are generated. Figure 2 shows 7 ps pulses measured by the standard intensity

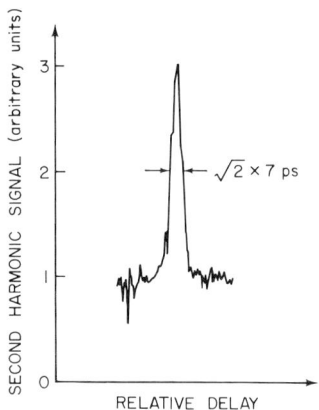

Figure 2. Autocorrelation trace of pulses from the laser. Gaussian pulse shape is assumed.

autocorrelation method with second harmonic generation. The pulse width is comparable to that obtained by a pulsed, passively mode-locked Nd:glass laser, yet with the advantages of high repetition rate, synchronizibility, and good reproducibility.

Two detailed observations suggest that, in addition to amplitude modulation, there is some frequency modulation in the mode-locking process in accordance with the FM mode-locking theory [6]: (1) the pulsewidth-bandwidth product is 1.1, slightly larger than the transform limit for an AM mode-locked laser; (2) the pulse repetition frequency (twice the modulator driving frequency) is about 6-8 KHz above the cavity round trip frequency when the shortest pulses are obtained.

The laser oscillator delivers 20 milliwatts of average power at 100 MHz pulse repetition rate, which suggests that a regenerative amplifier constructed from a similar system can provide microjoules of energy per pulse at kilohertz repetition rates. Furthermore, the Nd:Phosphate glass has enough bandwidth for 200 femtosecond pulses at 1.054 μm. Pulses can be further shortened by compression with an optical fiber and grating pairs.

We have also introduced a simple method to increase the pulse energy by mechanically chopping the Nd laser beam inside its cavity or the pump beam outside the cavity. Simultaneous mode-locking and relaxation oscillations occur regularly in both cases (Figure 3). The energy of the pulses in the

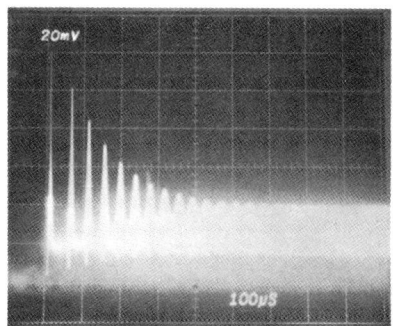

Figure 3. Simultaneous mode-locking and relaxation oscillations when the pump beam outside the cavity is chopped.

peak of the initial spike is increased over 40 times when the Nd laser beam is chopped inside the cavity, but the pulses are broadened to tens of picoseconds.

Acknowledgements

We would like to thank Dr. T. J. McIlrath, Dr. U. Hochuli, and Dr. G. L. Burdge for lending us equipment and Mr. D. Cooper for technical assistance. This work was supported in part by the Air Force Office of Scientific Research.

References

1. C. Kolmeder and W. Zinth, "Theoretical and Experimental Investigations of a Passively Mode-Locked Nd:Glass Laser", Appl. Phys. 24, 341 (1981).
2. L. S. Goldberg and P. E. Schoen, "Active-Passive Mode-Locking of a Nd:Phosphate Glass Laser Using #5 Saturable Dye", IEEE J. Quantum Electron. QE-20, 628 (1984).
3. T. Tomie, "Picosecond Pulse Generation by Self-Phase Modulation in an Actively Mode-Locked and Q-Switched Phosphate Glass Laser", Japan. J. Appl. Phys. 24, 1008 (1985).
4. S. A. Strobel, P.-T. Ho, C. H. Lee and G. L. Burdge, "Continuous wave mode-locked neodymium:phosphate glass laser", Appl. Phys. Lett. 45, 1171 (1984).
5. We are grateful to the Shanghai Institute of Optics and Fine Mechanics for the Nd:Phosphate glass slab.
6. D. J. Kuizenga and A. E. Siegman, "FM and AM Mode-Locking of the Homogeneous Laser - Part I:Theory", IEEE J. Quantum Electron. QE-6, 694 (1970).

# Part II

# Ultrafast Optical Generation and Measurement Techniques

# Fourier Transform Picosecond Pulse Shaping and Spectral Phase Measurement in a Grating Pulse-Compressor

*J.P. Heritage, A.M. Weiner, and R.N. Thurston*

Bell Communications Research, Inc., 331 Newman Springs Rd., Red Bank, NJ 07701, USA

We describe in this paper a technique for synthesizing arbitrarily shaped, Fourier transform limited picosecond and subpicosecond optical pulses [1-3]. We utilize these pulse shaping capabilities to perform a novel two slit temporal interference measurement of the phase spectrum of compressed pulses. In particular we have investigated the phase of the spectral components generated by self-phase modulation of a pulse distorted due to strong stimulated Raman scattering. We find that the red shifted frequency components are in phase over essentially the entire spectrum despite the fact that this region is dominated by pulse reshaping and depletion due to the stimulated scattering process.

The ability to control the amplitude and phase of the spectral components, and thus the temporal shape, of an ultrashort optical pulse would bring to the fields of ultrafast optical spectroscopy and wide bandwidth optical communications the flexibility long enjoyed in radio frequency electronics and nanosecond optics. Pulse shaping techniques which are successful in the nanosecond range, however, are difficult to apply to the picosecond and femtosecond regime. The very high speed of picosecond and femtosecond pulses makes it nearly impossible to tailor pulse shapes using electroptic or acoustooptic modulators. Passive techniques such as delay lines require interferometric precision. Techniques that employ scanning chirped beams and amplitude and phase masks leave the pulse substantially chirped. These techniques for pulse shaping, as well as others, are discussed in a recent review of the field [4].

Our Fourier transform pulse shaping approach overcomes these limitations. Pulse shaping is performed within a grating pair compressor, as depicted in Fig. 1. Because the optical spectral components are spatially dispersed after a single pass through the grating pair, spatial amplitude and phase filters may be introduced in this region to control the pulse's frequency spectrum and hence its temporal profile. It is important to emphasize that the synthesis of Fourier-transform-limited pulses requires that the various filtered frequency components be brought together at the same instant of time. We compensate the unavoidable temporal dispersion introduced by the grating pair by using a linearly chirped pulse (provided by self-phase modulation in an optical fiber) of precisely the correct chirp magnitude and sign. This configuration is immediately recognized as the well-known fiber and grating pair pulse-compressor. One significant difference is the addition of a second arm, formed by a beam splitter and a second return mirror, to the compressor. We place a spectral window [5] at this mirror to provide a background free compressed (but unshaped) pulse for high resolution crosscorrelation measurements of the complicated shaped pulses.

We have demonstrated our Fourier transform pulse shaper by fabricating a variety of pulse shapes that employ amplitude filters, phase filters and

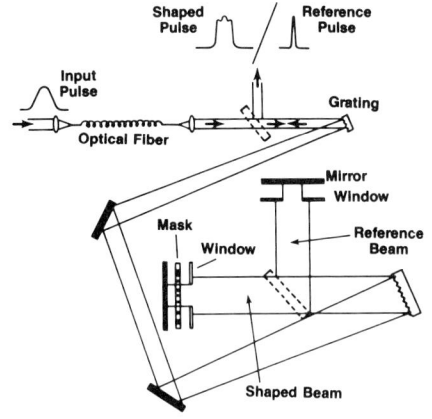

Fig. 1

Fiber and grating pulse-compressor modified for pulse shaping by addition of filters near turnaround. mirror. A second compressor arm provides a reference pulse for crosscorrelation measurements.

combined amplitude and phase filters. These reshaped pulses include a sequence of coherent pulses, an 'odd' pulse, a sequence of odd pulses [1], and a 'square' pulse [2]. We have also developed the theory of the Fourier transform pulse shaping, including the effects of finite spectral resolution [3]. All of our pulse shaping work was performed with a 1.064 μm modelocked Nd:YAG laser pulse train compressed from 75 psec to below 1 psec using a single mode, polarization holding, 400 meter fiber. The detailed experimental parameters are described in references 1 to 3.

We now turn our attention to an investigation of the optical phase spectrum. While our generated pulse shapes are in good agreement with calculations, none of the shapes presented so far have directly and sensitively tested the uniformity of the optical phase spectrum. The following discussion presents a simple new technique for measuring the compensated phase in a pulse compressor.

Our technique involves measuring the temporal beat note between two optical frequency components selected by a pair of narrow slits placed at the filter plane. The inset in Fig. 2 illustrates this arrangement; the

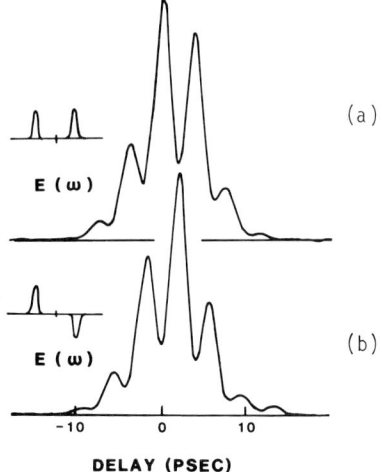

Fig. 2

Temporal beat pattern resulting from interference of two spectral components: (a) in-phase (b) 180 degrees out of phase.

traces in Fig. 2 are crosscorrelation measurements of the resulting interference. When the two components are in phase (upper trace), the resulting beat signal is proportional to the square of cos($\Delta\Omega t$), where $\Delta\Omega$ is the frequency spacing between the optical components. For any other relative phase, the position of the temporal beats will shift compared to the in-phase case. In particular when the phase change between the two interfering oscillators is $\pi$, the temporal intensity pattern is proportional to the square of sin($\Delta\Omega t$). This case is illustrated in the lower trace of Fig. 2. Here we have deliberately introduced a phase change of $\pi$ between the two oscillators by using a specially fabricated phase plate. The shift in phase is clearly evident and easily measurable.

We have applied this technique to investigate the phase spectrum of the compressed pulse under conditions of very strong stimulated Raman scattering (approximately 50% conversion in the 120 meter fiber). This fiber length exceeds the walk off length between pump and Stokes pulses but is sufficiently short that the influence of group velocity dispersion on the pump pulse itself can be neglected. Since the Stokes pulse propagates more rapidly than the pump (walk-off), the leading edge of the pump pulse is preferentially depleted and hence steepened. This steepening of the leading edge results in a nonsymmetrically broadened self-phase-modulation spectrum, which droops but extends further towards longer wavelengths [6], as shown in Fig. 3. The nonlinear chirp expected as a consequence of such nonsymmetric spectral broadening should prove detrimental for purposes of pulse compression and shaping. Experimentally we find that the pulse compression is fairly poor, even with significant spectral windowing. The pulses are about 2 picoseconds long but are accompanied by extensive wings and near-in structure.

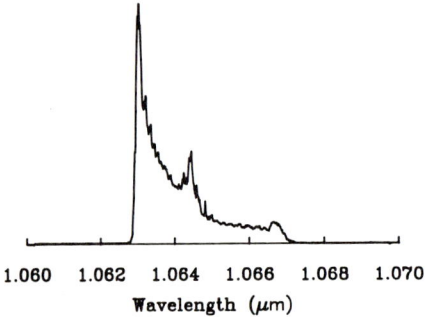

Fig. 3

Self-phase-modulation spectrum of the 1.0647 μm pulse under conditions of high efficiency stimulated Raman scattering.

To perform the phase measurements, we introduced a pair of slits into the filter plane, one centered at the optical carrier wavelength, the other at a slightly longer wavelength. After recording a crosscorrelation measurement of the beat profile, we performed a similar measurement on the short wavelength side of the spectrum, using an identical frequency spacing. This process was repeated for a sequence of some 15 spacings. We display the results of three of these measurement pairs in Fig. 4. In Fig. 4a, the measurement with the downshifted frequency (solid line) is virtually identical to that with the upshifted frequency (dotted line), indicating that for a wavelength spacing of 5 Angstroms the phase of the two outer frequency sources is identical. This behavior repeats itself until the separation is increased to 7 Angstroms, as displayed in Fig. 4b. In this case the downshifted frequency remains in phase but the upshifted pair departs by more than one half of a period. As we proceed from 7 Angstroms to 8 Angstroms

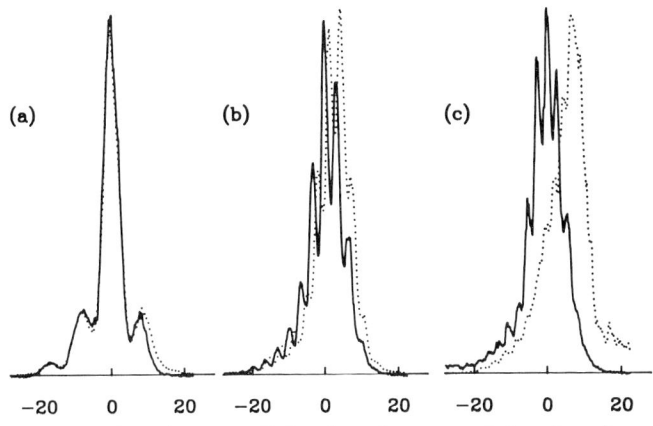

Fig. 4 Temporal beat pattern for downshifted (solid) frequency components and for upshifted (dotted) frequency components: (a) 5 Angstrom separation, (b) 7 Angstrom separation, (c) 8 Angstrom separation.

the upshifted phase varies so rapidly that it cannot readily be followed; simultaneously, a large time shift of the interference envelope appears. Note also that the fringe visibility has dropped dramatically for the upshifted frequencies indicating a loss of coherence. The downshifted wavelengths, however, remain coherent and in phase! Further measurements show that the downshifted components remain in phase out to the most extreme shift. This is quite a surprise, because it indicates that the phase spectrum is most uniform in the spectral region most affected by the stimulated Raman scattering.

In conclusion, we have described a technique for synthesizing arbitrarily shaped ultrashort pulses, by phase and amplitude masking in a fiber and grating compressor. We have applied this technique to measure the phase spectrum of the compressed pulse under conditions of strong stimulated Raman scattering. Efficient Raman conversion leads to a nonsymmetric self-phase-modulation spectrum which extends further towards longer wavelengths; surprisingly, the spectral field components remain in phase over a wide span of the downshifted frequencies. The method for characterizing the optical phase spectrum takes us one step further towards creating the highly complex tailored pulse shapes which we anticipate will find applications in picosecond and femtosecond spectroscopy and in communications systems based on frequency domain coding of individual ultrashort pulses.

1. J. P. Heritage, A. M. Weiner, R. N. Thurston: Opt. Lett. 10, 609 (1985)
2. A. M. Weiner, J. P. Heritage, R. N. Thurston: Opt. Lett. 11, 153 (1986).
3. R. N. Thurston, J. P. Heritage, A. M. Weiner, and W. J. Tomlinson: IEEE J. Quant. Electron. QE-22, 682 (1986)
4. C. Froehly, B. Colombeau, M. Vampouille: In Progress in optics XX, ed. by E. Wolf (Amsterdam, Holland 1983) p. 115
5. J. P. Heritage, R. N. Thurston. W. J. Tomlinson, A. M. Weiner, and R. H Stolen: Appl. Phys. Lett. 47, 87 (1985)
6. A. M. Weiner, J. P. Heritage, and R. H. Stolen: unpublished

# Picosecond Pulse Amplification Using Pulse Compression Techniques

D. Strickland, P. Maine, M. Bouvier, S. Williamson, and G. Mourou

Laboratory for Laser Energetics, University of Rochester,
250 East River Road, Rochester, NY 14623, USA

Nd:glass amplifiers have very good energy storage capabilities, but, the energy extraction is extremely inefficient for short pulse amplification. At relatively high peak intensities of 10 GW/cm$^2$, nonlinear phase shifts occur leading to beam wavefront distortion, which can result in filamentation and irreversible damage. In order that the peak intensity in the amplifier remain below this level, a picosecond pulse can be ampliified only to an energy density of 10 mJ/cm$^2$ which is more than two orders of magnitude less than the stored energy level of 5 J/cm$^2$. An amplification system, which uses an optical pulse compression technique and analogous to one used in the radar field, has been developed which circumvents this peak power limitation. With this technique, short pulses can be amplified to the high saturation energy levels, but, moderately low peak power levels are maintained in the amplifying medium. Over forty years ago, scientists working in the radar field overcame similar peak power limitations, by first frequency chirping a pulse and stretching it by passing it through a dispersive line prior to amplification, HUTTMAN [1] and CAUER [2]. By amplifying the longer pulses, much higher energies could be achieved. The echo is then passed through a matched filter which aligns all the frequency components in time such that the pulsewidth is compressed to a value approximately equal to $1/\Delta f$, where $\Delta f$ is the total chirped bandwidth.

To demonstrate the analogous technique in the optical regime, we have produced picosecond pulses at the tens of mJ energy level, STRICKLAND et al. [3]. The experimental system is shown in Fig. 1. A cw mode-locked, Nd:Yag laser produces low energy, 100–ps optical pulses at a 100–MHz repetition rate. The pulses are focussed into a 2.4–km long single-mode optical fiber. Figure 2 shows streak camera traces of the infrared pulses at the input and output of the fiber. Due to the combined effects of self-phase modulation and group velocity dispersion, the pulses at the output of the fiber are 330 ps in duration with a nearly linear frequency chirp over a 3.5–nm bandwidth, NAKATSUKA et al. [4]. The stretched pulses are amplified in a series of two silicate glass (Kigre Q-246) regenerative amplifiers. The silicate glass has a bandwidth of 20 nm and peaks at 1.062 μm. The fiber is isolated from the regenerative amplifier by a Pockels cell switch-out, which operates at a 5–Hz repetition rate. The switched-out pulses are then injected into the first regenerative amplifier by a second Pockels cell. Rather than using a separate $\lambda/4$ plate, we aligned the second Pockels cell to have a static $\lambda/4$ birefringence to frustrate lasing. We found that the original $\lambda/4$ plate acted as an etalon inside the cavity and caused double pulsing. After 100 round trips, the pulse has reached an energy of 2 mJ and is cavity

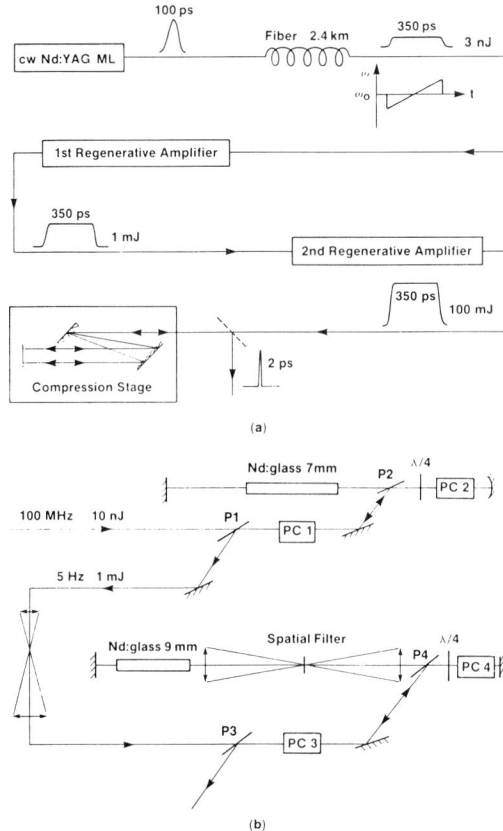

Fig. 1 (a) Block diagram of amplification and compression system. (b) Amplifier system configuration.

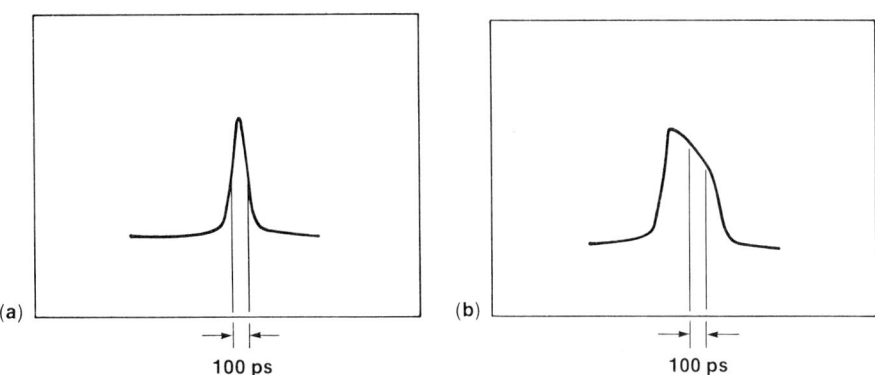

Fig. 2 (a) Streak camera traces of IR pulses at the A1 input. (b) Output of fiber.

dumped by the second Pockels cell. Before entering the second amplifier, the beam is up-collimated by a factor of 3 in order to fill the 9 mm rod. A third Pockels cell is used to isolate the first amplifier stage from the second. Pulses are injected into the second amplifier by a fourth Pockels cell operating at a 0.1-Hz repetition rate. A spatial filter is placed inside the second cavity, in order to maintain a good spatial profile. The pulses remain in the second regenerative amplifier for ten round trips and are then cavity dumped at an energy of 40 mJ. Because of gain saturation, the output pulse shape is not rectangular, but, the instantaneous frequency still exhibits a linear chirp across the pulse. The amplified pulses are then compressed by a double grating compressor, TREACY [5]. The gratings have a 1700 l/mm groove spacing and are used in a near Littrow configuration and spaced approximately 100 cm apart. The compressor is used in a double pass configuration in order to maintain a circular beam profile, DESBOIS et al. [6]. The overall grating compressor efficiency is 50 percent. Figure 3 shows a streak camera trace of a frequency doubled, amplified, and compressed pulse, with a reflection from an etalon. The streak camera trace shows that the amplified pulse was compressed to a value less than the streak camera limit of 2 ps.

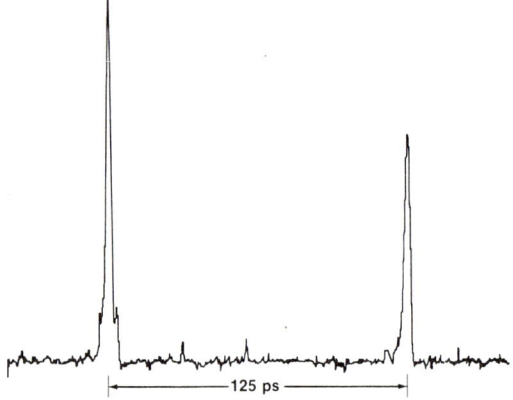

Fig. 3   Streak camera trace of amplified and compressed pulse with a reflection from an etalon. The pulses were frequency doubled in order to be detected by cathode. Pulsewidth is limited by the streak camera response of 2 ps.

We are presently developing an alternate system that should improve both the compressed pulsewidth and the repetition rate. The Nd:Yag oscillator will be replaced by Nd:YLF which has a broader gain bandwidth and can therefore permit shorter mode-locked pulsewidths. We have measured pulsewidths on the order of 35 ps from a cw mode-locked Nd:YLF oscillator, BADO et al. [8]. Considering other compression experiments using similar input pulsewidths, we expect to achieve compressed pulsewidths of approximately 0.5 ps, JOHNSON et al. [9]. Nd:YLF lases at 1.05 μm which matches the peak of the gain bandwidth of phosphate glass amplifiers. The stretched pulses from the Nd:YLF oscillator will therefore be amplified in phosphate glass amplifiers. The phosphate glass system can be operated at a higher repetition rate than the silicate glass system because the phosphate glass has both better thermal properties and higher gain. Using a Nd:YLF oscillator and a single phosphate glass regenerative amplifier we expect to produce pulses with energies of a few mJ and pulsewidths of 0.5 ps at a repetition rate of 30 Hz.

We feel that this completely solid state, short pulse system is a viable alternative to existing 10-Hz dye laser systems as an ultrafast spectroscopic source, [10]. The solid-state system is much less complicated and far more compact than the dye systems, and yet, produces comparable energies and pulsewidths. The peak power that can be produced with one regenerative amplifier is sufficient to produce a white light continuum and therefore provides a tunable source. Higher intensity levels may be achieved, because the beam quality of a solid-state system is superior to that of a dye system. We have measured the focal spot size of our amplified and compressed pulses to be within twice the diffraction limited spot size.

This technique is capable of being scaled to achieve extremely high peak powers. The bandwidth of Nd:glass amplifiers could allow 100 fs compressed pulsewidths and the stored energy in the amplifiers is on the order of 5 J/cm$^2$. Pulses with durations of a few hundred picoseconds can presently be amplified to the KJ level and, in principle could now be compressed with this technique to less than a picosecond. The dramatic improvement in efficiency and peak power that can be obtained with this technique is illustrated in Fig. 4. Curve (a) shows the energy that can presently be extracted from the Glass Development Laser at the Laboratory for Laser Energetics as a function of pulsewidth. The curve is determined by the peak power limitation of the glass amplifier. Curve (b) shows the performance improvement that this compression technique makes possible. For a picosecond pulsewidth, the efficiency is improved by more than two

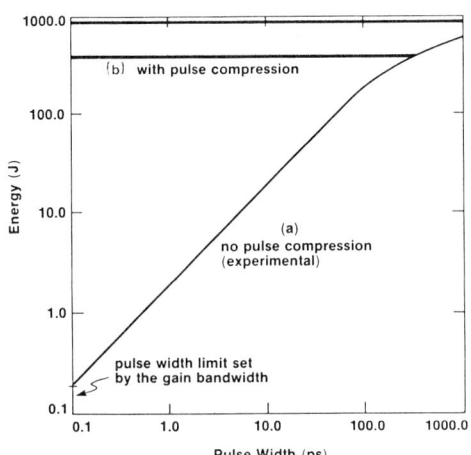

Figure 4  (a) Energy obtainable on the Glass Development Laser at the University of Rochester as a function of pulsewidth, curve (b) illustrates the performance improvements achievable by the pulse compression techniques.

orders of magnitude. This technique should allow for the first time peak powers of $10^{15}$ W, providing new possibilities to the field of matter under intense radiation, BOYER et al. [7].

Acknowledgment

This work was supported by the Laser Fusion Feasibility Project at the Laboratory for Laser Energetics which has the following sponsors: Empire State Electric Energy Reseach Corporation, General Electric Company, New York State Energy Research and Development Authority, Ontario Hydro, Southern California Edison Company, and the University of Rochester. Such support does not imply endorsement of the content by any of the above parties.

References

1. E. Huttman, German Patent No. 768068, issued March 22, 1940.
2. W.A. Cauer, German Patent No. 892,772, issued December 19, 1950.
3. Donna Strickland and Gerard Mourou, Opt. Comm. 56, 219 (1985).
4. H. Nakatsuka and D. Grischkowsky, Opt. Lett. 6, 13 (1981).
5. E.B. Treacy, IEEE J. Quantum. Electron. QE-5, 454 (1969).
6. J. Desbois, F. Gires, and P. Tournois, IEEE J. Quant. Electron. QE-9, 213 (1973).
7. K. Boyer and C.K. Rhodes, Phys. Rev. Lett. 54, 1490 (1985).
8. P. Bado and S. Coe, private communication.
9. A.M. Johnson and W.M. Simpson, J. Opt. Soc. Am. B 2, 619 (1985).
10. For a review of various amplification schemes see Picosecond Phenomena III, pages 10, 19 and 107. Edited by K.B. Eisenthal, R.M. Hochstrasser, W. Keiser, and A. Laubereau, Springer-Verlag, 1982.

# New Optical Design for a Jet Amplifier

*O. Seddiki, A. Goddi, R. Mounet, J.-F. Morhange, and C. Hirlimann*

Université P. et M. Curie, Laboratoire de Physique des Solides,
Spectroscopie des Solides, Associé au CNRS (LA 154),
4 place Jussieu, Tour 13-2, F-75252 Paris Cedex 05, France

It has been recognized during the past few years that it is possible to amplify light pulses using gain jets /1,2/. This has been shown to be of great importance in the femtosecond regime, where the length of the gain medium has to be minimized in order to avoid too much group velocity dispersion. We have designed a simple, compact, all reflective multipass amplifier.

The apparatus is depicted in Fig. 1. Two identical concave mirrors, with their centers placed at C1 and C2, are used to focus and recollimate the light beams. A flat mirror, orthogonal to the symmetry axis of the figure, reflects the beams onto the first concave mirror with a slight shift to insure the multiple passes of the light. By translating this mirror, in a direction perpendicular to its plane, it is possible to vary the number of passes. It appears, from symmetry considerations, that all the rays are concurrent at a same point. Although it is no longer true for beams, due to transverse spherical aberration, it has to be considered as an advantage for it results in a better depletion of the gain volume.

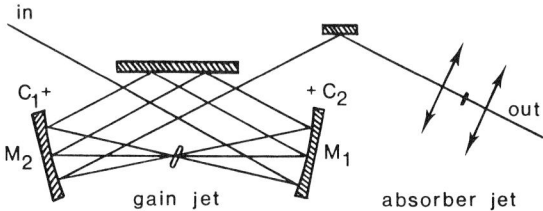

Fig.1 Optical scheme of the amplifier

We have tested our optical scheme by building up a laser preamplifier. The light pulses to be amplified are generated by a colliding-pulse mode-locked dye laser (CPM) and are 100 fs in duration, 50 pJ in energy /3/. The gain medium is excited by the doubled output (532 nm) of a 10 Hz repetition rate, Nd:YAG laser. The pumping pulse duration is 8 ns, slightly shorter than the 10 ns period of the CPM.

All the mirrors have enhanced silver coating and a reflectivity of 98%. The focusing mirrors are 150 mm in focal length. The flat mirror, 30 mm long, is placed in such a way that the beams are spread in an angle of $6°$. The pumping beam makes an extra $10°$ angle, in the horizontal plane, with the medium beam of the layer. A home-made stainless steel nozzle produces a 1.5 mm thick jet of dye dissolved in ethylene glycol. This jet is placed at the common focal point. Ten to twenty percent of glycerol is added to the solvent to enhance its viscosity /4/ in order to achieve a laminar flow and a good optical quality. Sulforhodamine 101 was chosen as a dye for its emission peaks at 608 nm, close enough to the 620 nm wavelength of the CPM laser. The pumping beam is focused, by means of a 30 cm focal length lens, down to a spot diameter of the order of 1 mm. A half-wavelength plate is used to turn the pumping electric field parallel to the amplified electric field.

The amplification part of the system is followed by a saturable absorber jet (Malachite green in ethylene glycol), placed in a 1:1, 3 cm focal length telescope. The low signal absorbance is set to 40.

Figure 2 shows the dependance of the total, 6 passes gain experienced by the amplified pulses, versus pumping energy. Three different regimes can be distinguished. Up to 2 mJ pumping energy, the total gain has a linear behavior. It then saturates up to 15 mJ where it rapidly drops down. The observed saturation reflects the competition between the amplification of the CPM pulse and the amplification of the spontaneous emission (ASE). The final drop in gain is probably due to the destruction of the dye molecules at high pumping energy. This curve shows that a total gain of 10000 can be achieved in the linear regime as was already observed in ref. /1/.

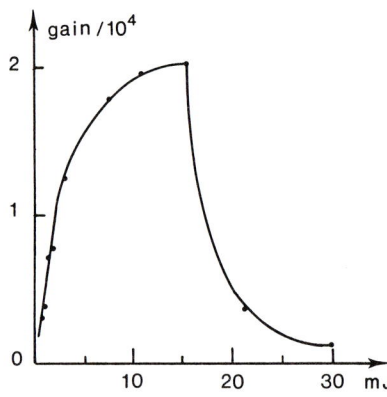

Fig.2 Total gain for 6 passes, versus the pumping energy

Time measurements showed that the duration of the pulses increases from 100 fs to 200 fs in the linear regime and for the 6 passes configuration. Further work is underway to check the ability of the pulses of being time recompressed in the linear amplification regime.

In a previous work /5/ it has been established that although the total gain increases with the number of passes, the mean gain per pass decreases. This can be understood as due to the growing ASE and also to the fact that the total path length for the CPM pulse comes closer to the spatial length of the pumping pulse when the number of passes is increased. We suggest attenuating the ASE by adding a proper saturable absorber dye to the gain medium.

Despite transverse aberration, due to the off-axis use of spherical mirrors, the optical quality of the amplified beam remains good, part of the distortion beeing cured by gain volume geometrical effect. Further improvement could be done by replacing the spherical mirrors by parabolic ones.

In conclusion, we have shown that our three mirror multipass laser amplifier is able to achieve good optical quality, high gain efficiency, in the femtosecond regime. This is done using only 10% of the pumping laser energy, giving us the opportunity to try further amplification by means of a multistage set up.

We would like to thank G. Mouget for setting up the synchronisation of the lasers and B. Barrau for realizing the mechanical parts of the system. This work has been performed under D.R.E.T. grant number 84-070.

1. W.H. Knox, M.C. Downer, R.L. Fork and C.V. Shank: Optics Lett. $\underline{9}$, 552 (1984).
2. T.L. Gustafson and D.M. Roberts. Optics Comm. $\underline{43}$, 141 (1982).
3. R.L. Fork, B.I. Green and C.V. Shank: Applied Physics Lett. $\underline{38}$, 671 (1981).
4. R.L. Fork and F.A. Beisser: Private Communication.
5. C. Hirlimann, O. Seddiki, J-F. Morhange, R. Mounet, and A. Goddi: To be published in Optics Commun.

# Fiber Raman Amplification Soliton Laser (FRASL)

M.N. Islam, L.F. Mollenauer, and R.H. Stolen
AT & T Bell Laboratories, Holmdel, NJ 07733, USA

We have demonstrated the first Fiber Raman Amplification Soliton Laser (FRASL) by synchronously pumping a loop of fiber with $\approx 10$ psec pulses from a modelocked color center laser. The FRASL output pulses are determined by pulse compression and soliton pulse shaping in the fiber, and have been to date as short as 240 fsec. The lasing wavelength can be time dispersion tuned[1]. The single cavity FRASL is a simpler configuration than the color center soliton laser[2], which requires feedback stabilization to maintain interferometric compatibility between its coupled cavities. Since the FRASL can be made to operate near $\lambda = 1.55\mu m$, it promises to be an attractive source for optical fiber communications or optical signal processing.

**Single Fiber FRASL**

The first experimental configuration for the FRASL is shown in Fig. 1. A dichroic beam splitter (highly reflective at the pump wavelength $\lambda_p$ and partially reflective at the stokes wavelength $\lambda_s$) couples in the pump and partially couples out the laser. The fiber is 500 m of single-mode, polarization-preserving, dispersion-shifted fiber[3] with a zero of dispersion near $1.536\mu m$ (lasing has been observed also using 100 m and 1.4 km fibers). A dispersion shifted fiber is required to increase the interaction distance between the pump and stokes pulses, and so that the pump can have normal dispersion while the stokes has anomalous dispersion. Although the threshold for lasing is a minimum when the

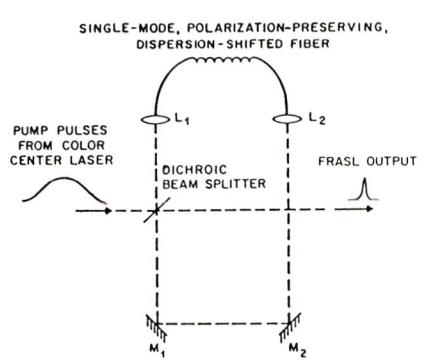

Figure 1. Experimental configuration for the first Fiber Raman Amplification Soliton Laser (FRASL).

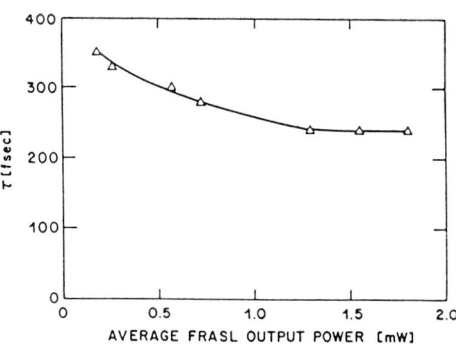

Figure 2. Pulse width versus average FRASL output power for a 500 m fiber and $\lambda_p \approx 1.47\mu m$.

pump and stokes pulses overlap over the entire length of fiber ($\lambda_p$ = 1.488 μm, $\lambda_s$ = 1.588 μm), we observe improved pulses when the pump is tuned to $\lambda_p \approx$ 1.47 μm. For the latter case, gain occurs over only about a fifth of the fiber length, and in the remaining length the soliton pulse shaping can take place without perturbation from loss or gain.

The range of pulse widths for $\lambda_p \approx$ 1.47μm is plotted versus average FRASL output power in Fig. 2. Close to threshold, approximately 350 fsec pulses are produced with a low, but broad, pedestal. The pedestal appears to be the non-soliton part which is dispersed out during the soliton pulse shaping.[4] As the power is increased, the pulse width narrows at first, consistent with the inverse proportionality between average power and pulse width for a fundamental soliton. However, the pedestal becomes more pronounced at higher powers, indicating an increased amount of non-soliton part for harder pumping. At sufficiently high pump powers, we observe both the first and second stokes wavelengths.

The spectral width for a 240 fsec hyperbolic secant pulse is about 44 cm$^{-1}$. This corresponds approximately to the measured bandwidth; therefore, at least the narrow part of the pulse is transform limited. The pulse width may be clamped at 240 fsec due to bandwidth limiting from the Raman gain itself. Also, we are able to time-dispersion-tune the laser over $\approx$ 80 cm$^{-1}$ by changing the cavity length of the FRASL. The FRASL output power near threshold is consistent with operation on a fundamental soliton.

Solitons can be amplified many-fold without a significant change in pulse shape if the gain occurs rapidly. For example, computer simulations of the nonlinear Schrodinger equation show that if the soliton is amplified sixteen-fold in a third of a period (the period $z_0$ is a characteristic length for a soliton), then the pulse width changes by less than three percent. Thus, the ideal FRASL should have amplification of the pulse in a distance short compared to $z_0$, followed by a length of fiber several $z_0$ long in which the soliton can readjust. Fibers with enhanced Raman cross-sections (e.g. $GeO_2$) could provide the necessary gain in a short distance.

## Two Fiber Configuration for FRASL

To incorporate these ideas, we designed and constructed a second configuration of the FRASL with distinct amplification and pulse shaping sections (Fig. 3). The two lengths of fiber are linked by a variable fiber coupler[5], which can transfer as much as 30 percent of the input. Now the pulse can be amplified to many times the soliton power in the first section, and then the coupler can be adjusted to yield close to the fundamental soliton power in the second fiber. The advantages of this second configuration are three-fold: (i) the FRASL output can be many times the fundamental soliton power; (ii) the pedestal should be reduced since the pulse shaping occurs at the correct power level; and, (iii) the alignment is simpler because of the increased number of accessible ports.

To start with, a 100 m gain section and a 500 m fiber for pulse shaping were used. Again, the FRASL can be time-dispersion-tuned; however, depending on where it is tuned, the pulse behavior is different. When the pump and

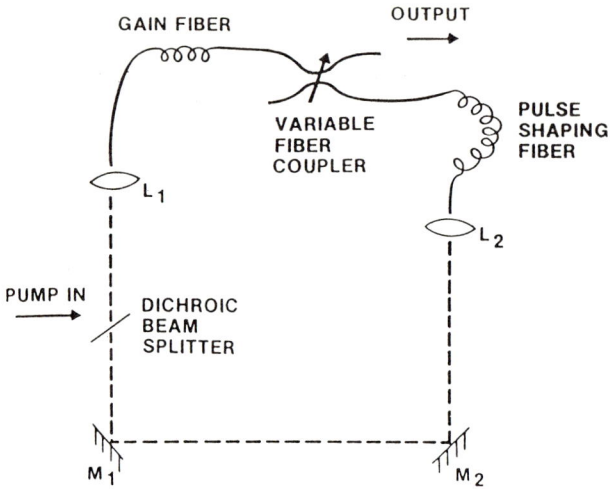

Figure 3. Second configuration of the FRASL with distinct amplification and pulse shaping sections.

stokes wavelengths are about 440 cm$^{-1}$ apart (near the peak of the Raman gain), we obtain a narrow 320 fsec pulse on top of a broad pedestal. As the coupler is detuned, the pulses widen while the pedestal decreases. Near threshold (coupling over $\approx$3%) we obtain 1.5 psec pulses with negligible pedestal (Fig. 4a). However, in this case the power in the second fiber is only about a tenth of the required soliton power.

On the other hand, if we lengthen the cavity and reduce the difference between pump and stokes wavelengths to $\delta\nu$=380 cm$^{-1}$, then at full coupling

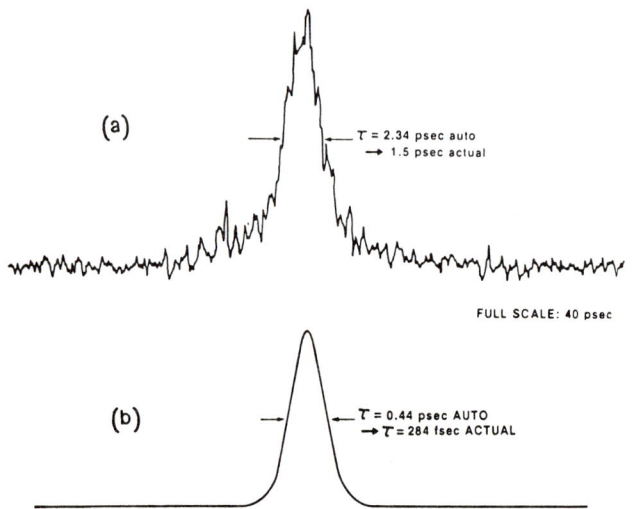

Figure 4. Autocorrelation of pulses from the laser of Fig. 3 with a 100 m gain section. (a) $\delta\nu$ = 440 cm$^{-1}$, coupler detuned to transfer $\approx$ 3%, laser near threshold; (b) $\delta\nu \approx$ 380 cm$^{-1}$, coupler transferring $\approx$ 30%.

we obtain pedestal-free 284 fsec pulses (Fig. 4b). The power in the pulse shaping fiber is approximately correct for a fundamental soliton. However, the soliton period $z_0$ is only 6.5 m, while the gain length $l_{gain}$ (limited by pulse walk-off) is about 52 m.

Since $z_0 \propto \tau^2$, the short pulses violate the criterion for the ideal FRASL that $(l_{gain}/z_0) \leq 1/3$. Apparently, the 100 m gain section has amplification as well as pulse compression and frequency shifting. In the first part of the 100 m, the pulse is amplified to a higher order soliton, which then compresses to a sub-picosecond pulse[5]. The recently discovered soliton self-frequency shift[6], which is proportional to $1/\tau^4$, then shifts the soliton to longer wavelengths. The resulting feedback in the laser is weak since the signal returning is at the wrong frequency.

To separate the amplification process from the other detrimental effects, we shortened the gain fiber to 50 m. The resulting pulses did broaden, but the link between pedestal and time-dispersion-tuning persisted. With the laser tuned for $\delta\nu \approx 440$ cm$^{-1}$, we obtain a narrow 0.64 psec pulse on top of a broad pedestal. When the laser is tuned to smaller $\delta\nu$, 0.85 psec pulses result with a much reduced pedestal. However, these pulses have excess bandwidth. For these broader pulses, further bandwidth limiting may be required to achieve transform limited pulses.

### Discussion

Although as Fig. 4 illustrates we have obtained attractive pulses from the FRASL, we have not as yet succeeded in separating the process of amplification from pulse shaping. To achieve this separation would require that $(l_{gain}/z_0) < 1/3$. If we have a long gain fiber, then pulse compression makes $z_0$ too short. If we shorten the gain fiber, then we end up with excess bandwidth. Somehow we must achieve stable operation on $\tau \geq 1$ psec pulses without generating unwanted bandwidth.

We do not yet fully understand the relation between time-dispersion-tuning and the level of pedestal. One possibility is that due to the soliton self-frequency shift, the broad pedestal is at a different frequency from the narrow peak. It also seems likely that the Raman amplification process chirps the stokes pulse. In the laser, Raman gain, time-dispersion-tuning and the soliton self-frequency shift complicate the usual self-phase modulation and group velocity dispersion of a soliton. The closure condition[4] requires that all these chirping mechanisms must balance in a round trip of the laser.

In summary, we have demonstrated the first fiber Raman amplification soliton laser. In the first configuration consisting of a beam splitter and a loop of fiber, we obtained pulses between 240 and 350 fsec with pedestals. In the second cavity consisting of a beam splitter and two lengths of fiber linked by a fiber coupler, we have achieved pedestal-free 284 fsec pulses. The aim is for a FRASL in which the gain length is a fraction of the soliton period $z_0$, while the pulse shaping occurs over several $z_0$.

**References**

[1] R.H. Stolen, Chinlon Lin and R.K. Jain, Appl. Phys. Lett. **30**, 340 (1977).

[2] L.F. Mollenauer and R.H. Stolen, Opt. Lett. **9**, 13 (1984).

[3] We thank J.R. Simpson, H-T Shang and A. Tomita for providing the fiber.

[4] H.A. Haus and M.N. Islam, IEEE J. Quantum Electron. **QE-21**, 1172 (1985).

[5] L.F. Mollenauer, R.H. Stolen and J.P. Gordon, Phys. Rev. Lett. **45**, 1095 (1980).

[6] F.M. Mitschke and L.F. Mollenauer, "Discovery of the Soliton Self Frequency Shift" (submitted to Opt. Lett.); J.P. Gordon, "Theory of the Soliton Self Frequency Shift " (submitted to Opt. Lett.); F.M. Mitschke and L.F. Mollenauer, postdeadline paper, 1986 Conf. on Ultrafast Phenomena, Snowmass, Co.

# Dispersion Compensated Fiber Raman Oscillator

J.D. Kafka, D.F. Head, and T. Baer

Advanced Product Technology Group, Spectra-Physics, Inc.,
Mountain View, CA 94040, USA

We have generated subpicosecond pulses from a synchronously pumped fiber Raman ring laser with a dispersion compensating delay line. The ring oscillator consists of 100 M of single mode fiber, a grating pair dispersive delay line, and an interference filter as a tuning element. (Fig. 1) The pump source is a CW modelocked Nd:YAG laser emitting 80 psec pulses at 1064 nm. We observe pulses at the the first Raman stokes band (1100 nm) with pulsewidths as short as 0.8 psec.

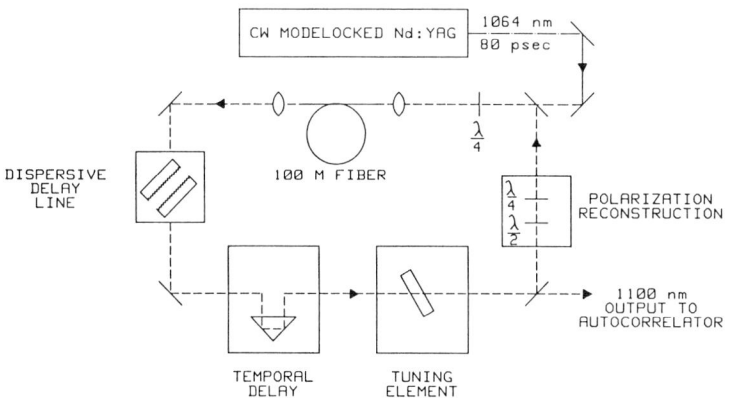

Figure 1 Schematic diagram of a dispersion compensated fiber Raman ring laser.

The fiber Raman laser is an ideal gain medium for a synchronously pumped source since it has high gain, which allows insertion of intracavity elements, and sufficient gain bandwidth to support subpicosecond pulses [1]. Two factors which potentially limit the pulsewidth from this oscillator are self-phase modulation (SPM) and group velocity dispersion (GVD) which take place in the fiber when a short pulse propagates. An elegant solution to this problem is to choose the oscillation wavelength such that the GVD of the fiber is slightly negative giving rise to soliton formation and ultrashort pulses [2].

An alternative solution is to incorporate a dispersive delay line which makes the GVD for a complete round trip in the ring oscillator slightly negative. This is accomplished by adjusting the spacing of the grating pair delay line to compensate for the linear dispersion in the fiber. The addition of the dispersive delay line should in principle allow the fiber Raman oscillator to support short pulses at any wavelength.

The operation of the dispersion compensated fiber Raman oscillator is similar to a synchronously pumped dye laser. Dispersion compensation is accomplished using a double pass grating pair with a grating separation of about 1 cm. The round trip time of the fiber ring oscillator must be matched to an integer multiple (approximately 40) of the pump pulse separation. An optical delay line is used to adjust the length of the oscillator to an accuracy of approximately 100 microns. The tuning element, which is a 10 nm bandpass filter, is necessary for short pulse generation. The tuning element allows the central wavelength to be selected as well as restricting the oscillating bandwidth.

Once the tuning element is inserted, the cavity length can be adjusted without the usual wavelength shift due to time dispersion tuning[1]. Moreover, as the length of the ring cavity is detuned from optimal, the shape of the output pulse autocorrelation varies in a manner similar to that of a synchronously pumped dye laser. Finally, a pair of waveplates are used to adjust the polarization of the 1100 nm light before it is reinjected into the fiber with a dichroic mirror. We find that correct adjustment of the polarization is necessary for short pulse operation and we are currently investigating polarization preserving fiber.

We have obtained pulses of 0.8 ps duration from the dispersion compensated fiber Raman oscillator with 20 mW of average power. The pulse autocorrelation is shown in Figure 2 with a baseline given for reference and a gaussian pulseshape assumed. Also shown in Figure 2 is the spectrum of the pulse at 1100 nm.

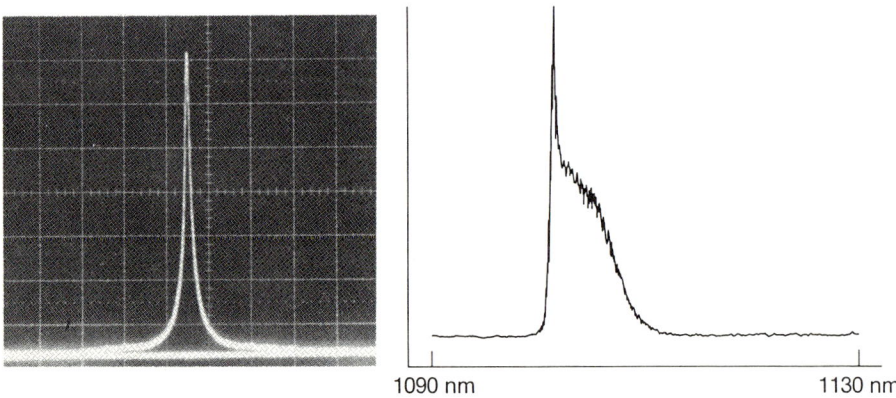

Figure 2 Autocorrelation and spectrum of the output pulse from the oscillator. The FWHM of the pulse is 0.8 psec assuming a Gaussian pulse shape.

At lower intensity the spectrum has the double peaked arch shape indicative of SPM [3]. When SPM occurs, the nonlinear index of the fiber causes the pulse to shift its instantaneous frequency first to the red and then to the blue. As the system is optimized, however, the intensity becomes sufficient for the pulse at 1100 nm to launch its own Raman band. This light travels faster than the 1100 nm pulse and preferentially depletes the leading edge. Since the leading edge of the pulse has been red shifted due to SPM, the red arch is depleted and the spectrum shown in Figure 2 results. The width of the spectrum is approximately 5 nm yielding pulses which are about two times the transform limit.

The dispersion compensated fiber Raman laser offers an attractive alternative to dye lasers or pulse compression for subpicosecond pulse generation. The output pulse characteristics do not appear critically dependent on pump source intensity and pulsewidth. The autocorrelation measurements indicate that the pulses have less energy in a background pedestal than compressed pulses. Finally, the fiber lengths and pump powers required can be made quite small.

References

1.  R. H. Stolen and C. Lin, CRC Handbook of Laser Science and Technology, ed. Marvin J. Weber, (CRC Press Inc., Boca Raton, FL) Vol. 1, p 265

2.  M. N. Islam and L. F. Mollenauer, Paper TuHH1, IQEC 1986, San Francisco, CA

3.  J. D. Kafka, B. H. Kolner, T. Baer and D. M. Bloom, Opt. Lett. 9, 505 (1984).

# 80-fs Soliton-like Pulses from an Optical Nonlinear Fiber Resonator

*B. Zysset, P. Beaud, W. Hodel, and H.P. Weber*

Inst. of Appl. Phys., University of Berne, Sidlerstr. 5, CH-3012 Berne, Switzerland

Generation of ultrashort pico- and subpicosecond pulses in the wavelength region between 1.3 and 1.55 µm is of importance for application in time resolved spectroscopy and in telecommunication. Very tempting seems the possibility of soliton pulse generation or narrowing with optical fibers. The soliton laser [1] and passive soliton-like pulse narrowing [2] have been demonstrated. Here we report on observations of frequency conversion and short, soliton-like pulse formation in a glass fiber nonlinear oscillator. Four-wave-mixing (FWM) and Raman interaction convert the pump light efficiently to a Stokes band. Short pulses of a minimum width of 80 fsec are formed out of this band through soliton pulse shaping mechanisms. The wavelength of these pulses is tunable in the range of 1.37-1.4 µm. The pulses from the resonator are compared with pulses generated in singlepass amplification, where a minimum pulse width of 160 fsec is found at a wavelength of 1.38 µm.

In our experiment we used a dye laser [3] to stimulate the nonlinear effects in the fiber. The dye laser is synchronously pumped by compressed 4 psec pulses of a cw-modelocked Nd:YAG laser [4]. Dye laser pulses of 300 fsec - 1 psec are obtained by choosing a proper bandwidth limiting element. In our experiment we worked with 1 psec pulses. The dye laser is tunable in the wavelength region between 1.25 - 1.35 µm. The average output power is 80 mW corresponding to a maximum peak power of 1 kW. These laser pulses are coupled into a single mode fiber to stimulate FWM and Raman scattering. Two different configurations were used: (i) single pass amplification to investigate the evolution of the spectrum and pulse duration as function of pump power and (ii) the fiber is placed inside a resonator to provide feedback thereby enhancing pulse shaping mechanisms and tunability. Linear and ring resonators were investigated. Coupling to the fiber is via IR antireflection coated microscope objectives to minimize losses which are estimated to 2-3 dB per roundtrip. The spectrum of the output as well as pulse duration are extremely cavity length dependent. Fine adjustments are made by moving the coupling port relative to the mirror for cavity length changes or by moving the objective relative to the fiber in order to change the coupling efficiency. No tuning or bandwidth limiting element is used in the cavity. Wavelength tuning is achieved by three different effects: (a) tuning the pump laser relative to the dispersion minimum of the fiber, (b) time dispersion tuning which is due to the difference in round trip time for different frequency components of the Stokes pulse, (c) an intensity-dependent frequency shift of the maximum gain of the FWM process as will be evidenced below. The interplay of these effects makes the wavelength tuning quite complicated in the experiment. However, it was possible to obtain short pulses in the region 1.37 - 1.4 µm. This range can probably be extended.

The non-polarization-preserving single mode fiber (Cableoptic, Cortaillod) used had a depressed cladding, a core radius of 4.1 µm, a cut-off wavelength of 1030 nm and a minimum dispersion wavelength $\lambda_o$ = 1.317 µm.

We first present spectral measurements of single pass amplification experiments in a 25 m piece of fibre. The spectra of the output was recorded as a function of pump power at different pump wavelengths. Examples thereof for the special case of $\lambda_p = \lambda_o$ are shown in Fig 1. Several features are remarkable: (1) The shift of the Stokes band differs from the ordinary Raman shift of 440 cm$^{-1}$ suggesting that we do not observe simple Raman amplification. (2) No Stokes band is observed if $\lambda_p$ is made smaller than $\lambda_o$, though the Raman process should be favoured as the effective interaction length increases due to the better match of the group velocities. However, efficient conversion is obtained if $\lambda_p \geq \lambda_o$. (3) The emission wavelength of the Stokes band is pump intensity dependent. (4) At high pump powers a second Stokes peak appears. (5) Very weak anti-Stokes components are observed with much smaller intensity than the Stokes band. The shift of the Stokes band with increasing intensity is shown in Fig. 2a. At a pump intensity of 0.17 GW/cm² a distinct peak at $\Omega$ = 150 cm$^{-1}$ appears. By increasing the pump level $I_p$ the Stokes band shifts rather rapidly away from the pump line. At $I_p$ = 0.47 GW/cm² $\Omega$ equals 310 cm$^{-1}$. At this power level a second Stokes component appears and the shift of $\Omega$ with pump power levels off. Obviously this spectral evolution can not be explained by a Raman process. The appearance of the anti-Stokes component indicates a FWM-process. For parametric FWM near a Raman resonance, theory [4] leads to a tentative explanation for the large intensity-dependent shift of the Stokes band. Accordingly, the maximum gain of FWM will occur for a certain optimum phase mismatch $\Delta k^{opt}$ which is proportional to the pulse intensity in the fiber. The normalized mismatch $\Delta k^{opt}/I_p$ is of the order of $-1\cdot 10^{-2}$ cm$^{-1}$/[GW/cm$^{-2}$]. The fiber itself provides this necessary phase mismatch. In Fig. 2b the calculated phase mismatch curve is plotted for Stokes shifts between 0-1000 cm$^{-1}$ for the fiber used. The mismatch is seen to be of the proper sign and smaller than $10^{-2}$ cm$^{-1}$ for frequency shifts $\Omega$ less than 300 cm$^{-1}$. It is also evident that an increasing $\Delta k^{opt}$ will increase the Stokes shift. Qualitatively, the observed intensity-dependent shift can therefore be explained. The suppression of the anti-Stokes component is explained by stimulated Raman absorption which is very efficient for frequency shifts close to Raman resonance and the high pump intensities used. Up to now we have only discussed spectra generated in single pass amplification. In case of the resonator, the situation is additionally complicated due to time dispersion and pulse shaping effects. However, the essential features of the spectra are similar in both cases.

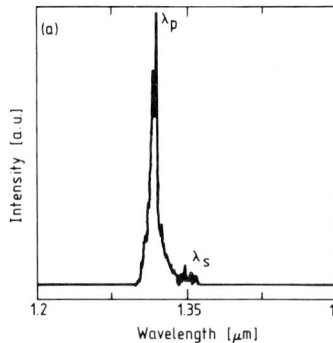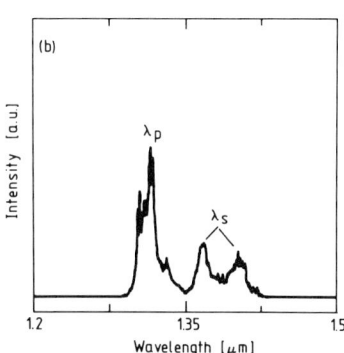

Fig. 1 Spectra generated by single pass amplification in 25 m of fiber at intensities of (a) 0.17 GW/cm², (b) 0.47 GW/cm². Pump laser wavelength is 1.317 μm

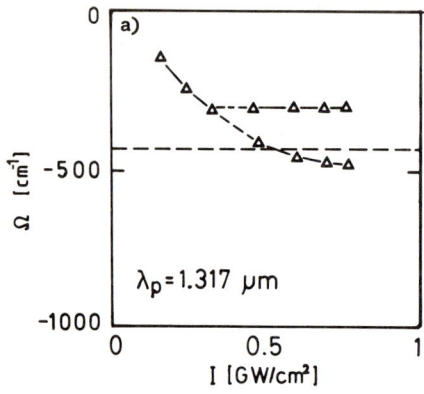
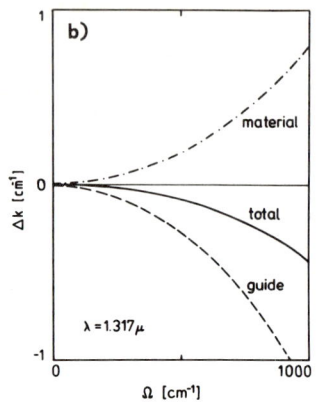

Fig. 2 (a) Shift of Stokes peaks relative to pump line as function of intensity. (b) Phasemismatch Δk of the fiber vs. Stokes shift Ω for pump at 1.317 μm

The duration of the Stokes pulses was measured by conventional background-free autocorrelation technique. The appropriate frequency components were filtered with selected interference filters. We generally found the pulse duration and the background energy to be strongly adjustment dependent. The procedure therefore was to set the dye laser to a given wavelength and adjust the nonlinear oscillator to maximum output at the Stokes wavelength. At this point the pulse of the Stokes band shows a intense, broad background of several picosecond duration with a short subpicosecond pulse superimposed, which was clearly identified as a pulse and not a coherence spike. Out of this situation short 80-100 fsec pulses are formed by length adjustments of the cavity. An example of such a pulse is shown in Fig. 3. The pump parameters were: Pp = 542 W, $\tau_p$ = 0.9 psec, $\lambda_p$ = 1.325 μm. Assuming a sech form the autocorrelation represents a 100 fsec pulse clearly documenting the pulse shortening mechanisms in the cavity. The noise upon this curve is mainly caused by the fluctuations of the pump. There is a weak broad background recognizable whose amount is strongly adjustment dependent. The measured spectrally integrated output power of the pulse and the background is 1259 W. This value can be compared to the peak power of a fundamental soliton with appropriate parameters $\lambda_s$ and $\tau_s$. This soliton power would be $P_s$ = 922 W. Comparing the two values one notes that the latter is somewhat smaller. It is supposed that the measured excess energy is due to the background in the pulse that associates the generated solitons. To judge this in more detail we have made several measurements of this kind and the results thereof are summarized in Fig 4. There, the measured spectrally integrated peak powers of the pulses are plotted versus a variable which comprises the soliton parameters $\lambda_s$ and $\tau_s$ and D the dispersion parameter of the fiber. The solid line represents the theoretically expected value for a soliton of order 1. All of the measured values are slightly higher but close to the theoretically expected value. A true identification of solitons should be done by either the measurement of the pulse shape or the transmission properties of the pulses in long lengths of fibers. This has not been done up to now but the above consideration strongly supports the assumption of soliton formation. Short pulse formation occurred also in single pass amplification. Such results are also included in Fig. 4. It was noticed, however, that these pulses are generally longer than from the resonator. The shortest pulses observed in this way had a duration of 160 fsec.

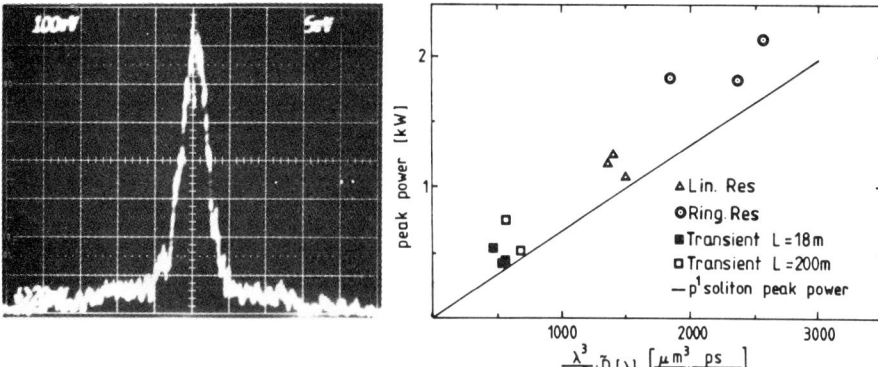

Fig. 3 Autocorrelation trace of soliton pulse. Pump at 1.325 μm, linear resonator, 18 m fiber. Scale is 220 fsec/div.

Fig. 4 Spectrally integrated peak power of pulses vs. soliton parameter. Symbols: experiment, solid line: theoretical soliton value

In conclusion, we have demonstrated FWM frequency conversion and short soliton-like pulse formation in a fiber resonator. Pumped by 1 psec pulses of a dye laser around the dispersion minimum of the fiber, Stokes pulses in the wavelength region 1.37-1.4 μm were obtained with a minimum width of 80 fsec.

We would like to thank L. Thèvenaz (University of Geneva) for helpful discussions and measurements of fiber parameters. Part of this work was sponsored by the R+D Department of the Swiss Post Office.

References

1. L.F. Mollenauer and R.H. Stolen: Opt. Lett. 9, 13 (1984)
2. L.F. Mollenauer, R.H. Stolen, J.P. Gordon, and W.J. Tomlinson: Opt. Lett. 8, 289 (1983)
3. P. Beaud, B. Zysset, A.P. Schwarzenbach, and H.P. Weber: Opt. Lett. 11, 24 (1986)
4. B. Zysset, W. Hodel, P. Beaud, and H.P. Weber: Opt. Lett. 11, 156 (1986)
5. A. Penzkofer, A. Laubereau, and W. Kaiser: Prog. Quant. Electr. 6/2, 55 (1980)

# The Stabilized Soliton Laser

F.M. Mitschke* and L.F. Mollenauer

AT & T Bell Laboratories, Holmdel, NJ 07733, USA

*In the soliton laser, feedback from a pulse shaping fiber controls output pulse width and shape. We describe servo control that locks to the required optical phase of that feedback, such that the laser produces an uninterrupted stream of pulses of uniform height and width. The stabilized soliton laser has produced pulses as short as 60 fs directly, and as short as 19 fs with compression in an external fiber.*

The soliton laser [1] consists of a synch-pumped, mode-locked color center laser, tunable in the 1.5 μm region, coupled to a second cavity, containing a single mode, polarization preserving optical fiber. (See Fig. 1.) Feedback from the fiber, where pulse compression and soliton formation take place, enables the color center laser to produce transform limited pulses of ~sech$^2$ intensity profile, with pulse width determined by the control fiber length. ($\tau$ scales as $\sqrt{L}$; see [1].)

Successful operation requires, of course, that the round trip time in the control cavity be matched to (or be an integral multiple of) that in the main cavity. But it is also required that the feedback from the fiber have the proper optical phase. Thus, without appropriate stabilization, mirror vibra-

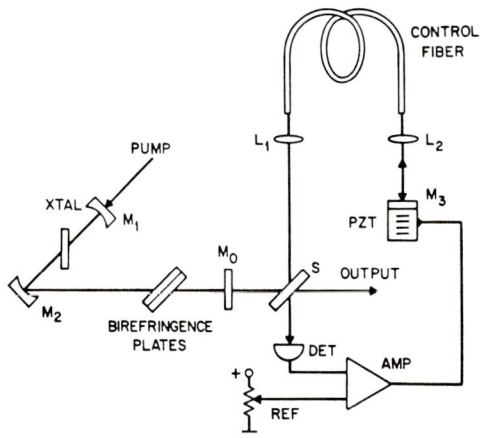

Fig. 1: Schematic of the soliton laser, with stabilization loop. Beam splitter S has $R = 30-50\%$. L: microscope objectives PZT: piezoelectric translator

---

* on leave from Institut für Quantenoptik, Universität Hannover, BRD

tion and drift of just a fraction of a wavelength cause the soliton laser action to flicker on and off at random.

Our stabilization scheme is based on the fact that the power in the control cavity varies with ϕ, the round trip optical phase shift in the control cavity, about as shown in Fig. 2.

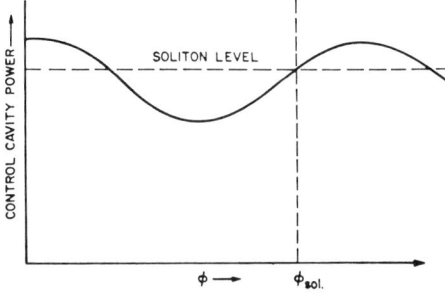

Fig. 2: Variation of control cavity power with round trip optical phase shift ϕ.

In addition, we find experimentally that soliton laser action is correlated with a well-defined level, lying somewhere within the middle of the range of cavity powers. The scheme of the stabilization circuitry shown in Fig. 1 is now easily understood. The error signal for the control of ϕ is generated by taking the difference between the detector signal (a measure of the control cavity power, and hence a measure of ϕ) and a reference voltage corresponding to the soliton level. The op-amp magnifies that difference and drives the PZT translator of the end mirror $M_3$. Thus, assuming correct choice of signal polarity, a closed negative feedback loop is formed.

The autocorrelation trace of Fig. 3 illustrates the stability and low noise of the laser's output resulting from the stabilization. Pulse intensity fluctuations are typically less than 1%, even in the face of ±5% or greater fluctuations in the pump laser intensity.

Although pulsing does not start by itself under cw pump conditions, we have discovered that once soliton laser action has begun, it will continue for a

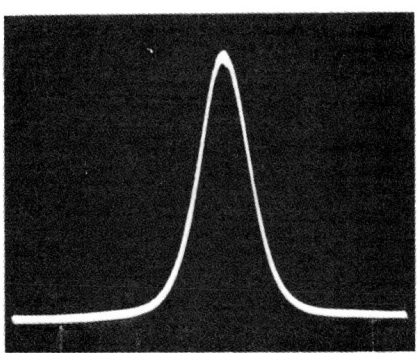

Fig. 3: Autocorrelation trace of the stabilized soliton laser output pulses. In this example, the FWHM of the pulses is 580 fsec (control fiber length $L = 1.6$ m).

while under cw pumping (up to a minute has been observed) until some large perturbation interrupts the pulse stream. Thus, the soliton laser is in fact passively mode-locked.

It was found experimentally in [1] that the laser operates on a so-called $N = 2$ soliton, i.e., a pulse that narrows and widens again, repeating its shape periodically as it travels down the fiber. The soliton period, $z_0$, scales with the square of the pulse width [2]. Thus, if there is a constant relation between $z_0$ and the control fiber length $L$, the width of the pulses formed by the soliton laser should scale with the square root of $L$. This dependence of the pulse width on $L$ has indeed been confirmed experimentally in [1]. The data there also fit the relation $L \approx z_0/2$, with the implication that in the stationary state, pulses returned from the fiber have substantially the same width as those launched into it.

Nevertheless, by duplicating control fiber conditions in a second, external fiber of adjustable length, we have been able to prove experimentally that stable solutions also exist for which $L$ is considerably less than $z_0/2$. Thus, pulses returned from the fiber are often significantly narrower than those circulating in the main cavity. This result shows that it is not possible to model the two cavity soliton laser in terms of an equivalent single cavity, as attempted earlier [3].

With a control fiber of $L \sim 20$ cm, the stabilized soliton laser produces pulses as short as $\sim 60$ fs. Compression of these in an external fiber has resulted in pulses of $\sim 19$ fs, or less than 4 optical cycles at the central wavelength of 1.52 μm (see Fig. 4). The bandwidth (FWHM) of such pulses is $\sim 520$ cm$^{-1}$, or $\sim 120$ nm. Pulse width is ultimately limited only by the bandwidth characteristics of the fiber dispersion; with special dispersion flattened fibers, it may be possible to produce pulses of little more than two optical cycles.

Finally, we note that the stabilized soliton laser has enabled discovery and exploration of a continuous down shift in optical frequency of the soliton, caused by a Raman self-pumping [4]. The effect is expected to allow for the

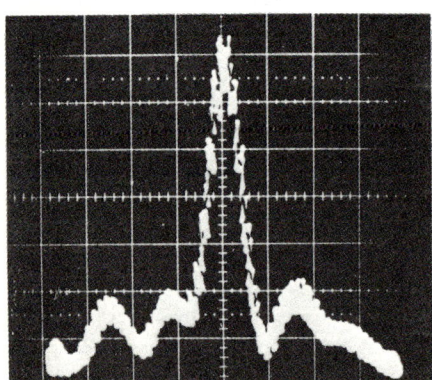

Fig. 4: Autocorrelation trace of soliton laser pulse after compression in an external fiber. Actual width, 19 fs FWHM. The autocorrelation width can be determined by counting ~6 interference fringes between half power points

generation of synchronized femtosecond pulses of different frequencies from a single soliton laser for pump-probe experiments.

1. L. F. Mollenauer and R. H. Stolen, "The Soliton Laser,"
   Opt. Lett. **9**, 13 (1985) and pp. 2-6 in Proc. of the
   IV International Conf. on Ultrafast Phenomena, June, 1984
   (Springer-Verlag, Berlin, 1984)

2. L. F. Mollenauer, R. H. Stolen, and J. P. Gordon, "Experimental observation of picosecond pulse narrowing and solitons in optical fibers," Phys. Rev. Lett. **45**, 1095 (1980)

3. H. A. Haus and M. N. Islam, "Theory of the Soliton Laser,"
   Proc. IEEE J.Q.E. **QE-21**, 1172 (1985)

4. F. M. Mitschke and L. F. Mollenauer, "Discovery of the Soliton Self Frequency Shift," submitted to Optics Lett.

# The Soliton Self Frequency Shift

*F.M. Mitschke, L.F. Mollenauer, and J.P. Gordon*

AT & T Bell Laboratories, Holmdel, NJ 07733, USA

*We describe experimental discovery of a continuous down shift in the optical frequency of a soliton pulse as it travels along the fiber. We show that the effect is caused by a Raman self-pumping of the soliton, by which energy is transferred from the higher to the lower frequency parts of its spectrum. In a several hundred meter long fiber, the net frequency shifts of subpicosecond pulses can amount to several percent or more of the optical frequency.*

The soliton self frequency shift represents a new and important effect for nonlinear pulse propagation in optical fibers, and it is one with practical implications for the measurement of ultrafast phenomena. We discovered the effect in experiments on the propagation of subpicosecond pulses in a several hundred meter length of single mode, polarization preserving fiber [1]. We had expected to observe only subtle temporal shifts, where the shifts were to result from higher order dispersive terms not normally included in the nonlinear Schrödinger equation. But the self frequency shift is qualitatively different, and the associated temporal shifts are much larger than those predicted.

In our experiments, the pulses are derived from a stabilized [2] color center soliton laser operating at about $\lambda = 1.5$ μm ($\nu_0 = 200$ THz). For fixed input pulse width $\tau_0$, we vary the power $P$ in the fiber by adjusting the input coupling efficiency, and we analyze both the shape and spectrum of the pulses emerging from the fiber by means of an autocorrelator and a scanning etalon, respectively.

For $P \ll P_1$, where $P_1$ is the soliton power, the emerging pulses are greatly broadened ($\tau \approx 45$ ps for $\tau_0 = 420$ fs), since the 392 m fiber is many soliton periods [3] long, and at $P = P_1$, the input pulse width is preserved. Such results are as expected. But for $P > P_1$, the pulse splits in two, with a separation as great as many tens of picoseconds. Such splittings are much larger than anticipated.

The splitting in time is accompanied by (and in fact is caused by) a large splitting in frequency, where a narrow peak remains at the input frequency $\nu_0$. A much broader peak, which corresponds to a pulse even shorter than the fiber input pulses ($\tau < \tau_0$), is strongly shifted to lower frequency ($\nu < \nu_0$). For fixed power and pulse width at input, and save for a anomalously large shift in the first part of the fiber, the down shift $\Delta \nu$ increases in direct proportion to the fiber length. We infer the continuous nature of the self frequency shift from that observation.

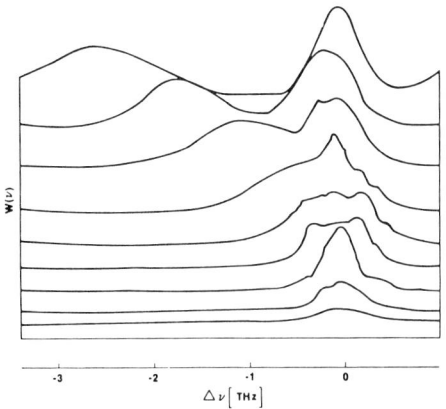

Fig. 1: Spectra at the output of a 52 m fiber for $\tau_0 = 475$ fs. All curves are to the same scale but displaced vertically in proportion to the time average power in the fiber.

The down shift $\Delta \nu$ also grows rapidly with increasing power. Fig. 1 shows several spectra as a function of the time average power in the fiber.

We interpret the above as follows: When pulses with $P > P_1$ are launched into the fiber, eventually a fundamental soliton is formed. Its width $\tau$ decreases, and its bandwidth correspondingly increases with increasing $P$. Any excess power not needed in the formation of this soliton is stripped off and becomes dispersion-broadened; it is this part that appears as the relatively narrow spectral peak at $\nu_0$.

The soliton part is subjected to the self shifting process, in which the higher frequency components of the pulse spectrum act as a Raman pump for its lower frequencies. This is possible because the broad Raman gain spectrum in fibers extends all the way down to zero frequency difference between pump and signal [4]. (At low frequencies, the gain is approximately proportional to the frequency difference.) Thus, the amount of frequency shift in a unit length of fiber depends strongly on the spectral width of the pulse; that is, the pulse rapidly "feels" more Raman gain as it becomes narrower and as its spectrum becomes correspondingly broader. Note, however, that the soliton continues to exist as a stable entity, in spite of the continuous transfer of energy to lower frequencies.

The behavior shown in Fig. 1 can now be understood, at least qualitatively, in terms of the decreasing pulse width of the emerging soliton with increasing power, and the rapid increase of shift with decreasing pulse width.

Raman effects can be introduced into the nonlinear Schrödinger equation by modification of the nonlinear term $|u|^2 u$ ($u$ is the pulse envelope function) to describe a delayed response. Nevertheless, calculation is easier to carry out by transforming to the frequency domain, and the self frequency shift itself can be treated as a perturbation. Fig. 2 shows the predicted shift [5] as a function of pulse width for fiber lengths of 1 m, $10^3$ m, and $10^6$ m, respectively. Note that the shift $\Delta \nu$ per unit length scales approximately as $\tau^{-4}$.

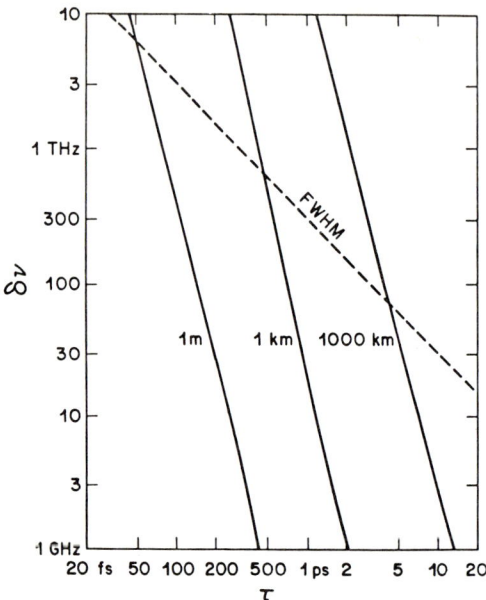

Fig. 2: Predicted soliton self frequency shifts as a function of pulse duration, for $\lambda = 1.5$ μm and $D = 15$ ps/nm/km, in lossless fibers. The pulse bandwidth is shown for comparison.

We have observed shifts of 8 THz for $\tau_0 = 560$ fs and $\tau = 260$ fs at the output of a 392 m fiber. For those same conditions, the theory [5] predicts about 4 THz. The agreement is about as good as could be expected in view of several uncertainties in pulse width and other parameters. Note that the 8 THz shift is already 4% of the optical frequency $\nu_0$. With pulses of $\tau = 120$ *fs*, we obtain a 10% shift, but in just 52 m of fiber. Comparison with the earlier experiment confirms the predicted $\tau^{-4}$ dependence to within the limits of experimental error.

It should be possible to use this phenomenon to derive pulses of different optical frequency from the same laser source for pump-probe experiments. The scheme becomes particularly attractive with pulses of just a few tens of fs, where a wavelength shift of hundreds of nm could be achieved in just a meter or so of fiber. (Pulse widths of ~60 fs have thus far been achieved directly from the soliton laser, and these have been reduced to ~19 fs with compression in a short length of fiber.) In that way, it may be possible to measure relaxation or other times to limits of resolution set only by the uncertainty principle.

1. F. M. Mitschke and L. F. Mollenauer, sub. to Opt. Lett.
2. F. M. Mitschke and L. F. Mollenauer, sub. to IEEE J. Quantum Electron.
3. L.F. Mollenauer, R. H. Stolen, and J. P. Gordon, Phys. Rev. Lett. **45**, 1095 (1980)
4. R. H. Stolen, C. Lee, and R. K. Jain, J. Opt. Soc. Am. *B1*, 652 (1984)
5. J. P. Gordon, sub. to Opt. Lett.

# Solitons at the Zero Dispersion Wavelength of Single-Mode Fibers

*P.K.A. Wai, C.R. Menyuk, H.H. Chen, and Y.C. Lee*

Laboratory for Plasma and Fusion Energy Sutdies, University of Maryland, College Park, MD 20742, USA

Summary

In low-loss, single-mode fibers, the maximum possible bit rate is determined by chromatic dispersion. At most carrier wavelengths, the pulse broadening due to dispersion can be adequately described by the second-order dispersion coefficient $\beta^{(2)} = \partial^2 \beta / \partial \omega^2$, where $\beta$ is the propagation constant. One way to avoid the spreading of the pulse transmitted by the fiber is to operate at the so-called zero dispersion wavelength, at which the second-order dispersion vanishes, $\beta^{(2)} = 0$. For pure silica fiber, this wavelength is 1.27 µm. However, even at this wavelength, it has been shown [1] that higher-order dispersion can cause significant pulse broadening. Another method to counter the dispersion, proposed by HASEGAWA and TAPPERT [2], is to make use of the nonlinearity of the refractive index, the Kerr effect, to balance the second-order dispersion. Soliton pulses could then be generated which propagate without dispersive broadening. Recent experiments [3] have shown the feasibility of this idea by demonstrating the propagation of solitons in the anomalous dispersion region of a single-mode silica fiber. However, whether such balance would occur between the nonlinearity and the higher-order dispersion at the zero dispersion wavelength has been unclear to date [4].

In this work, we show that solitons do exist at the zero dispersion wavelength; moreover, they exist at far lower power levels than at other wavelengths. At the zero dispersion wavelength the evolution of the pulse amplitude, $\phi(x,t)$, is described by the equation,

$$i \frac{\partial \phi}{\partial x} - \frac{i}{6}\beta^{(3)} \frac{\partial^3 \phi}{\partial t^3} + \frac{1}{2} \frac{\omega_0 n_2}{c} |\phi|^2 \phi = 0 , \tag{1}$$

where $n_2$ is the nonlinear index of refraction, $\beta^{(3)}$ is the third-order dispersion coefficient and $\omega_0$ is the carrier frequency. It is possible to show analytically that the soliton solutions to this equation, depending on the sign of $\beta^{(3)}$, must propagate at velocity greater or less than the group velocity. Since (1) is not integrable using spectral transform methods, it is solved numerically. We find that above a certain threshold power, it is possible to produce solitary wave solutions starting from several different initial profiles. We then numerically find the precise pulse shape of the solitary wave solutions by solving the ordinary differential equation which results from (1) when we demand $\phi(x,t) = \phi(t - x/v)$. A simulation of collisions, shown in Fig. 1, between two of these solitary wave solutions demonstrates that their shapes remain almost unchanged after collision, except for a phase shift. Hence, these solitary wave solutions are soliton-like. With some other initial profiles whose integrated power is larger, breather-like solutions, i.e., solutions whose shapes go through periodic contractions, are also observed.

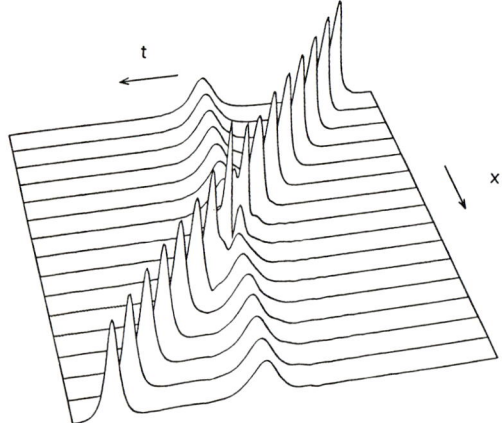

Fig. 1. Collision of two solitary wave solutions of (1)

Because of the smaller dispersion present at the zero dispersion wavelength, the minimum power required to produce these solitons and breathers is much smaller than at any other wavelength. For example, we consider a silica fiber of cross-sectional area 20 $\mu m^2$, and $\beta^{(3)}$ = 0.08 $(psec)^3$/km at the zero dispersion wavelength of 1.27 $\mu m$. The minimum power $P_0$ and the soliton pulsewidth $\tau$ (FWHM) is given by the following relation:

$$P_0(\tau)^3 \approx 0.24 \text{ W}(\text{psec})^3 . \tag{2}$$

Hence, for a 1 psec pulse, only 0.24 W is sufficient to produce a soliton, as compared to 1.6 W required for the soliton at $\lambda = 1.3$ μm in the anomalous dispersion region [5]. This difference becomes more dramatic for pulsewidths of 5 psec or more which are used in present experiments.

References

1. D. Marcuse, Appl. Opt. 19, 1653 (1980).
2. A. Hasegawa and F. Tappert, Appl. Phys. Lett. 23, 142 (1973).
3. R. Stolen, L. Mollenauer, and W. Tomlinson, Opt. Lett. 8, 186 (1983).
4. G. P. Agrawal and M. J. Potasek, Phys. Rev. A 33, 1765 (1986).
5. A. Hasegawa and Y. Kodama, Proc. IEEE 69, 1145 (1981).

# Active Mode-Locking of an InGaAsP Optical-Fiber Ring Laser

*G. Eisenstein, R.M. Jopson, M.S. Whalen, K.L. Hall, and G. Raybon*

AT & T Bell Laboratories, Crawford Hill Laboratory,
Holmdel, NJ 07733, USA

Mode-Locked InGaAsP optical-fiber composite-cavity lasers are compact sources which are capable of generating short pulses at high repetition rates. Such sources are important for high bit-rate communication, switching, signal processing, and sampling systems.

All previously reported lasers of this type used a linear configuration [1]. Here we describe for the first time active mode-locking in an InGaAsP optical-fiber ring laser [2]. The ring configuration has several potential advantages such as immunity to external reflections, the ability to insert additional intra-cavity elements, and the possibility of constructing a colliding pulse laser.

The ring laser described here is depicted schematically in Fig. 1. It consists of a 1.55-$\mu m$ traveling-wave optical amplifier [3] which is coupled via microlenses to a feedback loop containing a polarization maintaining fiber directional-coupler [4].

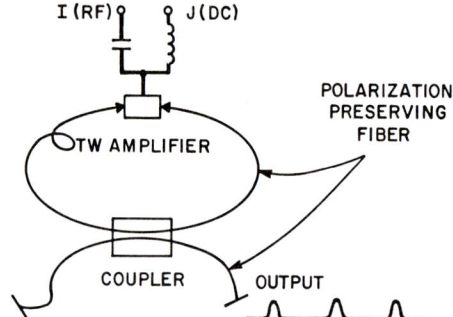

Fig. 1 schematic diagram

The polarization maintaining fiber together with the dichroism in the amplifier [3] act as an intra-cavity Lyot filter. This filter enables single frequency operation under cw drive conditions [2]. a similar filter obtained by proper choice of the fiber type and length may be used for spectral control and pulse shaping when the laser is mode-locked. The fiber length in the present laser is 1.66 m resulting in a fundamental resonance frequency of 125 MHz. the laser is mode-locked by driving the amplifier with a sinusoidal current at this frequency or a harmonic. Alternatively, the amplifier may be driven with a pulse train at the same frequency. The output pulses are detected using a fast InGaAs pin photodetector and a sampling oscilloscope. The impulse response of the detection system is 32 ps.

For sinusoidal drive at the low harmonics, the output pulses are rather wide. For example, Fig. 2a shows the pulses obtained at the second harmonic (250 MHz). The width of these pulses is approximately 200 ps. At such low frequencies, it is necessary to drive the laser with short electrical pulses in order to obtain short output pulses. Fig. 2b shows the pulses obtained when the laser is driven with 50 ps wide pulses from a comb-generator at 250 MHz. The observed pulse width is 34 ps which corresponds to an actual pulse width of 12 ps.

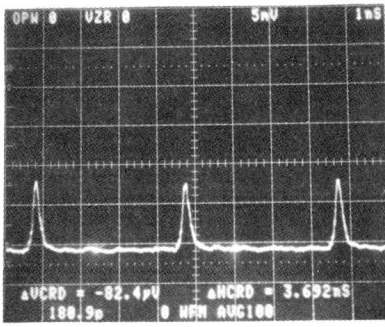
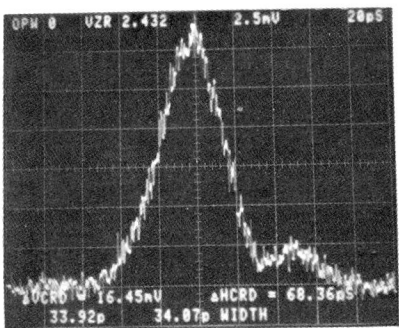

a - sinusoidal drive

b - pulsed drive

Fig. 2 : second harmonic - 250 MHz

Similar narrow pulse widths are obtainable with a sinusoidal drive at high frequencies. Fig. 3 shows the pulse obtained at the twenty-ninth harmonic (3.62 GHz). The observed pulse width is 34.5 ps corresponding to an approximate actual pulse width of 14 ps.

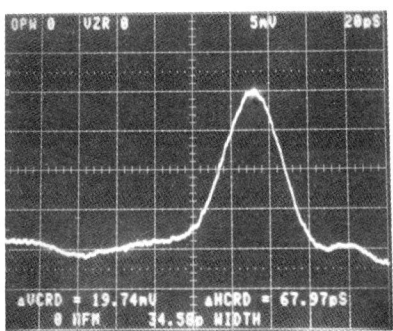
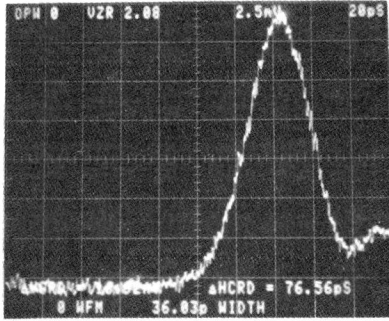

Fig. 3 : twenty-ninth harmonic - 3.62 GHz

Fig.4 : seventeenth harmonic - 2.12 GHz with a 10-15 dB intra-cavity loss

Finally, we examine the ability of the amplifier to overcome an intra-cavity loss. To this end, the fiber ring was broken and an air gap of approximately 75 $\mu m$ length was introduced. This gap represents a loss of 10-15 dB. Althogh the ring laser threshold current increased and its quantum efficiency decreased, it can still be mode-locked. Fig.4 shows the output pulse obtained at the seventeenth harmonic (2.12 GHz) under these conditions.

The observed pulse width is 36 ps corresponding to an actual pulse width of approximately 17 ps. This increase in pulse width is attributed to the lower Q of the cavity. Nevertheless, this measurement indicates that intra-cavity elements with 10-15 dB loss may be inserted inside the laser cavity while it could still be actively mode-locked.

**REFERENCES**

[1]   G. Eisenstein, R.S.Tucker, U. Koren, and S.K.Korotky, IEEE JQE QE-22, 142, (1986).

[2] R.M.Jopson, G.Eisenstein, M.S.Whalen, K.L.Hall, U.Koren, and R.J.Simpson, APL 48, 204 (1986).

[3] G.Eisenstein, and R.M.Jopson, Int. J. of Electron. 60, 113, (1986).

[4] C.A.Villaruel, M.Abebe, and W.K.Burns Electron. Lett. 19, 17, (1983).

# Femtosecond Resolved Fluorescence

W. Rudolph* and J.-C. Diels

Center for Applied Quantum Electronics, P.O. Box 5368,
North Texas State University, Denton, TX 76203, USA

Mode-locked dye lasers have been developed that provide an extremely stable and accurately characterized output [1]. The very high pulse to pulse reproducibility, combined with an accurate knowledge of the pulse waveform and phase, is an asset in any experiment involving nonlinear effect, and requiring precise temporal resolution. The high accuracy in pulse diagnostic led for instance to the determination of time constants much shorter than the pulse duration in an experiment of degenerate four wave mixing [2]. In general however, because of the low intensity of the source, this advantage is compromised against higher pulse energy by associating an amplifier with the source. It is demonstrated here that optimization and long integration times with simple conventional up-conversion techniques are sufficient to time resolve events as weak as the fluorescence of a dye with femtosecond resolution.

One of the two output beams of the ring dye laser [1] - each carrying an average power of 25 mW - is used to excite the sample, the other is used as a reference beam to gate the fluorescence (Fig. 1). A 20x microscope (aperture 0.4) objective is used to focus the beam 1 (Fig. 1) into the sample (a 200 μm thick jet of oxazine dye at 0.03 M/l in ethylene glycol). At that concentration, 70% of a focused beam of 25 mW is absorbed in 2 μm. For the particular oxazine dyes that we investigated, comparison of the absorption spectra of dilute and concentrated samples show no evidence of dimerization. The broadening of the signal due to the spatial spread of the

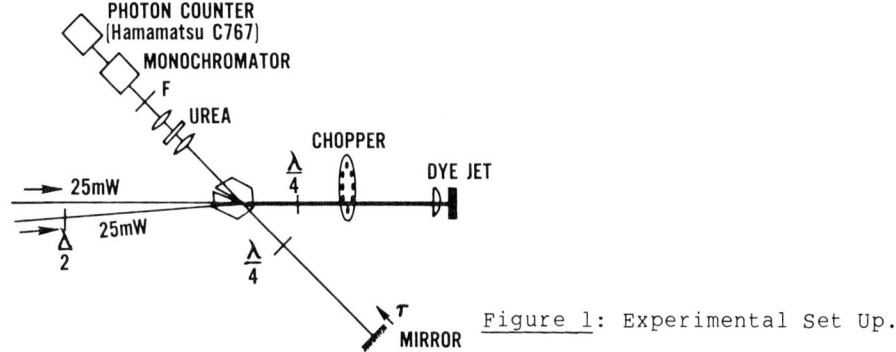

Figure 1: Experimental Set Up.

---

* Permanent address: Physics Dpt, FSU University, 69000 Jena, GDR.

source is thus less than 6 fsec. This configuration makes it also possible to have an absolute measurement of the "zero delay" point, by tuning the detection system to the second harmonic of the excitation wavelength. The wide angular aperture of the objective ensures that a large fraction (approximately 0.5%) of the fluorescence is collected. The fluorescence signal and the delayed reference beam are combined (cross polarized) by a calcite Glan Thompson prism (Fig. 1) and focused into a urea crystal, which is phase matched for generating the sum of the fluorescence and reference wavelengths. A combination of an interference filter and a spectrometer are used to minimize the background produced by the fundamental beams, and to select the wavelength range of interest. A photon counting instrumentation (Hamamatsu C767 counter with a C 716 preamplifier) in combination with a mechanical chopper in the signal beam is used to measure the background free upconverted signal (which is the difference between the total signal - chopper open - and the background - chopper closed - ). Since the signal is weak (of the order of one upconverted photon/$10^6$ laser pulses), a high rejection factor for the polarizing optics, a good quality crystal and negligible scattering are essential to eliminate the contribution of the reference beam to the background.

By tuning the detection to the second harmonic frequency of the laser beam, and orienting the jet at normal incidence to the pump beam, the reflection at the surface of the jet is measured, thus providing an absolute time reference for the optical delays.

The count rate of upconverted fluorescence photons is plotted in Fig. 2 as a function of reference delay, for the fluorescence of oxazine 170 perchlorate (Exciton oxazine 720) at the wavelengths of 660 and 650 nm (Fig. 2a), and nile blue at 660 nm (Fig. 2b). It is interesting to note that no wavelength dependence of the risetime can be resolved. The observed signal $S(\tau)$ is the correlation

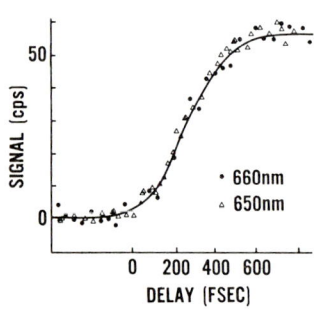
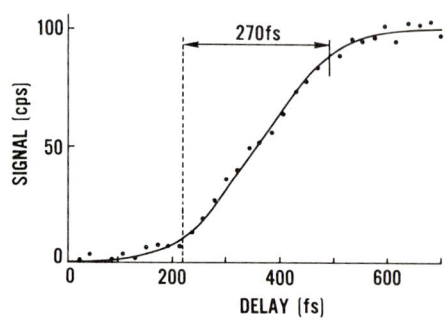

Figure 2a

Figure 2b

Up-converted fluorescence signal in oxazine 170, at 650 nm and 660 nm as a function of reference delay.

Up-converted fluorescence in nile blue at 660 nm as a function of reference delay.

$$S(\tau) = \int J(\tau-\tau')d\tau' \int I(t-\tau)F_\varrho(t)dt, \qquad (1)$$

where $F_\varrho(t)$ is the fluorescence signal, $I(t)$ is the pulse shape, and $J(\Delta\tau)$ is the jitter function between the two outputs of the ring dye laser. We have developed two methods to determine the shape from the pulse emitted by the laser. The first one is by successive iterative fitting of the pulse interferometric autocorrelation, intensity autocorrelation, and pulse spectrum [1]. The second method is by direct decorrelation of the pulse by itself after broadening through glass [3]. Both methods show an asymmetric pulse shape of 70 fsec FWHM, with a steep rise of 30 fsec. The jitter function is calculated by decorrelating the cross-correlation (which is a convolution product of the jitter function by the autocorrelation). We find that the relative timing of the outputs of the laser can be represented by a Gaussian distribution of 180 fsec FWHM. This very large jitter appears unavoidable when the laser is operating at average power levels exceeding 15 mW (a much smaller jitter was measured with the laser operating at power levels of a few mW [2]). Fourier transform techniques are again applied to deconvolute the measurement $S(\tau)$, knowing the pulse shape and the jitter function. The deconvoluted risetime of the fluorescence F (t) is 220 fsec for oxazine 170 (excitation wavelength: 617 nm; fluorescence wavelengths: 650 and 660 nm) and 110 fsec for nile blue (excitation wavelength 616 nm; fluorescence wavelength 660 nm). In the absence of any specific model to relate the excitation and the fluorescence, one can estimate the step function response of fluorescence by subtracting the pulse risetime from the date, yielding 190 fsec for oxazine 170, against 80 fsec for nile blue. It should be noted that the latter time matches the fast component of "equal correlation measurements" observed recently by ROSKER et al [4]. The fact that the risetime of fluorescence differs more than by a factor 2 between oxazine 170 and nile blue is rather surprising, in view of the structural similarity between the two dyes. The major conformation difference is an interchange of an H with a $C_2H_5$ group in going from oxazine 170 to nile blue, making structure of the latter more asymmetric.

The jitter between the two outputs of the dye laser is the major factor limiting the accuracy of the data. The use of two beams introduces large inaccuracies in the "zero delay" determinations, because of the difficulties in correcting simultaneously for pathlength variations due to angular dispersion and delay dispersion. The solution to these problems is to use the same laser pulse split between the reference and signal beams. The source should produce (in one beam) twice the single beam output of the ring laser. We demonstrated recently a hybrid antiresonant ring mode-locked dye laser, with the required characteristics (output power of 100 mW in a single beam, pulse duration of less than 70 fsec, pulse repetition rate 120 MHz), which is being used in further attempts to improve the resolution and absolute accuracy of the fluorescence measurements.

This work was supported by the National Science Foundation, under grant Nb ECS-8406985.

References

1. J.-C. Diels, J. J. Fontaine, I.C. McMichael, and F. Simoni, "Control and measurement of ultrashort pulse shapes (in amplitude and phase) with femtosecond accuracy", Applied Optics **24**, 1270-1282 (1985).
2. J.-C. Diels and I. C. McMichael, "Degenerate four wave mixing of femtosecond pulses in an absorbing dye jet", JOSA B, **3**, 535-543 (1986).
3. J.-C. Diels, N. Jamasbi, L. Sarger, "Design of a 50 fsec Linear Laser and Analysis of the Colliding Pulse Effect", "Ultrafast Phenomena V, Snowmass, Aspen, Co, paper TuD6 (June 1986).
4. M. J. Rosker, F. W. Wise, C. L. Tang, and A. J. Taylor, "Femtosecond Relaxation Dynamics of Large Organic Molecules", "Ultrafast Phenomena V, Snowmass, Aspen, Co, paper ThB4 (June 1986).

# Parametric Amplification Sampling Spectroscopy (PASS): A New Technique for Resolving Near-Infrared Luminescence on a Subpicosecond Time Scale

D. Hulin[1], A. Migus[1], A. Antonetti[1], I. Ledoux[2], J. Badan[2], and J. Zyss[2]

[1]Laboratoire d'Optique Appliquée, ENSTA, Ecole Polytechnique,
F-91120 Palaiseau, France
[2]Centre National d'Etudes des Télécommunications,
F-92220 Bagneux, France

In recent years, optical fiber communication requirements have aroused considerable interest in materials for optical devices operating in the 1.3-1.6 µm spectral region. Time-resolved spectroscopy is a fundamental tool for studying such material responses. We report here a new method, Parametric Amplification Sampling Spectroscopy (PASS), for time-resolving weak near-infrared emission with a subpicosecond accuracy. We have used this scheme to measure the ultrashort luminescence lifetime of an infrared dye and electronic transport properties of multiple-quantum-well structures (MQWS).

The principle of this new technique is based on the parametric amplification of an optical signal which overlaps with an intense ultrashort pump pulse inside a non-linear crystal. In this respect the new non-linear organic crystal NPP [for N - (4 nitrophenyl) - (L) - prolinol] has recently been shown [1] to lead to efficient gain for wavelengths between 0.9 and 1.6 µm. Due to the characteristics of the phasematching curve in this spectral region (quasi-vertical slope) a minimal group velocity dispersion is ensured which is fundamental for subpicosecond time resolution. Unlike other non-linear techniques such as up-conversion, the yield of this scheme is much greater than unity because any incoming signal photon overlapping with the pump pulse inside the crystal can produce as many as $10^4$ photons.

The time evolution of the signal of interest is resolved by variably delaying the arrival of the pump pulse on the non-linear crystal. Figure 1 shows the experimental configuration. The output of a colliding pulse mode-locked dye laser amplified at 10 Hz repetition rate (100 fs duration, energy per pulse up to 1

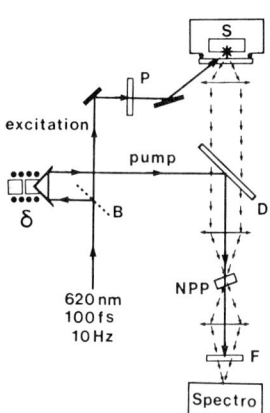

FIGURE 1. Experimental scheme; B is a beam splitter, P is a half-wave plate to adjust the polarization, S is the luminescent sample represented here inside a cryostat, D is a dichroic plate reflecting only the pump beam.

mJ, central wavelength at 620 nm) is split in two parts. The first one (excitation beam) is focused onto the sample, exciting the luminescence to be studied. The resulting signal is collected through a lens of large aperture and sent into the NPP crystal (thickness 1mm). The second beam (pump beam) propagates along an adjustable path before reaching the non-linear crystal where it spatially overlaps the signal beam. Its spot area ranges around 1 mm² and the corresponding incident intensity is a few GW/cm². The only part of the luminescence which experiences amplification is the one in temporal coincidence with the pump beam. After amplification in the NPP crystal, the signal is collected through a monochromator by a photodiode connected to a computer which controls the delay line. The zero delay determination is achieved by introducing a scattering medium into the sample holder and amplifying a scattered infrared beam used in place of the excitation beam.

The temporal resolution of the parametric amplification has been demonstrated to be less than 0.2 ps in the case of small input signal i.e. for no saturation of the gain. The experimental evidence for this feature has been obtained by recording the cross-correlation between the pump pulse and a pulse at 1.24 µm selected in a continuum generated by the other beam (same duration). The inset of Fig. 2 shows the result for a small incident signal intensity value. The velocity mismatch broadening is less than 100 fs. For a fixed pump value, we have checked experimentally that the amplified (output) signal is almost linear with the input signal intensity as long as saturation does not occur.

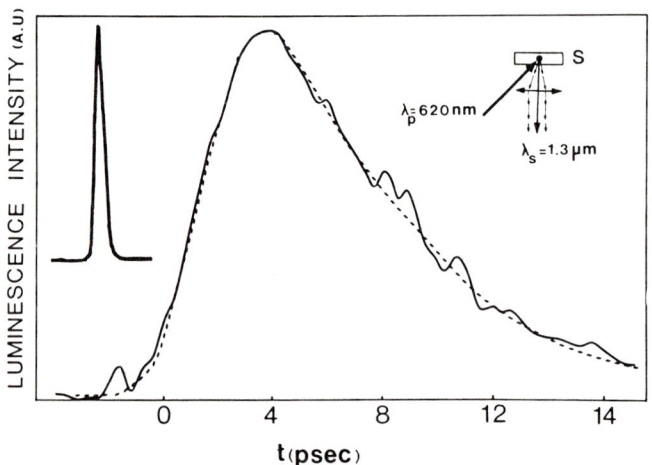

FIGURE 2. Time-resolved luminescence of an infrared dye at 1.3 µm obtained with the PASS.

Figure 2 shows an example of time-resolved luminescence. The emitting source is an infrared dye (number 26⁺HFB⁻ from Lambda Physics) dissolved in methanol. This dye which has recently been shown to operate as a gain medium in a subpicosecond synchronously pumped dye laser [2], was placed in a 1 mm thick cell and excited with the femtosecond laser at 620 nm. The luminescence was collected in the backwards direction. The PASS recorded at 1.3 µm exhibits an exponential decay which is best fitted with a time constant of 5 ps. This value is two times smaller than the one reported in [3] where the dye, dissolved in ethylene glycol, is excited at 1.06 µm. It should be noted that the rise of the signal is artificially lengthened due to the penetration depth of the pump pulse at 620 nm.

The near-infrared spectral region is very interesting for optical communications. Devices based on multi-quantum-well structures made of InGaAs show great promise in this area. The time-resolved luminescence of a InGaAs-InAlAs MQWS has been recorded with our PASS method. The sample is held at 15 K and is excited with the femtosecond laser pulse at 620 nm. Knowing the zero delay accurately, it can be seen in Fig. 3 that the rise of the photoluminescence signal is delayed by approximately 15 ps. Such a delay incorporates various contributions such as the migration of the carriers excited by the 2 eV photons from the InAlAs top layer (thickness 1 mm) followed by their trapping and cooling down in the quantum wells.

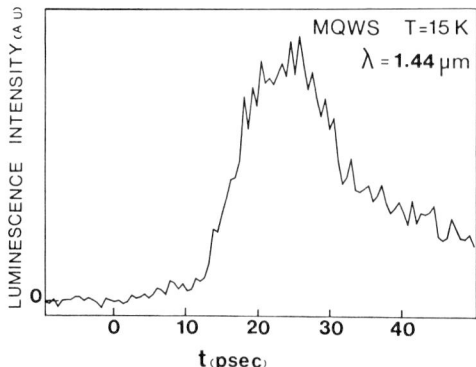

FIGURE 3. Time-resolved luminescence of a InGaAs-InAlAs multi-quantum-well structure at 15K excited with a 620 nm pump beam.

It should be pointed out that up to now only a few techniques have been developed for resolving fluorescence with a few picoseconds accuracy, namely streak-cameras, up-conversion and optical Kerr shutters. The latter method has been demonstrated to work at a subpicosecond level but the efficiency is poor. The first demonstration of the use of up-conversion on the femtosecond time scale can be found in this volume [4] but this experiment has been done at pump and luminescence frequencies very close to each other. In that case the severe problem of group velocity mismatch is avoided. On the other hand, the parametric amplification scheme allows a good velocity matching between the 0.62 µm pump pulse and near-infrared signals, yielding unique femtosecond resolution capabilities in this spectral domain.

1- I. Ledoux, J. Zyss, A. Migus, G. Grillon, A. Antonetti, Appl. Phys. Lett. in press
2- P. Beaud, B. Zysset, A.P. Schwarzenbach, H.P. Weber, Opt. Lett. 11, 24 (1986)
3- A. Seilmeir, W. Kaiser, B. Sens, K.H. Drexhage, Opt. Lett. 8, 205 (1983)
4- W. Rudolph, J.C. Diels, "Femtosecond-resolved fluorescence" (1986)

# Measurement of Optical Phase with Subpicosecond Resolution by Time Domain Interferometry

J.E. Rothenberg

IBM Watson Research Center, P.O. Box 218,
Yorktown Heights, NY 10598, USA

A new technique which determines the complete electric field of ultrashort optical pulses has been experimentally demonstrated. This general technique improves over previous cross-correlation[1] and (interferometric) auto-correlation[2] techniques in that it uniquely and completely determines the field's phase as well as its amplitude. The technique imposes no requirements on the pulse to be measured and only needs a synchronous probe pulse short enough to resolve all structure of interest.

The technique employed is straightforward. A pulse is sent into a Mach-Zehnder interferometer which has a Fabry-Perot etalon in one arm and is empty in the other. The Fabry-Perot etalon is chosen to be resonant with (some part of) the input pulse, thereby transmitting a monochromatic reference pulse. The finesse is made large enough so that the etalon cavity decay time (the inverse of the linewidth) is longer than the input pulse. The relative delay of the arms of the interferometer is adjusted to overlap the input pulse (which passes unaltered through one arm of the interferometer) and the nearly monochromatic pulse transmitted by the etalon in the other arm. The two pulses are combined by the output beam splitter of the interferometer which leads to interference (heterodyning) in the time domain. The intensity of the time domain interferogram is measured by standard intensity cross-correlation with a synchronous probe pulse short enough to resolve all structure of interest.

This technique is closely related to the self-induced heterodyne technique[3] which uses a resonant vapor instead of the Mach-Zehnder interferometer. In this previous technique the vapor provides the monochromatic reference pulse which is automatically copropagating (and thus interferes) with the input pulse. Unfortunately, this previous technique requires a vapor resonant at the frequency of interest and is only useful for pulses in which the amplitude varies much more slowly than the phase. The present interferometric technique has none of these drawbacks.

The technique was demonstrated experimentally using the apparatus shown in Fig. 1. The output of a synchronously pumped cavity-dumped dye laser was sent through a 3m single mode optical fiber. The frequency swept pulse emerged broadened from 6 to ~10 psec, and its spectral content broadened from 1 Å to ~55 Å. A portion of the frequency swept pulse was directed through the Mach-Zehnder interferometer, with the rest sent through a grating compressor. The compressed pulse, ~200 fsec in duration, was used to cross-correlate the output of the interferometer in a 200$\mu$ thick KDP crystal.

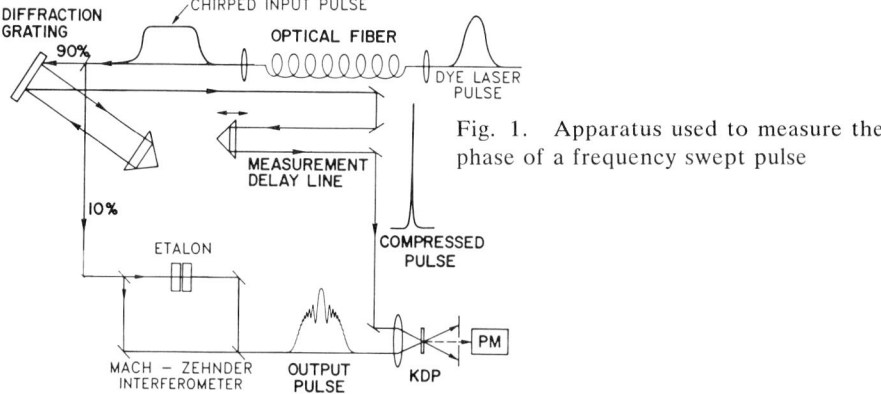

Fig. 1. Apparatus used to measure the phase of a frequency swept pulse

The input pulse to the interferometer is shown in Fig. 2(a), the output in Fig. 2(b), and the difference between these two in Fig. 2(c). Figure 2(c) is precisely one quadrature of the phasor representing the field relative to the "carrier" frequency of the resonant etalon. The intensity modulation shows the signature of a linear frequency sweep $(\propto cos(\phi(t)) = cos(at^2+\phi_0)$ ).

The phase can be determined from Fig. 2(a) by simply counting the extrema. For a pulse with smoothly varying amplitude the phase change from one extremum to the next is $\pi$ radians. In fact, one can interpolate between the extrema by assuming a $cos(\phi(t))$ dependence. Shown in Fig. 3(a) (solid line) is the relative phase $\phi(t)$ ($\phi_0 \equiv 0$) determined in this way from Fig. 2(c), where we have used the knowledge that there is a point of stationary phase at t=0, i.e. where the derivative of $\phi(t)=0$. The dashed curve represents the quadratic phase which a linear fre-

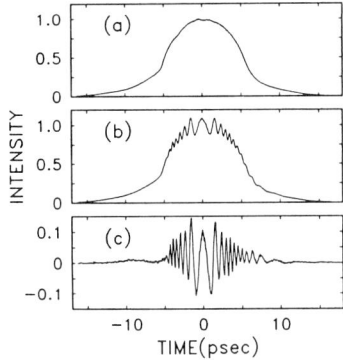

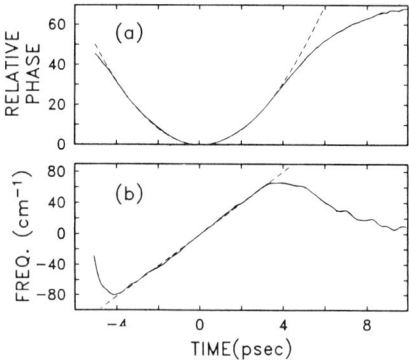

Fig. 2. (a) Input pulse to Mach-Zehnder interferometer; (b) Output pulse from the interferometer; (c) Difference between (b) and (a)

Fig. 3. (a) Relative phase (radians, solid curve) determined from Fig. 2(c) and quadratic phase (dashed) due to a 20 cm$^{-1}$/psec frequency sweep; (b) Instantaneous frequency (solid curve) obtained by differentiating (a) and dashed line representing 20 cm$^{-1}$/psec

quency sweep of 20 cm$^{-1}$/psec would produce. The instantaneous frequency is obtained by numerically differentiating $\phi(t)$ and is shown in Fig. 3(b) (solid line; the dashed line represents 20 cm$^{-1}$/psec). A linearly swept region as well as an asymmetric deviation from linearity at the ends of the pulse are readily apparent.

This technique does not give complete information on the phase of the pulse. The counting procedure only tells us phase differences, and not the direction of the phase change. For example, it could not predict the presence of the point of stationary phase. In addition, if the pulse amplitude changes rapidly the extrema can be shifted and thus give an erroneous phase.

These difficulties can be eliminated to a great extent by obtaining the other quadrature of the field. In this way we would have both the real and imaginary parts of the complex field $E(t)exp(i\phi(t))$. In practice this was achieved by rotating a microscope slide that was placed in one arm of the interferometer to give an additional $\lambda/4$ of delay. The two quadratures for a pulse with a somewhat larger frequency sweep are plotted in Fig. 4. Given both the real and imaginary parts of the field we can obtain $\phi(t)$ completely from the arctangent of their ratio without any assumptions or interpolation. In addition, because this is a ratio measurement the effects of a rapidly changing pulse amplitude are effectively cancelled out. It is helpful to plot one quadrature versus the other in the complex plane as is shown in Fig. 5. The phase is then simply the angle subtended by any point from the origin.

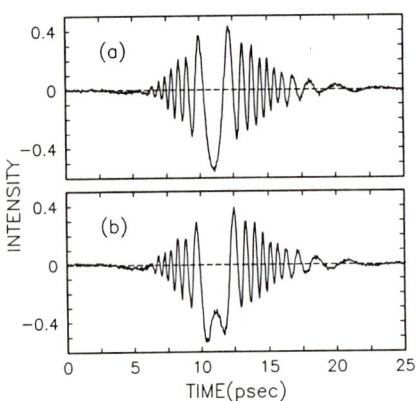

Fig. 4. Both quadratures of the field of a pulse with a somewhat larger frequency sweep

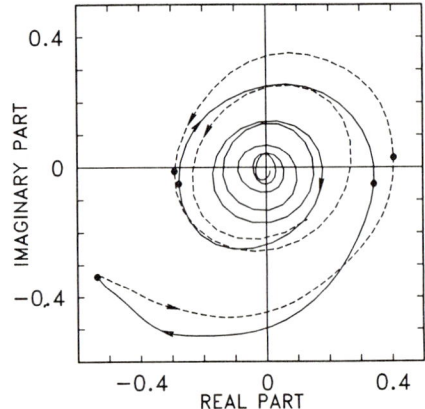

Fig. 5. The phasor representation of the field in the complex plane, i.e. a plot of one quadrature of Fig. 4 versus the other. The point of stationary phase as well as the adjacent minima and maxima are indicated by dots. The direction of time is indicated by arrows; the solid line is plotted for times before the point of stationary phase and dashed line for times after

Hence, we see the plot "winds" around the origin first in one direction then in the other (data for times before the point of stationary phase are plotted with a solid line, and after with a dashed line). Note that the point of stationary phase is directly observed as the point where the spiral reverses direction.

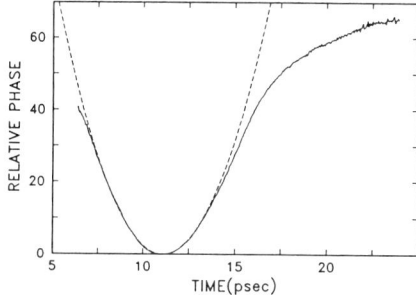

Fig. 6. (a) Relative phase (radians, solid curve) determined from Figs. 4 and 5, and quadratic phase (dashed) due to a 22 cm$^{-1}$/psec frequency sweep

The phase obtained directly from the ratio of the quadratures without any assumptions is shown by the solid line in Fig. 6, with the dashed line representing a parabolic phase corresponding to a 22 cm$^{-1}$/psec frequency sweep. It is interesting to note that the asymmetric disagreement near the pulse edges here is even more pronounced than in Fig. 3(a). A possible explanation is that this larger asymmetry is due to increased Raman competition because of the higher intensity pulse used in Fig. 4. We note that this asymmetry certainly decreases the compression efficiency and may contribute to the limits of compression observed recently.

The author gratefully acknowledges the help and cooperation of D. Grischkowsky throughout the course of this work.

1. H. Nakatsuka, D. Grischkowsky and A. C. Balant, Phys. Rev. Lett. 47, 910(1981).
2. W. Dietel, J. J. Fontaine and J. C. Diels, Optics Lett. 8, 4(1983).
3. J. E. Rothenberg and D. Grischkowsky, J. Opt. Soc. Am. B 2, 626(1985).

# Real Time Picosecond Optical Oscilloscope

J.A. Valdmanis

AT & T Bell Laboratories, 600 Mountain Ave., Murray Hill, NJ 07974, USA

We introduce a new, real time (i.e. non-sampling) measurement technique that is generally applicable to the characterization of picosecond and subpicosecond optical modulation phenomena. Here, we describe the technique as it is applied to the investigation of picosecond electrical transients. The system is based on the real time modulation of a relatively long, frequency swept (chirped) optical probe pulse, thus encoding temporal information as a function of frequency or wavelength. A spectrograph then converts the wavelength encoded temporal information to the spatial domain for readout. The method has the capability of analyzing picosecond signals on a single shot basis.

Figure 1 depicts the experimental arrangement for characterizing electrical signals, which is similar to that for conventional electro-optic sampling.[1] A femtosecond pulse laser output is split into two beams, a trigger beam and a probe beam. The trigger beam pulse is directed, via a fixed delay line, to the test device where it triggers the generation of the electrical transient to be characterized. The resulting electrical signal then propagates across a traveling wave Pockels cell and modulates the optical probe pulse. The probe beam pulse is directed through a dispersive medium, in our case a length of single mode polarization preserving fiber. Due to the large optical pulse bandwidth, dispersion readily lengthens the pulse and introduces a frequency sweep along the pulse in time, i.e. chirp. As the probe pulse propagates through the electro-optic medium its temporal profile is modulated in proportion to the profile of the synchronous electrical transient as depicted in Fig. 2. The fixed delay line is adjusted to ensure the electrical signal arrives within the optical measurement "window". Thus, the entire electrical signal is measured with every optical pulse. Optical phase compensation is included between the polarizers to ensure the modulator operates in the zeroth order of net phase retardance. The intensity modulated output pulse is directed into a spectrometer coupled to a detector array and spectrum analyzer for signal

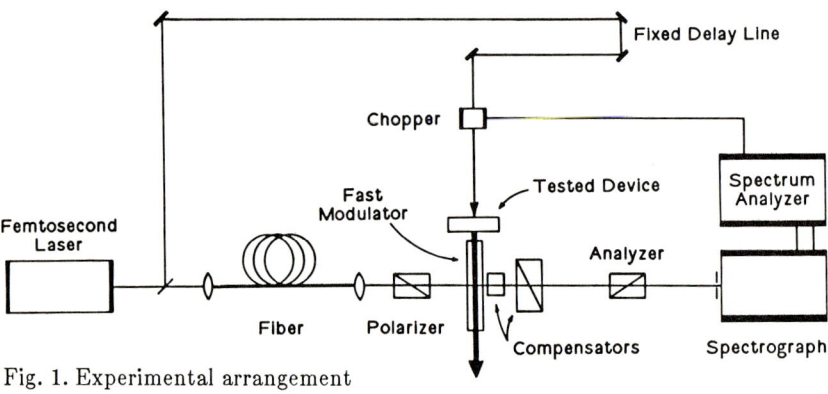

Fig. 1. Experimental arrangement

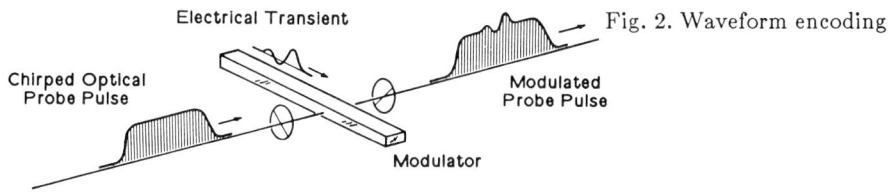

Fig. 2. Waveform encoding

retrieval. Spectra with and without modulation, as switched by the chopper, are subtracted in order to extract the waveform of interest.

70 fs optical pulses having a non-transform-limited bandwidth of ∼13 nm at 630 nm are generated at 100 MHz by a balanced colliding pulse dye laser.[2] Figures 3 and 4 show the autocorrelation trace and spectrum, respectively. The probe pulses are subsequently stretched to ∼300 ps in 50 m of optical fiber to provide for the measurement window. Because the dispersion of glass is very nearly linear over the pulse's bandwidth, the chirp of the probe pulse is also linear. No change between the fiber input and output spectrum was observed.

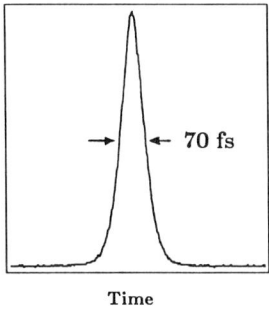
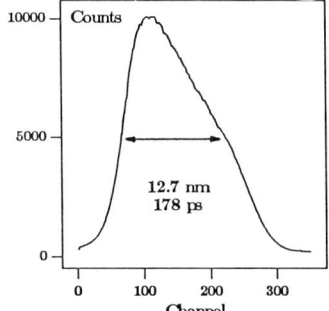

Fig. 3. Pulse autocorrelation          Fig. 4. Pulse spectrum

The temporal resolution of this system is ultimately determined by convolving the original pulse duration with a time given by the stretched pulse duration divided by the ratio of pulse bandwidth to spectrometer resolution. In this experiment, we used a 0.3 m spectrometer coupled to a 1024 element Reticon array and OMA-III spectrum analyzer. The system has a resolution of 0.04 nm and thus can resolve 325 points within the spectrum of the probe pulse, yielding a resolution of less than one picosecond. By stretching the pulse less, this limiting temporal resolution can be as short as the original pulse duration. Another temporal resolution limit arises from the time-wavelength duality of the chirped pulse. Additional frequencies, or wavelength's, (i.e. sidebands) introduced by the impressed modulation at a particular location in time (and hence wavelength) will appear to have come from "neighboring" times thus broadening the modulation. This temporal limit can be approximately given by the square root of the product of the stretched pulse duration and the original pulse duration.[3] In this example, this resolution is on the order of 3 ps. It is also interesting to note that as the chirped optical pulse impinges on the spectrometer grating (at 100 MHz), the diffracted light effectively reconstructs a wavefront that sweeps across the detector array, analogous to the electron beam in a real time oscilloscope or streak camera. However, in our optical "oscilloscope" the effective sweep speed can well exceed the speed of light.

The modulator arrangement was optically biased close to zero transmission in order to maximize the modulation depth of the transmitted light, even though the

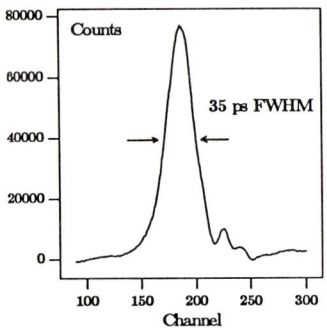
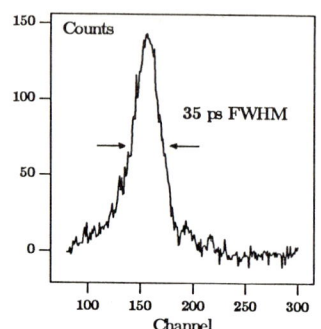

Fig. 5. 1000 scans of photodetector pulse with 35 GHz modulator. Scale is 1.15 ps/channel

Fig. 6. Single, "real time", scan of Fig. 5

overall intensity level at this point is significantly lower.[4] This is necessary because even near zero transmission, the modulation depth resulting from picosecond signals can often be less than $10^{-2}$ and the dynamic range of our multichannel analyzer is limited to $\sim 10^4$. By synchronizing the trigger beam chopper with alternating scans of the array detector system, background spectrum subtractions were accumulated "on the fly" to enhance the signal to noise ratio for small signals. Larger signals could be detected essentially "live" on a single scan basis. Typical scan times were 100 ms per scan.

We demonstrate an electro-optic application of the optical oscilloscope by measuring, in two configurations, the impulse response of an ion-bombarded GaAs photoconductive detector. As depicted in Fig. 2, the electrical signal propagates across a 250 micron thick balanced line traveling wave $LiTaO_3$ modulator with the resulting electric field along the c-axis. The chirped optical pulse is polarized at 45 degrees and propagates normal to the c-axis.

In the first configuration the detector is connected to the modulator via a short length of 35 GHz coaxial cable. Figure 5 shows the result of 1000 accumulated scans. The peak signal level is approximately one volt, and the pulse width is 35 ps, as limited by the cable. Figure 6 shows the same signal for a single scan, which is as it appears in real time on the analyzer screen. The electrical signals themselves are used to calibrate the temporal scale as well as to verify the linearity of the chirp. Figure 7 shows 6 overlayed pulses each delayed 30 ps from the

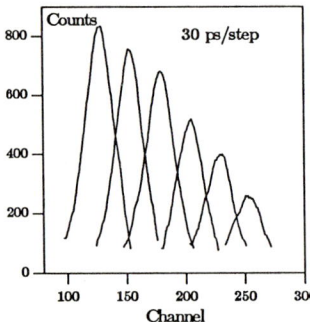
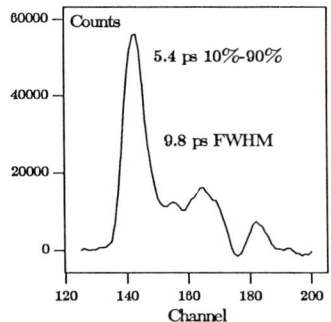

Fig. 7. Temporal calibration of scale to 1.15 ps/channel

Fig. 8. 1000 scans of photodetector pulse with hybrid modulator previous one by adjusting the fixed delay line. Using this technique, the temporal scale was established to be 1.15 ps per channel or 1.4 ps per Angstrom. The changing amplitude of the shifted pulses reflects the spectral pulse shape as a function of time.

In the second configuration the detector is directly attached to the end of the modulator crystal. This eliminates the bandwidth limiting effects of the coaxial cable. Figure 8 shows the resulting electrical waveform averaged over 1000 scans. The risetime of 5.4 ps is limited by the modulator configuration. The displayed waveform in the figure has been adjusted to compensate for the spectral shape of the measurement pulse even though, in this case, the effect is small.

This experiment demonstrates a new method of extracting picosecond or subpicosecond information from an optical modulation system in real time with the capability of operating on a single shot basis. In general, we anticipate the technique could be applied to a wide variety of ultrafast optical modulation situations including modulator testing, optical pulse probe experiments, and wavelength division multiplexing.

Many thanks are due Don Harter for key discussions, Peter R. Smith for supplying the photodetectors, and Princeton Applied Research for loan of the OMA-III.

[1] J.A. Valdmanis, G.A. Mourou, and C.W. Gabel, IEEE JQE, QE-17, 664 (1983)
[2] J.A. Valdmanis, R.L. Fork, and J.P. Gordon, Opt. Lett., 10, 131 (1985)
[3] J.A. Valdmanis, to be published.
[4] S. Williamson and G.A. Mourou, Proc. Conf. Psec. Elec. and Optoelec. 1985, Springer-Verlag, p.58

# Beam Overlap for Long Delay Lines Using Active Feedback

C. Doland, W.B. Jackson, and A. Andersson

Xerox Palo Alto Research Center, 3333 Coyote Hill Road, Palo Alto, CA 94304, USA

An active feedback system for long optical delay lines is described which stabilizes the beam position to a few microns for delays longer than 8 nanoseconds. Because the system is straightforward, improves the beam stability by over an order of magnitude, eliminates time-consuming delay line alignment, and reduces the linearity, quality, and hence, the cost of the delay line, the feedback system should find widespread application in experiments requiring long delays.

## 1. INTRODUCTION

Many ultrafast experiments utilize an optical delay line to create two beams with a variable delay between them. It is generally desirable for this delay line to have a maximum delay as long as possible in order to overlap the picosecond data with measurements obtained from fast photodiodes or photomultipliers.[1] A persistent problem is the alignment of the delay line to prevent motion of the delayed beam as the delay is changed.[2]

The origin of this problem can be understood from Fig. 1(a). A typical delay line consists of a retroreflector which can be moved along the beam axis on a translation stage. The retroreflector insures that the returning beam is parallel to the incident beam but the offset between the beams will vary if the axis of the translation stage and the beam are not colinear or there are nonlinearities in the motion of the stage. This motion causes the overlap of the pump and probe beams to vary since the sample is not precisely at the focus of lens L. Even if the system were correctly aligned, adjustments to the laser change the required alignment. The feedback system described below dramatically simplifies the operation and accuracy of long delay lines.

## 2. EXPERIMENTAL SETUP

The system, described in more detail in Ref. [3], is shown schematically in Fig. 1. A portion of the delayed probe beam is directed onto a two axis continuous position sensor[4] via a beam splitter. The retroreflector can be moved in the horizontal and vertical directions (both perpendicular to the beam axis) by a two-axis dc motor driven stage. The detector outputs generate error signals used to drive the motorized stages to correct for nonlinearities in the delay line.

A block diagram of the feedback electronics is depicted in Fig. 1 (b). The horizontal and vertical position signals from the photodetector are amplified with adjustable gains. The signals are then filtered to remove high frequency

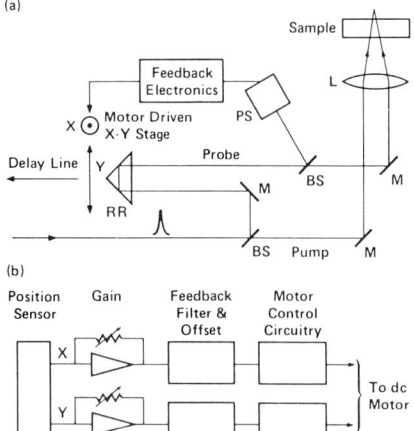
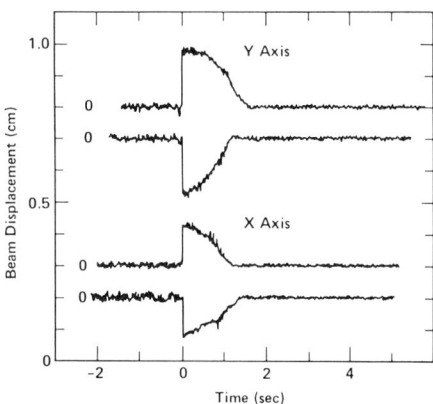

Fig. 1(a) Experimental setup where RR – retroreflector, M – mirror, BS – beam splitter, PS – position sensor, L – lens. The axis of delay line is along the probe beam axis. (b) Block diagram of feedback electronics.

Fig. 2. Large displacement response of feedback system versus time for a step function change in the beam position for both directions.

laser fluctuations and finally output to motor control circuits which drive the dc motors. The retroreflector is moved to reduce any tracking error to zero. It is important to have sufficient gain and correct filtration to minimize tracking errors.

## 3. RESULTS AND DISCUSSION

One of the important considerations for optimal feedback control is the correct gain and filtration to minimize tracking errors. The gain adjustment must be set as high as possible to minimize the beam position error but must not be so large as to result in oscillations. The gain was adjusted by observing the system response to a step function displacement. The beam position on the detector was altered by the insertion of an optical flat at an angle to the beam propagation direction, and the system was allowed to achieve a new equilibrium position. The beam position as a function of time was measured as the flat was abruptly removed. Results of this test for offsets in both directions along horizontal and vertical axes are shown in Fig. 2. The system was able to reach the new equilibrium in 1 to 2 sec limited by the slew rate of the motors. The gain was increased to a level just below the onset of overshoot.

The system was evaluated in several ways. First, the transmitted probe beam power passing through an limiting aperture slightly smaller than the beam size was measured. Relatively small deflections of this beam cause large changes in the probe power because the beam is moved off the pinhole. Ideally, the probe beam power should be constant as the delay is changed.

We measured the power through the pinhole both with and without the feedback for delays ranging from − 0.4 to 1.2 ns delay and from 0.6 to 2.2 ns.

The different delay ranges were obtained by translating the delay line along an optical rail. The beam was aligned on the pinhole at the beginning of the first scan. The results of these measurements appear in Fig. 3. The two scans without feedback differ significantly in the region where they represent the same delay. The difference between the scans indicates that alignment was not maintained between the two scans. The rapid falloff in power through the pinhole for delays larger than 1600 ps demonstrates that the beam was moving off the pinhole.

With feedback, the change in power through the pinhole is reduced to several percent per scan. The small remaining power drop is due to beam divergence causing increased power loss through the pinhole as the path length increases. The measurements are nearly identical in the regions of overlapping delay. The dramatic improvement demonstrates that the feedback system is effective in correcting for alignment errors or delay line irregularities.

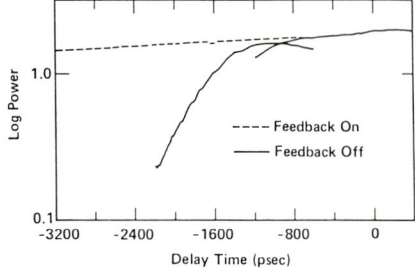

Fig. 3. Transmitted power of the probe beam through a slightly limiting aperture versus delay without feedback (solid) and with feedback (dashed).

Measurements were also made of the beam position determined by the position sensor outputs as the delay was varied. Typically, the beam did not wander more than ±5 µm over a 1600 psec delay (see Ref. [3] for further details). No alignment of the delay line was required other than to ensure that the delay line axis was nearly parallel to the probe beam. Similar results were obtained at different regions of the delay line.

## 4. CONCLUSIONS

The feedback system demonstrated in this paper has a number of important advantages over previous delay lines. The system markedly decreases the displacement of the delayed beam despite large misalignments of the delay line axis relative to the beam axis. Consequently, not only can the linearity and quality of the delay line track be substantially reduced, but there is virtually no need for any alignment of the delay line. Subsequent changes due to the alignment of the laser or optics is automatically compensated by the feedback system and the retroreflector. Because the feedback system is straightforward, lowers the system cost of the delay line, and improves the performance, it should find widespread use in experiments requiring long delay lines.

*REFERENCES*

1. See for example, D. R. Wake and N. M. Amer: Phys. Rev. B 27, 2598 (1983).
2. J. Strait. Ph.D. Thesis. (Brown University. May, 1985).
3. C. Doland, W. B. Jackson, and A. Andersson: (to be published).
4. The position sensor was a dual axis lateral cell made by Silicon Detector Corp. (Newbury Park, CA).

# Ultrashort Dye Laser Pulses
# Using the Sweeping Oscillator Method

*Y.H. Meyer, M.M. Martin, E. Bréhéret, and O. Benoist d'Azy*

Laboratoire de Photophysique Moléculaire du CNRS, Bât. 213,
Université Paris-Sud, F-91405 Orsay, France

Production of ultrashort pulses at various wavelengths is generally based on mode locking techniques and, for pratical applications, often requires the combined use of two expensive pump lasers, one CW, one pulsed. Other ways have been studied to produce short pulses using cavity transients, distributed feedback or cavity competition processes [1]. In this work we report progress in a new technique, recently proposed [2], for obtaining high power ultrashort pulses from dyes using a single standard nanosecond pump laser.

When a homogeneously broadened amplifying medium is used in an untuned laser cavity, spectral narrowing and sweep of the laser intensity can occur during each pulse [3]. This spectral evolution is due to interaction between the broadband intensity and the broadband gain which have related but distinct time-dependent spectra. As a result, a definite spectro-temporal evolution process develops which does not simply recopy the cavity transients such as relaxation oscillations or more complicated features related to the pump waveform. The sweep rate can be made very fast in short and low Q cavities. This spectral instability can be used to generate short pulses. It is sufficient to filter the broadband emission of a sweeping oscillator, in the adequate region, to obtain at once both a narrow line and a short pulse.

Bandwidth limited ultrashort pulses were obtained in a dye laser system with two sweeping oscillators in cascade (Fig. 1), pumped with a 15 ns pulse at 532 nm from a standard 10 Hz Q-switched Nd:YAG laser. The first sweeping oscillator $O_1$ is simply a 1 cm standard spectroscopy cell filled with a recirculating solution of R6G in ethanol 4 $10^{-4}$ M/l. Under transverse (or longitudinal) excitation, laser action occurs between the external uncoated faces of the cell. The collimated dye laser beam is diffracted with the grating $G_1$ and focused with a 50 cm lens on a longitudinally pumped amplifier A. The output is a 120 ps pulse when $G_1$ is tuned so that only wavelengths around 565 nm are amplified. When $G_1$ is tuned to the central wavelength of the broadband emission a pulse of several nanoseconds is obtained. The energy of the 120 ps pulse is

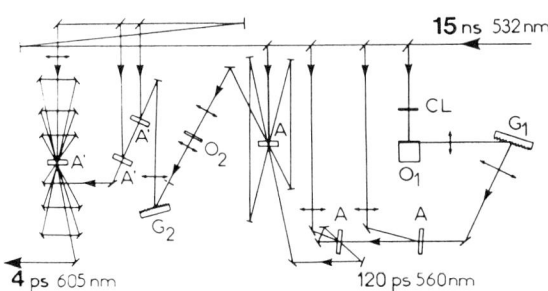

Fig. 1 Generation of picosecond single pulses by spectro-temporal selection starting from 15 ns excitation pulses

a few nJ and is further amplified in two multipass amplifiers to 20 µJ. Part of this energy is focused to pump longitudinally the second sweeping oscillator $O_2$ filled with Rhodamine 640 in ethanol 5 $10^{-3}$ M/l. The cavity length is 0.2 mm, and the mirrors are two standard R = 0.2 beam splitters (with antireflection coatings on the external face) separated by a 0.2 mm thick glass ring. The dye laser emission is collimated with a 20 mm lens and diffracted within a small spot < 1 mm in diameter, by the 1800 g/mm grating $G_2$. The second spectro-temporal selection is obtained from $G_2$ by amplification in the amplifiers A'. When $G_2$ is tuned in the region of 605 nm, i.e. in the blue wing of the broadband oscillator emission, so that only one cavity mode (FSR 0.7 nm) is amplified, a bandwidth limited pulse of 4 ps FWHM is obtained as measured with a streak camera (Fig. 2); this value is corroborated by the time average of 7.5 ps FWHM of the autocorrelation trace (non–collinear frequency doubling) and the measured spectral width of 3 $cm^{-1}$ FWHM. The energy of this pulse was above 10 µJ (3MW) using only a few tens of mJ of total pumping energy at 532 nm. Background energy due to nanosecond amplified fluorescence of A' was reduced to a few percent by the use of a saturable absorption cell with DODCI.

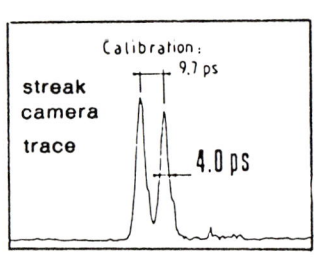
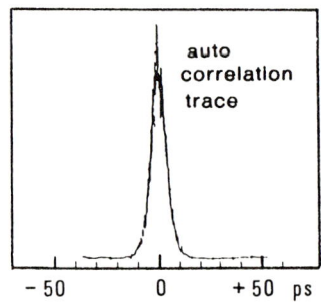

Fig. 2  Measurements of a single pulse duration (0.2 mm cavity) with a streak camera and of the average duration at 10 Hz by autocorrelation

Using a 100 µm cavity and a 7 µm thick tunable FP etalon (R = 0.95 $t_{cav}$ = 0.45 ps) as the spectral selector, we obtained an autocorrelation trace shape which corresponds to a 1.5 ps single-sided exponential pulse and to the measured linewidth of 3 $cm^{-1}$.

An interesting point in the process of intracavity spectro-temporal evolution is that, when started, the sweep will continue as long as enough photons in the cavity will remain to "remember" the spectral distribution resulting from the previous history, whatever the modification of the gain imposed by fast variations of the pump rate. Thus even for spiked pumping, which is often the case when the pump pulse comes from a multimode laser, the spectral evolution continues in the same direction. The complex time-frequency structure of the 0.2 mm cavity emission measured with a streak camera appears in Fig. 3 and clearly shows the spectral sweep within the set of well-resolved longitudinal modes. A single peak is emitted by a few modes in the band wings.

The quasi-periodic temporal substructure observed for some other mode is ascribed to combined effects of pump spikes and relaxation oscillations. A complete calculation of the multimode broadband emission as a function of time requires too much computing time. However some insight into the process can be obtained, assuming randomly phased modes, by means of the rate equation appro-

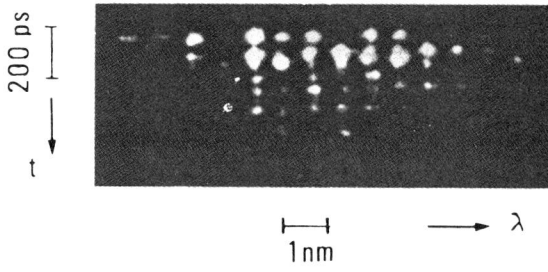

Fig. 3 Streak camera recording of the ultrafast spectro-temporal evolution within the multimode laser emission from 0.2 mm cavity

ximation extended to the spectral domain by the use of a sufficient number of coupled equations at different frequencies. The result of a calculation in the simplest case of a smooth pump pulse is shown in Fig. 4, which evidences the spectro-temporal evolution in a $\sim$ 0.2 mm cavity during the relaxation oscillations. Production of a short (5ps) and single pulse in the blue wing is predicted for pumping rates well above threshold.

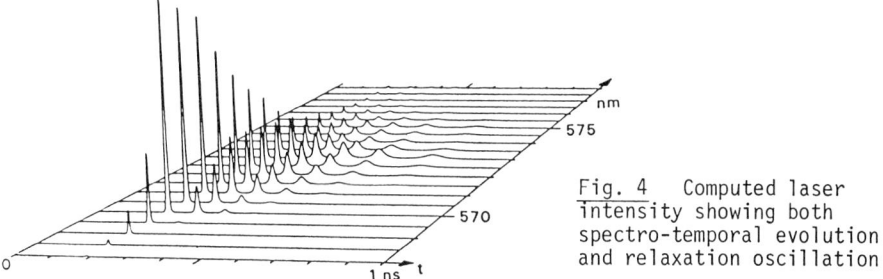

Fig. 4 Computed laser intensity showing both spectro-temporal evolution and relaxation oscillation

Experiments and calculations suggest that this method is able to produce pulse duration of the order of the roundtrip time. Thus subpicosecond pulses should be feasible with a cavity below 100 μm.

References

1. C. Lin, C.V. Shank, Appl. Phys. Lett. 26, 389 (1975)
   Z. Bor, B. Racz, A. Muller, Appl. Opt. 22, 3327 (1983)
   S. Szatmari, F.P. Schaefer, Appl. Phys. B33, 95 (1984)
2. M.M. Martin, E. Bréhéret, Y.H. Meyer, Opt. Commun. 56, 61 (1985)
3. Y.H. Meyer, P. Flamant, Opt. Commun. 19, 20 (1976)

# An Investigation on Ultrashort Light Pulse Generation by Travelling-Wave Amplified Spontaneous Emission

W. Lee, C. Ning, Z. Huang, and W. Wang

Laser Optics and Spectroscopy Laboratory, Physics Department,
Zhongshan University, Guangzhou (Canton), People's Republic of China

Travelling-wave amplified spontaneous emission (TW ASE) has been shown to be a promising method for ultrashort light pulse generation. *Bor* et al. [1], and *Szatmari* and *Schafer* [2] obtained a 6 ps TW ASE pulse with a 12 ps pumping pulse in a transverse travelling-wave pumping scheme. In a slightly different scheme, in which a gradient distribution of pumping energy was introduced, the authors [3] achieved a 50 ps TW ASE pulse using a nitrogen laser (700 ps) as the pumping source. Recently, we used a $2\omega$ mode-locked Nd:YAG laser (46 ps) as a pump source; 8–15 ps TW ASE output pulses were obtained.

Figure 1 shows the experimental set-up (see details in [3]). An output signal enhanced by 10:1 in the forward direction was observed by this transverse travelling-wave pumping scheme. However, the arrangement with a gradient distribution of pumping energy prevents the saturation of signal amplification as well as the long fluorescence tail. Experimental results for different different dyes and pump sources are given in Table 1.

To further understand the physical mechanism of shortening of TW ASE, we measured the output pulse shape for four different pumped lengths, as shown in Fig. 2. It shows clearly how a spontaneous emission signal is compressed when it travels from point A to point B.

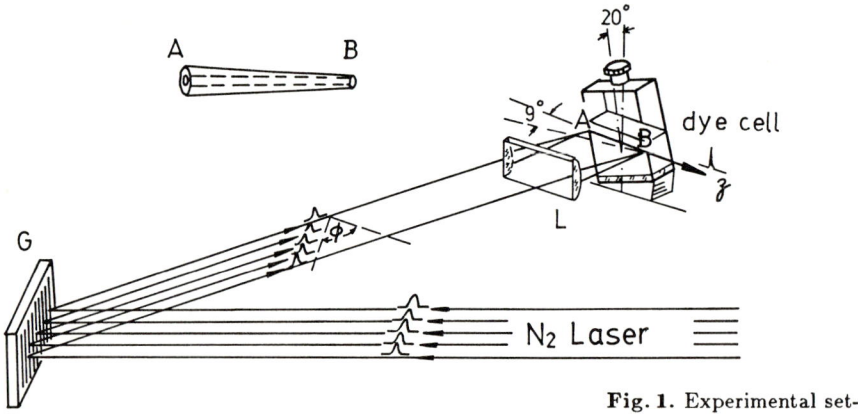

Fig. 1. Experimental set-up

Table 1.

| Dye | Pumped by $N_2$ Laser (700ps) | | | Pumped by $2\omega$ M-L Nd:YAG Laser (46ps) | | | | |
|---|---|---|---|---|---|---|---|---|
| | Coumarin 311 | Rh 6G | Rh B | Rh 6G | Rh 6G+ DODCI | Rh B | Rh B+ DODCI | DODCI |
| Concentration (M/L) | $5 \times 10^{-3}$ | $2.8 \times 10^{-2}$ | $2.5 \times 10^{-3}$ | $1 \times 10^{-2}$ | $1 \times 10^{-2}$ $1 \times 10^{-4}$ | $5 \times 10^{-3}$ | $5 \times 10^{-3}$ $1 \times 10^{-4}$ | $1 \times 10^{-3}$ |
| Output of signal: | | | | | | | | |
| pulsewidth | 45 ps | 55 ps | 45 ps | 15 ps | 11 ps | 10 ps | 8 ps | 8 ps |
| bandwidth | 68 Å | 107 Å | 57 Å | 140 Å | 210 Å | 140 Å | 270 Å | 160 Å |
| center of band | 458 nm | 591 nm | 600 nm | 579 nm | 592 nm | 612 nm | 614 nm | 638 nm |

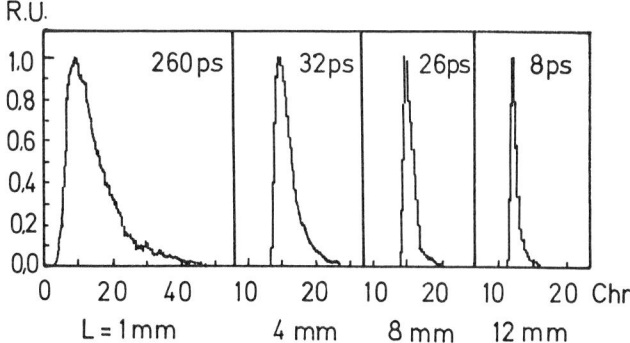

**Fig. 2.** TW ASE pulse shapes with different pumped lengths

In accordance with the above phenomena a mathematical model has been set up to simulate the temporal process of TW ASE generation [4]. The gain of the system is defined as

$$G(t, z) = I(t + z/V_g)/I(t, 0) . \qquad (1)$$

From the rate equations of the system we get the differential equation of $G$:

$$\frac{dG}{dt} = \sigma^e \sigma_p^a N \frac{P}{L\beta} \ln\left(\frac{\alpha}{\alpha - \beta z}\right) \exp\left[-(t - z/V_g)^2/T_G\right]$$
$$- (\sigma^e + \sigma_p^a) I(t, 0) G(G - 1) - \frac{G}{\tau_{\text{eff}}} (\ln G + \sigma^a N z) , \qquad (2)$$

where the function of pump pulse is assumed to be the Gauss profile and includes the description of the spatial delay; $\alpha$ and $\beta$ are parameters to describe the gradient distribution of the pumping energy; $\sigma^e$, $\sigma^a$ and $\tau_{\text{eff}}$ are the emission and absorption cross sections and the effective lifetime [5] of the dye molecules, respectively; $I(t, 0)$ is the spontaneous emission signal at point A. Using formula (2), we have calculated the shape and pulsewidth of output TW ASE of Rh 6G with different pump durations (Fig. 3). Where Fig. 3a is similar to the pump condition in [3], 3c is similar to Nd:YAG

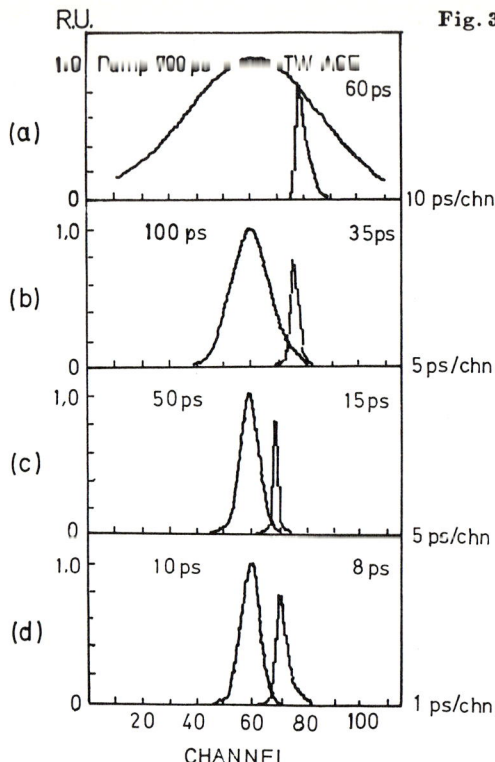

Fig. 3. TW ASE pulse shape depends on the pump duration

laser pumping and 3d is similar to that in [1]. The pulsewidths of ASE dependence on the concentration of dye C and pumped lengths L have also been calculated. Calculation results agreed with the experiments satisfactorily. The calculation also indicates that the ratio of pulse compression falls gradually when the pumping pulse duration is shorter than 50 ps because of the effect of the lifetime of the dye molecules and the saturation of the amplifying process.

The pulse emitted from three different dye solutions in three kinds of pumping schemes with the same pump energy were measured by a streak camera. The pulse shapes of TW ASE are shown in the upper row of Fig. 4, where a "seed" light signal starting from A is amplified and shortened step by step along the excitation volume as described above. The pulse shapes of a non-travelling-wave pump scheme are shown in the second row of Fig. 4. In this case the pump pulse arrives simultaneously at the whole pumped volume. Any "seed" signal of light generated anywhere inside the pencil-like region will be amplified in both directions. So, a longer ASE pulse appears. The pulse shapes of the transversal pumped with a standard laser cavity is shown in the lower row of Fig. 4. In comparison with the other two

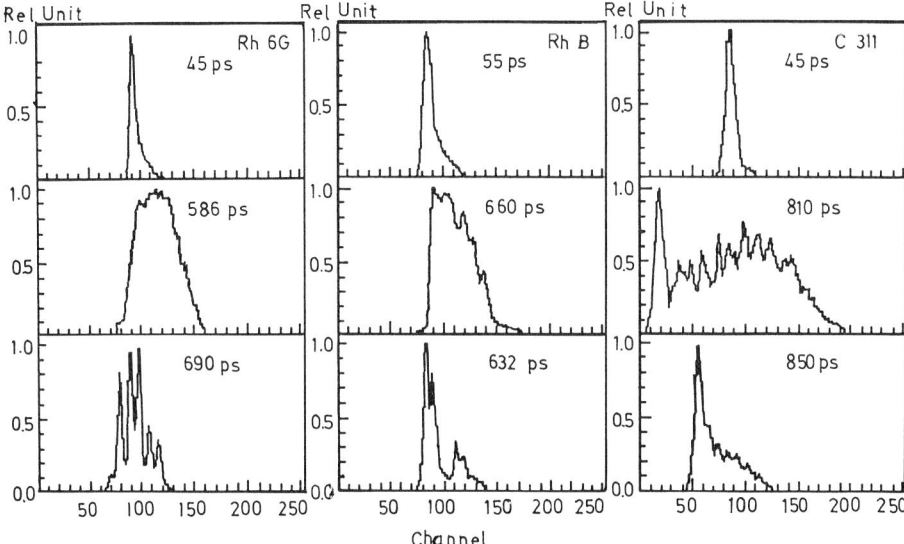

**Fig. 4.** Pulses emitted from three kinds of pumping scheme with Rh 6G, Rh B and C 311

pump schemes, we can see that the travelling-wave pump scheme is the only effective method which results in strong pulse shortening.

The spectrum of TW ASE from Coumarin 311 dye (see Fig. 5) has some clear fine structure. This is an interesting phenomenon and we are now making a further study of it.

Ultrashort light pulses with durations of 50 ps and 10 ps, respectively, pumped by a $N_2$ laser and a Nd:YAG laser, were obtained by our system. The wavelength can be easily changed by using different dyes and applying further narrowing with an additional spectral filter coupled to an amplifier, as shown in [2]. The efficiency of energy conversion is found to be 2%–5% for different dyes, and the variation of pulsewidth is 11%, it mainly depends on the stability of the pump source. Although the output energy here is rather low, it can easily be increased by using one- or two-stage amplifiers. On account of the above advantage, this system could be developed into a simple, low-cost, ultrashort light source.

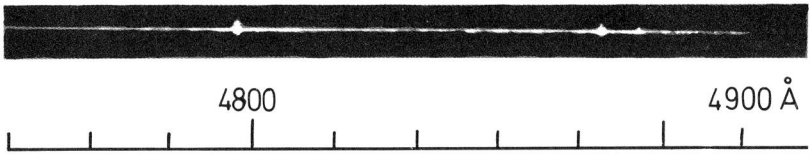

**Fig. 5.** The spectrum of TW ASE emitted from C 311

# References

1. Zs. Bor, S. Szatmari, A. Muller: Appl. Phys. B **32**, 101 (1983)
2. S. Szatmari, F.P. Schafer: Opt. Commun. **49**, 281 (1984)
3. Lee Wenchong, Ning Changlong, Huang Zuozhu: Appl. Phys. B **39** (1986)
4. Lee Wenchong, Ning Changlong, Huang Zuozhu: to be published
5. T.L. Koch, A. Yariv: J. Appl. Phys. **53**, 9 (1982)

# Part III

# Electrooptic Sampling Techniques

# Electrooptic Sampling
# of Gallium Arsenide Integrated Circuits

K.J. Weingarten, M.J.W. Rodwell, J.L. Freeman, S.K. Diamond, and D.M. Bloom

Stanford University, Edward L. Ginzton Laboratory,
Stanford, CA 94305, USA

1. Introduction

As GaAs integrated circuits (IC's) grow in speed and complexity, new methods to accurately measure the internal node response of these circuits are required. Sampling oscilloscopes have a 25 ps risetime and network analyzers can measure linear response to 100 GHz. However, both instruments load the test point with 50 Ω limiting their use to input or output ports. Electrooptic sampling was pioneered [1],[2],[3] to measure optoelectronic devices (photoconductors and photodetectors) with response times shorter than a sampling oscilloscope's resolution. To address the need to test internal nodes we have developed an electrooptic sampling system for probing signals directly within a GaAs IC [4].

2. Electrooptic Sampling System

Gallium arsenide is electrooptic; the electric fields from conductor voltages induce optical birefringence, causing a change in polarization to a suitably polarized probe beam passing through these fields. The polarization change can be detected by passing the probe beam through a polarizer and onto a photodetector. For the longitudinal probing geometry the change in intensity of the received beam is linearly proportional to the voltage [4] across the substrate at the test point. Probing geometries for measuring the voltage on various conductor types are shown in Fig. 1 [5],[6].

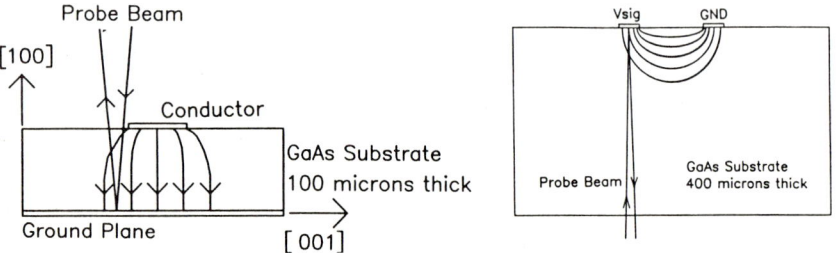

Figure 1: Cross sections of sampling geometries. Frontside (left) for probing microstrip transmission lines and backside (right) for planar transmission lines and wire interconnects

Figure 2 shows the sampling system schematic. A mode-locked Nd:YAG laser is driven at 82 MHz, producing optical pulses of 1.06 μm wavelength and 100 ps pulsewidth. A fiber-grating pulse compressor reduces the pulse duration to 2 ps [7] and a phase-lock-loop timing stabilizer reduces the laser timing jitter to 1 ps [8]. The probe beam passes through a polarizing beamsplitter and a $\lambda/4$ waveplate producing elliptically polarized light with its major axis 22.5° to the beamsplitter axes. A $\lambda/2$

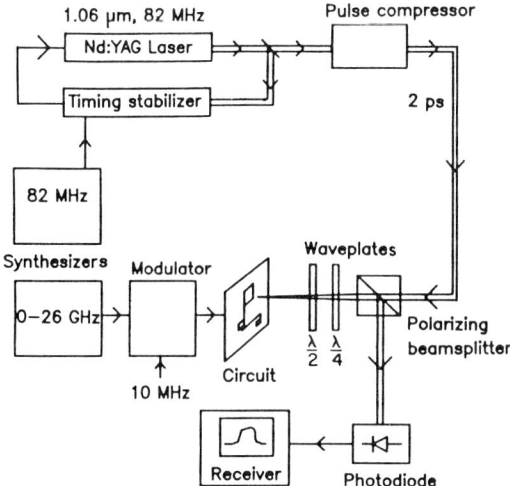

Figure 2: Schematic of the electrooptic sampler

waveplate rotates the axes of ellipse to 45° from the electrooptic crystal axes of the substrate. A microscope objective focuses the beam through the substrate adjacent to or on the conductor of interest for frontside or backside probing, respectively. The reflected light passes back through the waveplates and beamsplitter, producing linearly polarized light at 45° to the beamsplitter axes. The polarization component reflected from the beamsplitter is directed onto a photodiode connected to a receiver. The beam intensity at the photodiode varies linearly with the voltage on the conductor [4].

The system can be configured to emulate either a sampling oscilloscope or a network analyzer. To emulate a sampling oscilloscope, the microwave synthesizer is set to an exact multiple of the laser pulse repetition rate plus some small frequency offset, typically 10-100 Hz. The received intensity then varies in proportion to the signal voltage at this offset frequency. To improve the signal to noise ratio, the microwave signal is also pulse modulated at 10 MHz for synchronous detection. To use the sampler as a network analyzer, we remove the pulse modulation, offset the microwave frequency by 10 MHz and use a narrowband 10 MHz vector voltmeter.

The measurement bandwidth is set by the optical pulsewidth and the pulse-to-pulse timing jitter. The pulsewidth is measured with an optical autocorrelator; however, this does not uniquely determine the pulseshape. A more useful measure of the pulse's frequency content is the Fourier transform of its autocorrelation, the power spectral density (PSD) [9]. The compressed pulse autocorrelation, with a FWHM of 2 ps, has a -3 dB point of 100 GHz, slightly less than that of an ideal 2 ps FWHM gaussian pulse.

The timing jitter degrades the system performance in two ways, increasing the noise floor and decreasing the measurement bandwidth [8]. However, using a phase-lock-loop to stabilize the laser timing with respect to a stable microwave synthesizer, the r.m.s. jitter has been reduced to less than 1 ps with long term drift of about 0.5 ps/minute.

## 3. Circuit Applications

Since the sampling system can emulate either a sampling oscilloscope or a network analyzer, it is ideally suited for circuit measurements on GaAs IC's. Signal timing, risetimes, and propagation delays on a number of digital circuits have been measured [6],[10]. On an 8-bit multiplexer clocked at 2.6 GHz [11], signals on interconnects as narrow as 2 $\mu$m and the bottom of air bridge posts were measured, including the serial output word of the MUX and the timing of the 8-phase clock. To measure propagation delays, a test structure consisting a string of 20 inverters for average gate

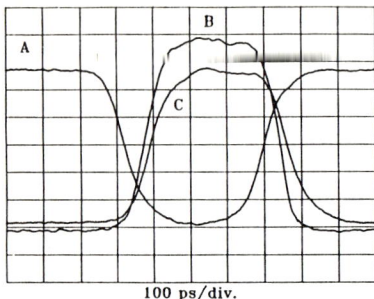

Figure 3: Propagation delays through one inverter of the 20-inverter test structure. A: Drive signal from previous inverter B: Delay through inverting FET, 60 ps C: Delay through buffer FET and diode level shifters, 15 ps

delay measurements with conventional electronics [12] was probed. By switching the first inverter with a microwave synthesizer at gigahertz rates, a repeating square wave rippled through the inverter chain. The probe beam was then positioned at nodes in the circuit to measure the delays between inverters and between the FET's internal to individual inverters (see Fig. 3).

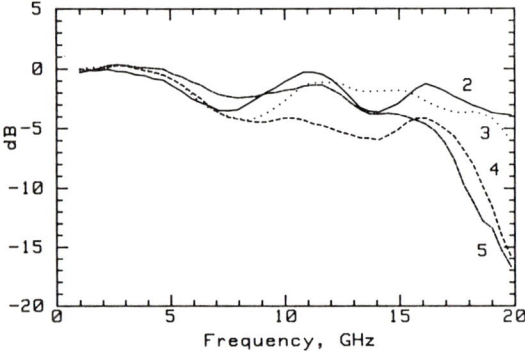

Figure 4: Small-signal voltage frequency response at the gates of a 5-FET distributed amplifier (gate 1 omitted for clarity)

More recently the system has been applied to the extensive characterization of microwave amplifiers [13]. Using the sampler in the network analyzer mode, the small-signal response at internal nodes was measured (see Fig. 4). Comparing the measured data to the simulated data, the process-dependent circuit parameters were modified to obtain a best-fit. Modeled gate termination resistance increased from 50 $\Omega$ to 80 $\Omega$, and the devices' $f_\tau$ decreased by 14%, for example. The large-signal, saturation behaviour of the amplifiers was studied using the system as a sampling oscilloscope. Harmonic distortion at fundamental frequencies as high as 21 GHz was observed.

In the network analyzer mode the sampler measures the vector voltage on the IC. A network analyzer, however, measures the forward and reverse traveling waves. By measuring the voltage along a conductor, the forward and reverse traveling waves can be calculated. Figure 5 shows an example of such a measurement on a transmission line with two different load terminations. For a line terminated in its characteristic impedance, the magnitude is constant and the phase varies linearly with slope $\beta$, the wavenumber, while for a perfect short or open, the magnitude varies sinusoidally. From

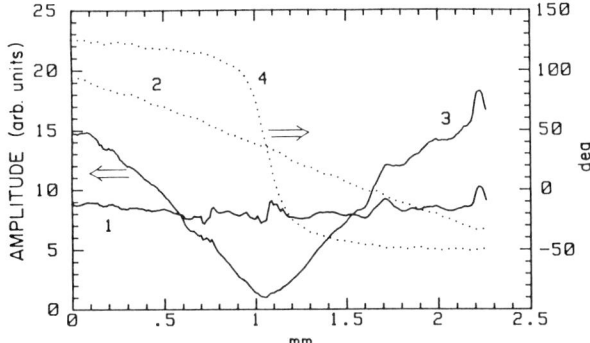

Figure 5: Length scan of microstrip transmission line for two load terminations at 16 GHz drive frequency: 1) magnitude and 2) phase with 50 Ω termination, 3) magnitude and 4) phase with short termination

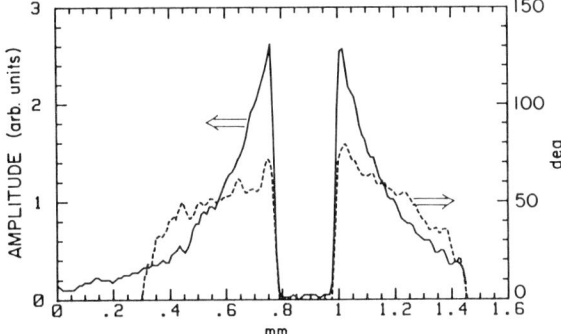

Figure 6: Width scan of microstrip transmission line at 31 GHz drive frequency. The conductor width is 220 μm and the substrate thickness 350 μm. The phase delay near the edge of the substrate is evidence of coupling to radiative modes

this data the reflection coefficient, loss, and dispersion can be calculated. Scanning perpendicular to the conductor allows for measurement of the potential distribution (Fig. 6). Note that the potential drops off in roughly the distance of substrate thickness, 350 μm, as expected. The phase shift near the edge of the substrate is evidence of coupling of the guided wave to higher-order leaky modes.

## 4. Conclusion

We have described an electrooptic sampling system designed for making ultrafast measurements on GaAs IC's. The system has a measurement bandwidth of 80 GHz, a timing drift of less than 1 ps, a noise floor of 300 μV (1 Hz), and a spatial resolution of 2 μm. Using the system as either a sampling oscilloscope or a network analyzer, we have measured signal timing, risetimes, and propagation delays in digital circuits with picosecond resolution, frequency response and harmonic distortion in microwave amplifiers to 20 GHz, and standing waves, propagation constants, and potential distributions of transmission lines to 40 GHz. The technique has a potential bandwidth extending past the millimeter-wave range (300 GHz), and future use of this system will concentrate on the study of new ultrafast GaAs circuits and devices.

## 5. Acknowledgements

The authors wish to thank Majid Riaziat of Varian Associates, Steve Swierskowski of Lawrence Livermore National Laboratory, and K. Reed Gleason of TriQuint Semiconductor for their invaluable assistance with their GaAs IC's. This work was supported by the Air Force Office of Scientific Research under contract number F49620-85K-0016. M. Rodwell, J. Freeman, and S. Diamond wish to acknowledge IBM fellowships.

## 6. References

1. J.A. Valdmanis, G. A. Mourou, and C.W. Gabel, IEEE J. Quant. Elec., **19**, 664 (1983)

2. B.K. Kolner, D.M. Bloom, and P.S. Cross, Elect. Lett., **19**, 574 (1983)

3. J.A. Valdmanis and G. Mourou, IEEE J. Quant. Elec., **22**, 69, (1986)

4. B.H. Kolner and D.M. Bloom, IEEE J. Quant. Elec., **22**, 79 (1986)

5. K.J. Weingarten, M.J.W. Rodwell, H.K. Heinrich, B.H. Kolner, and D.M. Bloom, Electron. Lett.,**21**, 765 (1985)

6. J.L. Freeman, S.K. Diamond, H. Fong, and D.M. Bloom, Appl. Phys. Lett.,**47**, 1083 (1985)

7. J.D. Kafka, B.H. Kolner, T.M. Baer, and D.M. Bloom, Optics Lett., **9**, 505 (1984)

8. M.J.W. Rodwell, K.J. Weingarten, D.M. Bloom, T. Baer, and B.H. Kolner, *submitted to* Optics Letters.

9. R.N. Bracewell, *The Fourier Transform and its Applications*, McGraw-Hill, New York 1978

10. M.J.W. Rodwell, K.J. Weingarten, J.L. Freeman, and D.M. Bloom, Electron. Lett., **22**, 499 (1986)

11. G.D. McCormack, A.G. Rode, and E.W. Strid, Proc. 1982 GaAs IC Symp., pp. 25-28

12. S. Swierkowski, K. Mayeda, and C. McGonaghy, Technical Digest, 1985 International Electron Devices Meeting, pp. 272-275

13. M.J.W. Rodwell, M. Riaziat, K.J. Weingarten, B.A. Auld, and D.M. Bloom, Technical Digest, 1986 IEEE MTT-S International Microwave Symp., pp. 333-336

# Picosecond Characterization of Ultrafast Phenomena: New Devices and New Techniques

D.R. Dykaar [1,2], R. Sobolewski [2], J.F. Whitaker [1,2], T.Y. Hsiang [2], G.A. Mourou [1], M.A. Hollis [3], B.J. Clifton [3], K.B. Nichols [3], C.O. Bozler [3], and R.A. Murphy [3]

[1] Laboratory for Laser Energetics,
[2] Department of Electrical Engineering, University of Rochester, Rochester, NY 14623, USA
[3] Lincoln Laboratory, Massachusetts Institute of Technology, Cambridge, MA 02139, USA

As switching speeds of microelectronic circuits increase, new problems arise in characterizing the real-time responses of these circuits. It is now typical for a new device to be faster than conventional electronic techniques can measure.

Direct measurement can, however, be achieved through the electro-optic sampling technique. First developed at Rochester in 1982, VALDMANIS et al. [1], this technique takes advantage of the advances made in ultrashort laser-pulse technology. This system makes use of a colliding-pulse mode-locked (CPM) laser which routinely emits pulses of less than 100–fs FWHM. These short optical time intervals permit a direct measurement of electrical signals with a duration of hundreds of femtoseconds.

Recently, this technique has been applied to the characterization of several new types of devices such as the two-dimensional electron gas field-effect transistor (TEGFET) and the metal semiconductor field-effect transistor (MESFET), MEYER et al. [2]. These devices had rise times of 16 ps and 25 ps respectively and, using conventional techniques, can only be tracted by indirect methods such as placing a series of similar devices in a ring oscillator and measuring the frequency of oscillation.

One device, however, resists even indirect attempts at characterization: the permeable-base transistor (PBT). Conceived at MIT Lincoln Laboratory in 1979, BOZLER et al. [3], the PBT presents unique problems in its manufacture. As shown in Fig. 1, a PBT is arranged similarly to a vacuum tube triode, except that here the "grid" has submicrometer dimensions. By interrupting the GaAs epitaxy process to fabricate the grating, contaminants are introduced, keeping yields low. As of this writing, only single devices have been fabricated, precluding the construction of ring oscillators. In addition, analog measurements have proven difficult because of the very high gain of the PBT. However, extrapolated frequency-gain measurements show a maximum frequency of oscillation in excess of 200 GHz, HOLLIS et al. [4]. The speed of this device results from advances in fabrication techniques that allow for extremely small dimensions in both the horizontal and vertical directions. These advances have led to a device that, until now, remained completely uncharacterized in the time domain.

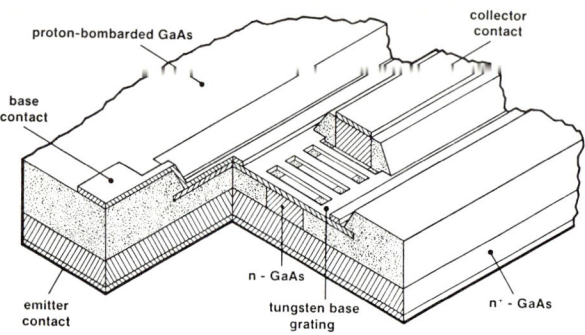

Figure 1  Geometry of PBT.

For electro-optic characterization, the PBT was built into the sampling geometry shown in Fig. 2. This geometry permits dc biasing as well as high-speed operation. The CPM laser-based sampling system measured the rise time, as shown in Fig. 3. Electromagnetic radiation from the input connection was thought to cause the negative prepulse. This feature did not scale in amplitude with PBT gain and has also been present in earlier measurements on TEGFET and MESFET devices. This measurement identifies, to the best of our knowledge, the currently fastest three-terminal room-temperature device.

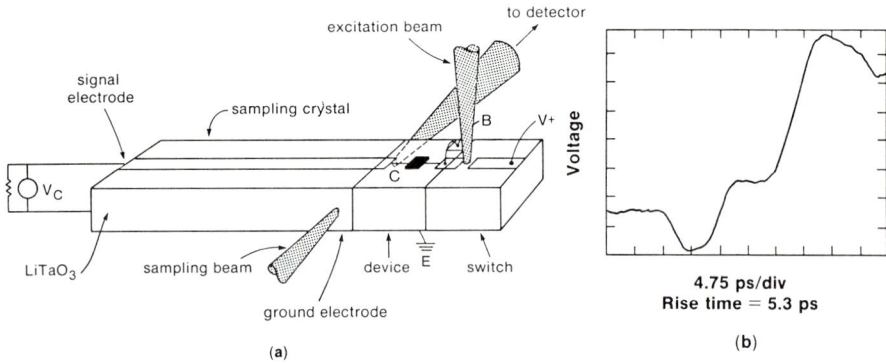

Figure 2(a) Sampling geometry for PBT characterization. Looping wire bonds were used for bias leads to prevent loading of high-speed signal paths. (b) Step response of PBT as measured by the electro-optic sampling system.

Clearly, with electrical rise times of a few picoseconds, it becomes necessary to consider the details of how these very fast signals are transmitted. In previous work, various geometries have been used to study propagation, KRYZAK et al. [5] and DYKAAR et al. [6]. In this work, the mechanisms contributing to signal distortion, namely frequency dependent absorption and modal dispersion, have been separated. This was accomplished by using a superconducting sampling system.

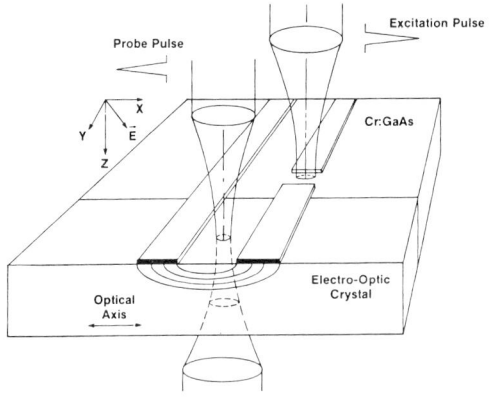

Figure 3  Superconducting sampling configuration.  In electrodes were evaporated directly across GaAs-LiTaO$_3$ interface.

As shown in Fig. 3, a GaAs/LiTaO$_3$ hybrid structure was fabricated. In operation, the entire unit was placed in a superfluid helium (T < 2 kelvins) environment.

The use of superfluid helium allowed the laser pulses to propagate into and out of the cryogenic environment, undistorted by fluid bubbling.  An early version of the sampler, using standard Pb-alloy technology for superconducting electronics showed a 5-fold increase in sampler performance to 1.0 ps, when cooled from room temperature.  A rise time of 360 fs was obtained using indium electrodes.  As shown in Fig. 4(a), the rise time dispersed to only 2 ps after 3 mm of propagation.  These results were modelled using a technique previously

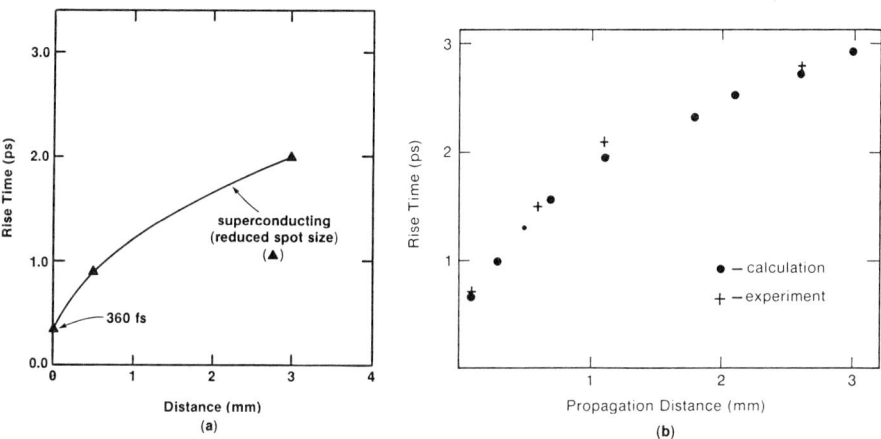

Figure 4(a)  Rise time versus distance for the superconducting sampler. (b) Early superconducting results modeled by only considering modal dispersion.

reported, WHITAKER et al. [7], with the only change in the model being to set the absorption of the transmission lines to zero. As shown in Fig. 4(b), the excellent fit suggests that for this system, only modal dispersion is present.

In conclusion, a new type of device has been studied using the electro-optic technique. Having a measured risetime of 5 ps, it is, to the best of our knowledge, the currently fastest three-terminal room-termperature device. In addition the electro-optic technique has been extended to include the cryogenic regime. This new capability has been utilized to study the propagation of femtosecond electrical signals.

## Acknowledgment

This work was supported by the Laser Fusion Feasibility Project at the Laboratory for Laser Energetics which has the following sponsors: Empire State Electric Energy Reseach Corporation, General Electric Company, New York State Energy Research and Development Authority, Ontario Hydro, Southern California Edison Company, and the University of Rochester. The PBT development is sponsored by the Defense Advanced Research Projects Agency and Department of the Air Force. Such support does not imply endorsement of the content by any of the above parties.

## References

1. J.A. Valdmanis, G.A. Mourou, C.W. Gabel: In Appl. Phys. Lett. 41, 211 (1982).
2. K.E. Meyer, D.R. Dykaar, and G.A. Mourou: In Picosecond Electronics and Optoelectronics, eds. G.A. Mourou, D.M. Bloom, and C.H. Lee (Springer-Verlag, Berlin, Heidelberg, New York, and Tokyo), 54 (1985).
3. C.O. Bozler, G.A. Alley, R.A. Murphy, D.C. Flanders, and W.T. Lindley: In Proc. 7th Bien. Cornell Conf. on Active Microwave Devices, 33 (1979).
4. M.A. Hollis, K.B. Nichols, R.A. Murphy, R.P. Gale, S. Rabe, W.J. Piacentini, C.O. Bozler, and P.M. Smith: In Technical Digest IEDM, 102 (1985).
5. C.J. Kryzak, K.E. Meyer, G.A. Mourou: In Picosecond Electronics and Optoelectornics, eds. G.A. Mourou, D.M. Bloom, and C.H. Lee (Springer-Verlag, Berlin, Heidelberg, New York, and Tokyo), 244 (1985).
6. D.R. Dykaar, T.Y. Hsiang, G.A. Mourou, Ibid., 249 (1985).
7. J.F. Whitaker, T. Norris, G. Mourou, T. Hsiang, IEEE MTT, to appear.

# Precise Measurement of Signal Propagation Characteristics in GaAs Integrated Circuits by Picosecond Electro-Optic Sampling

R.K. Jain, X.-C. Zhang, M.G. Ressl, and T.J. Pier

Amoco Research Center, Naperville, IL 60566, USA


A large number of picosecond optical and opto-electronic techniques have recently been proposed and demonstrated[1,2] for the characterization of high-speed GaAs integrated circuits (IC), including the use of electro-optic sampling[3] for the measurement of on-chip waveforms.[2] In this paper, we demonstrate the use of picosecond electro-optic sampling for the precise measurement of relative timing between waveforms at different points internal to a GaAs integrated circuit. In particular, we report precise pulse propagation and switching characteristics for both the "1"-to-"0" and "0"-to-"1" transitions in digital ICs of buffered FET logic design,[1,4] from which single gate propagation delays of 73 ± 3 picoseconds are inferred for the chosen inverter gates; in addition, temporal delays of less than 10 picoseconds have been measured for pulse propagation between diodes internal to such gates.

In contrast to the previous work[2] on electro-optic sampling in GaAs ICs, in which a single infrared probe pulse was synchronized to externally applied high-frequency electrical signals, we use a pair of synchronized picosecond optical pulses (in a manner similar to that used in conventional optical and optoelectronic sampling[3,5]), namely, a visible pulse (at $\hbar\omega > E_g$) to initiate an electrical transient within the circuit, and a subsequent precisely synchronized infrared pulse (at $\hbar\omega < E_g$) to probe the transient via the longitudinal electro-optic effect in GaAs. Such a two-pulse configuration is particularly useful for the measurement of precise temporal shifts between internal waveforms at different points within an IC, as demonstrated here.

Our electro-optic sampling measurements were performed on a variety of test device structures and circuits fabricated on (100)-oriented GaAs substrates. In each case, the two illuminating pulses were incident from opposite directions, with the 532 nm pulses ($\sim$ 2 ps wide) incident from above, and the 1.06 $\mu$m probe pulses ($\sim$ 3 ps wide) incident from below. The 1.06 $\mu$m beam is focused (1.5 $\mu$m diam.) on the reflective metallic interconnection line being probed, and polarization changes induced in the reflected beam (by transient voltages on the line) are measured by relatively standard signal detection techniques.[3,5] A plot of the detected signal versus the applied voltage indicated excellent linearity of the detection system over the entire voltage range of interest.

For brevity, we restrict this discussion to a single illustrative example, namely the measurement of propagation delays in logic gates in a high speed digital IC. The test circuit used for this measurement consisted of a string of 20 inverter gates of standard buffered FET logic design,[1,4] illustrated schematically in the cross-sectional view of Figure 1. Logic level electrical pulses with relatively fast risetimes for both the "1"-to-"0" and "0"-to-"1" transitions were easily induced at the output of these gates

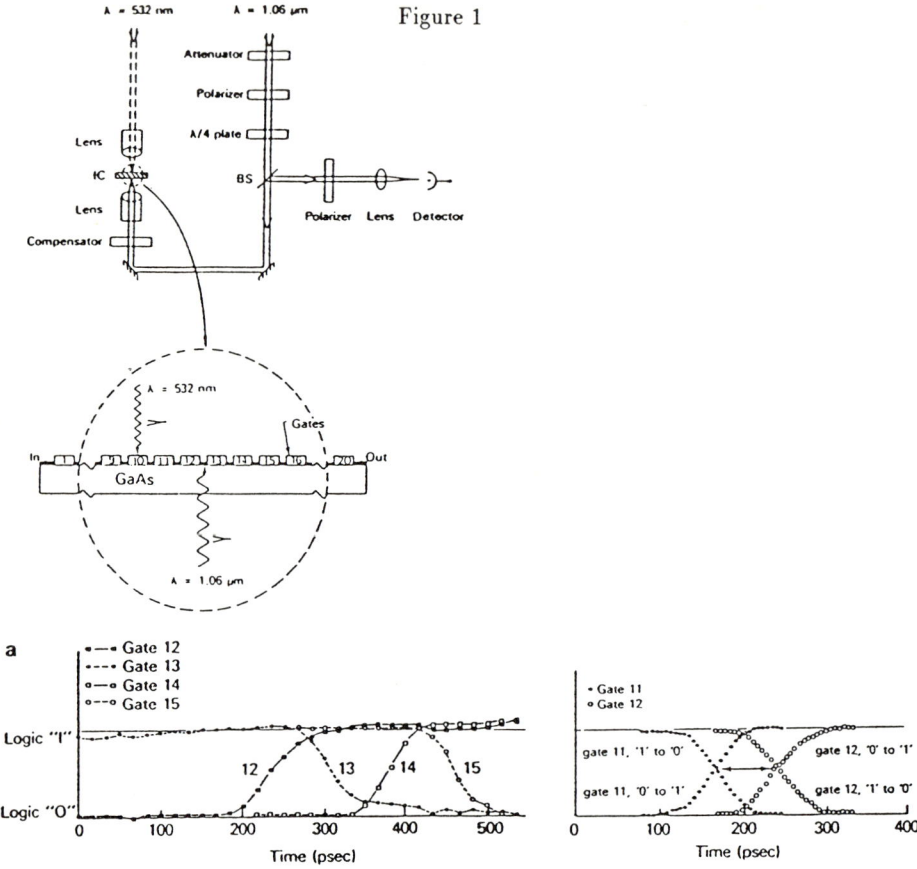

Figure 1

Figure 2

optically[1] by addressing appropriate field effect transistors within the circuit with ~ 2 ps, 532 nm pulses of ~ 6 pJ energy. For the data shown in Figure 2(a), the logic level at the input of the inverter string was set to logic "0" (-3V), and the logic state at the output of gate 10 was optically switched from logic "0" to logic "1" ($\cong$ 1V); the leading edges of the logic-switched waveforms at the output of several downstream gates (12 through 15) are shown in Figure 2(a). Likewise, leading edges of output waveforms for transitions of both polarities ("0"-to-"1" and "1"-to-"0") at the outputs of gates 11 and 12 are shown in Figure 2(b). Individual gate propagation delays of 73 ± 3 ps are obtained from such measurements, consistent with measurements performed by several other techniques.[1,4] More importantly, the measurements reported here represent the first clear experimental demonstration of the concept of propagation delay and how it relates to waveforms at the output of serially-connected logic gates. We have also measured waveforms and propagation delays at various points within the inverter, and delays smaller than 10 picoseconds have been measured for signal propagation between the various level-shifting diodes that comprise the gate.

Work currently in progress includes the use of electro-optic sampling in GaAs to study the precise effect of the fan-in and fan-out on gate propagation delay, and to map internal waveforms in more complex GaAs digital ICs such as sample-and-hold circuits, and gigabit rate multiplexers and demultiplexers.

References

1. R.K. Jain and D.E. Snyder, Opt. Lett., $\underline{8}$, 85 (1983); R.K. Jain, D.E. Snyder, and K. Stenersen, IEEE Electron Device Lett., $\underline{\text{EDL-5}}$, 371 (1984).

2. J.L. Freeman, S.K. Diamond, H. Fong, and D.M. Bloom, Appl. Phys. Lett., $\underline{47}$, 1083 (1985); K.J. Weingarten, M.J.W. Rodwell, H.K. Heinrich, B.H. Kolner, and D.M. Bloom, Electron. Lett., $\underline{21}$, 765 (1985).

3. D.H. Auston and A.M. Glass, Appl. Phys. Lett., $\underline{20}$, 398 (1972); J.A. Valdmanis, G. Mourou, and C.W. Gabel, Appl. Phys. Lett., $\underline{41}$, 211 (1982).

4. This GaAs-buffered FET logic circuit with MESFETs of 1-$\mu$m design rule was fabricated at the Lawrence Livermore National Laboratory by Steve Swierkowski and his colleagues, for measurement of average propagation delay per gate by conventional electronic means.

5. B.H. Kolner and D.M. Bloom, Electron. Lett., $\underline{20}$, 818 (1984).

# Propagation of Ultrashort Electrical Pulses on Superconducting Transmission Lines

*I.N. Duling III, C.-C. Chi, W.J. Gallagher, D. Grischkowsky, N.J. Halas*[**],
*M.B. Ketchen, and A.W. Kleinsasser*

IBM Thomas J. Watson Research Center, P.O. Box 218,
Yorktown Heights, NY 10598, USA

Subpicosecond electrical pulses have recently been generated on 5 micron coplanar aluminum transmission lines [1]. With the proper generation geometry these pulses can propagate as a single mode excitation of the transmission structure, which allows the pulse reshaping to be entirely due to the frequency dependent response of the materials of the transmission line. By Fourier analysis of the initial and propagated pulses, both the absorption and dispersion as a function of frequency can be obtained. In this manner, time domain spectroscopy is performed in the technically difficult regime from d.c. to 1 terahertz. This technique has been applied here to the study of superconducting transmission lines. The results clearly show the superconducting energy band-gap and the strong temperature dependence of the associated absorption edge.

The subpicosecond pulses are generated by the use of a fast-recombination-time photoconductive switch driven by subpicosecond laser pulses [1]. In this particular arrangement one line of the transmission structure is biased and the switch is used to short the two lines. This "sliding contact" method of excitation transfers charge from one line to the other providing a single mode excitation of the lines and producing an ultrashort pulse propagating in each direction. The transmission structure consisted of two lines 5 microns wide of 800 Angstrom thick niobium separated by 10 microns (Fig. 1). The photoconductive switch was implemented by depositing the transmission structure on ion-implanted silicon-on-sapphire, for which the carrier recombination time is approximately 500 fs. The niobium lines become superconducting at 9 degees Kelvin. The transmission lines are He vapor cooled and temperature regulated in an optical dewar. This arrangement allows good optical quality access to the photoconductive switches without resorting to the use of superfluid He. The opto-electronic sampling technique used is well suited to operation inside a dewar since all fast electrical signals (0-1 THz) are confined to the sample and only slow signals (0-1 KHz) must leave the dewar.

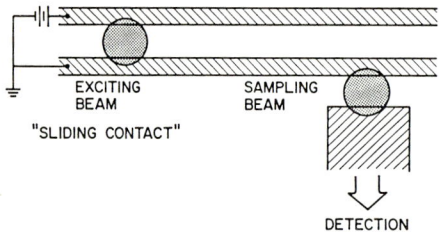

Figure 1 The transmission line geometry showing the movable excitation spot (sliding contact) and the sampling gap.

---

* This research was partially supported by the U.S. Office of Naval Research.
** Bryn Mawr College

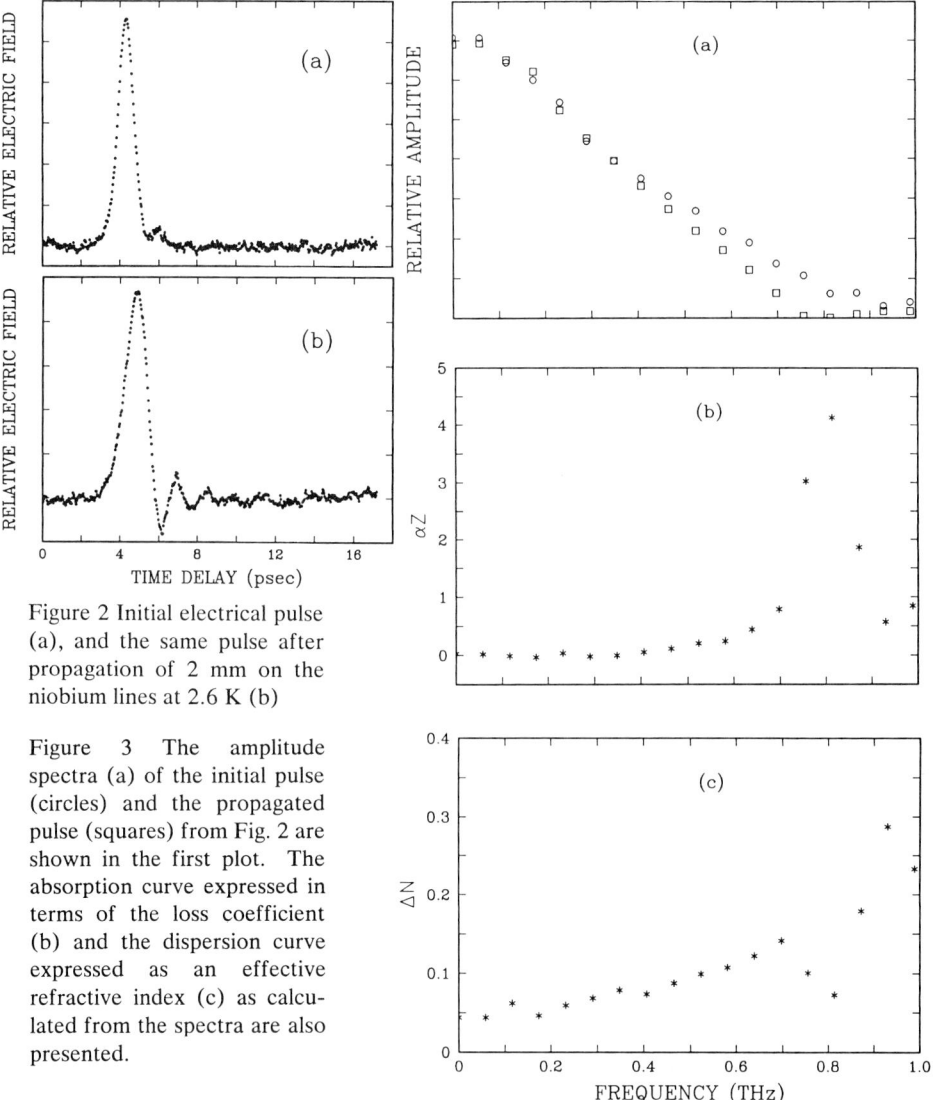

Figure 2 Initial electrical pulse (a), and the same pulse after propagation of 2 mm on the niobium lines at 2.6 K (b)

Figure 3 The amplitude spectra (a) of the initial pulse (circles) and the propagated pulse (squares) from Fig. 2 are shown in the first plot. The absorption curve expressed in terms of the loss coefficient (b) and the dispersion curve expressed as an effective refractive index (c) as calculated from the spectra are also presented.

The input and propagated pulses are shown in Fig. 2 for a temperature of 2.6 degrees Kelvin. The propagation distance was 2 mm. The observed change in shape of the pulse indicates that the structure has dispersion and absorption. By using the technique described briefly here and at more length by HALAS, et al. [2], the amplitude spectra and the resulting absorption and dispersion curves were calculated for the structure and are shown in Fig. 3. The results were normalized by assuming no absorption in the d.c. component.

It is expected that there will be a strong absorption for frequencies above the energy gap of the superconducting niobium, since these frequencies see a normal (resistive) transmission line. At room temperature the line resistance is 500 ohms/mm. The energy gap of niobium at 2.6 degrees Kelvin is approximately 3 meV which corresponds to a frequency of .72 THz. From Fig. 3 a strong absorption is evident at .7 THz and is consequently attributed to pair breaking in the superconducting niobium. The fall off of absorption after the initial absorptive peak is not physically significant due to the lack of signal to noise in this region.

The shape of the propagated pulse in Fig. 2 is in good qualitative agreement with that predicted by KAUTZ [3]. By using the absorption and dispersion predicted by the theory of Mattis and Bardeen, it is possible to model the propagation of wide-bandwidth electrical pulses on superconducting transmission lines. One result of this analysis is that the absorption increases by two orders of magnitude for frequencies above the band gap. This strong absorption would explain the loss of signal above the absorption edge.

In Fig. 4 the same procedure was followed to obtain the absorption spectra for two additional temperatures approaching the superconducting transition temperature. The closer the niobium is to becoming normal, the smaller the energy gap and the more the pulse will be attenuated. This is supported by the observed movement of the onset of absorption to lower frequencies as the temperature is raised. The predicted position of the absorption edge from the theory is .72, .64, and .49 THz for 2.6, 6.0, and 7.6 K respectively. These values fall above those obtained from the data indicating that only the beginning of the absorption edge is observed, which is further evidence of a strong normal state absorption.

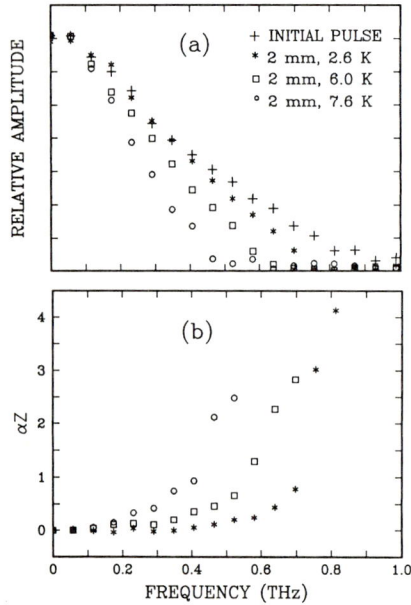

Figure 4 Amplitude spectra (a) and absorption curves (b) for the niobium transmission line at temperatures of 2.6, 6.0, and 7.6 degrees Kelvin showing the shift of the superconducting band edge.

Propagation of pulses for which the amplitude and phase can be recovered is proving to be a powerful technique for the evaluation of frequency dependent absorption and dispersion over the bandwidth of the pulse. In this case it has been applied to superconducting transmission lines and has proved to be a simple and direct way of measuring the energy gap of the superconducting material at a variety of temperatures. This example highlights the ability of this technique to perform time domain spectroscopy in the microwave frequency range.

The inherent underpinnings of this technique are equally important; the production of electrical pulses of subpicosecond duration, the single mode excitation of a coplanar transmission line, and the generation of this pulse in a guiding structure which allows long interaction lengths. All of these aspects combine to create a technique capable of examining a large variety of both spectroscopic and device oriented phenomena on a time scale rarely achieved in electrical measurements.

1. M.B.Ketchen, D.Grischkowsky, T.C.Chen, C.-C.Chi, I.N.Duling III, N.J.Halas, J.-M.Halbout, J.A.Kash, G.P.Li: Appl.Phys.Lett. 48(12), 24 March 1986 :see also JOSA A, vol. 2, 13 (1985).

2. N.J.Halas, I.N.Duling III, M.B.Ketchen, and D.Grischkowsky, conference proceedings CLEO 1986 p.328

3. R.L.Kautz, J.Appl.Phys. 49(1), January 1978.

# High Repetition Rate Electro-Optic Sampling with an Injection Laser

*A.J. Taylor, R.S. Tucker, J.M. Wiesenfeld, G. Eisenstein, and C.A. Burrus*
AT & T Bell Laboratories, Holmdel, NJ 07733, USA

The development of high-speed electronic and opto-electronic systems requires instrumentation to routinely characterize very high speed devices. We demonstrate that this type of measurement can be carried out using an electro-optic sampling system [1] with either a mode-locked or a gain-switched injection laser as the source of sampling pulses. We describe measurements of high repetition rate electrical waveforms [2] and high-speed long wavelength photodetectors [3]. This system is also suitable for noninvasive characterization of high-speed circuits [4] based on III-V materials technology.

The key element in this new electro-optic sampling system is an InGaAsP injection laser, at a wavelength of 1.3 or 1.55 μm, which produces 10-15 ps pulses in a mode-locked configuration and 18-20 ps pulses when gain-switched. An important feature of this laser is the wide tunability of its repetition rate which ranges from 0.1 GHz to 20 GHz. For measurements of electrical waveforms this allows both the laser and the test circuit to be driven by the same high frequency clock. In the mode-locked configuration, the tunability is possible because the fundamental resonance frequency of the laser can be tuned by adjusting the cavity length and then the laser can be operated at a high harmonic of this resonance frequency. For the gain-switched configuration, the repetition rate can be arbitrarily set to match the clock for the test circuit.

For measurements of external waveforms, the temporal resolution of the sampling system is degraded by jitter between the sampling pulses and the master clock. We have determined that the pulse-to-pulse timing jitter is < 1 ps in both mode-locked and gain-switched InGaAsP injection lasers. This jitter is due to phase noise on the microwave synthesizer and amplifier that drive the laser. For comparison the timing jitter in mode-locked Nd:YAG and dye lasers is typically 10-20 ps [4,5].

The sampling system employs the longitudinal electro-optic effect in the GaAs substrate of a microstrip line [4] which is probed by the below-bardgap radiation from the injection laser. The sampling system for the measurement of multigigabit voltage waveforms from an external circuit is illustrated in Figure 1. The laser and the circuit are driven by two microwave synthesizers, both at multigigabit rates, offset in frequency by $\Delta f$ (5 kHz) [4]. The arrival time of the optical pulse is thus scanned across the test waveform at a rate of $\Delta f$, enabling a real time display of the electro-optic signal on an oscilloscope. In Figure 2 we present such a real time electro-optic signal for the voltage waveform from an electrical comb generator (HP #33005C) driven at 2 GHz. The full width at half maximum (FWHM) of the electro-optic signal is 44 ps compared to 49 ps for the same waveform measured with a sampling oscilloscope.

The optical sampling source used in Figure 2 is a gain-switched InGaAsP injection laser with an 18 ps pulse width. The temporal resolution of this sampling system is 19 ps, and is dominated by the laser pulsewidth and the

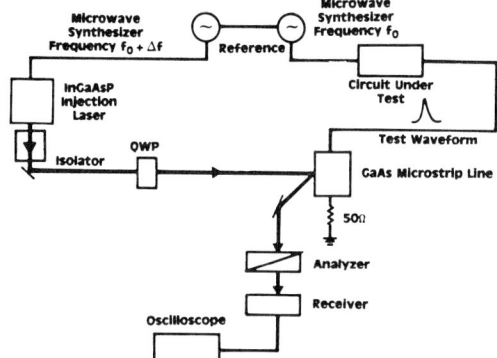

Fig. 1  Experimental set-up.

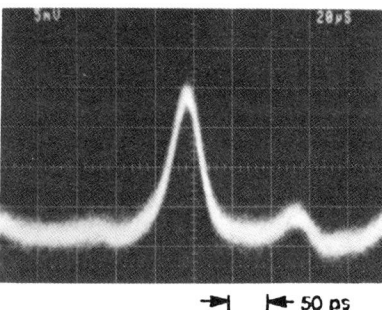

Fig. 2  Electro-optic signal from a comb generator.

6 ps optical transit time through the GaAs substrate. When we use a mode-locked injection laser with 10 ps pulses in place of the gain-switched laser, the system resolution becomes 12 ps [2]. In either case, the resolution of our system is superior to the 25 ps resolution of a sampling oscilloscope. The measured half-wave voltage is 6.3 kV at a wavelength of 1.3 µm. Our receiver, which consists of an InGaAs PIN photodiode followed by a FET amplifier, is shot noise limited. With typical average output powers from the laser, 0.5-2 mW, the minimum detectable voltage for a signal-to-noise ratio of 1 in a 1 Hz bandwidth is 1.5 mV/(Hz)$^{\frac{1}{2}}$.

A slightly different experimental set-up can be used for the evaluation of long wavelength photodetectors [3]. Pulses from the laser have the dual function of exciting the photodetector under test and sampling the resultant voltage waveform. A variable delay line scans the arrival time of the sampling pulse. The excitation beam is mechanically chopped and synchronous detection is employed. The time and frequency domain response of the photodetector can be obtained by deconvolving the measured autocorrelation from the measured electro-optic signal.

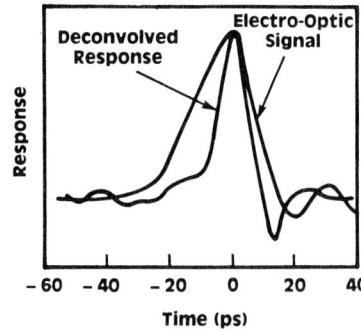

Fig. 3a  Response in time domain for InGaAs PIN photodiode in K-connector package.

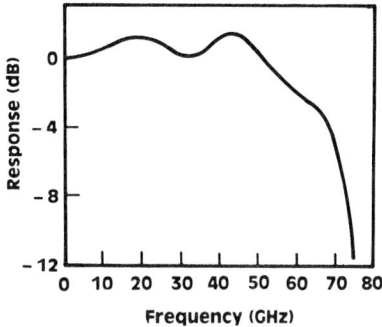

Fig. 3b  Response in frequency domain.

Figure 3 is the response in the time (3a) and frequency domain (3b) for an InGaAs PIN punch-through photodiode in an optimized coaxial package using a commercial K-connector. Parasitics are minimized and the photodiode is biased through an external bias tee placed after the sampler. The deconvolved time response reveals a 9 ps FWHM, and the frequency response exhibits a 3 dB rolloff around 67 GHz. This is the fastest response yet measured for a long wavelength PIN photodetector. We estimate a system resolution of 6 ps due to the optical transit time through the GaAs substrate, which may contribute to the observed high frequency rolloff. However, noise present in the measurement contributes ± 2 ps uncertainty to the deconvolved response.

For comparison with Figure 3, Figure 4 is the electro-optic signal and deconvolved response for a similar InGaAs PIN photodiode packaged in an HP coaxial microwave detector circuit [6]. This nonoptimized package contains a resistive impedance matching network and a bias-decoupling network. Hence, the impulse response has degraded to 19 ps FWHM. Figure 5 presents the deconvolved response for an InGaAs APD with separate absorption, grading, and multiplication regions (SAGM-APD) [7] operating with gains of 2.5 and 6. The FWHM of the corresponding impulse responses are 55 ps and 59 ps, corresponding to 3 dB bandwidths of 5.5 GHz and 5 GHz, respectively.

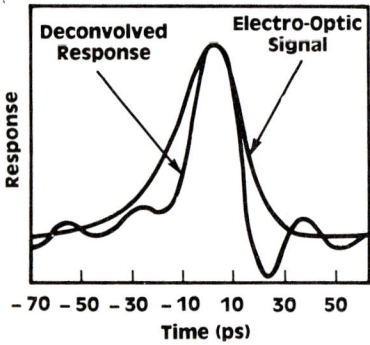

Fig. 4  Time domain response for InGaAs PIN photodiode in HP mount.

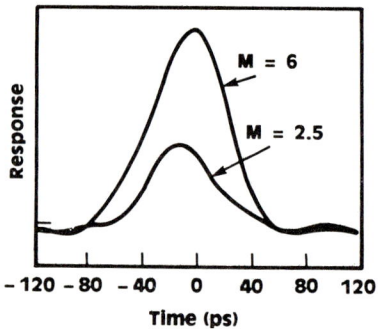

Fig. 5  Time domain response for InGaAs APD with gains of M = 2.5 and M = 6.

In summary, we have demonstrated the first electro-optic sampling system using InGaAsP injection lasers. The system has a resolution of 12 ps mode-locked and 19 ps gain-switched. By replacing the microstrip line with a GaAs circuit, this system could be used for noninvasive high-speed signal measurement internal to the circuit. Furthermore, the laser wavelength makes this system suitable for noninvasive characterization of optoelectronic integrated circuits based on InP/InGaAsP technology.

1. J. A. Valdmanis, G. Mourou and C. W. Gabel: Appl. Phys. Lett. 41, 211 (1982).
2. A. J. Taylor, J. M. Wiesenfeld, G. Eisenstein, R. S. Tucker, J. R. Talman, and U. Koren: Electron. Lett. 22, 61 (1986).
3. A. J. Taylor, J. M. Wiesenfeld, G. Eisenstein, J. R. Talman, and U. Koren: Electron. Lett. 22, 325 (1986).
4. B. H. Kolner and D. M. Bloom: IEEE J. Quant. Electron. QE-22, 79 (1986).
5. J. Kluge, D. Wiechert and D. Von der Linde: Optics Commun. 51, 271 (1984).
6. C. A. Burrus, J. E. Bowers and R. S. Tucker: Electron. Lett. 21, 262 (1985).
7. J. C. Campbell, A. G. Dentai, W. S. Holden, and B. L. Kasper: Electron. Lett. 19, 818 (1983)

# Picosecond Optoelectronic Sampling of Electrical Waveforms Produced by an Optically Excited Field Effect Transistor

*D.E. Cooper and S.C. Moss*

The Aerospace Corporation, Mail Stop M2-253, P.O. Box 92957, Los Angeles, CA 90009, USA

We have studied the electrical response of a submicron-gate GaAs FET to optical excitation by picosecond laser pulses. The optical excitation of FETs gives insight into the dynamical processes occurring within the device and is also of interest for evaluating the FET as a high-speed optical detector. Excitation by ultrafast optical pulses yields information on ultrafast carrier transport and offers an alternative to pulsing the device in the conventional electronic manner. Use of the FET as an optical detector is potentially interesting because, in contrast to the PIN diode, gain can be obtained without the extra noise associated with avalanche devices [1]. In addition, the output of a FET in an integrated circuit can be optically modulated, and in conjunction with electro-optic sampling this can be used for non-invasive diagnostics of fast GaAs ICs [2,3].

We characterize the electrical response of a FET to an optical pulse using photoconductive sampling of Auston switches [4] in microstrip transmission lines. The unpackaged FET (Avantek AT-12100, 0.3 micron gate) was bonded into a test fixture in the manner previously used for measurements of the electrical pulse response [5]. Sampling of the gate and drain waveforms was performed at photoconductive switches approximately 1 mm from the device. Optical pulses for exciting the FET and for photoconductive sampling were provided by a dye laser synchronously pumped by a frequency-doubled mode-locked Nd:YAG laser. The excitation beam was focussed onto the region around the FET gate. The shape of the gate and drain waveforms did not vary as the optical excitation beam was moved to different positions along the gate. The amplitude of the waveforms did vary showing three peaks corresponding to optical excitation at each of the three feeds from the gate pad. The laser spot size on the device was 10-15 microns, providing peak power densities of up to 10 MW/cm$^2$. The data, which consisted of electrical waveform amplitude as a function of sampling pulse delay, was acquired and processed by a microcomputer.

The sampled output of the gate of the unbiased FET in response to optical excitation is shown in Fig. 1a. The response is dominated by a sharp peak 70 mv in amplitude, followed by a long tail of lower amplitude. This signal is purely photovoltaic, since the FET is unbiased and the gate contact forms a Schottky diode. The peak is attributed to rapid sweepout of carriers created in the depletion region, and the tail is attributed to diffusive transport of carriers from the channel region (where the internal fields are much lower) into the depletion region. With the FET biased at an operating point of $V_{DS}$ = 2.5 V, $V_{GS}$ = -1.0 V, and $I_{DS}$ = 17 mA the gate response amplitude increases to 160 mV, and the initial peak becomes narrower (Fig. 1b). The changes are due to the increased size of the depletion region under bias, resulting in a greater collection of photo-generated carriers. The gate signal under bias is also purely photovoltaic, with no photoconductive gain.

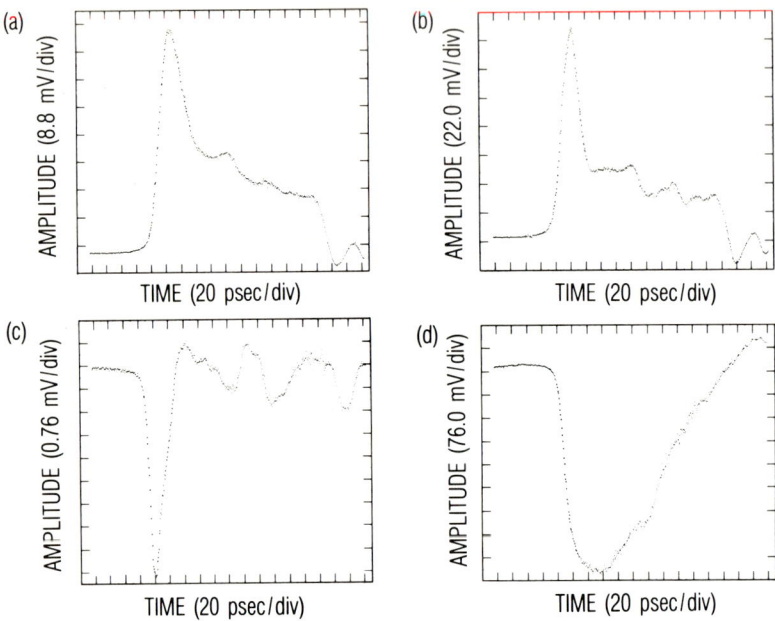

Fig. 1. FET response to optical excitation. Gate response (a) unbiased and (b) biased. Drain response (c) unbiased and (d) biased.

In contrast to the gate signal, the FET drain signal shows large variations with changes in both optical excitation intensity and FET biases. With no applied bias, the drain signal is also purely photovoltaic (Fig. 1c), and is similar to the gate signals. Short pulses (22 ps FWHM) are produced, but the amplitude is quite low (7 mv) and there is no long tail on the end of the pulse. These differences are attributed to the different geometries of the gate and drain: photogenerated carriers are produced on both sides of the gate and current flows towards the gate from both sides, producing a large prompt current pulse and the pulse tail as carriers continue to diffuse toward the gate. The drain current is produced when there is a difference between the current flowing towards the two sides of the gate. This difference is much smaller than the absolute magnitude of the current, and vanishes during the diffusive phase. The drain output under bias ($V_{DS}$ = 2.5 V, $V_{GS}$ = -1.0 V, $I_{DS}$ = 17 mA) is shown in Fig. 1d. Under bias the peak amplitude increases by a factor of about 90 due to photoconductive gain, completely obscuring the photovoltaic signal. The pulse has a sharp rise followed by a slower decay producing a pulse 132 ps wide. The decay of the electrical waveform is faster at higher drain bias or at lower optical excitation levels. We believe that the longer decays are produced by screening of the applied bias fields at high optical excitation levels resulting in slower transport and longer sweepout times.

The observed variations in FET output affect its performance as an optical detector. The fastest response is produced by the photovoltaic signal at zero bias, but there is no gain. With bias voltages turned on photoconductive gain is observed, but there is a loss of frequency response. There are still large high-frequency components in the sharp leading edge of the pulse, however. As excitation pulses optically generated within integrated circuits they can be used in conjunction with electro-optic sampling techniques [2] to study the high-frequency response of ICs. The amplitude of the largest observed pulses is close to that required to trigger logic level changes in digital ICs [3]. However, under the bias conditions which produce the largest transient gate and drain waveforms, both waveforms are superimposed upon a relatively large background with the same polarity as these fast transients. This background is most likely the result of long-lived transients due to the existence of trapping states and the resulting charge carrier dynamics at the substrate-channel interface [6]. At longer optical excitation wavelengths, these long-lived transients have been shown to have both the opposite polarity and larger amplitude than the fast transients. These changes in polarity, if rapid enough, could interfere with the ability to produce ultrafast logic level switching in digital ICs. In summary, the optical excitation of FETs yields insight into transport processes within the device and is a useful technique for integrated circuit diagnostics.

References

1. J. R. Forrest, F. P. Richards, and A. Perichon, "The Microwave MESFET Optical Detector," IEDM 1982 Technical Digest, p. 529.
2. B. H. Kolner and D. M. Bloom, "Electro-optic Sampling in GaAs Integrated Circuits," IEEE J. Quant. Electron. QE-22, 79 (1986).
3. X.-C. Zhang and R. K. Jain, "Measurement of On-Chip Waveforms and Pulse Propagation in Digital GaAs Integrated Circuits by Picosecond Electro-Optic Sampling," Electron. Lett. 22, 264 (1986).
4. P. R. Smith, D. H. Auston, A. M. Johnson, and W. M. Augustyniak, "Picosecond Photoconductivity in Radiation-damaged Silicon-on-Sapphire Films," Appl. Phys. Lett. 38, 47 (1981).
5. D. E. Cooper and S. C. Moss, "Picosecond Optoelectronic Measurement of the High-Frequency Scattering Parameters of a GaAs FET," IEEE J. Quant. Electron. QE-22, 94 (1986).
6. T. F. Carruthers, W. T. Anderson, and J. F. Weller, "Optically Induced Backgating Transients in GaAs FET's, "IEEE Electron Device Lett. EDL-6, 580 (1985). T. F. Carruthers, W. T. Anderson, and J. F. Weller, "Optically Induced, Spatially Resolved Backgating Transients in GaAs FET's, "IEDM 1985 Technical Digest, p. 106.

# Picosecond Electrical Pulses in Microelectronics

*P.G. May, G.P. Li, J.-M. Halbout, M.B. Ketchen, C.-C. Chi,
M. Scheuermann, I.N. Duling III, D. Grischkowsky, and M. Smyth*

IBM Thomas J. Watson Research Center, P.O. Box 218,
Yorktown Heights, NY 10598, USA

## 1. Introduction

In this paper we report on the photoconductive generation /1/ and on the propagation of picosecond electrical pulses on transmission lines of dimensions compatible with microelectronics applications. Various structures have been studied including pairs of coplanar aluminium lines of 1.2 μm width separated by 2.4 μm. These lines are deposited on either silicon on sapphire (SOS) substrates fabricated in a manner described previously /2/, or on polysilicon on quartz substrates. The silicon and polysilicon respectively act as photoconductors /3/ for pulse generation and sampling. In the latter case undoped polysilicon is first annealed to increase the carrier mobility and subsequently implanted in order to reduce the carrier lifetime. The results show the possibility of full integration of this technology with either bipolar or MOS silicon devices for in situ characterization.

## 2. Double Sliding Contact

For the experiments described here we use the "sliding contact" /2/ technique of excitation which has the advantage of a balanced excitation of the line. We have extended this technique further to the detection of pulses on coplanar transmission lines. Figure 1 shows the set up and results of an experiment where on the same transmission line a pulse is both generated and sampled. In order to detect the pulse on the line and not the integrated photocurrent from either illumination spot, the exciting and probing beam are chopped at two different frequencies and the signal is detected by a lock-in amplifier referenced at the sum of the two chopping frequencies, and connected at the end of the transmission line. Because of the inherent symmetry of the experiment, two pulses are seen on the traces corresponding to each of the spots acting as a generator site. The separation between the two peaks corresponds to twice the time taken for the electrical pulse to travel between the two spots. It is important to note that sliding contact sampling is a differential form of sampling and as will be described later in results on a crosstalk experiment it provides information that could not be obtained with the traditional side gap sampling technique.

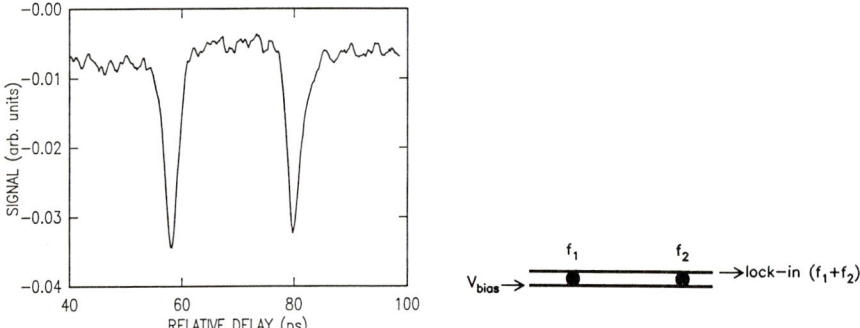

Fig.1 Schematic and result for double sliding contact measurement on 2.4 μm lines on polysilicon

## 3. Results and Discussion

The laser source used to drive the photoconductive switches is a Nd :YAG laser,acousto-optically mode-locked at 100MHz repetition rate.Its 73 ps pulses are compressed by passage through 40 meters of single mode optical fiber followed by a double-pass optical delay line composed of a pair of grazing incidence gratings.The compressed pulses are frequency doubled in a 0.5 cm KTP crystal to yield pulses of autocorrelated width 2 ps and average power 350 mW (5 W incident on fiber coupler).

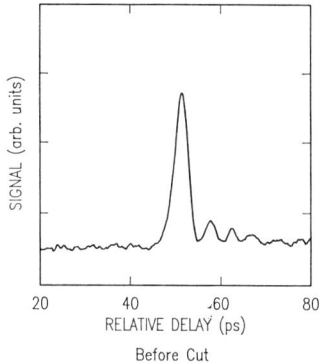
Before Cut

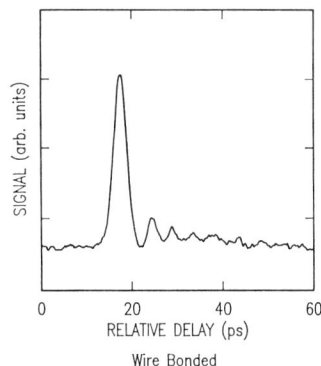
Wire Bonded

Fig.2 Pulse transmission through a wirebond

Figure 2 shows the results for the transmission of a 3 ps electrical pulse through a taper and wirebond.The coplanar lines are 10 $\mu$m wide, separated by 10 $\mu$m and taper out to 100 $\mu$m with the geometry and hence the characteristic impedance preserved.Two such tapered lines are connected by short lengths (250 $\mu$m) of 25 $\mu$m diameter aluminium wire.The electrical pulses are generated on one line by sliding contact and sampled on the other line with a side gap.With a bias voltage of 3 V the peak voltage of the generated pulses is estimated to be 200 mV.The ringing seen after the main pulse probably arises from the multiple reflections of a microwave pulse at the substrate-air interface.The results show that a 3 ps electrical pulse propagates virtually unaltered through both a taper and a short wirebond.The result is important for device characterization applications because it shows that with a little care, short electrical pulses can be delivered to a device from a discrete pulse generator and that its output can be connected to a discrete sampling site with minimum distortion to the waveforms.

Figure 3 shows the lay out and results for a crosstalk experiment where the sliding contact technique was used to generate pulses on one transmission line and to sample the pulses coupled on another line in close proximity. Treating the system as 4 coupled lines, modelling predicts that the waveform that is initially capacitively coupled to the upper two lines (ie. the component lines of the upper transmission line )are nearly identical and therefore a sliding contact sampling technique which measures the difference waveform should give a nullsignal.On the other hand a side gap measurement on the top line will give a signal.These predictions are confirmed by the experiment.If the coupled signals are measured at some distance from their generation point then in the sliding contact sampling case a small differentiated signal begins to appear (at about 2mm distance) and grow.This arises due to the different coupling in the system of the two top lines and hence pulses generated on each line will propagate at different velocities.

The double sliding contact technique is also particularly suitable for measuring the dispersion of electrical pulses down transmission lines, since there is total flexibility in choosing both the generation and sampling point.After 4mm travel for both 1.2 $\mu$m and 2.4 $\mu$m lines we have seen that electrical pulses of about 2 ps duration broaden by only a factor of approximately 50%.This result is the same

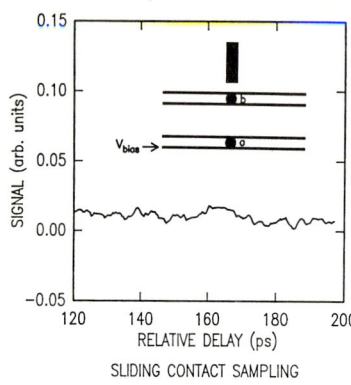 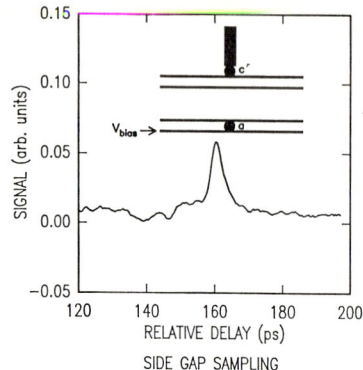

Fig.3 Coupled signal generated by sliding contact at "a" and sampled by sliding contact at "b" or by side gap at "c"

for both silicon on sapphire and polysilicon on quartz samples. This latter result is particularly encouraging as it was originally felt that the granular nature of the polysilicon surface would impair the deposition of micron structures.

4. Conclusions

It has been demonstrated that very short electrical pulses on aluminium micron lines can be propagated through impedance preserving tapers and through wirebonds - an important consideration regarding the use of photoconductive techniques in device characterization. A novel generating/sampling arrangement has been described. This double sliding contact technique allows differential sampling of the component lines of a transmission line. Further advantages of this technique include the freedom of choosing the desired sampling point on the transmission line and not having to design beforehand a sampling site; non-loading of the transmission line with spurious sampling sites; and a reduced capacitance at the sampling point which should yield a sharper sampling window.

References

1. D.H. Auston,in Picosecond Optoelectronic Devices, ed. by C.H. Lee (Academic, London, 1984).
2. M.B. Ketchen,D. Grischkowsky,T.C. Chen,C-C. Chi,I.N. Duling,N.J. Halas ,J-M. Halbout,J.A. Kash and G.P. Li,Appl. Phys. Lett. 48, 751, (1986);see also D.R. Dykaar,T.Y. Hsiang and G.A. Mourou,in Picosecond Electronics and Optoelectronics, ed. by G.A. Mourou,D.M. Bloom and C.H. Lee (Springer Verlag,1985),p 249;
3. W.R. Eisenstadt,R.B. Hammond,D.R. Bowman and R.W. Dutton, in Picosecond Electronics and Optoelectronics, ed. by G.A. Mourou,D.M. Bloom and C.H. Lee (Springer Verlag,1985),p 66;

# High Speed Circuit Measurements Using Photoemission Sampling

*J. Bokor, A.M. Johnson, R.H. Storz, and W.M. Simpson*
AT & T Bell Laboratories, Holmdel, NJ 07733, USA

We report the development of a new sampling technique for the measurement of high speed signals in electronic devices. The method is a hybrid between optoelectronic sampling[1] and e-beam probing[2] in that the signal waveform is measured by energy analyzing electrons ejected from the surface of a metallization line on the device into vacuum using a picosecond laser pulse to stimulate the electron emission. By exploiting the multiphoton photoelectric effect,[3] it is possible to use a visible wavelength laser to eject the electrons. Since the method involves the direct measurement of voltage on a metal line, it can be used on any type of electronic device, regardless of the electronic material being used to fabricate the device. The temporal resolution obtained in our preliminary experiments and reported here is better than 40 psec, and ultimate resolution of better than 10 psec should be achievable.

A short, visible laser pulse is focused onto the surface of a metal electrode on top of the circuit under test. In our experiments, we use a frequency doubled and fiber-grating pulse compressed cw mode-locked Nd:YAG laser.[4] This laser system produces pulses of duration less than 500 fsec at a wavelength of 532 nm, a repetition rate of 100 MHz and an average power of 200 mW. Under typical conditions, 30 mW is focused onto the sample surface using f/20 optics. This leads to peak pulse power on the surface of several hundred $MW/cm^2$ which is sufficiently high to produce easily detected multiphoton photoemission.[3] These electrons are accelerated by the electric field set up between the sample and an extraction grid. As the sample potential varies, so will the acceleration field between the sample and the extraction grid. This leads to a shift in the energy distribution of the electrons as they pass through the extraction grid. The magnitude of this shift is equal to the magnitude of the voltage change on the sample. By measuring the change in electron current which subsequently passes through a suitably biased retarding grid, the shift may be inferred and the sample voltage thereby measured.

For sufficiently short pulses, the time resolution of this technique is limited by the transit time of the electrons from the sample surface to the extraction grid. This is because the net acceleration is proportional to the time integral of the electric field over the duration of transit. (We note that this limitation applies equally to both photoemission sampling and electron beam probing.) For this reason, it is desirable to bring the extraction grid as close to the sample surface as possible and to use as high an extraction field as possible. The electron transit time, t, may be calculated as $t = 3.37 \times 10^{-9} (d/E)^{1/2}$ where d is the distance from the sample to the extraction grid in mm, E is the extraction field in V/mm, and t is in sec.

A very simple and compact parallel grid retarding field electron analyzer was used for the present experiments. A schematic diagram of the analyzer is shown in Fig. 1. The laser beam was focused through the back side of the analyzer as shown. A microchannel plate (MCP) detector with a 6 mm diameter center hole was used for this purpose. The collector grid also had a concentric 6 mm hole, but no special holes were needed in the extraction and retarding grids. The entire structure was held in a vacuum chamber at $10^{-6}$ torr pressure.

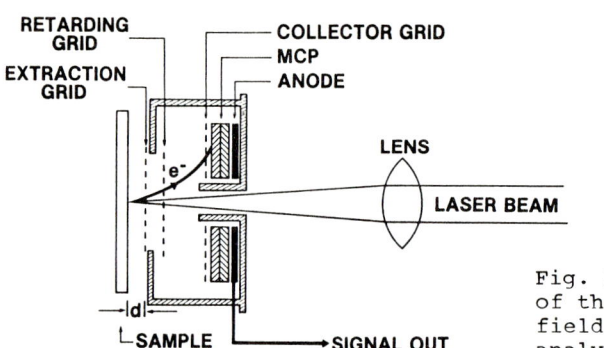

Fig. 1: Schematic diagram of the 3-grid retarding field electron energy analyzer.

We have tested the system on a simple microstrip sample consisting of a gold stripline on a high resistivity GaAs substrate. The work function for gold [5] is approximately 4.9 eV, hence three photons at 2.3 eV (532 nm) are required to overcome the work function and ejected electron kinetic energies of up to 2 eV are expected. The performance of the electron energy analyzer was tested by measuring the electron current detected by the microchannel plate as a function of sample voltage for various retarding grid bias voltages. The peak count rate from the channel plate under these conditions was typically $3 \times 10^5$ counts/sec. This corresponds to a photocurrent from the sample in the range of a picoampere. The retarding grid voltage necessary to extinguish the transmitted electron signal would ideally be equal to the sample voltage plus the approximately 2 V width of the initial energy distribution. In order to utilize the characteristics of this detector to achieve a linear voltage measurement with wide dynamic range a feedback circuit similar to that used in e-beam probing systems is used.[6] This circuit varies the retarding grid voltage as the sample voltage varies in order to maintain constant detector current.

Results obtained using the system to sample high speed pulse waveforms are now presented. A portion of the laser beam

was split off and passed through a variable optical delay line
to illuminate a high-speed radiation-damaged InP
photoconductor[7] with a previously measured impulse response
of 20 psec. The voltage pulse from this detector was then
launched down the gold stripline through 18 GHz sub-miniature
series A connectors and sampled by the photoemission sampler
after propagation down 1" of the stripline. Figure 2(a-c) shows
the sampled waveforms for various extraction fields at a d
spacing of 750 μm. For comparison, Fig. 2(d) shows the
waveform as measured by photoconductive sampling. For this
measurement, the voltage pulse was propagated down an identical
1" stripline to a CdTe photoconductive sampler with a
previously measured resolution of 10 psec.[8] By convoluting a
gaussian response function for the detector with the measured
pulse shape in Fig. 2(d) and fitting to the measured waveforms
in Fig. 2(a-c) we estimate the time resolution to be 121, 54,
and 39 psec, respectively. Based on the calculated transit
time, using a d spacing of 750 μm, we would expect the
corresponding values of 131, 65, and 46 psec. One possible
reason that the measured resolutions are better than those
calculated is that the extraction grid is somewhat flexible and
may have been attracted to the sample by the large applied
field, thus giving an actual d spacing that is smaller than
what was measured in zero field.

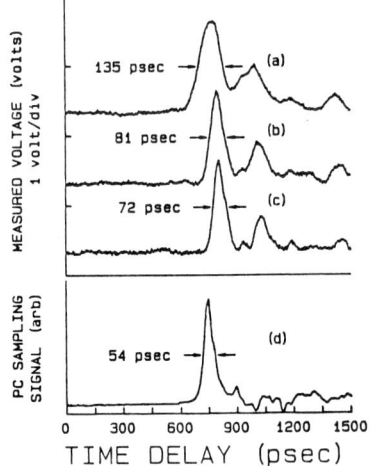

Fig. 2: (a-c) Measured
voltage waveforms.
Each curve was measured
using a different
extraction field:
(a) 666 V/mm;
(b) 2666 V/mm;
(c) 5333 V/mm. The
waveform in (d) was
measured using
conventional
photoconductive
sampling with a CdTe
sampler.

Experimentally, we found that the maximum extraction field
was limited to 5500 V/mm. Above this value, corona discharge
and electrical breakdown were observed. Improvement in time
resolution is still possible by further reducing the d spacing
at constant extraction field. In order to maintain a high
extraction field at the emission point, the d spacing must be
kept approximately 4-5 times the size of the grid holes. From
a consideration of gaussian beam focusing of the laser beam,
this leads to a lower limit on d of 20 μm which gives a time
resolution of 7 psec.

The photoemission sampled waveforms in Fig. 2 were taken with a 1 sec averaging time constant, and each waveform was averaged over 5 sweeps. From the noise level on the traces, we can estimate the present voltage sensitivity as approximately 40 mV/(Hz)$^{1/2}$. Further improvement in voltage sensitivity may be expected by improving the laser amplitude stability and/or increasing the modulation frequency to the 5 MHz range as is done in electrooptic sampling. We estimate that the shot noise limited sensitivity for this detector system is 1mV/(Hz)$^{1/2}$.

In summary, we have demonstrated a new technique for sampling high speed waveforms in electronic devices. Voltage waveforms are measured from metallization lines on the top of the device, thus the technique may be applied to electronic devices fabricated from any electronic material. The demonstrated time resolution is 39 psec and voltage sensitivity is 40 mV/(Hz)$^{1/2}$. Improvement of the performance to better than 10 psec time resolution and 1 mV/(Hz)$^{1/2}$ sensitivity is expected.

We would like to thank P. M. Downey for the loan of the InP photoconducting detector, D. W. Kisker for the preparation of CdTe films, and T. J. McIlrath for helpful discussions.

## REFERENCES

1. J. A. Valdmanis and G. Mourou, IEEE J. Quantum Electron. QE-$\underline{22}$, 69 (1986); B. H. Kolner and D. M. Bloom, IEEE J. Quantum Electron. QE-$\underline{22}$, 79 (1986).
2. E. Menzel and E. Kubalek, Scanning $\underline{5}$, 103 (1983).
3. S. I. Anisimov, V. A. Benderskii, and G. Farkas, Usp. Fiz. Nauk, $\underline{122}$, 185 (1977) [Sov. Phys. Usp. $\underline{20}$, 467 (1978)].
4. A. M. Johnson, R. H. Stolen, and W. M. Simpson, Appl. Phys. Lett. $\underline{44}$, 729 (1984).
5. American Institute of Physics Handbook, (McGraw-Hill, New York, 1963).
6. A. Gopinath and C. C. Sanger, J. Phys. E: Sci. Instrum. $\underline{4}$, 334 (1971).
7. P. M. Downey and B. Schwartz, Proceedings of the Society of Photo-Optical Instrumentation Engineers $\underline{439}$, 30 (1983).
8. D. H. Auston, A. M. Johnson, P. R. Smith, and J. C. Bean, Appl. Phys. Lett. $\underline{37}$, 371 (1980).
9. A. M. Johnson, D. W. Kisker, W. M. Simpson, and R. D. Feldman, in Picosecond Electronics and Optoelectronics, G. A. Mourou, D. M. Bloom, and C. H. Lee, eds., (Springer-Verlag, New York, 1985), p. 188.

# Photoemissive Sampling of Picosecond Electrical Waveforms

*A.M. Weiner, R.B. Marcus, P.S.D. Lin, and J.H. Abeles*

Bell Communications Research, Inc., Navesink Research and Engineering Center, Red Bank, NJ 07701, USA

Ultrafast optics is becoming increasingly important for probing high-speed devices as electronic switching times drop deep into the psec range. Established optoelectronic sampling techniques utilize either the electro-optic effect /1,2/ or photoconductive switching /3/. Here we describe a new approach for contactless probing of high-speed electrical waveforms, in which fsec laser pulses stimulate photoelectron emission from metallic interconnection lines on the surface of the device under test /4/. The potential at the emitting surface is derived from an energy analysis of the photoelectrons. This photoemissive sampling technique may be applied to circuits and devices on any semiconductor.

The experimental arrangement is sketched in Fig. 1. 80-fsec, 3-nJ pulses are generated at a frequency of 1 MHz from a cavity-dumped, colliding-pulse-modelocked (CPM) ring dye laser /5/. These pulses are admitted through a window into the vacuum station and focused by a 40x, 10.1 mm working distance microscope objective to a spot size of a few microns on a 50 Ω gold microstrip transmission line on a semi-insulating GaAs substrate. A high-speed photodetector is used to generate psec electrical pulses, which are fed through into the vacuum and coupled onto the microstrip line. A planar copper anode mounted 6 mm above the sample surface functions as electron spectrometer and detector; a wire screen 3.5 mm below the anode is used to establish an extraction field. The anode contains a 3/8 inch diameter hole, covered by a transparent conducting film of SnO on glass for optical access. The light incident on the photodetector is chopped; the resultant modulation in anode current is measured with a 1 MΩ sampling resistor and a lock-in amplifier.

The work functions of gold (∼5 eV) and other metallizations used on GaAs and silicon are sufficiently high that ultraviolet light is required for single photon photoemission. In order to operate at visible wavelengths,

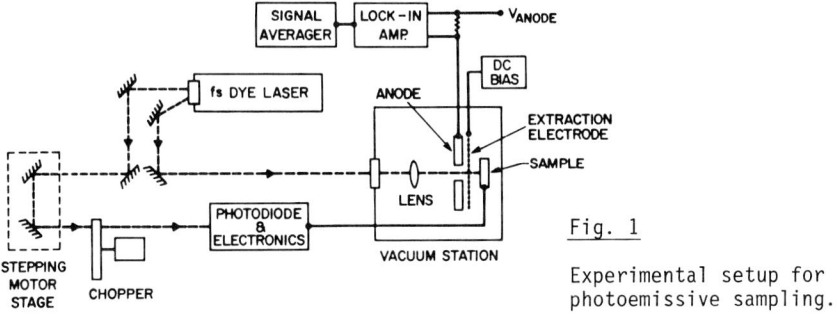

Fig. 1

Experimental setup for photoemissive sampling.

we utilize a multiphoton photoelectric effect. For gold, three 2-eV photons from the CPM laser are required for emission of a single photoelectron. A plot of photocurrent vs. optical intensity, shown in Fig. 2, exhibits the expected cubic power law dependence. Average photocurrents of up to 100 pA are produced by the 3-nJ, cavity-dumped pulses incident on a smooth gold surface.

We have also observed three photon photoemission using the less energetic pulses from an undumped CPM laser, by utilizing a surface-enhanced multiphoton photoelectric effect from roughened Au surfaces. 50-pJ pulses at a 100 MHz repetition rate produce currents as large as several nA; the photoelectron yield is enhanced by at least four orders of magnitude compared with emission from smooth gold. Roughening has been achieved in a variety of ways: by electrochemical treatment, by depositing gold on microlithographically patterned Si /6/ and unpolished GaAs substrates, and by focusing the optical beam on the edge of a metallization line. To our knowledge this work constitutes the first report of a surface-enhanced multiphoton photoelectric effect. The enhancement of Raman scattering /6,7/ and other nonlinear optical phenomena on roughened metallic surfaces has been well documented.

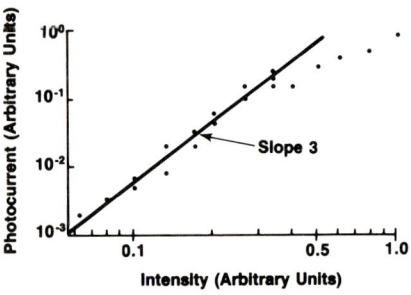

Fig. 2

Anode current plotted vs. optical intensity, for gold deposited on microlithographically patterned silicon.

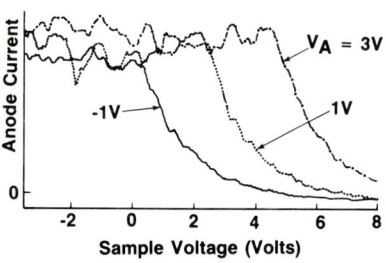

Fig. 3

Anode current plotted vs. sample potential, for a 40 V/cm extraction field (10 V grid voltage) and for various anode potentials.

The electron spectrometer is operated with the wire grid at a positive potential in the range 10-3000 V and the anode at a slightly negative potential. The grid provides an extraction field which accelerates photoelectrons away from the sample; in the region between the grid and the anode, electrons are decelerated. Figure 3 shows the anode current plotted as a function of the static voltage applied to the test sample, for various fixed anode potentials. The anode current increases with increasing negative bias of the emitting surface; measurements are sensitive to sample potential differences of a few tens of millivolts (with one second averaging time).

Time-resolved measurements were performed using 90 ps, -850 mV pulses, which were generated by a comb generator and coupled onto a 50 Ω microstrip transmission line on unpolished GaAs. Time-resolved photoemission traces are shown in Fig. 4 for extraction fields of 40, 400, 1600 and 8000 V/cm, with an anode bias of -1.5 V. At low extraction fields the waveforms display a long leading edge (at 40 V/cm the 10-90% rise time is 1.08 nsec) due to the limitation on temporal resolution to be discussed below. With an increasing extraction field, the waveforms become progressively more symmetrical; at 8000 V/cm the rise time is 136 ps and the FWHM 151 ps. The fall times of all traces are roughly the same (∼100 ps) and are comparable to the fall time measured with a sampling oscilloscope.

The temporal resolution of photoemissive sampling is governed by the effect of a changing sample potential on the photoelectrons during their travel toward the anode. The electric field is established primarily by the extraction grid and is sensitive to the sample potential only in the immediate vicinity of the sample (within a distance comparable to the spacing between the signal line and the nearest rf ground plane). We define an "effective" transit time as the time required for electrons to traverse the region of space in which the field significantly depends on the sample potential. Electrons leaving the sample in advance of a signal pulse by one effective transit time or less are affected by the signal while those leaving after the pulse are unaffected; this broadens the leading edge of the recorded waveform but leaves the trailing edge undistorted. The effective transit time T is given by:

$$T = \frac{2m}{eE} \left[ \sqrt{seE + U_0} - \sqrt{U_0} \right] , \qquad (1)$$

where E is the extraction field, m and e are the electron mass and charge, s is the spatial extent of the region influenced by the sample potential, and $U_0$ is the initial electron kinetic energy in the direction of the extraction field. With $U_0$ = 1 eV and s = 600 μm (the spacing between the signal line and ground plane is 300 μm), the calculated transit times for extraction fields of 40 V/cm and 8 kV/cm are 712 and 88 ps, respectively, in general agreement with the data shown in Fig 4. These data demonstrate that temporal resolution is improved by using high extraction fields.

The effective transit time may also be reduced by using a small spacing between signal line and ground plane. To demonstrate this point, we performed photoemissive sampling measurements of 50 ps electrical pulses coupled onto a 20-μm, 50 Ω gold coplanar stripline fabricated on smooth GaAs. The results

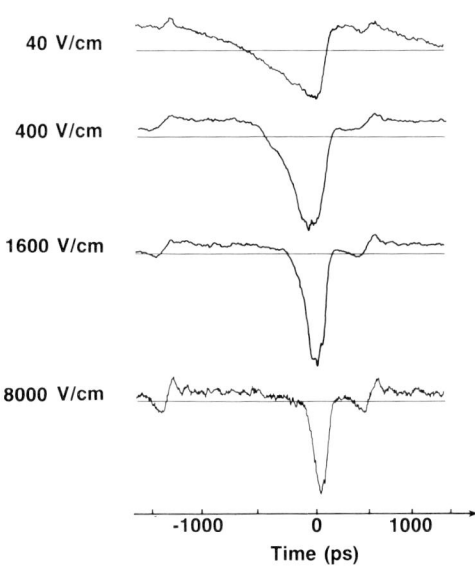

Fig. 4

Time-resolved measurements of 90 ps electrical pulses on a 50 Ω microstrip transmission line on unpolished GaAs, for various extraction fields. The anode voltage is -1.5V.

are displayed in Fig. 5a and 5b, respectively, for extraction fields of 100 V/cm and 1.1 kV/cm. Measurements on the microstrip transmission line on unpolished GaAs taken under similar conditions are also plotted. The temporal resolution obtained with the 20 μm coplanar stripline is enhanced dramatically compared to that obtained with the microstrip.

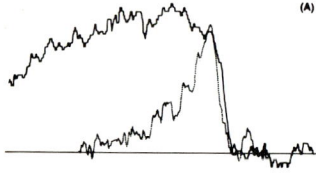

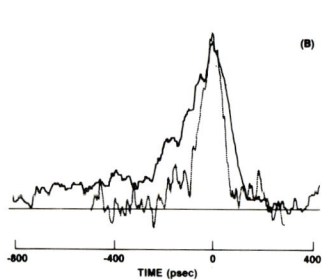

Fig. 5

Time-resolved measurements of 50 ps electrical pulses on a 20 μm coplanar stripline (dotted curve) and on a microstrip transmission line (solid curve). The extraction field is (a) 100 V/cm for both lines, (b) 1.1 kVcm for the coplanar line, 1.6 kV/cm for the microstrip line.

Further reductions in the effective transit time should be possible. For example, in the case of a 5-μm coplanar line, the effective transit time for an electron accelerated by a field of 100 kV/cm is only 3 psec. Under such conditions photoemissive sampling should provide an important new tool for contactless probing of high-speed electrical waveforms on circuits and devices on any semiconductor.

## References

1. J. A. Valdmanis and G. A. Mourou, IEEE J. Quantum Electron. QE-19, 664 (1983).
2. B. H. Kolner and D. M. Bloom, Electron. Lett. 20, 818 (1984).
3. D. H. Auston, A. M. Johnson, P. R. Smith and J. C. Bean, Appl. Phys. Lett. 37, 371 (1980).
4. R. B. Marcus, A. M. Weiner, J. H. Abeles and P. S. D. Lin, Appl. Phys. Lett., in press.
5. R. L. Fork, B. I. Greene and C. V. Shank, Appl. Phys. Lett. 38, 671 (1981).
6. P. F. Liao, J. G. Bergman, D. S. Chemla, A. Wokaun, J. Melngailis, A. M. Hawryluk and N. P. Economou, Chem. Phys. Lett. 82, 355 (1981).
7. M. Fleischmann, P. J. Hendra and A. J. McQuillan, Chem. Phys. Lett. 26, 163 (1974).

# Nonlinear Responses of Picosecond Photodetectors to Photogenerated Carriers

*T.F. Carruthers and J.F. Weller*

Naval Research Laboratory, Washington, DC 20375, USA

We report a simple means of measuring device response times which yields results similar to cross-correlation techniques on linked pairs of photodetectors but which avoids the bandwidth limitations imposed by interconnecting transmission lines and largely eliminates the need for elaborate device mounting procedures. Measurements are therefore possible on devices, such as those which must be mounted singly or which are embedded in an integrated circuit without direct electrical access, whose intrinsic response speeds were previously not observable.

In our experiment a train of picosecond optical pulses is split into two beams, which are chopped at different frequencies; one beam is delayed a variable time $\tau$ with respect to the other, and the beams are focused to overlapping spots on a photosensitive device. The lowest order *nonlinear* component of the device's response is extracted by detecting the component of its electrical output at the sum and difference of the chopping frequencies.

If the electrical output signal $I(t)$ of a photodetector is assumed to depend only upon the instantaneous number $N(t)$ of photogenerated carriers present in the device, the similarity of the nonlinear response to cross-correlation measurements is easily demonstrated. The two optical beams, separated by a delay time $\tau$, generate carriers of number $N_1$ and $N_2$. The device response $I(N)$ can be expanded in a Taylor series, retaining terms only to second order:

$$I(N) = N\frac{dI}{dN} + \frac{1}{2} N^2 \frac{d^2I}{dN^2} + \cdots = I(N_1) + I(N_2) + N_1 N_2 \frac{d^2I}{dN^2} + \cdots \qquad (1)$$

The last term of (1) is the only one containing frequency components at the sum and difference of the two chopping frequencies. The correlation signal $Q(\tau)$ is the time average of this term, so long as the time scales of the device response and the pulse repetition rate are widely separated, so that

$$Q(\tau) \approx \frac{d^2I}{dN^2} \int_{-\infty}^{\infty} dt\, N_1(t)\, N_2(t+\tau) \approx \frac{d^2I/dN^2}{(dI/dN)^2} \int_{-\infty}^{\infty} dt\, I_1(t) I_2(t+\tau) \qquad (2)$$

to second order in N. If $N_1$ and $N_2$ are identical except for their time delay, $Q(\tau)$ is the autocorrelation function of the device response. Unless stated otherwise, the nonlinear device responses reported here reflect a sublinear dependence of the output signal upon incident optical energy, corresponding to a negative second-order Taylor coefficient.

We have studied a number of different types of photodetectors and, where possible, compared their nonlinear responses with other response time measurements. Figure 1 presents the responses of an ion bombarded InP

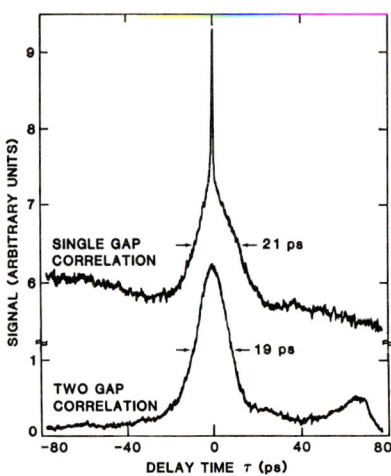

Fig. 1 Comparison of one- and two-gap correlation techniques on an ion bombarded ($10^{12} B^+/cm^2$ at 200 keV) InP photoconductive detector.

planar photoconductive sampling gate to 1.0 ps, 620 nm optical pulses; the two traces correspond to nonlinear response (single gap) and the more traditional cross-correlation response (two gap) measurements.

The two gap correlation measurement yields a central peak with a full width at half-maximum (FWHM) of 19 ps and a secondary peak corresponding to reflections from microstrip discontinuities at a relative time delay of 65 ps. The single gap response exhibits a coherence spike at $\tau=0$, due to the fact that the two beams are overlapping, and a correlation response with a FWHM of 21 ps on top of a broad background signal. No signal corresponding to reflections of electrical transients off sampler boundaries is evident in the single gap response trace.

We have investigated a number of planar photoconductive detectors with response times ranging between 200 and ~ 1 ps and have found the same features to be present in all cases: a central coherence spike, a broad background, and a signal corresponding to the photoconductor's response time closely matching the signal FWHM measured in cross-correlation experiments.

The response of a nominally 100 GHz GaAs Schottky-barrier photodiode [1] was found, under appropriate illumination conditions, to be sufficiently fast to monitor the quality of our 1.5 ps pulses. Figure 2 compares the response of the photodiode with that of a KDP second harmonic generating (SHG) crystal when the photocarriers are injected into a region of the diode which has been damaged by proton bombardment [2]. The device bandwidth under these illumination conditions is between 150 and 200 GHz, corresponding to a device response time somewhat briefer than an electron transit time through the 0.3 μm active region.

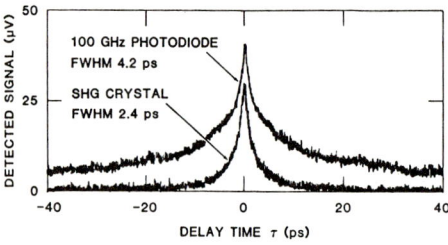

Fig. 2 The nonlinear response of a 100 GHz Schottky-barrier GaAs photodiode compared with the output of a noncollinear beam SHG pulse monitoring system.

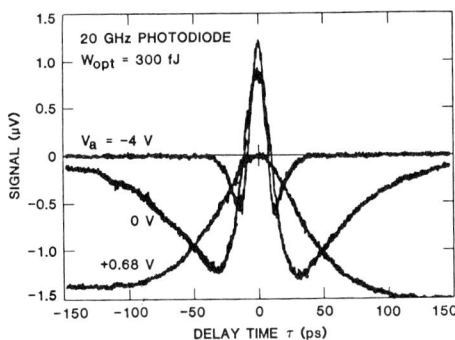

Fig. 3 The nonlinear response of a 20 GHz Schottky-barrier GaAs photodiode at 4 V reverse bias, zero bias, and 0.68 V forward bias.

Photodetectors which are physically more intricate exhibit more complicated responses which are outside the range of validity of the simple model presented above. Figure 3 shows the nonlinear photoresponse of a nominally 20 GHz GaAs photodiode [4] at varying bias conditions. Under reverse and zero bias the central peak about $\tau=0$ corresponds as usual to a mutually attenuating effect of the optical pulses upon one another. At longer time delays, however, the nonlinear response signal changes sign. An optical pulse, therefore, at first attenuates and then enhances the device's sensitivity to a subsequent pulse. At a forward bias of 0.68 V the optical pulses elicit no response at all when they arrive simultaneously, but produce a large (but slow) signal when they are separated by ~ 100 ps. (This bias corresponds to a crossing point of the dark and illuminated current-voltage curves, and the device would be expected to be insensitive to light here [5].) Details of the bias and intensity dependence of the photodetector's response lead us to tentatively attribute the delayed enhancement in sensitivity to the saturation of surface recombination sites by photocarriers and the consequent increase in collection efficiency of the device to a subsequent pulse.

The GaAs photodiodes are on loan to us from Hewlett-Packard Laboratories.

References

1. S. Y. Wang and D. M. Bloom, Electron. Lett. 19, 554 (1983).
2. T. F. Carruthers and J. F. Weller, Appl. Phys. Lett. 48, 460 (1986).
3. T. F. Carruthers and J. F. Weller, IEEE IEDM Tech. Dig. (Washington, DC, December 1985) p. 483.
4. S. Y. Wang, D. M. Bloom, and D. M. Collins, Appl. Phys. Lett. 42, 190 (1983).
5. D. K. Donald (unpublished).

# Direct Generation of Picosecond to Subpicosecond Optical Pulses Using Electrooptic Modulation Methods

*T. Kobayashi, A. Morimoto, T. Fujita, K. Amano, T. Uemura, and T. Sueta*

Electrical Engineering, Engineering Science, Osaka University,
1-1 Machikaneyama, Toyonaka, Osaka 560, Japan

The electrooptic methods to generate ultrashort pulses [1,2] have better controllability and reliability than the passive modelocking and the nonlinear fiber pulse compression [3]. With these methods, the temporal pulse position is electrically controllable and the pulsewidth does not depend on laser power and linewidth. Here, we report two new results about purely electrooptical short pulse generation; [I] 0.8 picosecond pulse generation from a cw laser source by using a Fabry-Perot modulator (the pulsewidth is the shortest ever reported with an electrooptic technique), and [II] direct pulse compression of cw light in picosecond range.

## 1. 19GHz-0.8 ps Pulse Generation Using a Fabry-Perot Electrooptic Modulator

As shown in Fig.1, an electrooptic phase modulator is placed in the optical resonator to form a Fabry-Perot modulator. The modulator is driven by an rf signal synchronized to the cavity round trip. Then the modulation frequency $f_m$ is equal to $nc/2L$ (L: cavity length, n: integer). Under this condition, the transmittance T of the modulator is the same as the static case, i.e.

$$T(t) = \frac{T_{max}}{1 + (2\mathcal{F}/\pi)^2 \sin^2[\theta_0 + \Delta\theta \sin(2\pi f_m t)]}, \quad (1)$$

where $\mathcal{F}$ is the finesse, $\Delta\theta$ is the phase-modulation index, and $\theta_0$ is the optical bias [4]. The transmittance T(t) shows periodically sharp peaks. Accordingly ultrashort light pulses much shorter than the modulating period can be generated from cw light. The pulsewidth is about $1/(2\mathcal{F}\Delta\theta f_m)$ for $\theta_0 = 0$.

In our experiment, the Fabry-Perot modulator was formed by a 16cm long confocal optical cavity with an electrooptic crystal $LiTaO_3$ (1.5x2.8x8mm) at the center of the cavity ($\mathcal{F} \sim 40$). The crystal was mounted at the end of a microwave waveguide, and was driven by a 9.35GHz high-power microwave source, while the light source was a continuous argon ion laser ($\lambda = 514.5$nm, single longitudinal mode). The phase-modulation index was $0.65\pi$ radian. Figure 2 shows an example of the output spectrum measured with a scanning Fabry-Perot interferometer. The spectral shape, which shows the wide exponential tails on

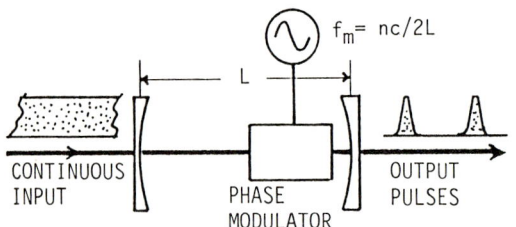

Fig.1 A scheme of ultrashort pulse generation using a Fabry-Perot modulator.

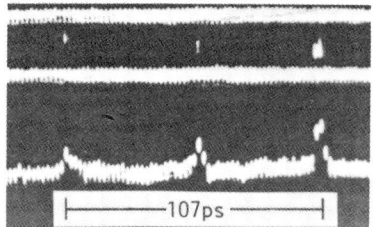

Fig. 2  An example of the output spectrum. It shows the wide exponential tails of 280GHz FWHM.

Fig.3  Streak trace of the output pulses. The repetition rate is 18.7GHz. The pulsewidths are limited by the resolution 9ps of the streak camera.

both sides, agrees with that calculated theoretically. The spectral width is 280GHz. The theoretical time-bandwidth product of the pulses is about 0.22. Using this relation, the output pulsewidth is estimated to be 0.79ps. This pulsewidth is the shortest one generated with a purely electrooptic means. Figure 3 shows an example of the streak camera trace. Ultrashort pulses are obtained at the repetition rate 18.7GHz (= $2f_m$), though the observed pulsewidths are limited by the resolution 9ps of the streak camera.

## 2. Electrooptic Pulse Compression in Picosecond Range for a CW Laser

A pulse compression method utilizing electrooptic phase modulation can be applied not only to forced modelocked lasers [5,6] but also to cw lasers [1]. In picosecond or shorter range, however, any successful result has not been reported. This is the first successful result in picosecond range.

When the sinusoidally phase modulated optical field (modulation index $\Delta\theta$, modulation frequency $f_m$) passes through the dispersive element (medium / circuits) having the phase delay $\theta(\omega) = \theta(\omega_0) + (\omega-\omega_0)\tau(\omega_0) + \frac{1}{2}(\omega-\omega_0)^2 \partial\tau/\partial\omega$ (see Fig.4(a)), the output optical field becomes

$$E_0(t) = E_0 e^{j\{\omega_0 t - \theta(\omega_0)\}} \sum_n J_n(\Delta\theta) e^{-j[n\omega_m\{t-\tau(\omega_0)\} + n^2 f_m^2 \pi(\partial\tau/\partial\nu)]}, \quad (2)$$

where $\tau(\omega) = \partial\theta/\partial\omega$ is the group delay in the dispersive element. Thus $\partial\tau/\partial\nu$ corresponds to the group delay dispersion. The output shapes of the optical power ($\propto |E_0(t)|^2$), calculated by eq.(2), are shown in Fig.4(b). It is seen from the figure that sharp pulses are formed at the adequate amount of dispersion. The analysis shows that for the optimum dispersion condition, the peak power of the pulse is enhanced by the factor of $3.14 \times \Delta\theta^{0.6}$ and the pulsewidth is roughly $0.7/(2\Delta\theta f_m)$, (for $\Delta\theta \leq 10\pi$).

In the experiment, a cw argon laser was used as a light source and the electrooptic phase modulator was similar to that described in 1. A combination of right angle prism and grating (2400 lines/mm), was used as a dispersive circuit . The modulation index was measured to be $1.3\pi$ (spectral width $\sim$75GHz), which was limited by electric discharge on the surface of $LiTaO_3$. Figure 6 shows the streak trace of the output pulse train. Its period and pulse width are measured to be 107ps (=$1/f_m$) and 12ps, respectively. The real pulse width is evaluated to be 8ps taking account of the time resolution

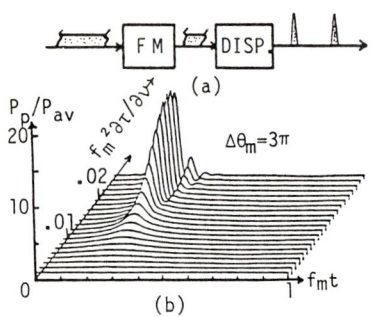

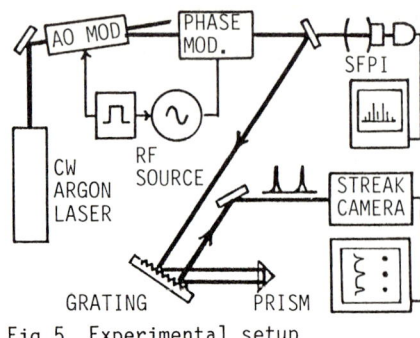

Fig.4 Pulse growth from a phase-modulated light in the dispersive element.

Fig.5 Experimental setup

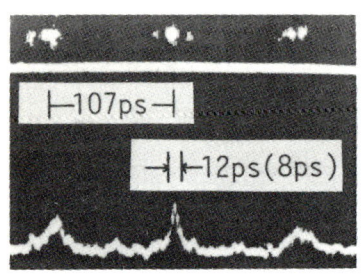

Fig.6 Streak trace of compressed pulses.

of the streak camera 9ps. This result agrees well with the theoretical one, 9ps. The time-bandwidth product of the pulses is about 0.6 (almost transform limited). The power efficiency (average power of output pulses /average power of input) was only a few percent, which was mainly limited by the diffraction efficiency of the grating. This method is, however, essentially reactive and should be possible to have high power efficiency (near 100%) by using a highly efficient diffraction grating or a low loss single-mode optical fiber as a dispersive element. If the optimum-designed phase modulator were provided, one magnitude of increase in the modulation index and consequently the compression to subpicosecond pulses would be expected using the same rf source.

## References

1. T. Kobayashi and Tadasi Sueta: CLEO '84 Anaheim, WG-1 (1984).
2. A. Morimoto, S. Fujimoto, T. Kobayashi, and T. Sueta: Ultrafast Phenomena IV, ed. by D. H. Auston and K. B. Eisenthal (Springer-Verlag, Berlin, 1984) p.84.
3. W. H. Knox, R. L. Fork, M. C. Downer, R. H. Stolen, C. V. Shank and J. A. Valdmanis: Appl. Phys. Lett., 46, 1120 (1985).
4. T. Kobayashi, T. Sueta, Y. Cho, and Y. Matsuo: Appl. Phys. Lett., 21, 341 (1972).
5. J. A. Giordmaine, M. A. Duguay, and J. W. Hansen: IEEE J. Quant. Electron., QE-4, 252 (1968).
6. M. A. Duguay and J. W. Hansen: Appl. Phys. Lett., 14, 14 (1969).

# Elimination of Dynamic Flash in a Picosecond Streak Image Tube

*Huanwen Zhang*

Xian Institute of Optics and Precision Mechanics, Academia Sinica, Xian, Shaanxi, People's Republic of China

## 1. Abstract

The cause of dynamic flash in a picosecond streak image tube has been analyzed. The flash is known to be produced by an electrical discharge between the deflector and the anode cone aperture of the tube. Material with a small reflective coefficient has been chosen for the deflectors. This measure was taken to reduce reflected photoelectrons and other stray electrons. Dynamic flash in the tube has been effectively eliminated.

## 2. Discussion

A model WS-306 picosecond streak tube with a curved cathode and a curved grid mesh was used in a model BWS-5K picosecond streak camera, both of which have been developed in our Institute. A time resolution of 2 ps has been achieved with this system and image curvature has been eliminated [1,2].

It has been found, in the course of making dynamic measurements on the picosecond streak tube, that a strong background flash appeared on the screen when the sweep circuit was laser triggered, thereby applying the sweep voltage ramp to the deflectors. The effect was also present when the shutter aperture was closed and no light was present on the photocathode. The position of the dynamic flash on the screen was related to the direction of the predeflection. If predeflection was to the left, the flash appeared on the right part of the screen and vice versa. The intensity of the dynamic flash depends on the streak speed: the lower the speed, the stronger the flash.

Why would a dynamic flash appear in a tube which is good under static conditions? According to our analysis, it has been confirmed that the origin of the flash is an electrical discharge between the anode aperture and the deflectors in the tube. The construction of the tube and the voltages applied to the electrodes are shown in Fig. 1. When static predeflection voltages of +1.2kV and -1.2kV were applied to the left deflector (lower plate in Fig. 1) and the right deflector (upper plate), respectively, the discharge did not appear, because the electric field between the deflectors and the anode aperture was not high enough. However, when the streak voltage of 5kV was

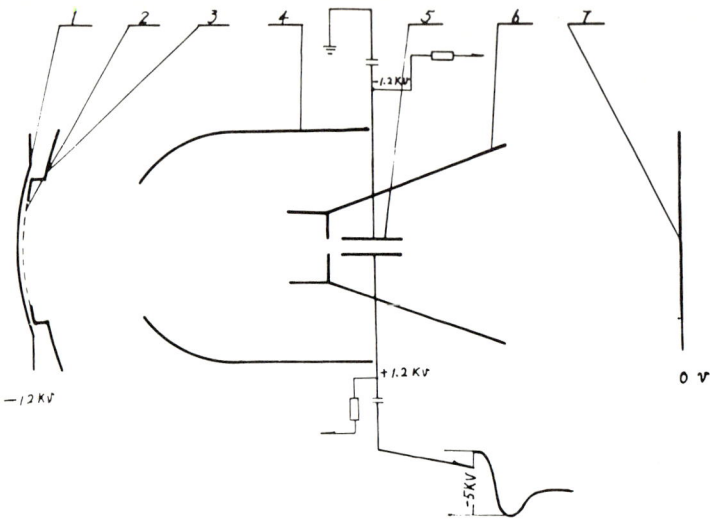

Fig. 1. Structure of the tube and voltages applied on electrodes of the tube. 1. Photocathode; 2. Mesh; 3. Shutter; 4. Focus; 5. Deflector; 6. Anode; 7. Screen.

applied to the left deflector, the intensity of the electric field between the left deflector and the anode aperture could be as high as 1.9 kV/mm. The electrodes were also made of very thin metal plates. Under these conditions, the discharge would take place. The light produced by the discharge would irradiate the photocathode from which photoelectrons would be emitted. These photoelectrons would go into the deflection region and strike the right deflector. The section of the deflector against which the photoelectrons would strike can be calculated according to

$$L = \left(\frac{4Vd/2}{U}\right)^{1/2}, \qquad (1)$$

where L stands for the distance between the entrance of the deflectors and the first part of the section of the deflectors against which the photoelectrons strike; d represents the distance between the two deflectors; V denotes the anode potential with respect to the cathode, and U is the voltage between the two deflectors.

According to our calculations, the deflected photoelectrons would strike the right deflector in a region 8mm from the exit end. The reflected electrons in the region of deflection were deflected to the right and directed to the right part of the screen. It is worth while pointing out that, because of their low energy, the secondary electrons produced by the photoelectrons striking the deflector cannot excite the fluorescent screen.

The above analysis explains the reason for producing a flash on the right portion of the screen when predeflection is to the left. It can also explain the relationship between the flash intensity and the streak speed. When the streak speed is low, the flash is strong since additional electrons would be intercepted by the screen.

The photoelectrons excited by the signal light would not produce a dynamic flash for the following reasons:

(1) Sweeping the whole screen only requires a voltage of approximately 1 kv; namely, only the central and most linear part of the 5kV sweep voltage ramp.

(2) The duration of the signal light is of the order of ps, whereas the electrical discharge time is of the order of ns. In addition, the light emitted by the discharge would illuminate the whole photocathode, whereas the signal light would illuminate the photocathode through a slit measuring a few micrometers in width. It can be seen that the photoelectrons emitted by the signal light would be much lower in number than those emitted by the discharge light.

(3) Because of limited dynamic range, the slit image of the laser pulse from an etalon on the screen would become weaker with the increase of the sweep voltage, until it cannot be recorded.

According to the above analysis, in order to eliminate dynamic flash, a special material with a small reflective coefficient for electrons has been selected for the deflectors. The reflective coefficient for electrons with high energy ($>$ 1 keV) is larger than for low energy electrons, since they can be directly reflected by the nuclei. The reflective coefficient will increase with the atomic number of the atoms of the materials. For the heavy elements, the reflective coefficient can reach as high as 50% when the energy of the primary electrons is larger than 11 keV.

## 3. Conclusion

In addition to selecting a material with a small reflective coefficient, measures were taken to screen stray electrons and electrons reflected from the rear of the deflectors. As a result, the dynamic flash has been effectively eliminated.

## 4. Acknowledgements

The author wishes to thank Professor H. Niu for his collaboration. The author also wishes to thank Miss Guo Lihua for the typing, and Zhou Jiyinglam for the drawing.

## 5. References

1. H. Niu, H. Zhang, "Picosecond Streak Image Tube," SPIE, Vol. 491, <u>High Speed Photography</u>, p. 669, (1984).

2. X. Hou, H. Niu, *et al.,* Picosecond Streak Camera System, SPIE, Vol. 491, <u>High Speed Photography</u>, (1984).

# Part IV

# Nonlinear Optics and Continuum Generation

# Parametric Chirp Reversal and Enhancement: Application in Femtosecond Optics

A. Piskarskas, D. Podenas, A. Stabinis, A. Umbrasas, A. Varanavichius,
A. Yankauskas, and G. Yonushauskas

Vilnius University, Laser Research Center, 232054, Vilnius,
Lithuanian SSR, USSR

## 1. Introduction

Three-wave interaction of phase modulated pulses is of considerable interest in femtosecond nonlinear optics and devices. We concentrate our attention on the phenomenon of parametric chirp reversal (PCR) in real time and parametric chirp enhancement (PCE) due to three-wave interaction. The PCR of light signals by four-wave parametric interaction has been discussed earlier in theoretical work [1,2], where it was proposed to use the phase conjugation of spectral components for the compensation of phase distortion introduced by the group velocity dispersion (GVD). The main demand for the realization of the PCR is $\Delta\nu \ll 1/\tau_R$, where $\tau_R$ is the response time of the nonlinear interaction in the medium and $\Delta\nu$ is the frequency deviation. It is obvious that the predominant contribution of the slow components in the third-order nonlinearity limits the frequency band of PCR in isotropic media. On the other hand, three-wave parametric processes are related to electronic nonlinearity, the response time of which is nearly a femtosecond. Thus, using $\chi^{(2)}$ the phase conjugation of spectral components of extremely wide optical range is possible [3]. It is necessary to emphasize that broad spectral bandwidth of parametric amplification realized in a number of crystals (CDA, KDP, etc.) solves the problem of amplification of weak phase-modulated picosecond signals (e.g. coming from optical fibers) for 5–6 orders and enables one to obtain two phase-conjugated light pulses with opposite chirps. The PCR by paramagnetic amplification allows one to use in pulse compressors both media with negative and those with positive GVD and to achieve femtosecond pulses with power exceeding gigawatts. Furthermore, the phase conjugated pulses with the linear chirp open up new possibilities in four-photon-phase spectroscopy, dynamic holography of space-time events, as well as in systems of optical data processing (e.g. for the formation of correlation signals).

Second, we report the new effective method of parametric chirp enhancement (PCE) [4] by chirp conversion in the process of parametric generation.

## 2. Parametric Chirp Reversal

We have reported [5,6] about phase self-conjugation (passive PCR) of picosecond pulses excited in an optical parametric oscillator (OPO) and we have proposed that it should be used for the elimination of phase modulation in order to generate band limited pulses.

In the present paper we report a study of an active PCR which occurs during the injection of a picosecond signal with a linear chirp into an optical parametric amplifier (OPA). The solution of the corresponding truncated equations [7] for a given pump field and for group matching of the signal and idler waves can be written

$$A_1(\eta, z) = \frac{A_{10}(\eta)}{2} \exp\left[\sigma \int_0^1 a_{30}(\eta - \nu_{31}z)dz\right],$$

$$A_2(\eta, z) = A_1^*(\eta, z) , \qquad (1)$$

where $A_j = a_j \exp(i\phi_j)$ is the complex wave amplitude, $\eta = t - (z/u_1)$, $\nu_{31} = 1/u_3 - 1/u_1$, $u_j$ is the group velocity of the wave, and $\sigma$ is the coupling coefficient. We use the notation

$$A_j(\eta, z)\bigg|_{z=0} = A_{j0}(\eta).$$

We assume $A_{10} \neq 0$, $A_{20} = 0$, and $1 \gg L_{n\ell}$, where $\ell$ is the length of the nonlinear medium, $L_{n\ell} = 1/[\sigma A_{30}(0)]$. We see that the phase modulation of the signal pulse is not distorted ($\phi_1 = \phi_{10}$), and a pulse is generated at the idler frequency with a conjugate phase ($\phi_2 = -\phi_{10}$). It is thus possible to conjugate optical signals with either random or regular phase modulation; specifically, it is possible to reverse a linear chirp.

The experiments have been carried out according to a nearly collinear parametric amplification scheme (Fig. 1). The single pump pulse of 5 ps duration with an energy of 3–4 mJ has been formed by the passively mode-locked Nd-phosphate glass laser (1). The signal pulse with linear positive chirp has been produced in the single-mode optical fiber (4) with the length of 1.3 m. For that purpose the laser pulse with energy $\sim 1 \,\mu$J ($\lambda = 1.054 \,\mu$m) has been coupled into a fiber with efficiency of about 40%. The combined effects in the optical fiber i.e. self-phase modulation and GVD have caused the output pulse to be broadened both in time and spectrum. Typically, the pulse is broadened up to $\sim 10$ ps and the bandwidth is increased up to $\sim 400$ cm$^{-1}$. CDA crystal has been chosen as the non-linear medium in order to realize broad band parametric amplification ($1 = 4$ cm, 1st phase-matching type. A 90° phase-matching angle allows one to obtain a large bandwidth of amplification (up to 2000 cm$^{-1}$ in degenerate case [3]). The

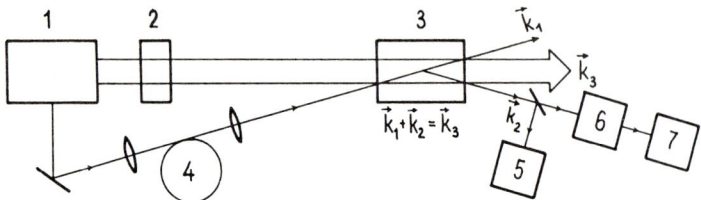

**Fig. 1.** Experimental arrangement: 1 – picosecond phosphate glass laser, 2 – second harmonic generator (KDP), 3 – parametric amplifier (CDA), 4 – optical fiber, 5 – dynamic interferometer, 6 – compressor, 7 – pulse durationmeter (streak-camera, autocorrelator)

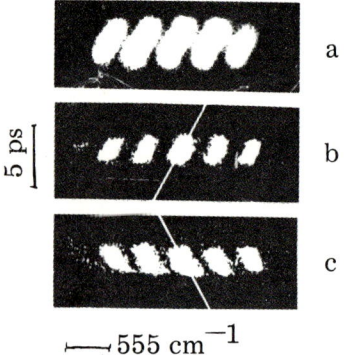

**Fig. 2.** Dynamic interferograms of the pulses: at the output of the fiber (a), the signal (b) and idler (c) at the output of the parametric amplifier

signal and pump pulses have been directed into OPA through the time delay matching lines and an energy gain by a factor $\sim 10^4$ has been achieved ($E_s$, $E_i = 0.5$ mJ). Since the pump pulse "cuts out" a corresponding spectral band from the signal pulse, the spectrum of signal and idler pulses decreases up to $\sim 200$ cm$^{-1}$ after amplification. The temporal phase behaviour of pulses has been investigated by the dynamic interferometry technique involving a Michelson-type interferometer and an "Agat SF-3" streak-camera. Figure 2 shows dynamic interferograms of light pulses. The inclination of the interference pattern is proportional to $\Delta\nu/\tau$. The direction of the inclination depends on the sign of the chirp.

Figures 2b and 2c confirm that signal and idler pulses are phase-conjugated. Compression of the idler pulse to 280 ps (Fig. 3) and 1 GW power in a positive GVD medium (KRS crystal) has been achieved.

## 3. Parametric Chirp Enhancement

PCE takes place in OPO pumped by chirped pulses. This effect may be explained by the analysis of OPO tuning curves (Fig. 4). The curves show the dependence of OPO output wavelength on pump wavelength. This depen-

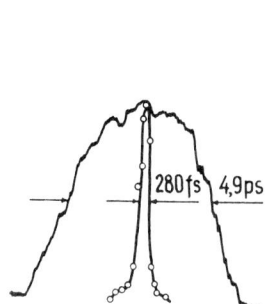
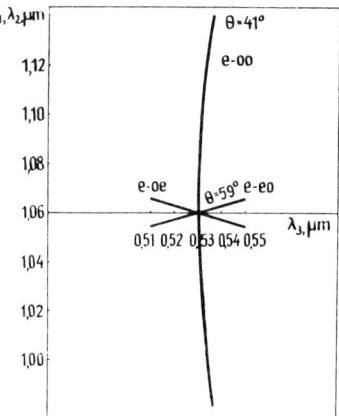

**Fig. 3.** Pulse microchronogram before and autocorrelogram after pulse compression

**Fig. 4.** Tuning curves of KDP OPO pumped by $\lambda = 0.53\,\mu m$

dence is very strong for phase-matching conditions of the first type in KDP pumped by the second harmonic of a Nd:glass laser. It means that a small change in pump wavelength results in a strong change of signal wavelength. If it takes place during the chirped pump pulse the chirp of OPO output becomes magnified and enhanced. This effect does not depend on pump power, and is determined only by crystal dispersion. PCE of two orders of magnitude is possible. Besides this, signal and idler pulses have reversed chirps.

It is easy to show that the deviations of signal $(\delta w_1)$ and idler $(\Delta w_2)$ pulses depend upon the pump deviation $\Delta w_3$ as follows [4]:

$$\Delta w_1 = p_1 \Delta w_3, \quad \Delta w_2 = p_2 \Delta w_3, \qquad (2)$$

where $p_1 = \nu_{32}/\nu_{12}$; $p_2 = 1 - p_1$ is the chirp enhancement coefficient, $\nu_{32} = 1/u_3 - 1/u_2$ and $\nu_{12} = 1/u_1 - 1/u_2$ is the group velocity mismatch.

If $|\nu_{32}|\nu_{12}|$, the chirp of OPO radiation may be enhanced in comparison with pump chirp. If $u_1 \approx u_2$ it is necessary to take into account GVD.

Let us evaluate $p_1$ for some crystals. If $\kappa = w_2/w_3$ we obtain $p_1 = a/(1-2\kappa)$, where $a = \nu_{31}(w_3 d^2 k_1/dw_1^2)^{-1}$.

For $\lambda_3 = 0.53\,\mu m$ we obtain $p_1 = 0.54/(1-2\kappa)$ for LiNbO$_3$ (Fig. 5) and $p_1 = 0.16/(1-2\kappa)$ for KDP.

Both regular phase and random phase spectral components are generated in OPO [8]. Parametric amplification bandwidth when the OPO is pumped by a band-limited pulse (or instantaneous bandwidth of OPO with chirped pump):

145

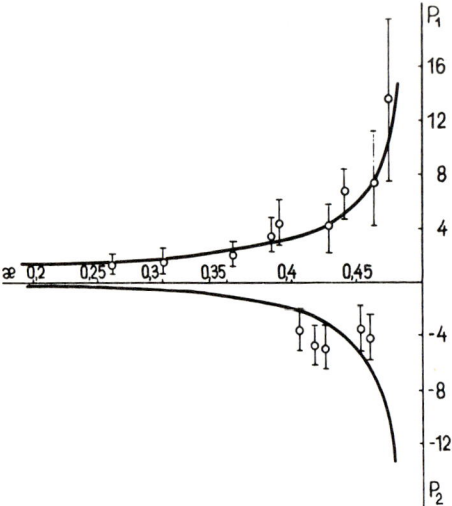

**Fig. 5.** Dependence of chirp enhancement coefficient $p_1$ and $p_2$ upon $\kappa$ in the LiNbO$_3$ OPO pumped by $\lambda = 0.53\,\mu$m (solid curves: calculation, circles: experimental results)

$$\Delta w_{\text{amp}} = 4\sqrt{\ln 2}/|\nu_{12}|\sqrt{1 L_{n\ell}} \qquad (3)$$

Experimental data show that in a superfluorescence OPO usually $\Delta w_1, \Delta w_2 \ll \Delta w_{\text{amp}}$. During pulse compression with linear chirp, components of random phase distort this process. The role of random components decreases by decreasing noise spectral bandwidth. In order to reduce the negative role of parametric noise the synchronously pumped OPO is preferable in which a considerable increase of effective crystal length takes place. The last is responsible for the narrowing of OPO radiation instantaneous bandwidth (3).

Experiments have been carried out with a synchronously pumped temperature tunable LiNbO$_3$ OPO. Figure 6 shows dynamic interferograms of pump and OPO signal pulse ($\lambda = 0.99\,\mu$m) at the beginning, in the middle and at the end of a train. Relatively long pulse train (~30 pulses) and pulse duration (~60 ps) enables one to clarify chirp evolution in the OPO. Pump chirp is caused by self-phase modulation in the laser rod. OPO regular chirp appears in the middle of the pulse train. At the end of the pulse train, narrowing of the instantaneous OPO pulse spectrum and increase of the pump chirp are responsible for the final formation of linear chirp up to ~0.5 cm$^{-1}$/ps. This corresponds to pump chirp enhancement of 7 times. Due to OPO pulses shorter than pump pulses, the nonlinear region of pump chirp is "cut out" and OPO chirp linearizes.

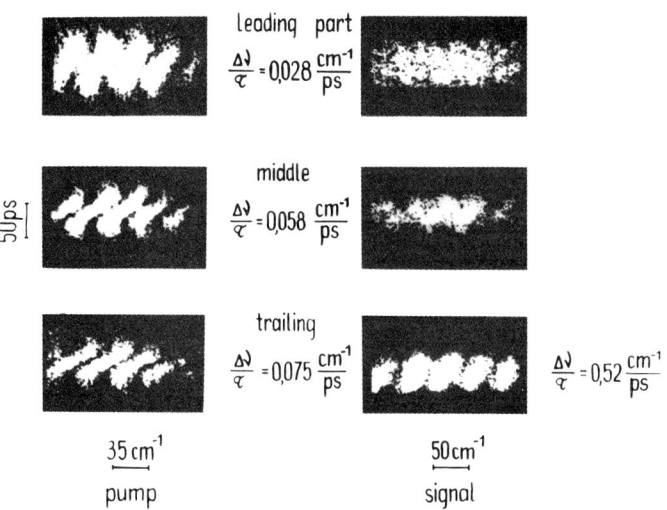

**Fig. 6.** Chirp evolution of pump ($\lambda = 0.53\,\mu$m) and OPO pulse ($\lambda = 0.99\,\mu$m) along the train

A similar experiment has been provided by using a second harmonic train of a passively mode-locked LaBe ($La_2Be_2O_5:Nd$) laser as a pump. Shorter pulse durations (10 ps) and higher intensities led to stronger pump chirp (0.8 cm$^{-1}$/ps at the end of the train). Experimentally noticeable chirp enhancement coefficient $p_1 = 10$ has been achieved at $\kappa = 0.476$. Figure 7 shows the results of pulse compression in a positive GVD medium. KRS-6 crystal has been used due to its high dispersion. After passing through KRS-

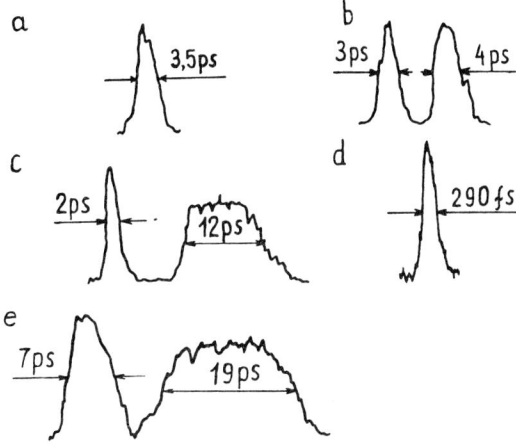

**Fig. 7a–e.** Pulse compression results in positive GVD compressor. Idler and signal pulses at the compressor input (**a**); at the compressor output after passing various distances in KRS crystal (**b**), (**c**) and (**e**) measured by streak-camera. Autocorrelogram of optimally compressed idler pulse (**d**)

6, signal ($\lambda_1 = 1.02\,\mu$m) and idler ($\lambda_2 = 1.12\,\mu$m) pulses have separated in time due to different wavelengths (b). Crystal length at which idler pulse optimally compresses is ~35 cm. At the same distance the signal pulse spreads considerably due to positive chirp (c). The idler pulse starts to spread at a longer distance (e) after being compressed. More accurate measurements of idler pulse duration by a second-order correlation technique show that it is compressed to 280 fs (d).

## 4. Conclusions

We have presented the new light phase nonlinear conversion phenomena caused by three-wave interaction. The main results are as follows: chirp reversal, chirped pulse parametric amplification and chirp enhancement due to second-order susceptibility. Apparently, these results may be applied in further development of femtosecond lasers.

**References**

1. J.H. Marburger: Appl. Phys. Lett. **32**,372 (1978)
2. A. Yariv, P. Fekete, D. Pepper: Opt. Lett. **14**,52 (1979)
3. R. Danielius, A. Piskarskas, V. Sirutkaitis, A. Stabinis, A. Yankauskas: JETP Lett. **42**, 122 (1985)
4. A. Piskarskas, A. Stabinis, A. Yankauskas: Kvantovaya elektronika **12**, 1781 (1985)
5. A. Varanavichius, R. Grigonis, A. Piskarskas, A. Stabinis, A. Yankauskas: Sov. Tech. Phys. Lett. **6**, 624 (1980)
6. R. Grigonis, A. Yankauskas, A. Piskarskas, A. Stabinis, A. Varanavichius: In Proc. 2nd Intern. Symp.: Ultrafast Phenomena in Spectroscopy, **1**, 150 (Reinhardsbrunn, Jena 1980)
7. S.A. Akhmanov, A.S. Chirkin, K.N. Drabovich, R.V. Khokhlov, A.P. Sukhorukov: IEEE J. Quantum Electron. **QE-4**, 598 (1968)
8. A. Piskarskas, A. Stabinis, A. Yankauskas: Kvantovaya elektronika **11**, 2375 (1984)

# Supercontinuum Generation in Gases: A High Order Nonlinear Optics Phenomenon

*P.B. Corkum*[1], *C. Rolland*[1], *and T. Srinivasan-Rao*[2]

[1] National Research Council of Canada, Division of Physics, Ottawa, Ontario, Canada, K1A 0R6
[2] Brookhaven National Laboraory, Upton, NY 11973, USA

The recent development of high power, ultrashort pulse sources has created renewed interest in the interaction between intense laser radiation and free atoms and molecules. Not only is it feasible to apply laser fields that approach, or exceed, the strength of the atomic field as seen by the outer electrons, but it is also possible to apply these fields nonadiabatically using ultrashort pulses. Recent experiments have uncovered a number of new phenomena, including energetic electron production from rare gas atoms, [1,2] multiply charged ions, [3] and inner shell excitations.[4]

From a theoretical perspective, when the laser field approaches one atomic unit, it can no longer be considered as a perturbation to the atomic field. Thus non-perturbative theoretical approaches are being developed to treat both atoms [5] and molecules [6] in strong fields.

Until now, experiments have been restricted to isolated atoms. However, because the theories of nonlinear optics and multiphoton ionization are so interrelated, we should expect these new phenomena to have optical signatures. In addition to their intrinsic interest, nonlinear optics experiments can add a new perspective for judging emerging theories of high intensity laser processes. Clearly, there is a new class of experiments to be performed using ultrahigh power, ultrashort pulses.

We describe an experiment performed in high pressure gases (1 atm < p < 40 atm) with a 2 psec or 70 fsec 0.6 μm pulse focused to a peak intensity of $I \lesssim 10^{13}$ W/cm$^2$. The maximum intensity exceeds that in which multiphoton ionization is observed in longer pulse experiments (~25 psec) in Xe using either 1.06 μm or 0.53 μm radiation. It is approximately the intensity at which tunnel ionization is observed with nanosecond 10 μm pulses in very low pressure Xe.[7] Furthermore, at such intensities in 0.53 μm and 1.06 μm experiments, high energy electrons (up to 20 excess photons absorbed) are observed from Xe.[2]

The experiment was performed by focusing the $5 \times 10^9$ W, 70 fsec pulse or the $0.4 \times 10^9$ W, 2 psec pulse produced by an excimer pumped dye laser [8] with an F/170 lens into a high pressure gas cell. The output was monitored (i) by eye, (ii) on a 10 nm/mm spectrograph and recorded on an OMA or (iii) on a 10Å/mm spectrograph and recorded photographically.

Seven gases were investigated, Ne, Ar, Kr, Xe, $N_2$, $H_2$ and $CO_2$. Our primary experimental observation was that supercontinua (Figs. 1 and 2) are produced in all of the above gases except Ne. Supercontinuum generation appears to be a near universal response of high pressure gases to high intensity radiation.[5] However, it has never been predicted nor observed previously. (Working independently, P. Sorokin's group at IBM

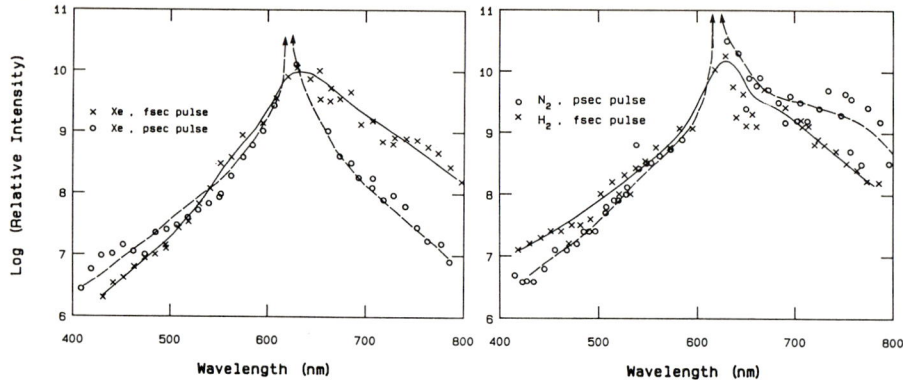

Fig. 1: Continuum Spectra from Xe
× P = 30 atm, τ = 70 fsec
○ P = 15 atm, τ = 2 psec

Fig. 2: Continuum Spectra
× P = 38 atm $H_2$, τ = 70 fsec
○ P = 40 atm $N_2$, τ = 2 psec

has recently observed supercontinuum from high pressure gases.[9] Their work has expanded on their previous observation of spectral broadening of picosecond UV pulses when focused in room air.[10]) Supercontinuum generation itself is not a new phenomenon.[11] The inadequacy of standard theories of continuum generation [11,12] for the case of high pressure gases is discussed in a related paper and the reader is referred to that work for details.[13]

Comparing Fig. 1 and Fig. 2 we see a remarkable similarity of the spectra for both picosecond and femtosecond pulses and for both molecular and rare gases. (Although the continuum intensity is in arbitrary units in both figures, the units are the same for all of the curves plotted.) The anti-Stokes component is nearly the same for all gases and pulse durations investigated and has a minimum wavelength of $\lambda_{min} \sim 300$ nm. We have not investigated the Stokes component as thoroughly, but there are significant variations between gases and pulse durations. In most cases, the Stokes component contains more energy than the anti-Stokes. At least with the femtosecond pulse in $CO_2$ the radiation extends beyond 1.3 μm, the limit of a S-1 photocathode.

All continua showed a distinct threshold. Below the continuum generation threshold there was little or no broadening. The threshold value of the gas pressure multiplied by the laser power required for supercontinuum generation is plotted in Fig. 3 as a function of the inverse of the laser power. The plotted data are for femtosecond pulses.

Self-focusing has a similar threshold dependence to that shown in Fig. 3. For self-focusing the pressure-power product is given by:

$$(\underline{P}\ p)_{TH} = \pi\varepsilon_0 c\lambda^2\ \eta_2^{-1}, \qquad (1)$$

where $\underline{P}$ is the laser power, p the gas pressure, and $\eta_2$ is the second order nonlinear refractive index at atmospheric pressure. Equation 1 indicates that each gas has a characteristic pressure-power product at the self-focusing threshold. Figure 3 shows a similar behaviour for the threshold of continuum generation. Taking the ratio between the values of $(\underline{P}\ p)_{TH}$ in Fig. 3, we find that all ratios agree within 20% with the ratio

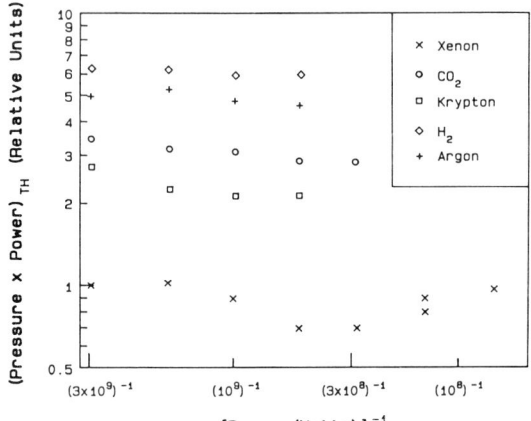

Fig. 3: Pressure-power product at threshold, plotted as a function of the inverse of the laser power for different gases

of hyperpolarizability.[14] Finally, the absolute value of the critical power for self-focusing agrees within ~20% with the threshold power for continuum generation. Only when $N_2$ and $CO_2$ were illuminated with picosecond pulses was the continuum threshold different (lower) than the self-focusing threshold. There is evidence that the Raman active mode in these molecules is responsible for the lower threshold.

Since the maximum intensity of the unattenuated femtosecond pulse, even in the absence of self-focusing, exceeded $10^{13}$ W/cm$^2$, some plasma must be produced.[15] One might expect plasma formation to be important in the spectral broadening process leading to continuum generation.[12] Plasma production is rejected as the direct cause of the supercontinuum (for detailed discussion see Ref. 12). However generation of even modest plasma densities can limit the size of the focal spot.[13]

If we assume that continuum generation requires self-focusing, at least in the rare gases, [16] what role can it play? Self-focusing, by sweeping the focal spot, allows the laser field, as seen by the interacting gas molecules, to build up on a much shorter time scale than the pulse duration.[17] The maximum intensity is limited by the refractive index change that results from the build-up of the total plasma density ($N_e$) in the focal region. Equation 2 gives the conditions required for the plasma nonlinearities to compensate for the electronic nonlinearities and, therefore, limit the maximum intensity:

$$\eta_2 E^2 \simeq \frac{N_e}{2N_c} = \frac{1}{2N_c} \int_{-\tau/2}^{\tau/2} \frac{dN_e}{dt} dt \quad . \tag{2}$$

In Eq. 2 $N_c$ is the critical electron density and all other symbols have their usual meaning. Equation 2 indicates that a shorter effective pulse duration $\tau$, allows a higher ionization rate $dN_e/dt$. A larger ionization rate occurs with stronger electric fields which enhance the nonlinearity of the interaction.

In conclusion we suggest that experiments analogous to those being performed on isolated atoms and molecules will be an important new direction in nonlinear optics. In one such experiment we have observed supercontinuum generation from gases. Self-focusing correlates with continuum generation in nearly all cases. We propose that self-focusing allows the field to be applied in a less adiabatic manner and therefore increases the nonlinearity of the interaction.

References

1.  P. Agostini, F. Fabre, G. Mainfray, G. Petite and N.K. Rahman, Phys. Rev. Lett. 42, 1127 (1979).
2.  L.A. Lompré, A. L'Huillier, G. Mainfray and C. Manus, J. Opt. Soc. Am. B2, 1906 (1985).
3.  A. L'Huillier, L-A. Lompré, G. Mainfray and C. Manus, in Laser Techniques in the Extreme Ultraviolet, edited by S.E. Harris and T.B. Lucatorto, AIP Conference Proceedings No. 119, AIP, New York (1984), p. 79.
4.  C.K. Rhodes, Science 229, 1345, (1985).
5.  See for example Multiphoton Ionization of Atoms, edited by S.L. Chin and P. Lambropoulos (Academic Press, New York, 1984).
6.  See for example T. Tung Nguyen-Dang and A.D. Bandrauk, J. Chem. Phys. 79, 3256 (1983); J. Chem. Phys. 80, 4926 (1984); A.D. Bandrauk and T. Tung Nguyen-Dang, J. Chem. Phys. 83, 2840 (1985).
7.  S.L. Chin, F. Yergeau and P. Lavigne, J. Phys. B: At. Mol. Phys. 18, L213 (1985).
8.  C. Rolland and P.B. Corkum, accepted for publication, Opt. Commun.
9.  P.P. Sorokin - private communication.
10. J.H. Glownia, G. Arjavalingam, P.P. Sorokin, and J.E. Rothenberg, Opt. Lett. 11, 79 (1986).
11. J.T. Manassah, M.A. Mustafa, R.R. Alfano and P.P. Ho, IEEE J. Quantum Electron. QE-22, 197 (1986).
12. N. Bloembergen, Opt. Commun. 8, 285 (1973).
13. P.B. Corkum, C. Rolland and T. Srinivasan-Rao, submitted for publication, Phys. Rev. Lett.
14. H.J. Lehmeier, W. Leupacher and A. Penzkofer, Opt. Commun. 56, 67 (1985).
15. P. Lambropoulos, Phys. Rev. Lett. 55, 2141 (1985).
16. In preliminary experiments we have unsuccessfully attempted to produce supercontinua with our psec pulse using small F# optics while maintaining the pressure-power product below threshold.
17. J.H. Marburger and W.G. Wagner, IEEE J. Quantum Electron. QE-3, 415 (1967).

# New Excitation and Probe Continuum Sources for Subpicosecond Absorption Spectroscopy

*J.H. Glownia, J. Misewich, and P.P. Sorokin*

IBM Thomas J. Watson Research Center, Yorktown Heights, NY 10598, USA

1. Introduction

There appears to be an increasing trend in photochemistry to try to characterize more fully the primary photodissociative or photoreactive process, portions of which can occur on a picosecond, or even femtosecond, time scale. Recent experiments[1,2] have shown that it is possible to find instances where the growth of simple fragments (e.g., I atoms, OH radicals) resulting directly from photolysis can be monitored in real time with the use of tunable probe lasers. However, it is also of interest to observe the spectrum of the reacting parent compound while it moves on an excited state potential surface. Here one can possibly rely on absorption of an ultrafast broadband probe continuum to monitor electronic transitions. Alternatively, molecular vibrations of the photoexcited compound can be probed by absorption of an ultrafast continuum in the IR. In either case, a method for simultaneously generating both ultrafast, energetic UV excitation pulses and suitable ultrafast probe continua, with precise control of the timing between pump and probe, becomes essential. In this paper we describe an apparatus that conveniently provides such pulses in some of the frequency ranges desired. Subpicosecond pulses at both 308nm and 248nm, with energies of several millijoules at each wavelength, are available from this apparatus for purposes of excitation. A simultaneously generated ultrafast UV probe continuum pulse spans the range ~450nm to ~235nm. An ultrafast IR continuum pulse extends from 2.2-2.7$\mu$m and can be upconverted to the visible for ease of detection. Since both the 248nm and probe continuum pulses are derived by nonlinear techniques from the 308nm pulses, the arrival times of all pulses at a sample should, in principle, be precisely controllable.

2. Generation of Ultrafast Continua in High Pressure Gases

We recently reported[3] the generation of ultrafast continua in several high pressure gases (Ar, $H_2$, and $CO_2$ ), induced by the application of focused 308nm pulses of ~300fs duration and ~10GW power. Ultrashort continuum pulse generation in gases is a phenomenon that has only recently been observed. Our motivation to study continuum generation in high pressure gases was based upon our previously reported finding[4] that gentle focusing in air of ~1.5mJ, ~350fsec, 308nm pulses produces ~ 1000$cm^{-1}$ spectral broadening of these pulses. P. Corkum et al.[5] have independently observed ultrashort continuum generation in high pressure gases, both with the use of focused 625nm, 70fs, 5GW pulses, as well as with 2psec pulses of the same frequency.

We found the Ar continuum to have the broadest spectrum of the three gases studied and hence recommend its use for a probe continuum. The Ar gas (at ~33atm) was contained in a 1m long cell. To generate the continuum, ~300fs, 2.5mJ pulses at 308nm, produced in a single-stage XeCl amplifier, were focused 20cm before the exit window of the cell with a 1.5m lens. The output beam of the Ar cell, which typically displayed on the order of 10 filaments in the near field, was directed via mirrors through a path length of several meters onto the 50$\mu$ slit of a 1.5m spectrograph equipped with an unintensified Tracor-Northern OMA detector. Sufficient light was found to be present to enable single-shot time-resolved absorption spectroscopy to be carried out over the range ~450nm to ~235nm with good spectral resolution. Although the single-shot power spectra were observed to contain one or more regular modulation frequencies, it should be possible to ratio these out by means of a reference arm involving a second OMA. Possible sources of the modulation frequencies are discussed in Ref. 3.

In $H_2$ both stimulated Raman scattering (SRS) and continuum generation were observed to be present (Fig. 1). The continuum is more heavily weighted to wavelengths shorter than 308nm, resulting in its largely being absent in the region of the higher $H_2$ Raman Stokes bands, while contributing significant fraction to the total intensity in the region of the $H_2$ Raman anti-Stokes components. At lower pressures the contribution of the continuum to the total spectrum drops. Filamentation is observed in the output beam at the higher $H_2$ pressures. We have not as yet determined whether there exists a lower pressure regime in which SRS occurs without self-focusing or self-phase modulation.

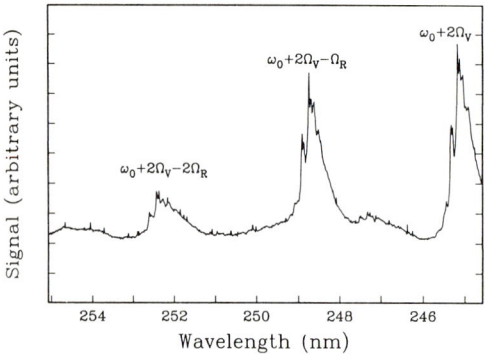

Fig. 1: $H_2$ continuum plus Raman beam spectrum in second anti-Stokes region. Multi-shot average (128 shots), 65 cm cell, $P_{H_2}$ ~18 atm.

## 3. Amplification of Ultrashort Pulses at 248nm

For a pump wavelength of 308nm there occurs a coincidence between one of the Raman shifted frequencies in $H_2$ and the peak of the KrF B → X excimer transition. Consider specifically the ideal case in which the pump radiation $\omega_0$ is circularly polarized. At sufficiently high pump powers, one can expect both stimulated rotational Raman scattering ($\Delta v = 0$, $\Delta J = 2$) and stimulated vibrational Raman scattering ($\Delta v = 1$, $\Delta J = 0$) to occur, producing first Stokes waves at $\omega_S^R = \omega_0 - \Omega_R$ and $\omega_S^V = \omega_0 - \Omega_V$, respectively. By a resonantly enhanced four-wave mixing process, light at a new frequency $\omega_U = \omega_0 - \omega_S^V + \omega_S^R = \omega_0 + \Omega_V - \Omega_R$ can then be generated. The same basic mixing process can then be repeated, this time with $\omega_U$ as the third wave, to produce light at $\omega_{U'} = \omega_0 - \omega_S^V + \omega_U = \omega_0 + 2\Omega_V - \Omega_R$. Using the ortho-$H_2$ values $\Omega_V$ = 4155.2 cm$^{-1}$, $\Omega_R$ = 587.0 cm$^{-1}$, and choosing a pump wavelength $\lambda_0$ = 308.0 nm(air), one finds $\lambda_{U'}$(air) = 248.8 nm, sufficiently close to the known wavelength of the KrF excimer transition to make this process a candidate for generation of a seed pulse suitable for amplification in a KrF gain module.

Figure 1 shows that a band corresponding to $\omega_0 + 2\Omega_V - \Omega_R$ is observed in the spectrum of the output beam from the $H_2$ cell. For Fig. 1 a hydrogen pressure of ~18atm was used in a cell 65cm long. The ~2.5mJ, ~300fs, 308nm linearly polarized pump beam was focused into the cell with a 0.75m quartz lens. Nonlinear birefringence associated with self-focusing in the gas evidently provided the circularly polarized component of light required to drive the stimulated rotational Raman scattering process. The output beam from the $H_2$ cell was observed to be no longer linearly polarized.

We proceeded to collimate the output beam from the $H_2$ cell and then to send it in a single pass through a Lambda-Physik EMG200 excimer gain module. Amplification was readily achieved, with measured amplified output pulse energies of ~7mJ. Isolation of the amplified short pulse energy from that of the ASE originating in the KrF excimer module was readily achievable by virtue of the much lower divergence of the amplified short pulse. Autocorrelation traces, based upon two-photon ionization in NO, showed the 248nm amplified beam to have an actual pulse width of ~450fsec, distinctly greater than the ~320fsec 308nm pulsewidth measured the same day.

## 4. Multiplexing the 308nm Pulses

At first glance, it might appear that a third excimer gain module would be required in an actual pump-probe apparatus capable of simultaneously providing a subpicosecond photolysis output at either 308nm or 248nm and an ultrafast Ar probe continuum. However, this is actually not the case. Because the efficiency of second harmonic generation in the KDP crystal that generates the 308nm seed pulse is only ~10%, enough 616nm light remains to generate a second UV seed pulse, one having almost the same energy as the first. If the two seed pulses are spaced by two or three nanoseconds, there is sufficient time for complete repumping of the XeCl B state, according to Ref. 6. This we confirmed by generating an additional ~2.5mJ, ~300fsec, 308nm pulse with the use of a simple scheme[3] in which the two pulses are injected into the excimer amplifier with orthogonal polarizations.

## 5. Generation and Upconversion of an Ultrafast IR Continuum (Subpicosecond TRISP)

Generation of powerful subpicosecond IR pulses at ~2.4 $\mu$m was accomplished by stimulated electronic Raman scattering (SERS) in Ba vapor[7]. As pump pulses we again utilized amplified ~350 fsec, 308 nm pulses. Although application of ~20 nsec XeCl laser pulses to Ba vapor is known to produce SERS only on the $6s^2\ ^1S_0 \rightarrow 6s5d\ ^1D_2$ transition, with a Stokes output near 475 nm, we find, by contrast, that with ultrashort 308 nm excitation, SERS occurs only on the $6s^2\ ^1S_0 \rightarrow 6s7s\ ^1S_0$ transition, with a Stokes output peaked near 2.4 $\mu$m. The reason for this switch is conjectured in Ref. 7. The 2.4 $\mu$m SERS output is highly photon efficient, with measured IR output pulse energies of ~0.4 mJ for ~5 mJ UV input pulses. The SERS threshold was less than 1 mJ. Spectrally, the IR output was found to be a continuum, extending from 2.2 - 2.7 $\mu$m, provided the 308 nm pump pulses were spectrally broadened[4] in air by self-phase modulation. Without spectral broadening of the pump pulses, the IR continuum was found to extend from 2.3 - 2.5 $\mu$m.

With the use of a polished Si wafer, the horizontally polarized ultrashort IR pulses ($\nu_{IR}$) were collinearly combined with the vertically polarized ~15 nsec pulses ($\nu_L$) from a tunable narrow-band Furan 1 dye laser. With the timing between the two sources adjusted so that the subpicosecond IR pulses occurred within the 15 nsec long dye laser pulses, both beams were sent into a Rb upconverter, where the dye laser beam induced SERS on the Rb 5s→6s transition, producing a narrow-band, vertically polarized, Stokes wave $\nu_S$. Horizontally polarized, visible continuum pulses at $\nu_L - \nu_S \pm \nu_{IR}$ were then observed to emerge from the Rb cell when $\nu_L$ was tuned to phase match either upconversion process.

In Fig. 2 portions of two (lower sideband) upconverted spectra are superimposed. In one case, the IR was passed through an empty 20 cm cell; in the other case the IR was passed through the same cell filled with 200 Torr of CO. A surprise finding is the observed increase in upconverted signal at the peaks of the CO 2-0 bands. This we explain as follows. Under the conditions of Fig. 2 the upconverted signal was very heavily saturated by the subpicosecond IR pulse, that is, too few photons at $\nu_L$, $\nu_S$ were available during the actual IR pulse to allow efficient upconversion of the latter. Therefore, a decrease in transmitted light due to molecular resonance ab-

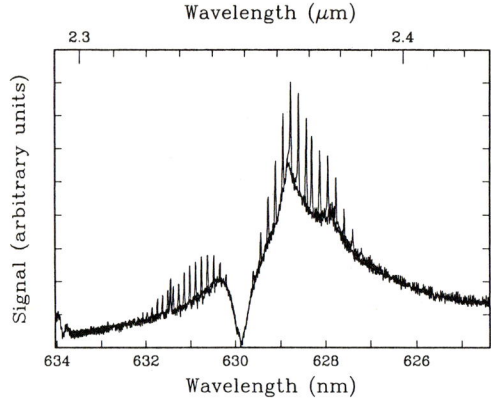

Fig. 2: Superimposed upconverted TRISP spectra (cell empty and cell filled with 200 Torr CO gas). Each spectrum is the average of 128 shots. Prominent absorption at 629.8 nm due to Rb excited state (5p) atoms.

sorption during the IR pulse did not result in a measurable decrease in upconverted signal. However, the coherently reemitted light of the molecules[8], occurring for a time on the order of $T_2$ after the IR pulse, when the upconverter is no longer saturated, was able to be efficiently upconverted, resulting in the observed peaks. Positive IR absorption was observed in the upconverted spectra when the subpicosecond IR probe beam was attenuated.

An experiment is planned to measure directly the $\tilde{B}$ to $\tilde{A}$ non-radiative decay in DABCO vapor[9]. This experiment requires both the ultrafast 248nm excitation and ultrafast 2.2-2.7$\mu$m probe sources described above.

References
1. N. F. Scherer, F. E. Doany, A. H. Zewail, and J. W. Perry: J. Chem Phys. 84, 1932 (1986).
2. J. L. Knee, L. R. Khundkar, and A. H. Zewail: J. Chem Phys. 83, 1996 (1985).
3. J. H. Glownia, J. Misewich, and P. P. Sorokin: submitted to JOSA B.
4. J. H. Glownia, G. Arjavalingam, P. P. Sorokin, and J. E. Rothenberg: Opt. Lett. 11, 79 (1986).
5. P. B. Corkum: private communication, October 1985; P. B. Corkum, C. Rolland, and T. Srinivasan-Rao: preprint of summary submitted to this conference.
6. P. B. Corkum and R. S. Taylor: IEEE J. Quantum Electron. QE-18, 1962 (1982).
7. J. H. Glownia, J. Misewich, and P. P. Sorokin: submitted to Opt. Lett.
8. H. -J. Hartmann and A. Laubereau: J. Chem. Phys. 80, 4663 (1984).
9. J. H. Glownia, G. Arjavalingam, and P. P. Sorokin: J. Chem. Phys. 82, 4086 (1985).

# Induced Phase Modulation and Spectral Broadening of a Weak 530-nm Picosecond Pulse by an Intense 1060-nm Picosecond Pulse in Glass

*R.R. Alfano, Q.X. Li, T. Jimbo, J.T. Manassah, and P.P. Ho*

Institute for Ultrafast Spectroscopy and Lasers and
Photonic Engineering Laboratories, Departments of Physics and
Electrical Engineering, The City College of New York, NY 10031, USA

Techniques to control and transfer the spectral bandwidth of ultrashort laser pulses are important for many scientific and technological applications. It is well known that when an intense pulse propagates through a medium, it causes a refractive index change. This induces a phase change of the optical electric field which will in turn cause a frequency sweep within the pulse envelope. This process is called <u>self phase modulation (SPM)</u>[1-4]. When a weak probe pulse at a different frequency propagates into a medium whose index of refraction has been disrupted from equilibrium by a strong pulse, the phase of the probe pulse can be modulated by the time variation of the index of refraction originating from the primary intense pulse. This process is defined to be <u>induced phase modulation (IPM)</u>[5-7] which results in a spectral broadened probe pulse.

We report on the first observation of the IPM of a weak second harmonic generated by the presence of a primary intense pulse in a BK-7 glass. The primary strong pulse at $w_1$ induced the refractive change causing a phase change and a frequency sweep of a weak probe pulse at $2w_1$. The enhancement of the bandwidth of the weak pulse at $\lambda_2$ by propagating an intense laser pulse at $\lambda_1$ in condensed media is attributed to IPM. This new effect has technological importance in communications and signal processing by allowing pulse coding in different frequency regions and noise generation in glass fibers.

Experimentally, a single 8ps laser pulse at 1060 nm generated from a mode-locked glass laser system was used as the pump beam to initiate the induced spectral broadening process. A second harmonic pulse was used as the probe beam. Pulses at the primary 1060 nm and the second harmonic 530 nm wavelengths were weakly focused into a 9 cm long BK-7 glass. The incident laser beam size at the sample site was about 1.5 mm diameter. A weak supercontinuum signal was measured when only the 530 nm laser pulse or the 1060 nm laser pulse was incident into the sample. An enhanced signal was obtained when both 530 nm and 1060 nm laser pulses were sent through the sample at the same time.

In this experiment, the second harmonic 530 nm laser pulse intensity was kept nearly constant with pulse energy of about 80 µJ. The primary 1060 nm laser pulse energy was a controlled variable changing from 0 to 2 mJ. Neutral density filters were used to adjust the input laser intensity of the 1060 nm pulse. Output signals were sent through filter sets to remove the incident laser pulses and were different for the Stokes and anti-Stokes measurements. Signals were also directed into a spectrometer with an optical multichannel analyzer to measure the spectral intensity distribution. A typical experimental result of SPM and IPM spectra of the weak probe pulse are shown in Fig.1 (a) and (b), respectively.

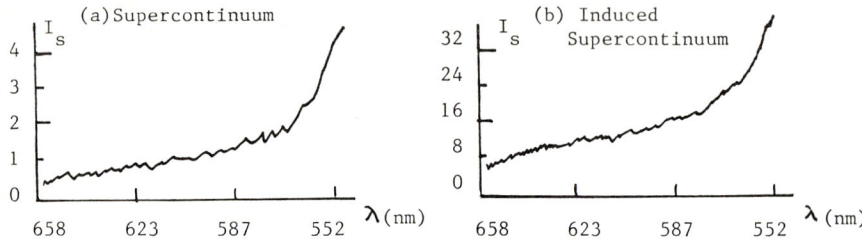

Fig. 1. The Stokes side of the supercontinuum spectral distribution. The vertical axis is calibrated as a relative intensity scale. (a) 530 nm alone. (b) Both 530 nm and 1060 nm pulses used. The 530 nm laser pulse energy was kept at about 80 µJ.

The 530 nm laser pulse generated a weak supercontinuum while the intense 1060 nm laser pulse served as a catalyst to enhance the induced supercontinuum of the 530 nm pulse. The entire Stokes spectrum of the induced supercontinuum was about 10 times greater than that of the SPM spectrum generated by the 530 nm pulse alone. The supercontinuum signal generated by the 1060 nm pulse alone in this 550 nm - 660 nm spectral region was less than 1% of the total signal. Signal shapes in Fig.1 have a similar spectral distribution, however, the IPM spectrum was about ten times stronger than the SPM signal. In addition, we have studied the supercontinuum signal enhancement as a function of the addition of the 1060 nm laser pulse energy where the 530 nm pulse energy was set at 80±15 µJ. The IPM spectrum increased linearly as the added 1060 nm laser pulse energy was increased from 0 to 200 µJ. When the 1060 nm pulse energy was over 1mJ, the signal enhancement reached a plateau and saturated at a gain factor of about 11 times over the supercontinuum intensity generated by the 530 nm pulse alone.

The output signal was imaged onto the slit of a spectrograph to separate contributions from two possible mechanisms : phase modulation (PM) and four photon parametric generation (FPPG) of the supercontinuum from the spectral spatial distribution. Films were used to measure the distribution of the supercontinuum spectrum. Geometrical blocks were arranged in the path for the selection of a particular process. The induced spectrum displayed in Fig.2 shows a similar spatial spectral distribution of the conventional supercontinuum.

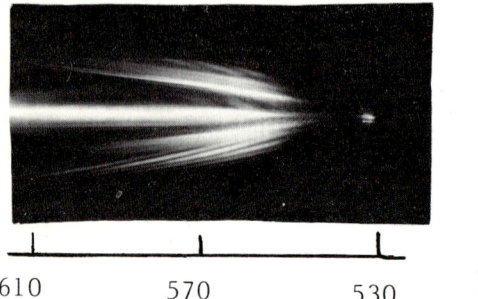

Fig 2. Spectrograph of the Stokes side of the induced supercontinuum. The collinear profile arising from IPM has nearly the same spatial distribution as the incident pulse. Two wings at noncollinear angles correspond to FPPG arising from the phase-matching condition of the generated wavelength emitted at different angles from the incident laser beam direction.

Using a photomultiplier system and spatial filtering, quantitative measurements of signal contributions from the collinear IPM and the noncollinear FPPG parts were obtained[5]. There was little gain from the contribution of the FPPG process over the entire added pulse energy dependent measurement. The main enhancement process of the induced spectral broadening is consequently attributed to IPM.

A Hamamatsu model #C1587 streak camera with 2 ps resolution was used to measure the temporal distribution of the signal. The duration of induced supercontinuum with a selected 10nm bandwidth was measured to be about the same as the incident laser pulse.

From our previous analysis[4,7], the solution of the electric field of the nonlinear wave equation in this system can be expressed as

$$E = E_0 [a\, e^{i\alpha}\, e^{i(kz-wt)} + \delta b\, e^{i\beta}\, e^{2i(kz-wt)}],$$

where $E_0$ is the electric amplitude of the primary incident pulse; $\delta$ is the relative strength of the SH signal to the primary frequency signals; a and b are the pulse shape function of the primary and the SH signal; and $\alpha$ and $\beta$ are the phase of these two pulses, respectively. Assuming $\delta \ll 1$ and the group velocity to be constant for wavelengths from 400 nm to 1060 nm in glass, b, $\alpha$ and $\beta$ can be derived from Eqs. 28 and 29 of Ref. 7 and knowledge of $E_0$, the pulse duration, the nonlinear index of refraction $n_2$, and the index of refraction of the glass.

Neglecting fast oscillatory terms in $\chi^3$, quasilinear partial differential equations have been obtained from the nonlinear Maxwell equation. Analyzing the partial differential equations with the assumption that the primary and SH lasers have the same duration gives $\beta = 2\alpha$ and a = b. Thus, the primary pulse only modulates the SH pulse through the index of refraction change. There is no direct energy transfer from the primary to the induced supercontinuum. Furthermore, the induced spectral distribution of the SH pulse has the same spectral distribution as the supercontinuum of the primary pulse. When the factor $n_2 E_0^2 z/(c\tau)$ from both the primary and the SH lasers are small which is a good approximation in our experiment, the supercontinuum generated by 530 nm should have the same spectral distribution as the supercontinuum generated by 1060 nm. This means the spectral distributions of the supercontinuum and the induced supercontinuum are geometrically similar, as shown in Fig. 1.

This research is supported by NSF, AFOSR, Hamamatsu and PSC/BHE.

References
1. R.R. Alfano and S.L. Shapiro, Phys. Rev. Lett.,24, 584,592,1217 (1970)
2. T.K. Gustafson, J. Taran, H.A. Haus, J. Lifsitz and P.L. Kelley, Phys. Rev. 177, 306 (1969).
3. R.H. Stolen and C. Lin, Phys. Rev. A17, 1448 (1978).
4. J. Manassah, M. Mustafa, R. Alfano and P. Ho, IEEE, J. QE-22, 197 (1986)
5. R.R. Alfano, Q. Li, T. Jimbo, J. Manassah and P.P. Ho, "Induced Spectral Broadening in Glass ..." Opt. Lett. Sept. (1986)
6. R. Alfano, Q. Wang, T. Jimbo and P.P. Ho, "Induced Spectral Broadening in ZnSe ..." submitted for publication.
7. J.T. Manassah, M.A. Mustafa, R.R. Alfano and P.P. Ho, Phys. Lett.,113A 242 (1985)

# The Observation of Chirped Stimulated Raman Scattered Light in Fibers

*A.M. Johnson, R.H. Stolen, and W.M. Simpson*

AT & T Bell Laboratories, Holmdel, NJ 07733, USA

Stimulated Raman scattering (SRS) limits the maximum achievable power in the fiber-grating compression of optical pulses [1]. We have recently experimentally investigated SRS in single-mode fibers in the regime where group velocity dispersion (GVD) causes large Raman Stokes pulse walkoff from the pump pulse [2]. From measurements of pulse spreading we inferred that the Raman pulse was generated with a large frequency chirp. The present work presents a direct measurement of the chirp of the stimulated Raman pulse. This is not the first observation of chirped SRS in a fiber [3], but it is the first measurement in a length of fiber only a few walkoff lengths long. The distinctive features of the present results are that over this short length of fiber, subsequent modifications of the Raman pulse by self-phase modulation (SPM) or GVD are minimal and the pump pulse is only minimally chirped. The wings of a spectrally broadened pump pulse are thus not the source of the chirping of the Raman pulse.

In the previous experiments [2], the source of the Raman pulse and the initial Raman pulsewidth were extracted from high-speed pulse propagation measurements. We found that the Raman pulse was produced within the first 3-4 walkoff lengths, where a walkoff length is defined as the distance in which the Raman signal passes through one pump pulsewidth. The peak of the Raman pulse is generated about two walkoff lengths into the fiber for conversion near 20% and moves closer to the input for higher pump powers due to pump depletion and further from the input for lower powers. The initial Raman pulsewidth is about the same as that of the initial pump pulse and could be shorter for very high pump powers and be much longer for very low powers.

Extraction of an initial pulsewidth from the Raman pulsewidth data [2] requires knowledge of the pulse spreading. This could be obtained directly by comparing the Raman output pulsewidth for different fiber lengths longer than the Raman pulse generation distance. For 35% conversion of 36 psec pulses at 532 nm we measured a spreading of 1.66 psec/m using fiber lengths of 34 m and 101 m. This is larger than a predicted value of 1.16 psec/m based on a dispersion of 436 psec/nm-km (0.071 in dimensionless units) and a measured spectral halfwidth of 90 cm$^{-1}$. This value of the dispersion is corroborated by excellent agreement with pump pulse and Raman pulse separation data for various fiber lengths. If we assume that the pulse is produced with no initial chirp we can approximate the Raman pulsewidth by a Gaussian convolution of the initial pulsewidth with the measured spreading coefficient over the appropriate length of fiber:

$$T_p = \{ (t_o)^2 + ([\Delta t/\Delta \ell]L)^2 \}^{1/2} , \qquad (1)$$

where $T_p$ is the Raman pulsewidth after propagation down a length L of fiber with a spreading coefficient of $[\Delta t/\Delta \ell]$ (psec/m) and an initial pulsewidth $t_o$. This Gaussian approximation leads to highly unphysical results. For example, measurements from the 101 m fiber predict an intial pulsewidth of 102 psec, while similar data from the 34 m fiber predict an initial pulsewidth of 58 psec. Not only should the initial pulsewidth be the same in both fibers but it has to be less than 50 psec which we measured out of a 20 m fiber.

Another approach is to assume that the Raman pulse is generated with a chirp. By assuming an initial linear chirp the pulsewidth out of the fiber is the sum of the initial width and the subsequent spreading:

$$T_p = t_o + [\Delta t/\Delta \ell]L \ . \tag{2}$$

Using the measured pulse spreading coefficient and output pulsewidths for the 34 m and 101 m fiber lengths, Eqn. [2] predicts an initial pulsewidth of 32 psec for both fiber lengths. For a pump pulse of 36 psec duration and 35% Raman conversion an initial Raman pulsewidth of 32 psec is quite reasonable. This result implies that at the generation site, a 32 psec Raman pulse is produced with a linear chirp of 90 - 100 $cm^{-1}$ across the half power points of the measured spectrum. We then proceeded to measure this chirp directly.

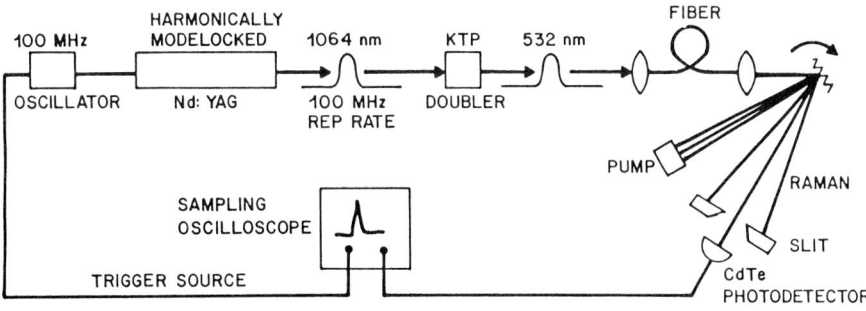

Fig. 1: Experimental setup for the measurement of the chirp on the Raman spectrum.

We investigated the Raman chirp with the experimental arrangement indicated in Fig. 1. The pump pulses consisted of 36 psec transform-limited pulses at 532 nm produced by a frequency-doubled and harmonically modelocked Nd:YAG laser [4]. The Raman chirp measurements were made with a 21 m (3.4 walkoff lengths) polarization-preserving fiber with a core diameter of 4.1 microns. The Raman spectrum was dispersed by a grating and imaged onto a high-speed polycrystalline CdTe photodetector [5]. The transient signal was coupled into a sampling oscilloscope triggered by the 100 MHz rf signal used to drive the acousto-optic modulator of the Nd:YAG laser. The system response of the sampling oscilloscope, photodetector combination was 38 psec (FWHM). The experiment consisted of scanning the dispersed Raman spectrum across a slit (15 $cm^{-1}$ resolution) on the photodetector.

161

A clear variation in the relative delay of the peak of the Raman pulse as a function of frequency shift from the pump, indicative of a chirped spectrum, is displayed in Fig. 2. The difference in delay was 80 psec over the extremes of the spectrum. The sense of the chirp was identical to the positive chirp the pump pulse would experience in a longer length of fiber by the action of SPM. The measured frequency chirp is displayed in Fig. 3. The chirp was not linear over the entire spectrum. For comparison we fit a linear chirp to the experimental data over the spectral region between the half power points. A 3.2 $cm^{-1}$/psec chirp over an initial Raman pulsewidth of 32 psec yields a chirp between half power points of 102 $cm^{-1}$ in excellent agreement with the pulse spreading results. In addition to the chirp there was a clear variation in the pulsewidth across the spectrum (Fig. 2).

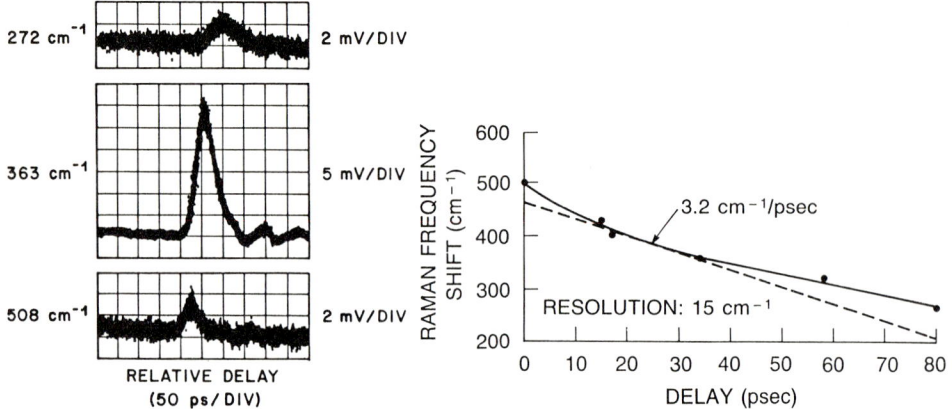

Fig. 2: Sampling oscilloscope response of the grating dispersed Raman spectrum as a function of the frequency shift from the pump, with a resolution of 15 $cm^{-1}$.

Fig. 3: Measured chirp of the stimulated Raman pulse.

There are several possible sources of a chirp on the Raman pulse. These include GVD, self-phase modulation, cross-phase modulation from the pump pulse and direct transfer of the pump chirp to the Raman pulse [6]. By using a fiber short enough that the pulse doesn't travel very far after it is generated, all these sources of chirp should be negligible except for direct transfer of the pump chirp. The chirp of the first Stokes pulse would then be expected to comparable to the pump chirp [6] but we find a Raman chirp at least three times the pump chirp at the source of the Raman pulse. The implication of the experimental results is then that there is an additional source of Raman chirp related to pulse walkoff as the Raman pulse is produced.

This is the first observation of chirped SRS near the generation point in a single-mode fiber. Direct measurements of the chirp in the regime of normal GVD have revealed an "up" chirped spectrum, which suggests the possibility of external grating compression of the Raman pulse.

References

1. A. M. Johnson, R. H. Stolen, and W. M. Simpson, Appl. Phys. Lett., 44, 729 (1984); D. Grischowsky and A. C. Balant, Appl. Phys. Lett., 41, 1 (1982).
2. R. H. Stolen and A. M. Johnson, IEEE J. Quantum Electron., November, 1986.
3. E. M. Dianov, A. Ya. Karasik, P. V. Mamyshev, G. I. Onishchukov, A. M. Prokhorov, M. F. Stel'makh and A. A. Fomichev, Sov. Phys. JETP Lett., 39, 691 (1984).
4. A. M. Johnson and W. M. Simpson, J. Opt. Soc. Am. B 2, 619 (1985).
5. A. M. Johnson, D. W. Kisker, W. M. Simpson, and R. D. Feldman, in *Picosecond Electronics and Optoelectronics*, ed. by G. Mourou, D. Bloom, and C. Lee, (Springer, New York, 1985) p. 188.
6. V. N. Lugovi, Sov. Phys. JETP, 44, 683 (1976).

# Observation of 7.2-THz Beats Between the D-Lines of Atomic Rb

*J.E. Golub and T.W. Mossberg*

Department of Physics, Harvard University, Cambridge, MA 02138, USA

Over the past few years, dramatic advances in short pulse technology have pushed studies of dynamical processes to ever shorter time scales. Unfortunately, short pulse technology remains beyond the means of many investigators. A simple and clever alternative approach to achieving high temporal resolution was recently proposed by several groups [1-4]. In this approach, time resolution is limited only by the coherence time of the excitation light, rather than by its duration. Since laser light of short coherence time (i. e. large bandwidth) is much more easily generated than the corresponding short pulse, this result is of great importance.

Schemes [1-4] for obtaining high temporal resolution using broadband light typically involve the generation of a four wave mixing (FWM) signal with two or three beams derived from the single output pulse of a broadband laser. In the present experiment, we study transient three input-beam FWM phase matched in the folded boxcars configuration. Material relaxation and beat signals are studied by monitoring the intensity of the FWM signal as a function of the delay, $t_{21}$, between pulses 1 and 2. The third pulse acts as a probe.

Our results are well described by a simple linear systems theory [4-6] which relates the population transfer out of the material ground state, $\Delta n$, to the (spatially varying) power spectrum of pulse 1 plus pulse 2, $S(r,\omega,t_{21})$, and to the homogeneous line shape of the transition excited, $h_1(\omega)$, according to

$$\Delta n \propto \int d\omega\, S(r,\omega,t_{21})\, h_1(\omega), \qquad (1)$$

where $S(r,\omega,t_{21}) = 4S_o(\omega)\cos^2\{[\omega t_{21}+\varphi(r)]/2\}$; $S_o(\omega)$ is the power spectrum of a single excitation pulse; $\varphi(r) = (k_2-k_1)\,r$ is a spatially varying phase; and $k_i$ is the wavevector of pulse i.

Equation 1 describes a spatial population grating whose $t_{21}$-dependent magnitude can be sampled via scattering of the third input laser pulse. When two material transitions are simultaneously excited, $h_1(\omega)$ in Eq. 1 must be replaced by $h_1(\omega) + h_2(\omega)$, where $h_2(\omega)$ corresponds to the second transition. Gratings corresponding to $h_1$ and $h_2$ oscillate in relative spatial phase according to $(\omega_1-\omega_2)t_{21}$, where $\omega_i$ is the frequency of transition i, thereby modulating the magnitude of the overall grating [6]. This oscillation introduces beats in the FWM signal as a function of $t_{21}$.

In Fig. 1, we show the results of a measurement of the FWM signal versus $t_{21}$ in Rb. With the laser operated over the full gain profile of methyl DOTC, the 5 nsec excitation pulses were resonant with both D-lines. The trace shows a pronounced oscillation of period 139 ∓4 femtoseconds. This period corresponds well with the 240 cm$^{-1}$ energy difference between the Rb $^2P_{1/2}$ and $^2P_{3/2}$ states. To our knowledge, this is the highest frequency material-specific beat yet observed. To obtain the time resolution necessary to observe these beats, it was necessary to use a relatively small (10 mrad) angle between the beams. Under these conditions, detection of the FWM signal becomes somewhat difficult.

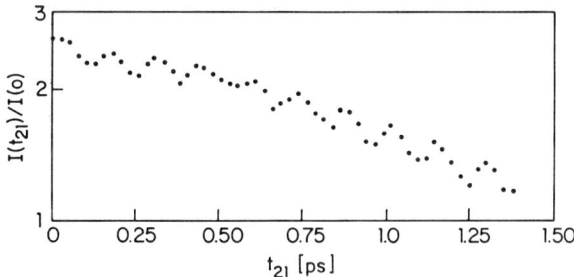

Fig. 1. FWM intensity versus delay $t_{21}$ showing 7.2-THz oscillations

When the dye laser is tuned to excite a single D line, the FWM signal exhibits exponential decay as a function of $t_{21}$. As discussed elsewhere [1,2,4], the exponential decay constant is determined primarily by the homogeneous dephasing time, $T_2$, of the transition excited. By introducing argon perturber gas at pressures of one to several atmospheres, we observe picosecond exponential decay of the FWM signal and can deduce Rb-Ar collisional broadening cross sections. Our values for the cross sections agree well with previous measurements performed with traditional spectral line broadening techniques. These are apparently the first time domain relaxation measurements of the rubidium-rare gas system.

We thank the Joint Services Electronics Program (N00014-84-K-0465) and the National Science Foundation (PHY-85-04260) for financial support. We thank J. Rothenberg for a useful loan of equipment.

## References

1. S. Asaka, H. Nakatsuka, M. Fujiwara, and M. Matsuoka, Phys. Rev. A 29, 2286 (1984).
2. N. Morita and T. Yajima, Phys. Rev. A 30, 2525 (1984).
3. R. Beach and S.R. Hartmann, Phys. Rev. Lett. 53, 663 (1984).
4. H. Nakatsuka, M. Tomita, M. Fujiwara, and S. Asaka, Opt. Comm. 52, 150 (1984).
5. A.M. Weiner, S. De Silvestri, and E.P. Ippen, J. Opt. Soc. Am. B 2, 654 (1985)
6. J.E. Golub and T.W. Mossberg J. Opt. Soc. Am. B 3, 554 (1986).

# Coherent Multiphoton Resonant Interaction and Harmonic Generation

*A. Mukherjee*[1], *N. Mukherjee*[1], *J.-C. Diels*[1], *and G. Arzumanyan*[2]

[1]Center for Applied Quantum Electronics, P.O. Box 5368,
 North Texas State University, Denton, TX 76203, USA
[2]Research Institute of Condensed Media Physics,
 Yerevan State University, Kievian Str. 1A, SU-375028 Yerevan, USSR

The presence of multiphoton resonances enhances the non-linear susceptibility responsible for harmonic generation, but also reduces the maximum achievable conversion efficiency [1] because of resonant losses resulting in a non-uniform phase—matching condition over the length of the sample. The absorption losses can be avoided by using coherent interaction of properly phase-shifted and time delayed pulse sequences propagating through several absorption lengths [2]. We demonstrate experimentally for the first time enhancement of harmonic generation through coherent propagation of a pair of pulses under condition of multiphoton resonance. An overall energy conversion efficiency of one percent has been obtained, which is the theoretical limit for the system under investigation [1, 2].

This experiment requires phase coherent picosecond pulses with an energy density in excess of 50 $mJ/cm^2$ over a distance of several cm at a wavelength of 571 nm, with a flat and uniform wavefront. The energy density is required to significantly excite the 4s level of lithium (with two photons absorbed from the 2s ground level). The interaction length is needed to observe significant two-photon absorption at a pressure sufficiently low to neglect collisions during the interaction. As a succession of two pulses propagates through a medium which exhibits a resonance (single or multiphoton), the resonant part of the susceptibility will depend on the relative phase of the two pulses. We measured the transmitted fundamental and the third harmonic energy as a function of the relative phase and delay between a pair of pulses.

The pulses from the oscillator are amplified to approximately 1 mJ at 10 pps by a frequency-doubled Nd:YAG laser pumped 3-stage amplifier [3]. The phase coherence of the pulses as well as the uniformity of the wavefront is monitored by a Mach Zehnder interferometer, which serves also as pulse sequencer. To adequately define the relative phase of the pulses transmitted by the interferometric delay line, their spacing is monitored with an accuracy better than $1/20^{th}$ of the wavelength, by sending a He-Ne laser beam through the same Mach Zehnder interferometer, and measuring the successive constructive-destructive interferences. The two-pulse sequence is sent through the resonant medium - a heat pipe containing lithium and magnesium vapor for phase matching the third harmonic. Interference filters are used to separate the third harmonic. The transmitted fundamental as well as the initial pulses are recorded in second harmonic.

For an interpulse delay of 3 psec, Fig. 1 shows the incident and transmitted energy (measured in second harmonic) versus phase. As expected, a maximum transmission is observed for a phase difference

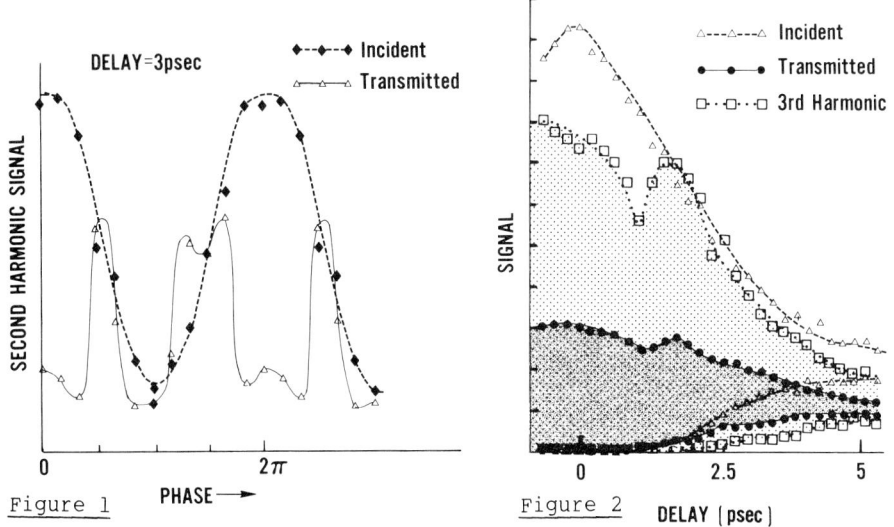

Figure 1

Figure 2

of $\pi/2$, when the second pulse recuperates all the energy lost by the first one through two photon stimulated emission.

Figure 2 is obtained by plotting the maxima and minima of the phase dependencies (as in Fig. 1) as a function of interpulse delay. The dashed and solid lines correspond to the envelopes of the second order interferometric autocorrelations [4] of the pulse before and after the heat pipe. The dotted lines are the maxima and minima of the phase dependence of the third harmonic, hence would be the envelope of a third-order interferometric autocorrelation in the absence of coherent effects. The "dip" near a delay of 2 psec observed in the upper envelopes is a very sensitive function of the frequency of the pulses, as illustrated by a plot of the envelopes of the transmitted signal and third harmonic versus delay, calculated for various detunings. Far below resonance, there is no absorption (right side of Fig. 3), therefore, the transmitted autocorrelation is the same as

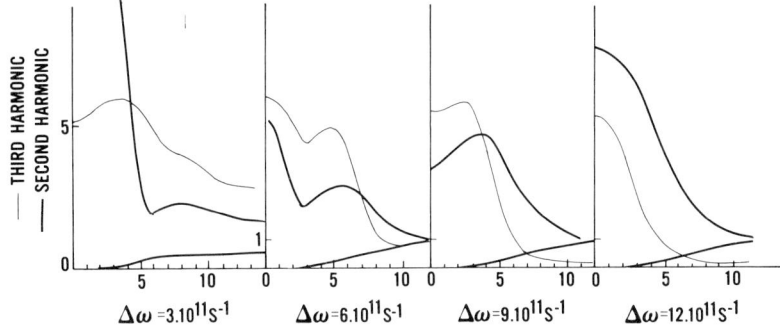

Figure 3

that of the pulse, with a peak to background ratio of 8 to 1. The third harmonic follows a third-order interferometric autocorrelation, with a peak to background ratio of 32 to 1. Closer to (zero field) resonance, the coherent superposition of the two fields is just sufficient to Stark shift the levels into resonance, hence the absorption dip at zero delay. At different detunings the dynamic Stark shift produced by the coherent superposition of the two fields brings the atom into resonance at different inter-pulse delays, causing the dip to move to different parts of the autocorrelation trace.

Coherent enhancement of third harmonic is demonstrated in fig. 4, which shows the envelopes (maxima and minima of the phase dependence) of the third harmonic signal versus delay, in a phase-matched mixture of lithium and magnesium (10 torr lithium; 20 torr magnesium). Since a maximum conversion efficiency of 6 % was predicted for a **plane wave** [2], the maximum of 1 % for a Gaussian beam is close to the theoretical limit. It should be noted that, in the case of incoherent interaction, the third harmonic energy should be 32 times larger at zero delay than at 2 psec delay. therefore, the peak efficiency of 1 % corresponds to a forty-fold enhancement by coherent interaction.

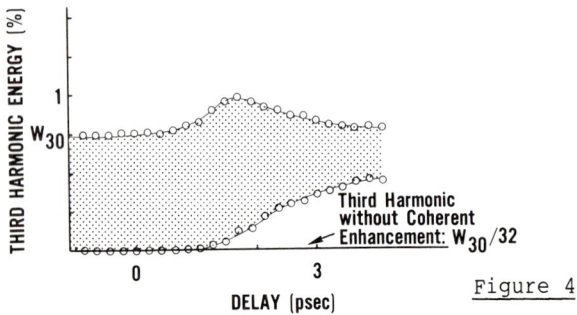

Figure 4

We have demonstrated the possibility of controlling phased pulse sequences in the visible. Extension of this technique to the infrared has important applications in selective excitation of molecules [5].

This work was supported by the National Science Foundation, under grant Nb ECS-8406985.

References.

1. J.-C. Diels, Proceedings of the $6^{th}$ Vavilov Conf., **1**, 206 (1979).
2. J.-C. Diels and A. T. Georges, Phys. Rev. **A19**, 1589-1606 (1979).
3. H. Vanherzeele, H. Mackey, J.-C. Diels, Appl. Opt. **23**, 2056 (1984).
4. J.-C. Diels, J. J. Fontaine, I.C. McMichael, and F. Simoni, Applied Optics **24**, 1270-1282 (1985).
5. J.-C. Diels and S. Besnainou, "Multiphoton Coherent Excitation of Molecules", submitted to J. Chem. Phys. (1986).

# Ultrafast Chaos from Semiconductor Lasers

*Y. Cho[1], T. Umeda[1], I. Jun Cha[1], M. Koishi[2], and M. Miwa[2]*

[1]The Institute of Scientific and Industrial Research,
 Osaka University, Mihogaoka 8-1, Ibaraki, Osaka, 565 Japan
[2]Hamamatsu Photonics K.K., Ichino-cho 1126-1, Hamamatsu, 431-32 Japan

## 1. Introduction

Since the chaos study started from fluid dynamics, its relevant time regime was originally in milli-second range. Even after chaos entered into the laser field [1], experimental works have mainly been done in up to micro-second range since the lasers used in those experiments were mainly gas-phase lasers of IR [2] and FIR range [3] and solid-state lasers [4] whose characteristic relaxation times were in milli-second to micro-second ranges. If a semiconductor laser could be brought into chaotic region, considering its carrier relaxation time of sub-nano-second range, it is expected to obtain chaos in the 100-ps regime, and it can be called "ultrafast chaos" even though it is only in 100-ps time regime.

Meanwhile, it is commonly recognized that the noise level of semiconductor lasers rises sharply in the presence of even a small amount of reflection of their output to their own output facet. This brought severe difficulties for semiconductor-laser applications as a stable light source such as for the high-speed optical communication and for video disks. However, the origin of this semiconductor laser noise caused by the reflection has not yet been fully understood. On the other hand, we pointed out that a semiconductor laser could be brought into chaotic oscillations by an application of delayed feedback [5]. Those chaotic oscillations were observed directly on their waveforms in nano-second time regime using a high-speed photo-detector and a single-scan 1 GHz oscilloscope [6]. Here, as an extension of the above experiment, we report ultrafast chaos observed using a streak camera and a discussion on its connection to the semiconductor laser noise caused by the feedback.

## 2. Experimental

Tested semiconductor lasers ($\lambda=0.78$ μm) include different types depending on manufacturers. To obtain the delayed feedback, the output light emitted from one facet was focussed by an objective of x20 onto a plane mirror placed at distances 6 - 30 cm and fed back on this facet, forming an external cavity as shown in Fig.1. Reflections obtainable were estimated to be up to 2 %. Output from another facet was fed to a streak camera system (Hamamatsu Photonics C1587). In the external cavity an air-gap etalon was inserted to keep the oscillation in a single device mode. An ND filter was inserted in front of the camera to suppress the reflection from a window face of the streak tube. All possible care was taken to avoid extrinsic stochastic perturbations.

An example of waveforms reproduced from streaked and filed digitized data is shown in Fig.2. The averaged global period of these waveforms was equal to the feedback delay time, and could be shortened by shortening the delay time down to 400 ps, which was limited by the optics configuration used in the experiment. Periods of many rapid fluctuations are around 200 ps, presumably corresponding to the characteristic fluctuation period of carriers in the laser. The relative fluctuation amplitude $<\Delta P>/P$ could easily be brought up to unity, where P is the averaged power level and $<\Delta P>$ is the RMS value of fluctuation.

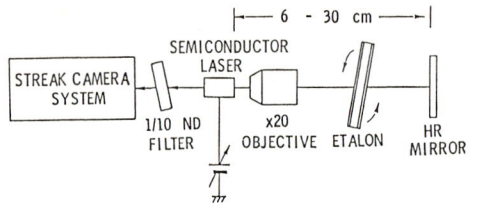

Fig.1 Experimental setup

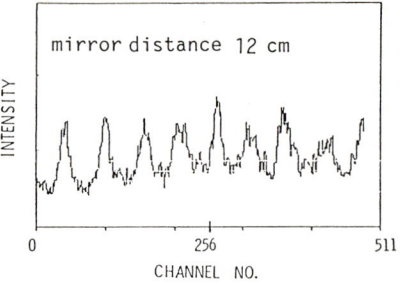

Fig.2 An example of observed waveforms, time/channel is 12.5 ps

## 3. Computer Simulations

Computer simulations have been tried on a set of equations based on the Maxwell-Bloch equations [7] (i.e., Lorenz equations [8]) modified so as to include particular characteristics of semiconductor lasers, such as carrier density dependent complex refractive index change [5], inhomogeneous broadening [9], additive noise due to spontaneous emission. Used equations are

$$\dot{E} = -\kappa \cdot ( E + E_i - P ) - i \cdot \kappa \cdot \Delta\omega \cdot ( E + E_i ) ,$$

$$\dot{P} = - P + D \cdot ( E + E_i + E_f ) + \nu^2 S - i \cdot ( \kappa - 1 ) \Delta\omega P ,$$

$$\dot{D} = b \cdot ( r - D ) - \text{Re} [ P \cdot ( E + E_i + E_f ) ],$$

$$\dot{S} = - S - P ,$$

where E is the electric field, P is the polarization, D is the inversion (i.e., carrier density), S is the off-center component of polarization, $\kappa$ is the field relaxation rate, b is the inversion relaxation rate, r is the pumping parameter, $\nu$ is inhomogeneous saturation parameter, $\Delta\omega$ is the cavity detuning from the line center, and the time as well as all the relaxation parameters are normalized to the polarization relaxation rate. $E_f$ is the feedback field $E_f(t)= f \cdot [E(t-\tau)]+E_f(t-\tau)]$, where f is the amplitude feedback factor and $\tau$ is the delay time. $E_i$ is the Langevin field due to spontaneous emission.

An example of simulated waveforms for a set of appropriate parameter values for a semiconductor laser is shown in Fig.3. A fairly good resemblance to the observed waveform can be seen.

## 4. Dimensionality Tests

Dimensionality of the observed waveforms were checked on filed digitized data in accordance with the correlation dimensionality test procedure [10]. An example of obtained $\ln C_n(d)$ vs. $\ln d$ plot averaged from more than ten shots is shown in

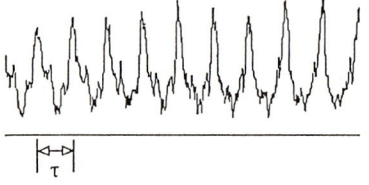

Fig.3 An example of simulated waveforms. Used parameter values are;
$\kappa=0.1$, $b=10E-4$, $r=1.4$, $\nu^2=0.2$, $\Delta\omega=0$, $f^2=0.16$, $\tau=800$,
spontaneous emission factor=10E-4

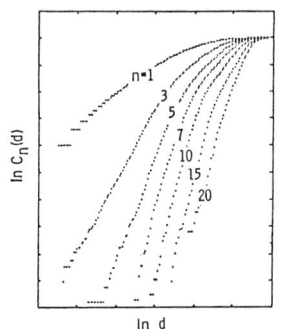

Fig.4  ln $C_n(d)$ vs. ln d plot taken from observed data

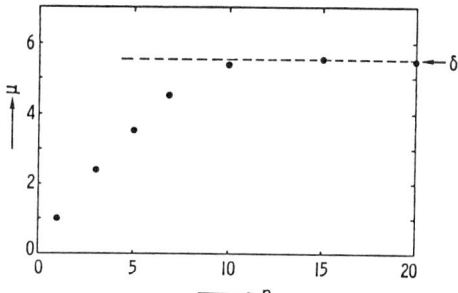

Fig.5  μ vs. n plot for Fig.4 data

Fig.4, where d is the distance between vector points in an embedding n-dimensional space deduced from those digitized waveform data and $C_n(d)$ is the accumulated probability falling inside a distance d. Slopes μ of linear portion of each curve are plotted against n in Fig.5. Saturated value of μ indicates a kind of dimension δ (correlation dimension) of a strange attractor to be estimated existing behind observed waveforms. We found a rather low dimension of around δ=5 as shown in Fig.5. The same testing procedure was applied also on simulated waveforms and the result also gave a dimension of around 5. These facts ensure the observed streaked waveforms are not stochastically induced noise, but chaos.

## 5. Relative Intensity Noise

The noise level of semiconductor lasers are customarily evaluated in terms of relative intensity noise, RIN(dB/Hz) = $10 \log_{10}[(1/\Delta f)(<\Delta P>/P)]$, where $\Delta f$ is the bandwidth. Assuming a bandwidth of the streak camera system to be 20 GHz (corresponding to 50 ps resolution determined by a slit opening of ∼100 μm), a value of $<\Delta P>/P=1$ on the streaked waveform gives RIN=-103 dB/Hz. Values of RIN around -110 to -120 dB/Hz are typically observable values in semiconductor lasers exposed to reflected outputs. Observed $<\Delta P>/P$ values of 0.5 to 0.05 on the streaked waveforms correspond to these typical RIN values.

## 6. Conclusion

In conclusion, using a streak camera, we could observe ultrafast chaotic waveforms in the 100 ps time regime from semiconductor lasers with delayed feedback. We believe that the origin of noise increase of semiconductor lasers when exposed to reflected outputs is, at least partly, the chaotic oscillations of semiconductor lasers due to delayed feedback.

1. H.Haken, Phy.Lett. 53A, 77 (1975).
2. F.T.Arecchi, G.L.Lippi, G.P.Puccioni, J.R.Tredisse, Optics Commun. 51, 308 (1984).
3. C.O.Weiss, W.Klische, P.S.Ering, M.Cooper, Optics Commun. 52, 405 (1985).
4. K.Otsuka, IEEE Jour.Quant.Electron. QE-15, 655 (1979).
5. Y.Cho, T.Umeda, 13th IQEC, WEE2 (1984).
6. Y.Cho, T.Umeda, Optics Commun. 60, (Accepted) (1986).
7. R.Graham, Phys.Lett. 58A, 440 (1976).
8. E.N.Lorenz, Atmos.Sci. 20, 20 (1963).
9. R.Graham, Y.Cho, Optics Commun. 47, 52 (1983).
10. P.Grassberger, I.Procaccia, Phys.Rev.Lett. 50, 346 (1983).

# Part V

# Applications to Semiconductors, Quantum Wells, and Solid State Physics

# Thermodynamics and Kinetics of Melting, Evaporation and Crystallization, Induced by Picosecond Pulsed Laser Irradiation

*F. Spaepen*

Division of Applied Sciences, Harvard University,
Cambridge, MA 02138, USA

## 1. Introduction

The irradiation of a strongly absorbing solid surface with a short laser pulse is a means of concentrating thermal energy in a very thin layer. The timescale considered in this paper is greater than 1 ps, which is longer than the time constants for most electronic relaxation processes, so that only thermal effects need be considered [1]. In a strongly absorbing solid (metal, UV on semiconductors) the absorption depth, $\alpha^{-1}$ ($\alpha$: absorption coefficient), is less than the thermal diffusion length during irradiation, $l_T = (D_{th} t_p)^{1/2}$ ($D_{th}$: thermal diffusivity; $t_p$: pulse length). For a 30 ps pulse on Iron, the example used in all further discussion, these lengths are 200 Å and 500 Å, respectively. A general discussion of the two regimes corresponding to the relative values of $\alpha^{-1}$ and $l_T$ can be found in a review paper by Bloembergen [2]. It is worth noting that for $t_p < 1ps$, $l_T < \alpha^{-1}$, so that shortening the pulse further no longer leads to a steeper temperature gradient and a higher quench rate.

If the fluence is low enough so that no visible permanent surface damage occurs (typically around 100 mJ/cm$^2$), a liquid layer of thickness $l_T$ with a temperature of several thousand degrees is created during the irradiation. Further melt-in, to a maximum thickness d of a few thousand Å, occurs until the crystal-melt interface temperature drops to the equilibrium melting temperature, $T_m$; this stage lasts on the order of 100 ps. Upon further cooling, the interface temperature drops below $T_m$, and the crystal regrows. The thermal parameters of this cooling process can be estimated simply by a dimensional analysis based on the thickness of the molten layer, d, and the temperature of the melt, $T_m$, as shown in Table 1 [3].

Table 1. Thermal Parameters in Melt Quenching of Metals with a 30 ps Laser Pulse. From [3]

| | | |
|---|---|---|
| Melt temperature | $T_m$ | $= 10^3$ K |
| Melt thickness | d | $= 10^{-7}$ m |
| Temperature gradient | $\nabla T = T_m/d$ | $= 10^{10}$ K/m |
| Cooling rate | $\dot{T} = D_{th} \nabla T/d$ | $= 10^{12}$ K/s |
| Melt lifetime | $\tau = T_m/\dot{T}$ | $= 10^{-9}$ s |
| Isotherm velocity | $u_T = \dot{T}/\nabla T$ | $= 100$ m/s |
| Heat-flow limited crystal growth velocity | $u_h = \kappa \nabla T/ \Delta H_{c,v}$ | $= 230$ m/s |

The metastable phases formed in this process are: the crystal overheated above $T_m$, the liquid overheated above the boiling point, $T_b$, the liquid undercooled below $T_m$, and, possibly, new metastable crystalline phases or glasses. This paper gives a concise review of the thermodynamic stability of these phases, and the kinetics of the transitions between them.

## 2. Kinetics of Crystallization

The theory of the motion of interphase boundaries has been reviewed in detail in a number of papers [4]. The results for crystallization, which are easily generalized for other transitions, can be summarized as follows. The crystal growth velocity can be written as

$$u = f k \lambda \left[ 1 - \exp\left(-\frac{\Delta G_c}{RT_i}\right) \right] , \qquad (1)$$

where k is the atom jump frequency across the crystal-melt interface, f is the fraction of interface sites that can incorporate a new atom, $\lambda$ is the interatomic distance, $\Delta G_c$ is the difference in molar free energy between crystal and melt, and $T_i$ is the interface temperature. If the undercooling of the interface, $\Delta T_i = T_m - T_i$, is small, (1) can be linearized to

$$u = k \lambda (\Delta T_i / T_i) . \qquad (2)$$

The crystal growth rate also depends on the rate at which the heat of crystallization, $\Delta H_{c,v}$ per unit volume, can be removed. Since this must occur by heat flow down the temperature gradient at the interface, $\nabla_i T$, the heat flux, J, must obey the relations

$$J = u \Delta H_{c,v} = \kappa \nabla_i T , \qquad (3)$$

where $\kappa$ is the thermal conductivity. Equation (3) can be turned around to define a heat flow-limited velocity:

$$u_h = \kappa \nabla_i T / \Delta H_{c,v} , \qquad (4)$$

which, for the example used here, is also listed in Table 1. Combining (2) and (4) then gives for metals

$$u_h / k \lambda = (T_m - T_i)/ T_i . \qquad (5)$$

The maximum growth velocity of a crystal, $u_{max} = k\lambda$, is determined by the nature of the atomic rearrangements that control the growth process. In pure metals, dilute alloys or compounds with simple crystal sturctures, k can be taken as the frequency of thermal vibration (~ Debye frequency), so that $u_{max} \sim u_s$, the speed of sound in the liquid, which is around 3000 m/s for Fe. This is called the *collision-controlled* regime, because the growth occurs by simple impingement of the liquid atoms onto the crystal. This is confirmed by direct measurements of the growth velocity of pure metals and disordered alloys (Au, Cu, Cu-Au) by transient reflectivity [5] which yield large values (>100 m/s) that can only be accounted for by a collision-controlled mechanism. From (5) and the values from Table I, the interface in Fe is found to be undercooled by only 128 K. An undercooling of at least 295 K is required for homogeneous crystal nucleation to be appreciable [6].

In more concentrated alloys or in compounds with a more complicated crystal structure, where a change of the nearest neighbor configuration, i.e. diffusion, is necessary, the jump frequency can be taken as $D_l/\lambda^2$, and

$u_{max,D} = D_l/\lambda$, which is on the order of 10 m/s. This is the *diffusion-controlled* regime. Table I shows that for picosecond laser quenching $u_{max,D} \ll u_T, u_h$, so that this type of growth can easily be suppressed. The interface temperature drops far below $T_m$, and glasses can be formed. This is confirmed by numerous experiments on binary metallic systems (e.g., Ni-Nb [7]), in which glasses have been formed by ps irradiation at compositions where the intermetallic crystalline compounds have a large unit cell, or in which long-range chemical order must be established.

## 3. The Overheated Crystal

Melting of a crystal is a first-order transition, which usually occurs heterogeneously, i.e., by nucleation of the liquid and motion of a liquid-crystal interface. Since for most materials $\sigma_{cv} > \sigma_{lv} + \sigma_{cl}$ ($\sigma_{ij}$: surface tension of the interface between phases i and j; c: crystal; v: vapor; l: liquid), there is no barrier to nucleation of the liquid at the surface of the pure crystal [8]. It should be kept in mind, however, that if the crystal is covered with a high melting layer (e.g., an oxide) and the surface tension of the resulting interface is low, nucleation of the melt may require some overheating of the crystal.

The motion of the crystal-melt interface is governed by the kinetics described in Section 2. During the initial stages of the melt-in phase the temperature gradients are an order of magnitude greater than during regrowth. This means that the heat-flow limited crystal growth velocity, given by (4) is also an order of magnitude greater than the value listed in Table I, or about 2000 m/s. For pure metals, where $u_{max} = u_s \sim 3000$ m/s, (5) gives for the overheating: $\Delta T_i = 0.66\, T_i$, which for Fe corresponds to an overheating of several thousand degrees. For materials where melting is diffusion-limited, the overheating is even higher.

The overheating required to make homogeneous nucleation of the melt (i.e., in the interior of the crystal) the dominant mechanism can be estimated from classical nucleation theory [8], which gives the following expression for the *steady state* nucleation frequency (number of nuclei per unit volume and time):

$$I = N k \exp\left( - \frac{16\pi}{3} \frac{\sigma^3}{\Delta G_v^2 k_B T} \right), \qquad (6)$$

where N is the number of atoms per unit volume, k is the interfacial jump frequency, $\Delta G_v$ is the free energy difference per unit volume, $\sigma$ is the crystal-melt interfacial tension, and $k_B$ is Boltzmann's constant.

The minimum nucleation frequency required can be estimated as: $I_{min} = \alpha/a^2 t_p$, where a is the minimum lateral distance between nuclei for inducing melting throughout the layer. Using $a = 1\mu m$ gives a value for $I_{min}$ of $10^{30}$ m$^{-3}$s$^{-1}$. (It should be kept in mind that the analysis is very insensitive to the choice of $I_{min}$). Using a conservatively high estimate for $\sigma$, 0.3 N/m [6], an overheating of 600 K is found to be sufficient to exceed $I_{min}$ in Fe. The transient time for establishing the steady-state nucleation rate [9] can be estimated as the time required to assemble a critical nucleus of radius $r^* = -2\sigma/\Delta G_v$, which is $r^*/k\lambda$; for the example of Fe overheated by 600 K, this is only a few ps, so that homogeneous nucleation of the melt can be considered possible here.

For systems where the melting kinetics are relatively slow (i.e. the jump frequency, k, scaling both nucleation and growth is small) it is possible to overheat far enough to reach the stability limit of the crystal. This is the temperature at which the shear modulus, $\mu$, vanishes, and at

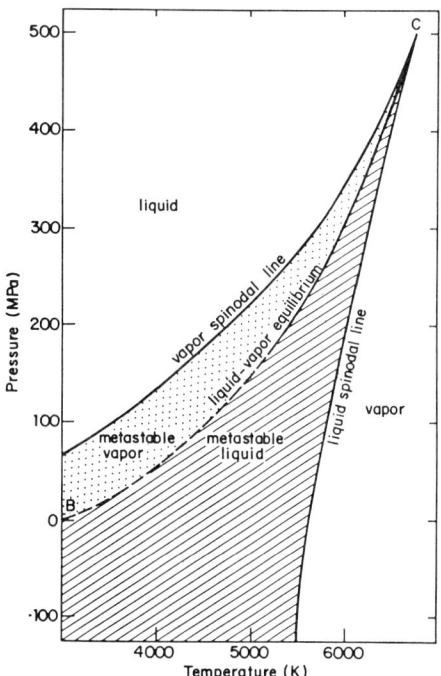

Fig. 1. P-T diagram for Fe, calculated from the van der Waals model and the critical parameters [14].

which melting can occur homogeneously by unlimited thermal shear displacements throughout the crystal. An extrapolation based on typical values of $d\mu/dT$ [10] sets this stability limit at about $2T_m$.

## 4. The Overheated Liquid

That liquids can be overheated considerably above their melting temperature is a well-known experimental fact [11], which has been qualitatively understood since van der Waals (Figure 1). The liquid can be overheated into the metastable region, where nucleation of a vapor bubble is necessary to induce the transformation, up to the spinodal line, where the liquid-vapor surface tension, $\sigma_{lv}$, and hence the nucleation barrier, vanish. Figure 1 shows that at atmospheric pressure (0.1 MPa), this instability occurs at 5700 K.

A calculation of the rate of homogeneous bubble nucleation, based on the classical expression similar to (6), taking into account the roughly linear decrease of $\sigma_{lv}$ with temperature to the spinodal [12], shows that the temperature of the overheated liquid must be within a few hundred degrees from the spinodal for the nucleation rate to exceed $I_{min} = 10^{30}$ $m^{-3}s^{-1}$, estimated above. An estimate of the transient time for this transformation is $r^*/\Gamma \lambda^3$ ($\Gamma = p/\sqrt{2\pi mkT}$, the impingement rate of the vapor atoms), which is 7 ps for the present example. Given that this is a lower limit (perfect sticking is assumed in calculating $\Gamma$), and that the extreme overheat required for nucleation is only present during a fraction of $t_p$, the transient effects probably enhance the stability of the overheated liquid.

In a typical ps melt quenching experiment the fluence is not high enough to reach the extreme temperatures required for bulk vaporization.

Only surface evaporation needs to be considered, which can be estimated conveniently from the impingement rate, $\Gamma$, of the equilibrium vapor phase. It has been shown that the loss of mass and energy by this mode is negligible in ps experiments [13].

The permanent surface damage (holes) sometimes observed in ps experiments is therefore not due to direct evaporation, but to *mechanical* displacement of the liquid by the recoil pressure of the evaporating atoms. This is confirmed by the morphology of the displaced material. Direct evidence of the recoil pressure can be obtained by observing the vibration of a metallic reed pulse irradiated at one end and clamped at the other [14].

The author's work in this area is supported by the Office of Naval Research under contract number N00014-85-K-0684.

## References

1. For a review of these relaxation processes, see W.L. Brown: Mat. Res. Soc. Symp. Proc. 23, 9 (1984).
2. N. Bloembergen: In *Laser-Solid Interactions and Laser Processing*, ed. by S. D. Ferris, H.J. Leamy and J.M. Poate (AIP, New York 1979) p. 1.
3. F. Spaepen and C.J. Lin: In *Amorphous Metals and Non-Equilibrium Processing*, ed. by M. von Allmen (Les Editions de Physique, Les Ulis, France 1984), p. 65.
4. F. Spaepen and D. Turnbull: In *Laser Processing of Semiconductors*, ed. by J.M. Poate and J.W. Mayer (Academic, NY 1982), p. 15.
5. C.A. MacDonald, A.M. Malvezzi and F. Spaepen: Mat. Res. Soc. Symp. Proc. 51, (1986) in press.
6. D. Turnbull: J.Appl. Phys. 21, 1022 (1950).
7. C.J. Lin and F. Spaepen: Acta Metallurgica 34, in press.
8. D. Turnbull: Sol. St. Phys. 3, 225 (1956)
9. K.F. Kelton, A.L. Greer and C.V. Thompson: J. Chem. Phys. 79, 6261 (1983). This paper also contains an extensive review of earlier work.
10. H.J. Frost and M.F. Ashby: *Deformation-Mechanism Maps* (Pergamon, NY 1982).
11. R.E. Apfel: J. Acoust. Soc. Am. 49, 145 (1971).
12. J.W. Cahn and J.E. Hilliard: J. Chem. Phys. 28, 258 (1958).
13. C.J. Lin: Ph.D. thesis, Harvard University (1983).
14. Landolt-Börnstein Tables, Vol. II, Part 1, Mechanisch-Thermische Zustandgrössen, ed. by H. Borchers et al. (Springer, Berlin 1971) p. 332.

# Investigation of Nonthermal Population Distributions with 10-fs Optical Pulses

*C.V. Shank, R.L. Fork, C.H. Brito Cruz\*, and W. Knox*
AT & T Bell Laboratories, Holmdel, NJ 07733, USA

An ultrashort femtosecond pulse is a unique tool for exciting a nonequilibrium population distribution in a solid or large molecule. The optical energy must be put into electronic or vibronic energy levels in a time short compared to the interaction with the thermal bath. This time scale is on the order of 100 femtoseconds or less.

Advances in femtosecond measurement techniques have made experiments of the type just described feasible. A new high repetition rate amplifier using a copper vapor laser has been developed [1]. This system is capable of producing femtosecond optical pulses at a repetition rate of 10 kHz with energy of a few microjoules. These amplified optical pulses have been successfully compressed to 8 femtoseconds [2] using a short 7 mm piece of optical fiber and a grating pair compressor.

Time resolved absorption spectra are usually measured using the "white light" continuum technique. Interestingly, the spectral extent of an 8-femtosecond optical pulse is nearly 100 nm, making this a nearly ideal light source for measuring spectra. Since the pulse is well characterized, there is minimal frequency sweep.

For a large dye molecule in solution at room temperature, the ground and excited singlet states are strongly broadened vibronic levels. These levels are perturbed by interaction with the thermal bath. The thermal bath can exchange energy with these levels by both intramolecular and intermolecular processes. Vibrational modes in the molecular backbone of these large molecules can form a thermal bath within the molecule itself. It is also possible for intermolecular energy transfer to take place on a somewhat longer time scale by collisions, dipole-dipole interaction etc. In the experiment described here we use a 60-femtosecond optical pulse to excite a band of states approximately 8 nm in width and observe the temporal evolution of the absorption of excited molecules with a second 10-femtosecond optical pulse.

The experimental apparatus is arranged to perform a pump-probe type measurement. The amplified 60-femtosecond optical pulse is split into two parts. One part is used to optically excite the sample while the other part is compressed with a fiber and grating pair to 10 femtoseconds. The shorter pulse is then used to probe the absorption spectrum by passing through the excited sample into a spectrometer and diode array.

---

\*On leave from Instituto de Fisica-UNICAMP
13100, Campinas, S.P., Brazil

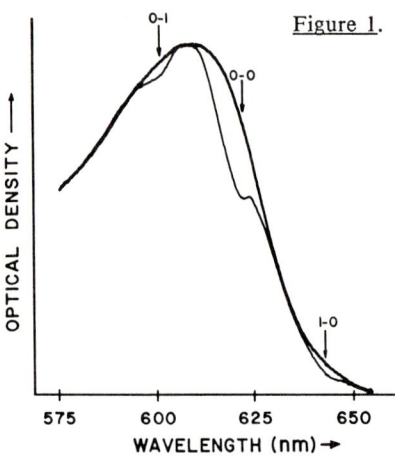

Figure 1. Plot of cresyl violet absorption before and after excitation.

The result for the molecule cresyl violet dissolved in ethylene glycol is shown in Fig. 1 where we plot the optical density, $\alpha$, as a function of wavelength before and after excitation. The pump and the probe are coincident in time. In the wavelength range near the pumping frequency a decrease in absorption or spectral hole burning is observed. This is to be expected since we excite molecules that are resonant with the pump. To our surprise we see two additional holes 600 $cm^{-1}$ above and below the pumping frequency.

These replica holes can be explained by investigating the vibrational substructure in the optical absorption spectrum. The wavelength of the pumping pulse is very close to the 0-0 transition. The measured Raman spectra of the molecule cresyl violet [3] reveal a strong vibrational mode at $\omega_v$ = 590 $cm^{-1}$. We attribute the apparent hole at the pump frequency, $\omega_p$, to both the depletion of the 0-vibrational level of the ground state and the increase in the population of the 0-vibrational level of the excited singlet state. In addition, the depletion of the 0-vibrational level in the ground state also causes a bleaching of the transition to the first vibrational level of the excited singlet, which appears blue shifted at the frequency $\omega_p + \omega_v$ and is labeled 0-1 in the figure. The transient excess population of the 0-vibrational level of the excited singlet similarly gives rise to a bleaching of the transition to the first vibrational level of the ground state. This appears as the red shifted hole at the frequency $\omega_p - \omega_v$ which is labeled 1-0.

The dynamics of the evolution of the nonthermal population distribution has been investigated by measuring differential spectra as a function of time delay between the pumping and probing pulse. These spectra are obtained by chopping the pumping beam at 10 Hz and accumulating the output from the detector array in the computer in phase with the chopper. In Fig. 2 we have plotted the change in optical density $(-\Delta\alpha)$ for a range of relative delays between the pump and probe. At the earliest time we see the hole at the pumping frequency with the two replica holes positioned 590 $cm^{-1}$ above and below the pumping frequency. As time progresses the holes disappear and the differential absorption spectrum approaches what appears to be the expected spectrum for a thermalized population distribution in about 150 femtoseconds.

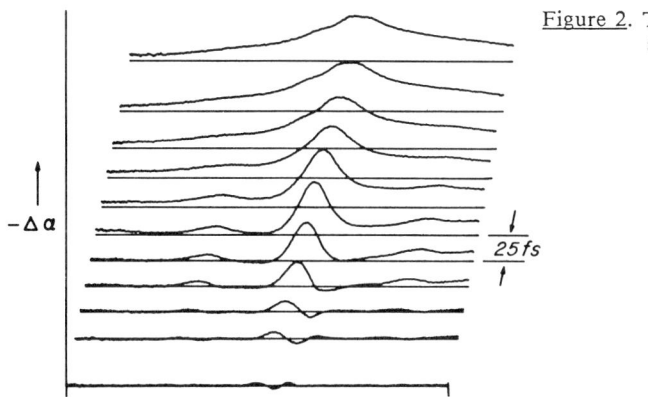

Figure 2. Time resolved differential spectra for cresyl violet.

In Fig. 3 we plot normalized differential spectra ($-\Delta\alpha/\alpha$) for the dye molecule nile blue. In this case the pumping wavelength 618 nm is to the blue edge of the fundamental absorption. Again we see a hole burnt at the pumping frequency. In addition, we see a broad hole at the long wavelength side of the absorption spectrum. This is attributed to bleaching of transitions that terminate on an excited ground state. Note the overshoot of the bleaching at the long wavelength edge as the nonthermal population distribution thermalizes.

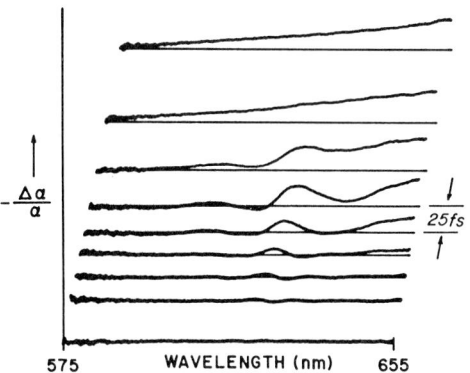

Figure 3. Time resolved differential spectra for nile blue.

In conclusion, we have demonstrated a new technique for measuring time resolved spectra using compressed 10-femtosecond optical pulses. We have observed spectral hole burning in dyes and have observed the time evolution to the thermalized spectrum.

1. W.H. Knox, M.C. Downer, R.L. Fork, and C.V. Shank, Optics Lett., 9, 552 (1984).

2. W.H. Knox, R.L. Fork, M.C. Downer, R.H. Stolen, C.V. Shank, and J. Valdmanis, Appl. Phys. Lett., 46, 1120 (1985).

3. W. Werncke, A. Lau, M. Pfeiffer, H.J. Weynmahn, G. Hunsalz and K. Lenz, Optics Commun., 16, 128 (1976).

# Superheating During Ultrafast Laser Heating of Semiconductors

*D. von der Linde, N. Fabricius, B. Danielzik, and P. Hermes*

Fachbereich Physik, Universität-GHS-Essen,
D-4300 Essen 1, Fed. Rep. of Germany

## 1. Introduction

When a short laser pulse is incident on some strongly absorbing semiconductor or metal, the absorbed optical energy is thermalized in a time typically of the order of a picosecond. Thus, for pulse durations much longer than a picosecond, the interaction of the laser pulse with the material can be described as a very rapid thermal heating process [1]. The thickness of the heated surface layer is typically of the order of several hundred Angstroms, being determined by the penetration depth of the light and/or by the thermal diffusion length. It follows that with incident pulse energies of about 0.1 $J/cm^2$ surface temperatures well in excess of the melting temperature of many materials can be achieved.

The solid-to-liquid phase transition usually starts from nucleation centers at the surface and proceeds by propagation of the solid-liquid interface into the bulk [2]. The velocity of propagation of the melt-front is a function of the temperature difference between the interface temperature $T_i$, and the equilibrium melting temperature $T_m$. For conventional heating conditions $T_i - T_m$ is usually very small, e. g. $(T_i-T_m)/T_m \ll 1$. However, extremely fast heating rates can be achieved with the use of ultrashort laser pulses. Under these conditions the solid-liquid interface temperature during melting may greatly exceed $T_m$, and the solid may be driven into a metastable, highly superheated state.

Indication of superheating has been observed in a number of recent laser heating experiments [3-5], most notably perhaps in the picosecond electron diffraction experiments of Williamson and Mourou [5]. Here we describe measurements of the surface temperature of crystalline GaAs during heating with picosecond laser pulses. We believe that our results provide clear evidence of superheating of solid GaAs by several hundred degrees Kelvin [6].

## 2. Experimental

For the determination of the surface temperature of the laser-heated material we measure the velocities of the particles which evaporate or sublimate from the hot surface. We make use of the fact that in a thermal evaporation process the velocity distributions of the emitted particles are given by Maxwell distributions with a temperature corresponding to the surface temperature.

The GaAs samples mounted in an ultrahigh vacuum chamber with a pressure of about $10^{-10}$ bar are irradiated with the laser heating pulses at an angle of incidence of 45 degrees. A quadrupole mass spectrometer (QMS) positioned in the direction of the surface normal at a distance of L=9.2 cm serves as a detector of the emitted particles. We measure the time of flight from the surface of the sample to the detector [7]. The times of flight are obtained from the measured time between the laser pulse and the output pulses of the QMS, after properly taking into account the particle drift times in the QMS, which are small in comparison with the flight times.

We use laser heating pulses of 25 ps duration at a wavelength of 532 nm. The pulses are obtained from the second harmonic of an actively and passively mode-locked Nd-YAG laser. As will be discussed below we also use heating pulses of 10 ns duration from a frequency-doubled, single transverse and longitudinal mode Q-switched Nd-YAG laser.

The laser-heated area on the surface of the sample is characterized by a Gaussian intensity distribution of the laser beam which has a 1/e diameter of 0.74 mm. To detect the onset of melting we monitor the optical reflectivity of the surface with the help of a cw krypton laser beam (wavelength 647 nm) focussed to a spot of 0.1 mm in the center of the laser-heated area. Melting of the surface is associated with a strong increase of the optical reflectivity [8], because molten GaAs is metallic. The time resolution of the reflectivity measurement is 0.8 ns.

Since the number of atoms for a single laser pulse is not sufficient to establish the complete velocity distribution, the data of many laser pulses must be accumulated. The sample is raster-scanned during the experiment to provide a fresh sample area for each laser pulse.

## 3. Velocity Distributions

An important result of both ns and ps laser heating experiments is that for a rather wide range of laser fluences the observed velocity distributions correspond to thermal Maxwell distributions. Figure 1 depicts as an example the measured time-of-flight distributions of arsenic atoms for four different fluence values (picosecond heating). The distributions are normalized to unity. We note, however, that the total number of emitted particles is strongly increasing with the laser fluence. The dotted curves in Fig. 1 represent time-of-flight curves corresponding to Maxwell velocity distributions. It can be seen that the Maxwellians with the indicated temperature values provide an excellent fit of the measured data (solid curves).

The fact that the measured distributions can be represented by Maxwell distributions is indeed noteworthy. Since there is no spatial and temporal resolution in our present velocity measurement, our data represent spatial and temporal averages. Generally speaking, an average over different Maxwell distributions is not expected to produce a Maxwellian. However, because the number of emitted particles is a strongly increasing function of the temperature, it is very likely that the properly weighted average is in

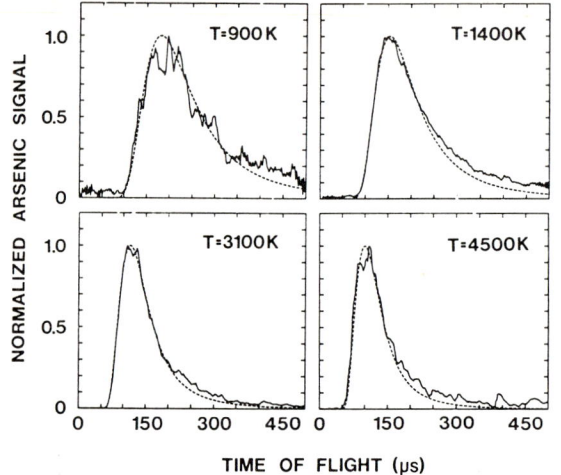

Fig.1:

Time of flight distribution of As atoms for different laser energies.
a) 27 mJ/cm$^2$, b) 36 mJ/cm$^2$
c) 59 mJ/cm$^2$, d) 87 mJ/cm$^2$.

fact dominated by the Maxwell distribution corresponding to the maximum temperature in space and time. The observation that the measured distributions are Maxwellian is therefore taken as strong evidence that it is a good approximation to interpret the experimental temperatures as the maximum surface temperature.

## 4. Results of the Temperature Measurements

Before discussing ultrafast laser heating with picosecond pulses it is instructive to consider first the results of nanosecond experiments. In Fig. 2 the temperature obtained from the measured velocities of Ga atoms is plotted as a function of the laser fluence for ns heating. The temperature is seen to increase continuously from room temperature (lower dashed line)

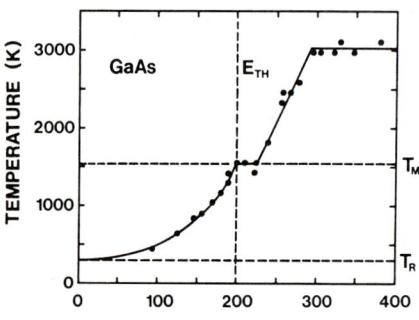

Fig. 2: ENERGY DENSITY (mJ/cm²)
Measured surface temperature vs laser energy for ns laser pulses.

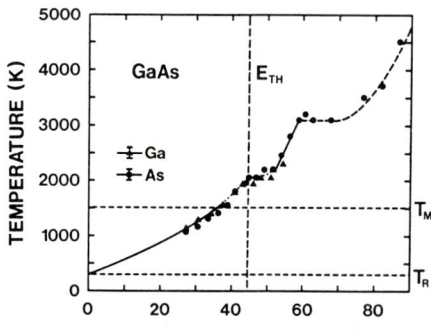

Fig.3: ENERGY DENSITY (mJ/cm²)
Measured surface temperature vs laser energy for ps laser pulses.

to the melting temperature (upper dashed line), which is reached at $E_{TH}=200$ mJ/cm$^2$. An important point to note is that for a further increase of the fluence the temperature remains constant at the value $T=T_m$ up to $E=1.1E_{TH}$. For still higher fluences the temperature rise is resumed.

The nanosecond measurement is interpreted as follows. The abrupt break of the tempoerature curve at $T=T_m$ is clear evidence of the onset of melting. For $E_{TH} < E < 1.1E_{TH}$ the incident laser energy is consumed as latent heat of melting for the formation of a liquid surface layer. The temperature rise for $E > 1.1E_{TH}$ indicates a further heating of the liquid layer, when the thickness of the developing film exceeds the optical skin depth or the thermal diffusion length. The onset of melting at 200 mJ/cm$^2$ is confirmed by the independent optical reflectivity measurements, which show that the reflectivity begins to increase at $E=E_{TH}$, and reaches the maximum around $E=1.1E_{TH}$.

Results of temperature measurements for picosecond laser heating are plotted in Fig. 3, which shows data both from Ga and As velocity measurements. The temperature increases with fluence reaching the nominal melting point $T_m=1511$ K at 37 mJ/cm$^2$. However, comparing with the nanosecod data, the striking difference is that there is no break of the temperature curve at $T=T_m$. Rather, the temperature continues to rise without any sign of a discontinuity of the slope. The change of the optical reflectivity which indicates the development of a molten surface layer is observed to set in at a fluence 45 mJ/cm$^2$, substantially greater than the fluence that is required to reach the melting temperature. We estimate that the liquid layer must exceed a thickness of at least several nm to be detectable by the measurement of the optical reflectivity.

## 5. Discussion

The kinetics of the melting process have been discussed elsewhere in this book by Spaepen [9]. According to this model, formation of the melt-front starts at the surface, and propagation sets in as soon as the surface temperature exceeds the melting temperature. The velocity of propagation of the melt-front increases with the degree of superheating of the interface, $T_i-T_m$. The interface temperature $T_i$ is determined by the energy balance between heating and the consumption of latent heat of melting associated with the progressing melt-front. When the heating rate is much faster than the consumption of latent heat, the interface temperature must go up.

We believe that the continuous temperature rise at $T=T_m$ observed for picosecond laser heating is direct evidence of such a situation in which the heating rate is so great that the usual pinning of the interface temperature near the equilibrium melting point -which has been observed for nanosecond heating, see Fig. 2- is overcome. Our picosecond data indicate that $T_i$ is driven substantially above $T_m$.

It should be emphasized that for $T > T_m$ our data represent the surface temperature $T_s$ of the liquid overlayer and not directly the liquid-solid interface temperature. To infer the proper interface temperature the possible temperature drop across the liquid must be taken into account. To illustrate this point let us consider Fig. 4 which shows calculated temperature profiles from a detailed computer simulation. Although the values of the temperature-dependent material parameters for GaAs are not well known it turns out that during the initial stage of melting the temperature gradient of the liquid is very small. This result holds for rather general conditions, independent of the detailed values of the material parameters. Figure 4 shows the calculated temperature versus depth for different times during and after the laser heating pulse for a relatively larger fluence corresponding to 2.6 times the fluence necessary to reach the melting point. The dotted vertical lines mark the positions of the melt-front.

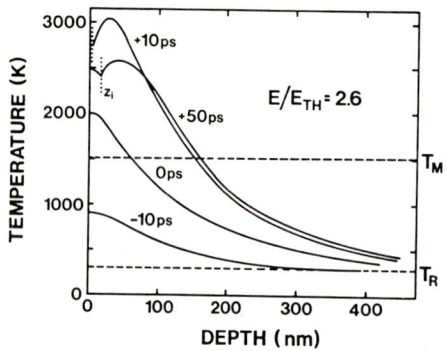

Fig. 4:

Calculated temperature profiles at different times after excitation.

On the depth scale of Fig. 4 the interface position t=0 (maximum of the laser pulse) is indistinguishable from the surface, and the temperatures of the surface and the interface are the same, $T_i=T_s=2000$ K. For t=10 ps the melt-depth is about 4 nm; the temperature difference $T_s-T_i$ is still very small. Note that the temperature gradient in the solid near the interface is positive: As a consequence of the consumption of latent heat there is a temperature minimum at the interface and heat transport occurs from the strongly superheated solid to the interface. The maximum temperature of the solid reaches 3100 K indicating that superheating in the bulk is still much stronger than at the interface. For t=50 ps, finally, the melt-depth is about 20 nm, and yet the temperature drop across the film is only about 100 K.

These results encourage us to interpret the measured surface temperatures as a good approximation of the liquid-solid interface temperature. For example, our picosecond temperature data give a surface temperature of 2000 K at the threshold of the formation of an optically detectable liquid layer, where the difference $T_i-T_m$ has been shown to be very small. The picosecond temperature data therefore indicate a superheating of the liquid-solid interface in GaAs of about 500 K.

## 6. Conclusions

We have used measurements of the velocity distributions of evaporated atoms to determine the surface temperature of laser-heated GaAs. For ns heating the observed liquid-to-solid transition temperature agrees with the equilibrium melting temperature, $T_m$=1511 K. Analysis of our picosecond data indicates strong superheating of the solid by several hundred degrees K.

References

1. N. Bloembergen, in Laser Solid Interactions and Laser Processing, ed. by S. D. Ferris, M. J. Leamy, and J. M. Poate, (American Institute of Physics Conference Proceedings No. 50, New York, 1979), p. 1
2. S. Spaepen and D. Turnbull, in Laser Annealing of Solids, ed. by J. M. Poate and J. M. Mayer, (Academic Press, New York, 1982), p. 15
3. P. H. Bucksbaum and J. Bokor, Phys. Rev. Lett. 53, 182 (1984)
4. M. O. Thompson, P. H. Bucksbaum, and J. Bokor, Mat. Res. Soc. Symp. Proc. 35, 181 (1985)
5. S. Williamson, G. Mourou, and J. C. Lee, Mat. Res. Soc. Symp. Proc. 35, 87 (1985)
6. N. Fabricius, P. Hermes, D. von der Linde, A. Pospieszczyk, and B. Stritzker, Sol. State Comm. 58, 239 (1986)
7. B. Stritzker, A. Pospieszczyk, and J. A. Taggle, Phys. Rev. Lett. 47, 356 (1981)
8. D. H. Auston, C. M. Surko. T. N. C. Venkatesan, R. E. Slusher, and S. A. Golovchenko, Appl. Phys. Lett. 33, 437 (1978)
9. See article by F. Spaepen, in this volume

# Non-equilibrium Carriers in GaAs: Secondary Emission During the First Two Picoseconds

*J.A. Kash and J.C. Tsang*

IBM Thomas J. Watson Research Center, P.O. Box 218,
Yorktown Heights, NY 10598, USA

The relaxation of non-equilibrium ("hot") carriers in semiconductors is one of the problems intensively studied using ultrashort laser pulses[1]. For these studies, the laser is commonly used to first inject the carriers (with a strong pump pulse) and then at some later time a property of the semiconductor is measured which depends on the injected carriers, often using a second laser pulse to probe the material. In this paper, we will show how an ultra-sensitive multichannel optical detector, the imaging microchannel plate photomultiplier (MCP) can be used to study the relaxation of hot carriers by monitoring emission which would otherwise be impossible to detect. In particular, we have measured the time-resolved Raman spectrum of the nonequilibrium LO phonons generated as the hot carriers lose energy to the lattice and have also measured hot band-to-band luminescence emitted by these carriers at very high energies and at very short times, before substantial cooling takes place.

1. Experiment

The experiment, shown in Fig. 1, resembles the standard pump-probe setup except for the triple spectrometer and MCP. The laser system is a synchronously pumped dye laser operating at 580 nm, with the pulses compressed to 0.7 psec (autocorrelation FWHM) in a fiber optic pulse compressor. Slits are placed in the grating stage to reject the frequency tails of the pulse which would otherwise obscure the weak emission signals. The laser bandwidth is 30 cm$^{-1}$, about a factor of two from the transform limit. Substantially broader bandwidth (as from shorter laser pulses) would seriously degrade the spectral resolution needed for these experiments[2]. Pump and probe beam are orthogonally polarized to avoid coherent artifacts. The samples in the experiments described here are

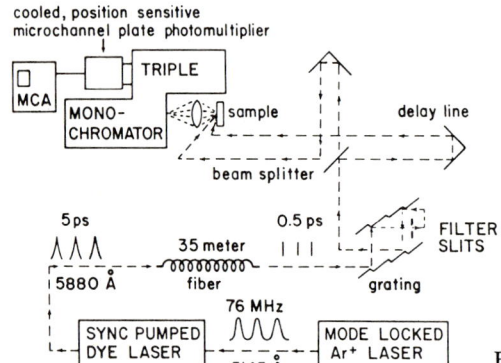

Fig. 1. Experimental apparatus

nominally undoped epitaxial layers of GaAs on semi-insulating substrates, kept at room temperature. The imaging MCP detector (ITT F4146M) has a photocathode response (MA3) out to 900 nm and a resistive anode. A Surface Science Model 2401 Position Computer is used to determine both the arrival of a photon and the wavelength of the photon to within 0.3 nm. When used for spectroscopy, this detector is equivalent to having 400 small photon counting photomultiplier tubes (effective photocathode dimension 3mm by 60$\mu$m) at the exit of the spectrometer, each with a dark count of 0.01 cps. Further details about the detector are given elsewhere[3]. The extremely low dark count, the linearity, and the multichannel nature of the detector are all important in the detection of weak signals in the presence of possible background signals and laser fluctuations. The only other detector which approaches these specifications is the much more complex charge coupled device (CCD)[4], but the present level of CCD readout noise ($\sim$40 electrons/pixel) and so-called "radiation fails" are a significant limit to the available sensitivity. It should be noted, however, that the 10 microsecond dead time of the MCP precludes its use with low repetition rate pulsed lasers (<10kHz). Also, the MCP has no intrinsic time resolution, so time-dependent phenomena must be monitored with a pump-probe technique.

## 2. Carrier Relaxation in GaAs

When electrons and holes are optically injected into GaAs by a short pulse laser with photon energy well above the direct gap, the initial monoenergetic carrier distribution is rapidly changed by both carrier-carrier scattering and electron-LO phonon scattering. The effect of each mechanism on the carrier distributions is very different. Electron-LO phonon scattering causes the carriers to lose energy to the lattice. On the other hand, carrier-carrier scattering redistributes energy within the carriers and produces a carrier distribution characterized by a well-defined temperature for the electrons and the holes that can be different than the lattice temperature. In addition[5,6] $\tau_{e-LO} \simeq 165$ fsec and is independent of carrier concentration for $n < 10^{18}$cm$^{-3}$, while $\tau_{c-c}$ depends[5] on concentration n roughly as $\tau_{c-c} \simeq 2 \times 10^4$ sec /n cm$^3$. Thus, the relative importance of each mechanism depends upon the carrier concentration. The initial relaxation of the optically injected carriers is dominated by $\tau_{e-LO}$ at low n, and by $\tau_{c-c}$ at high n. The two rates are equal at $n \simeq 10^{17}$cm$^{-3}$.

## 3. Time Resolved Raman Scattering

We have measured $\tau_{e-LO}$ directly using time resolved Raman scattering[6]. The experiment is indicated schematically in Fig. 2. Carriers are injected with a subpicosecond pump pulse. The non-equilibrium phonons generated as these carriers relax to the band edge are then sampled by measuring the spontaneous Stokes or anti-Stokes Raman emission excited by a delayed probe pulse. For these experiments, the carrier concentration is low enough to ignore carrier-carrier scattering. We observe (Fig. 3) a non-equilibrium phonon population which increases for 2 psec and then decays. A simple analysis of the cascade process[6] leads to $\tau_{e-LO} \simeq 2$ psec / 12 = 165 fsec. Previous time resolved Raman scattering in GaAs[7] used a conventional photomultiplier. The MCP improves signal/noise by about 100. This improvement is needed in order to measure the creation of the phonons, as shorter laser pulses are necessary. Unfortunately, such short pulses have a substantially broader bandwidth ($\sim$1 nm) which lowers the peak signal $\sim$10 times, making our experiment extremely difficult with a conventional photomultiplier.

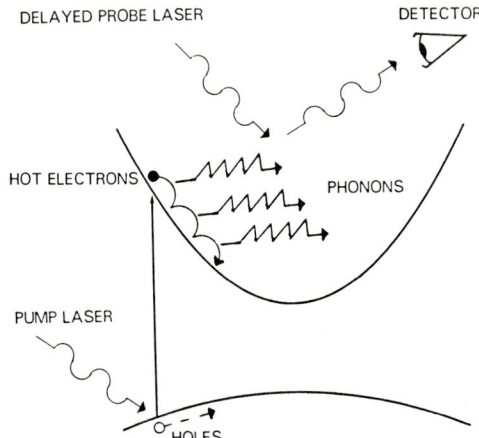

Fig. 2. Schematic of time resolved Raman scattering experiment

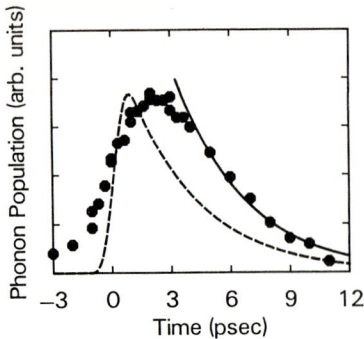

Fig. 3. Non-equilibrium phonon population vs. time. The dashed curve indicates the system response and the solid curve shows the 3.5 psec decay of the phonons at room temperature

## 4. Band-to-band Recombination Luminescence

At carrier concentrations $n > 10^{17}$ cm$^{-3}$ the carriers continue to lose energy to the phonons. However, the simple cascade pictured in Fig. 2 is no longer valid since carrier-carrier scattering also becomes important. Here we observe broad band emission at energies up to and even above (i.e. "anti-Stokes") the laser energy. This emission, shown in the inset to Fig. 4, is "hot" luminescence resulting from band-to-band carrier recombination. Previous studies of this emission[8-10] have concentrated on emission at energies of 1.8 eV and below, where the emission is fairly intense. With the additional sensitivity of the MCP, we are able to make measurements up to 2.4 eV, i.e. 0.3 eV *above* the laser photon energy. As we will show, unique information is contained in the anti-Stokes emission relating to the thermalization of the carriers. We analyze the emission using two techniques. The first is picosecond luminescence correlation[8], in which we measure the time-integrated emission from the sample at various photon energies as a function of delay between two subpicosecond laser pulses. This correlation technique, which relies upon the bimolecular nature of band-to-band recombination, measures how long electrons remain in high energy states. The results are shown in Fig. 4. As has been shown previously, the correlation time decreases with increasing emission energy. For the anti-Stokes emission we find FWHM correlation times ranging between 1.5 and 3 psec, and we believe that still shorter times could be measured at higher energies. This shows that the anti-Stokes emission comes out only during the first picosecond. Thus it is a sensitive probe of the first picosecond, a regime which has yet to be probed by direct time-resolved luminescence techniques.

For a single laser pulse which generates time dependent electron and hole distributions $f_e$ and $f_h$, the time-integrated emission at photon energy $E = E_{gap} + E_e + E_h$ is

$$I(E) \propto \int_0^\infty f_e(E_e, t) f_h(E_h, t) dt \quad ,$$

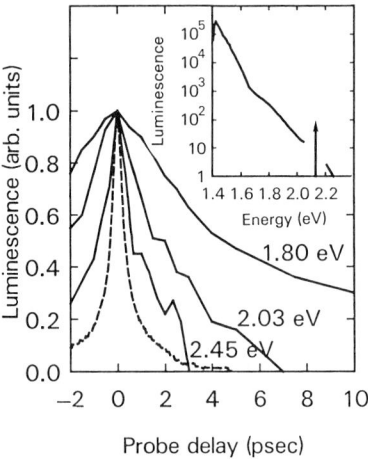

Fig. 4. Picosecond luminescence correlation data for several emission energies E. Pump and probe each inject $n \simeq 5 \times 10^{17} \text{cm}^{-3}$, except for E=2.45 eV, where $n \simeq 10^{18} \text{cm}^{-3}$. The dashed curve is the laser autocorrelation. (Inset) Hot luminescence spectrum vs. emission energy for $n \simeq 4 \times 10^{17} \text{cm}^{-3}$, normalized against system response. The spectrum could not be measured near the laser energy, indicated by the arrow

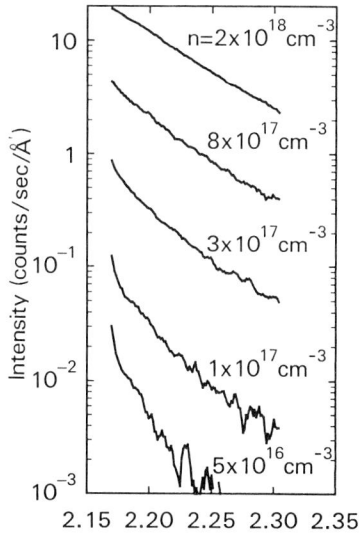

Fig. 5. Anti-Stokes hot luminescence spectra for several values of n

where $E_{gap}$ is the bandgap of GaAs and $E_e$ and $E_h$ are the electron and hole energies measured with respect to the band extrema. Examination of this equation shows that the *shape* of I(E) will not change if the time evolution of $f_e$ and $f_h$ does not vary with n except for amplitude. In this case, I(E) will simply increase quadratically with n. On the other hand, if the shape of I(E) is observed to change with n, then the distribution functions must be changing with n. Such changes should be observed when carrier-carrier scattering is just beginning to be important, but not yet dominant.

If we measure the anti-Stokes spectrum arising from single pulse excitation as a function of injected carrier density (Fig. 5), we see that for $n > 3 \times 10^{17} \text{cm}^{-3}$ the spectrum simply increases quadratically in intensity with n but the shape remains the same. At lower densities however, the shape of the spectrum changes with decreasing n. These changes, which could not be measured with a conventional detector, result from the decreasing rate of carrier-carrier scattering as n decreases. For $n > 3 \times 10^{17} \text{cm}^{-3}$, $\tau_{c-c} \ll \tau_{e-LO}$, and a thermalized carrier distribution is obtained via multiple carrier-carrier scattering events before the carriers lose any energy to the lattice; thus the shape of the spectrum becomes constant. At lower n the carriers lose some energy to the lattice before acheiving a thermal distribution, resulting in changes in the spectral shape with density as the relative importance of carrier-carrier scattering and carrier-phonon scattering changes. At lower n longer times are required to thermalize the carriers, and so the temperature of the initial thermalized carrier distribution is lower. Since a thermal distribution has an exponential tail, the slope of the time-integrated emission at the highest energies (i.e. the anti-Stokes side) is dominated by the highest temperature achieved by the distribution. Thus a steeper slope is observed at lower n. In addition, there cannot be substantial numbers of electrons at energies higher than the creation energy unless some electrons gain energy from carrier-carrier scattering.

## 5. Conclusions

We have shown that a sensitive optical detector, the imaging microchannelplate photomultiplier, can be used in conjunction with ultrashort laser pulses to study rapid emission processes far too weak to be measured with conventional detectors. In particular, we have measured the electron-LO phonon scattering time in GaAs to be 165 fsec. We have also observed the competition between electron-phonon scattering and carrier-carrier scattering which occurs at carrier densities greater than $10^{17} cm^{-3}$. Further improvements in multichannel detectors will allow additional experiments. Charge coupled device (CCD) detector improvements offer great hope. With quantum efficiency 5 to 10 times that of photocathode-based detectors and the ability to detect many photons simultaneously (i.e. no dead time), CCD's will allow the study of weak emission excited by low repetition rate sources, such as amplified femtosecond pulses.

## References

1. See for example: <u>Proceedings of the Fourth International Conference on Hot Electrons in Semiconductors</u>, ed. by E. Gornik, G. Bauer, and E. Vass, Physica <u>134B</u> (1985).
2. Sudhanshu S. Jha, J.A. Kash, and J.C Tsang: Phys. Rev. <u>B</u>, to be published.
3. J.C. Tsang: In <u>Dynamics on Surfaces</u>, ed. by B. Pullman, J. Jortner, A. Nitzan, and B. Gerber (Reidel, Dordrecht, 1984) p. 379.
4. Cherry A. Murray and C.B. Dierker: J. Opt. Soc. Am. <u>A</u>, to be published.
5. E.M. Conwell and M.O. Vassel: IEEE Trans. Electron. Devices <u>13</u>, 22 (1966).
6. J.A. Kash and J.C. Tsang: Phys. Rev. Lett. <u>54</u>, 2151 (1985).
7. D. von der Linde, J. Kuhl, and H. Klingenberg: Phys. Rev. Lett. <u>44</u>, 1505 (1980).
8. D. von der Linde, J. Kuhl, and E. Rosengart: J. Luminescence <u>24</u>, 675 (1981).
9. Kathleen Kash, Jagdeep Shah, Dominique Block, A.C. Gossard, and W. Weigmann: Physica <u>134B</u>, 189 (1985).
10. J.F. Ryan, R.A. Taylor, A.J. Turberfield, Angela Maciel, J.M. Worlock, A.C. Gossard, and W. Weigmann: Phys. Rev. Lett. <u>53</u>, 1841 (1984).

# Ultrafast Carrier Dynamics in GaAs and $Al_xGa_{1-x}$ As

*W.Z. Lin[1], J.G. Fujimoto[1], E.P. Ippen[1], and R.A. Logan[2]*

[1] Department of Electrical Engineering and Computer Science and Research Laboratory of Electronics, Massachusetts Institute of Technology, Cambridge, MA 02139, USA

[2] AT & T Bell Laboratories, Murray Hill, NJ 07974, USA

The dynamics of excited carriers in GaAs are relevant to both electronic and optoelectronic devices. Recent advances in the generation of ultrashort optical pulses permit the investigation of these processes on a femtosecond time scale[1-3]. In this paper we describe the investigation of femtosecond absorption recovery dynamics in GaAs and $Al_xGa_{1-x}As$ using pulses as short as 35 fs generated by a colliding pulse modelocked (CPM) ring dye laser[4]. Pump probe measurements reveal the presence of a two component carrier relaxation process. In GaAs, the initial ultrafast process occurs on a time scale comparable to the pulse duration.

Experiments were performed with pulses from a CPM dye laser incorporating a multiple prism arrangement for control of dispersion[5]. The use of prisms both inside and outside the laser cavity permitted the independent control of pulse duration and bandwidth as well as frequency chirp. Pump probe measurements were performed in thin cladded layers of GaAs grown by liquid phase epitaxy. Absorption saturation dynamics were investigated high above the bandgap using a photon energy of 2 eV in samples at room temperature.

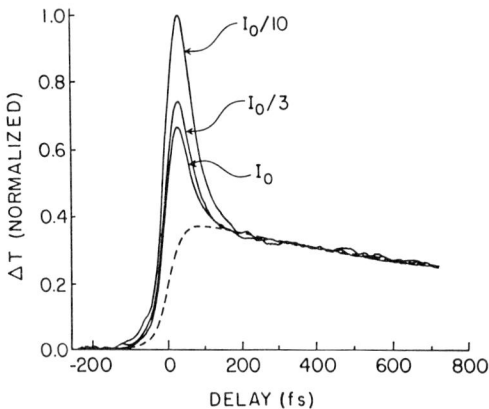

Fig. 1. Femtosecond absorption saturation measurements with excited carrier densities of $10^{18}$, $3 \times 10^{17}$, and $10^{17}$ cm$^{-3}$.

Figure 1 shows experimental results obtained for three different carrier densities ranging from $10^{17}$ to $10^{18}$ cm$^{-3}$. The pump and probe beams were orthogonally polarized. Careful measurements with different polarizations confirmed that this configuration minimized the contribution from coherent artifacts. The data shown indicate an initial rapid recovery of the absorption, followed by a slower recovery with a ~1.5 ps time constant. The initial process arises from a relaxation mechanism which occurs on a time scale comparable to or shorter than the 35 fs pulse duration. The varia-

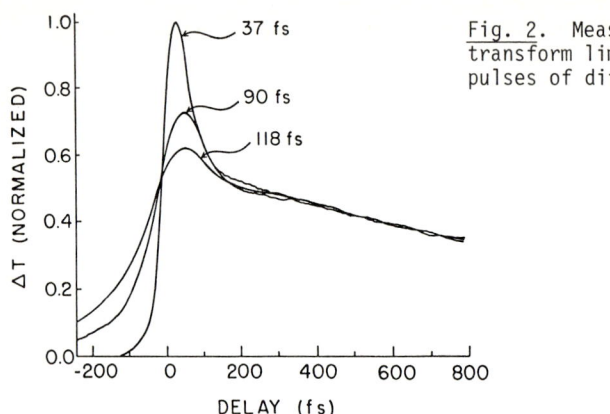

Fig. 2. Measurement with nearly transform limited (zero chirp) pulses of different durations.

tion in amplitude of the initial rapid peak indicates that the relaxation time decreases with increasing carrier density. A more accurate fitting of the data was performed by first subtracting the effects of the slow relaxation component as modelled by the convolution of the pulse intensity autocorrelation with a 1.5 ps exponential response time. The remaining fast component was then fit by time constants of 13 fs, 17 fs, and 30 fs at carrier densities of $10^{18}$, $3 \times 10^{17}$, and $10^{17}$ cm$^{-3}$ respectively.

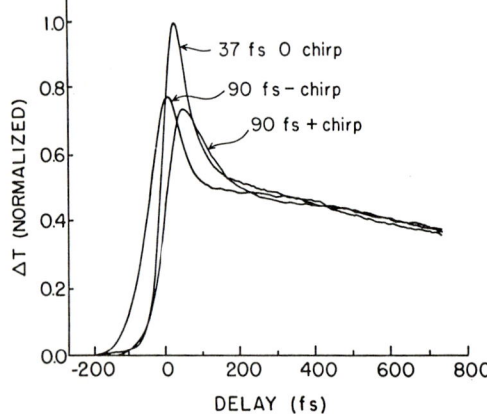

Fig. 3. Measurement with pulses of identical bandwidth by varying chirp.

The validity of the use of peak height as an indicator of relaxation time was confirmed by measurements using different pulse durations as shown in Fig. 2. The decrease in amplitude of the initial rapid transient is consistent with an initial rapid response time which is shorter than the pulse duration. Special care was taken to use pulses of zero chirp. Experiments performed for different chirped pulses with identical spectral bandwidths are shown in Fig. 3. The pulse chirp is shown to be an important parameter in influencing the measured response function. The use of negative chirp (frequency decreasing with time) results in a narrow feature which is shifted to earlier times; in contrast, positive chirp yields a broad feature shifted to later times. This behavior can be explained by the presence of a rapid carrier relaxation process which causes excited carriers to relax in energy and hence induces absorption saturation at lower energies at later times.

For GaAs, the observed optical response is dominated by the electrons because of the large effective mass of the holes. These electrons are initially excited at energies 0.5 eV in the band, high enough to allow scattering to the L valleys. The carrier density dependence of the initial rapid relaxation process suggests that it is dominated by carrier-carrier scattering. This mechanism in conjunction with carrier-phonon scattering causes the establishment of a quasi-equilibrium within the carrier distribution. The longer time scale of 1.5 ps is commensurate with LO phonon emission which produces a cooling of the excited carrier distribution. It is important to note that absorption saturation measures the scattering time out of the initially excited states rather than the relaxation time of the distribution.

In order to further investigate the dynamics of these carrier relaxation mechanisms, we have performed similar measurements on samples of $Al_xGa_{1-x}As$ for mole fractions of x = 0.1, 0.2, 0.3, and 0.4. With increasing concentration of Al, the bandgap at the Γ and L points increase. However, for photon energies of 2 eV, L valley scattering is always allowed.

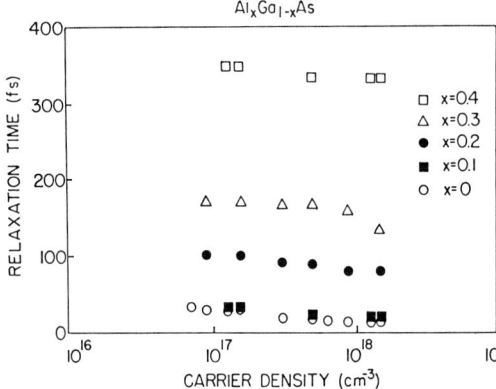

Fig. 4. Initial relaxation times as a function of excitation density in $Al_xGa_{1-x}As$.

Figure 4 shows the experimentally measured initial relaxation times observed as a function of carrier density and x. Measurements indicate that this mechanism slows down significantly as the bandgap is increased. The decrease in carrier scattering rate may be due to a number of factors including the decrease in the density of allowable scattering states as carrier energies approach the band edge, the decreasing effective temperature of the distribution, the increase in screening effects, and possible changes in intervalley scattering rates. Measurements performed closer to the band edge indicate the presence of band filling, however, the time constant of the slow carrier cooling process remains relatively constant.

In summary, we have shown that variable pulse durations and chirp on a femtosecond time scale can be a valuable technique for the study of ultrafast processes. Investigations of GaAs indicate the presence of a rapid carrier density dependent initial relaxation process occurring on a time scale of 10-35 fs as well as a slow 1.5 ps carrier cooling process. Studies in $Al_xGa_{1-x}As$ indicate that the time scale of this initial process slows significantly for increasing x. The dynamics of excited carrier relaxation processes are relevant to understanding the nonequilibrium behavior of excited carriers in optical and electronic materials and devices.

This work was supported in part at MIT by AFOSR Grant 85-0213 and by a Grant from the ITT Corporation. We thank S. Brorson for helpful discussion and technical assistance. WZL is visiting from the Zhongshan University, Guangzhou, People's Republic of China.

REFERENCES

1. C. V. Shank, R. L. Fork, R. F. Leheny, J. Shah: Phys. Rev. Lett. 42, 112 (1979)
2. A. J. Taylor, D. J. Erskin, C. L. Tang: J. Opt. Soc. Am. B2, 663 (1885)
3. J. L. Oudar, D. Hulin, A. Migus, A. Antonetti: Phys. Rev. Lett. 55, 2074 (1985)
4. J. A. Valdmanis, R. L. Fork, J. P. Gordon: Opt. Lett. 10, 131 (1985)
5. R. L. Fork, O. E. Martinez, J. P. Gordon: Opt. Lett. 9, 150 (1984).

# Subpicosecond Optical Non-linearities in GaAs Multiple-Quantum-Well Structures

D. Hulin[1], A. Antonetti[1], A. Migus[1], A. Mysyrowicz[1], H.M. Gibbs[2], N. Peyghambarian[2], W.T. Masselink[3], and H. Morkoç[3]

[1]Laboratoire d'Optique Appliquée, ENSTA, Ecole Polytechnique, F-91120 Palaiseau, France
[2]Optical Sciences Center, University of Arizona, Tucson, AZ 85721, USA
[3]Coordinated Science Laboratory, University of Illinois at Urbana-Champaign, Urbana, IL 61801, USA

We have investigated the dynamics of optical non-linearities observed in multiple-quantum-well structures (MQWS) of GaAs on a subpicosecond time scale. In this paper, we discuss the results obtained at low temperatures and present as an application the demonstration of an ultrafast all-optical logic gate with subpicosecond switch-on and -off times, operating both at room and low temperature.

The principle of the experiments makes use of the pump and probe technique. The transmission of a broad-band probe pulse of duration typically 150 fsec is recorded at different delays after the irradiation of the sample with a pump pulse of adjustable wavelength and of similar duration [1]. In this way, a picture of the evolution of the absorption spectrum of the sample is obtained with subpicosecond accuracy.

If the pump central wavelength is selected to correspond to a subband-to-subband transition inside the GaAs wells, a fast bleaching of the exciton resonance is observed. This effect occurs in less than 150 fsec, which corresponds to the time resolution of the experiment. A study of the exciton bleaching in function of pump-generated pair density shows that the exciton bleaching is gradual but is not accompanied by any measurable shift (see fig. 1).

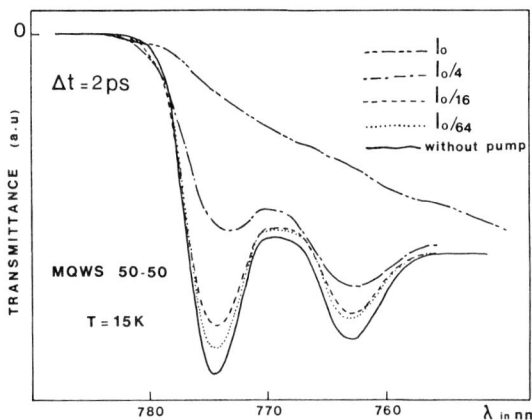

FIGURE 1. Absorption spectra of a GaAs-AlGaAs MQWS at 15K in presence of free electrons and holes (recorded 2 psec after band to band excitation) for different pump intensities.

This constancy of the exciton absolute energy in the presence of free carriers is similar to what is observed in bulk crystals. It is explained by the quasi-perfect compensation between two different effects of the carriers acting upon the excitons : a screening of the electron-hole Coulomb interaction in the exciton (leading to a blue shift since the exciton binding energy is decreased), and a reduction of the forbidden gap energy in presence of excited free carries (band gap renormalization) resulting in a red shift of the exciton position.

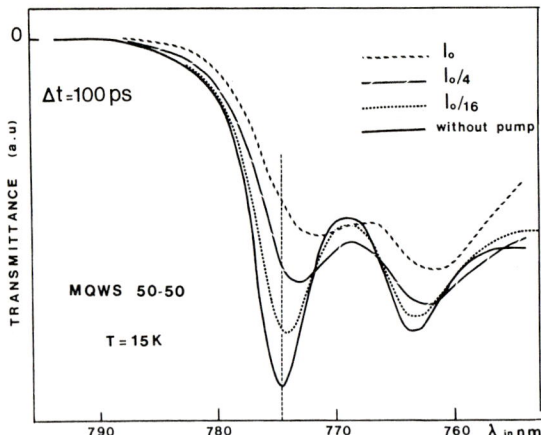

FIGURE 2. Absorption spectra of a GaAs-AlGaAs MQWS at 15K in presence of excitons (recorded 100 psec after band to band excitation) for different pump intensities.

If the pump laser wavelength is now adjusted inside the lowest exciton absorption line, a different behaviour is observed. There also, at sufficiently high densities of photogenerated (bound) electron-hole pairs, a reduction of exciton oscillator strength is occurring. However, simultaneously a high energy shift of the exciton resonance is observed [2], but only in narrow wells [3]. A similar behaviour is observed at long delay times ($\gtrsim$ 100 psec) after non-resonant excitation (see figure 2). In this case, the initially generated free carriers pair together to form a high density gas of free excitons. This blue shift is clearly related to the density of excitons and to the well size, being apparent only in wells with thickness less than 100 Å. This exciton shift in the presence of other excitons is explained by the fact that in two-dimensional systems at high densities, the attractive part of the potential between the bound particles is no more compensated by the repulsive part due to the Pauli exclusion principle acting upon the individual electrons and holes forming the excitons. This imbalance is manifest because the long-ranged, Van der Waals like attraction between excitons is strongly reduced in two-dimensional (2-d) systems. It has been shown by Schmitt-Rink et al.[4] that in a 2-d system, the magnitude of the blue shift for a given resonance should be directly proportional to the amount of exciton bleaching, since both processes have their origin in the Pauli exclusion principle. This relation is verified experimentally as shown in figure 3 in the case of a 40 Å GaAs well, 100 Å AlGaAs barrier structure, both for the heavy-hole and light-hole exciton.

In a third series of experiments we have explored the time-resolved changes of absorption of GaAs MQWS with the pump laser tuned below the lowest exciton resonance, in the transparency region of the medium. Typical results of

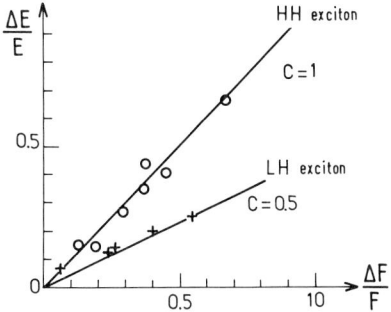

FIGURE 3. Fractional change of exciton binding energy of heavy hole (HH) and light hole (LH) as a function of relative decrease of exciton oscillator strength. The well is 40A thick.

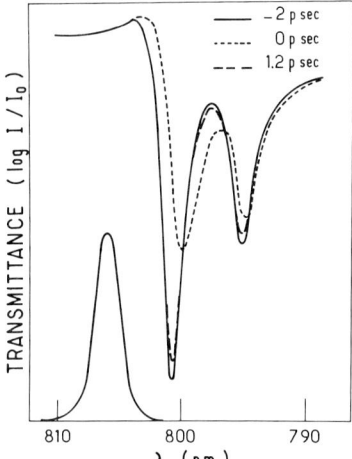

FIGURE 4. Subpicosecond time-resolved absorption spectrum of a GaAs-AlGaAs MQWS with well and barrier thickness of 100A, recorded at different delays from a non-resonant pump pulse at 807 nm.

the induced non-linear response in the exciton region are shown in figure 4, recorded at the time of maximum of pump intensity and 1 ps later. A large blue shift is visible together with partial bleaching, both for the heavy hole and light hole exciton in presence of the pump electromagnetic field. The recovery of the absorption spectrum is almost complete after the end of the excitation, indicating that no real particle (exciton or charged carrier) has been created during the process. In contrast to the resonant case described above, the exciton blue shift observed here is not sensitive to the well dimension, and can be observed even in bulk GaAs. The shift magnitude is larger for smaller detuning of the pump frequency with respect to the excitonic resonance and for larger pump intensities [5]. We attribute this blue shift to a pump-field-induced optical Stark effect. This effect is well known in atomic spectroscopy but has not been observed so far in condensed matter, except recently in a particular case [6].

Non-linearities based on the optical Stark effect are very interesting from a point of view of applications, since they occur only during the presence of the pump field. They are therefore well adapted to the operation of devices with ultrafast response times. As an application of this effect, we have demonstrated the operation of an all-optical logic gate (NOR, OR) with subpicosecond switch-on and switch-off times [7]. The device consists of a GaAs superlattice sample (well size = 100 Å ; barrier size 25 Å) between two 90 % reflecting gold mirrors. The

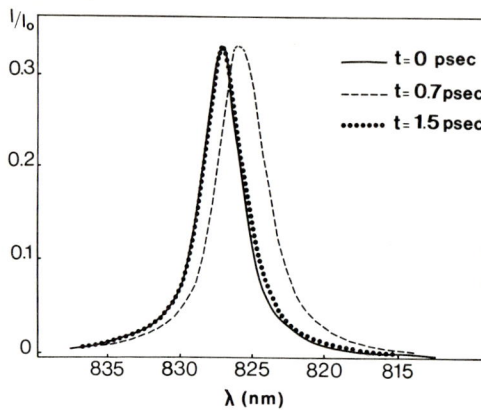

FIGURE 5. Time-resolved transmission spectra of the Fabry-Perot etalon for a subpicosecond excitation at 827 nm. The pump spectrum is broader than the FP mode. Zero delay is chosen just before the onset of the signal. This experiment has been performed at 150K, but the same results have been obtained at room temperature (with less efficiency due to a larger detuning).

pump field induced changes of exciton lines lead to a change of refractive index at longer wavelengths, which in turns induces a displacement of the modes of the non-linear etalon. This performs the logic function on a probe beam set at the peak of an perturbed mode. In view of the small thickness of the etalon (1.3 µm) and taking into account an instantaneous non-linearity the intrinsic response time is limited by the build-up time of the fields in the cavity, of the order of 250 fsec. Results shown in figure 5 are in agreement with this estimate.

In summary, we have used time-resolved detection techniques to study the dynamics of optical non-linearities in GaAs multiple-quantum-well structures. A high energy shift of the exciton resonance occurs in two different experimental situations. The first one takes place in presence of a large exciton population and is due to the two-dimensional character of very thin wells. The second is observed during the presence in the sample of a high electromagnetic field tuned in the non-absorbing region of the optical spectrum. This virtual excitation implies no real creation of excited particles. We have taken advantage of this last feature to demonstrate the operation of an optical logic gate with subpicosecond on and off switching times.

REFERENCES

1- A. Migus, A. Antonetti, J. Etchepare, D. Hulin, A. Orszag, J. Opt. Soc. Am. 2, 584 (1985)
2- N. Peyghambarian, H.M. Gibbs, J.L. Jewell, A. Antonetti, A. Migus, D. Hulin, A. Mysyrowicz, Phys. Rev. Lett. 53, 2433 (1984)
3- D. Hulin, A. Mysyrowicz, A. Antonetti, A. Migus, W.T. Masselink, H. Morkoç, H.M. Gibbs, N. Peyghambarian, Phys. Rev. B 33, 4389 (1986)
4- S. Schmitt-Rink, D.S. Chemla, D.A.B. Miller, Phys. Rev. B 32, 6601 (1985)
5- A. Mysyrowicz, D. Hulin, A. Antonetti, A. Migus, W.T. Masselink, H. Morkoç, Phys. Rev. Lett. 56, 2748 (1986)
6- D. Fröhlich, A. Nöhte, K. Reimann, Phys. Rev. Lett. 55, 1335 (1985)
7- D. Hulin, A. Mysyrowicz, A. Antonetti, A. Migus, W.T. Masselink, H. Morkoç, H.M. Gibbs, N. Peyghambarian, submitted to Appl. Phys. Lett. .

# Picosecond Relaxation of Nonthermal Wannier Excitons in GaAs

*L. Schultheis*[1], *J. Kuhl*[1], *A. Honold*[1], *and C.W. Tu*[2]

[1]Max-Planck-Institut für Festkörperforschung,
D-7000 Stuttgart 80, Fed. Rep. of Germany
[2]AT & T Bell Laoratories, Murray Hill, NJ 07974, USA


The energy and momentum relaxation of optically excited free electron-hole pairs occurs via carrier-carrier scattering within a few hundred femtoseconds /1/. In contrast, little is known about the fast (momentum) relaxation processes of the excitonic groundstate. This lack of knowledge is partly due to the polariton character of excitons in a bulk semiconductor which impedes optical dephasing experiments in the time as well as in the frequency domain. A further problem especially for nonlinear experiments is the high absorption coefficient ($10^5 cm^{-1}$) at the excitonic resonance leading to reabsorption of the nonlinear signal and necessitating such a high excitation level that exciton-exciton interaction hinders the interpretation of the experimental data /2/.

We overcome these difficulties by using an optically thin layer of GaAs ($L_z$=190nm). The excitonic motion perpendicular to the layer is quantized and polariton transport effects are negligible. Thus, time-resolved Degenerate Four-Wave-Mixing (DFWM) can be applied to study the ultrafast relaxation of excitons. For our experiments we used two synchronously pumped dye lasers with Styryl 9. The autocorrelation width was about 3.7ps, the width of the power spectrum 0.9meV, the jitter between the two dye laser pulses less than 2ps.

The phase relaxation of nonthermal excitons, coherently excited by a short optical pulse and subjected to collisions with noncoherent excitons is explored by means of a pump and probe experiment employing time-resolved DFWM: Two weak probe pulses with identical frequencies, tuned into the excitonic resonance, are self-diffracted off an orientational grating, thus probing the excitonic phase coherence. A much stronger pump pulse advanced by 20ps creates additional excitons which are already dephased when the DFWM of the coherent exciton ensemble starts. Figure 1 shows the self-diffracted intensity versus the delay $\tau_{12}$ between the two probe DFWM pulses for three exciton densities at 2K. Without an additional background exciton density (pump pulse blocked) the self-diffraction of the two weak probe pulses directly reveals the intrinsic excitonic phase coherence time $T_2$. The fit to the experimental diffraction curve obtained by solving the optical Bloch equation in the small signal limit and by using Gaussian pulse shapes matched to the autocorrelation trace yields $T_2=(7\pm0.5)ps$.

This surprisingly short excitonic phase coherence time is attributed to scattering with acoustic phonons and to scattering

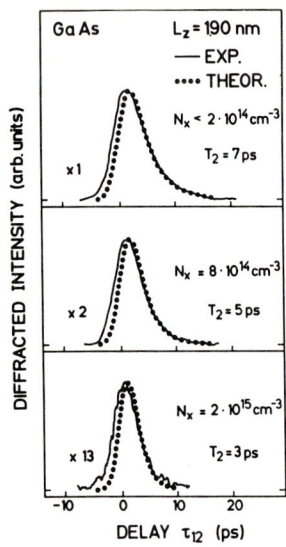

Fig.1: Experimental and theoretical diffraction curves for various background exciton densities $N_X$

with residual impurities. This is confirmed by an observed linear dependence of $T_2$ on the temperature.

Also shown in Fig.1 are the diffraction curves for background exciton densities of $8*10^{14}cm^{-3}$ and $2*10^{15}cm^{-3}$. Fits to the experimental curves yield phase coherence times of $(5\pm0.5)$ps and $(3\pm0.5)$ps. This result demonstrates that exciton-exciton collisions destroy the excitonic phase coherence even at relatively low densities indicating efficient exciton-exciton interaction. These results are consistent with the corresponding density dependence of the absorption linewidth. Time-resolved studies of the transmission spectra confirm the correspondence of line broadening and reduced phase coherence time as a function of the exciton density.

In conlusion we have explored the phase coherence to study the ultrafast dynamics of nonthermal excitons in a thin GaAs slab. Time-resolved DFWM with an additional pump pulse for the generation of incoherent excitons is used to study the dependence of the excitonic phase coherence on the exciton density. For low exciton densities the phase coherence is limited by acoustic phonon scattering and residual impurity scattering whereas at high exciton densities a further loss of phase coherence due to mutual exciton-exciton collisions is observed.

REFERENCES

1. J.L. Oudar, D. Hulin, A. Migus, and F. Alexandre, Phys.Rev.Lett.55, 2074 (1985)
2. Y. Masumoto, S. Shionoya, and T. Takagahara, Phys.Rev.Lett.51, 923 (1985)

# Picosecond Observation of the Photorefractive Effect in GaAs

*A.L. Smirl*[1], *G.C. Valley*[1], *M.B. Klein*[1], *K. Bohnert*[2], *and T.F. Boggess*[2]

[1] Hughes Research Laboratories, 3011 Malibu Canyon Road, Malibu, CA 90265, USA
[2] Center for Applied Quantum Electronics, North Texas State University, Denton, TX 76203, USA

For more than 15 years, photorefractive materials such as $LiNbO_3$, $BaTiO_3$ and $Bi_{12}SiO_{20}$ have been widely investigated for applications in holographic storage, optical data processing, and phase conjugation [1]. Previously, however, transient studies in these materials have been limited to time scales of ns or longer and to investigations of the photorefractive effect where the space charge field is produced between one-photon-ionized donors or acceptors and charged traps [2-4]. Although picosecond pulses and trains of pulses have been used to provide the large fields necessary for two-photon ionization of donors in $LiNbO_3$ [5,6] and potassium tantalate niobate (KTN) [7], no subnanosecond time resolution of the photorefractive processes was reported. More recently, the photorefractive effect has been demonstrated in the semiconductors GaAs, InP, and CdTe on time scales from 250 $\mu$s to steady state [8-11]. Here, we report (to the best of our knowledge) the first investigations of the photorefractive effect on picosecond time scales.

In our experiments, we measured the energy transferred between two 1.06 $\mu$m pulses that were spatially and temporally coincident in a GaAs sample. The laser source for our experiments was an actively and passively mode-locked Nd:YAG laser operating in the fundamental transverse mode. A single pulse was switched from the train of pulses and amplified. The pulse was then divided into two parts and recombined at a small angle in the sample. The optical pulse width was measured to be 43 ps (full width at $e^{-1}$ of the intensity). We wish to emphasize that the crystal studied here is the same one that was used by KLEIN [9] in his cw experiments. The crystal cross section was 6 mm x 5 mm, and the thickness was 4 mm. The crystal orientation was identical to that used by KLEIN [9]. That is, the grating wave vector (sample surface) was along the (100) direction; both pulses were polarized along (110); and the surface normal was parallel to ($\bar{1}$10). The beam diameters at the sample surface were 5 mm (full width $e^{-1}$ of the maximum intensity), as determined by pinhole scans.

We measured the beam coupling as a function of the ratio of the energies in the two pulses, the total fluence, time delay between the two pulses, and crystal orientation. We found the direction of energy transfer between two equal beams to depend definitively on crystal orientation, an unambiguous signature of the photorefractive effect. Moreover, we measured photorefractive gains for weak picosecond signal pulses of a few percent using pump fluences of less than 0.2 mJ/cm$^2$ (i.e., 2pJ/$\mu$m$^2$). At these low fluences and times too short for significant carrier-trap recombination to occur, the space charge field is established between ionized donors and mobile free carriers, in contrast to the cw photorefractive effect where the space charge field is between immobile ionized donors and immobile traps. As a consequence, for picosecond pulses, the space charge field in this regime is limited by the number of available donors; whereas, for cw radiation the limiting field is determined by the trap density. The former

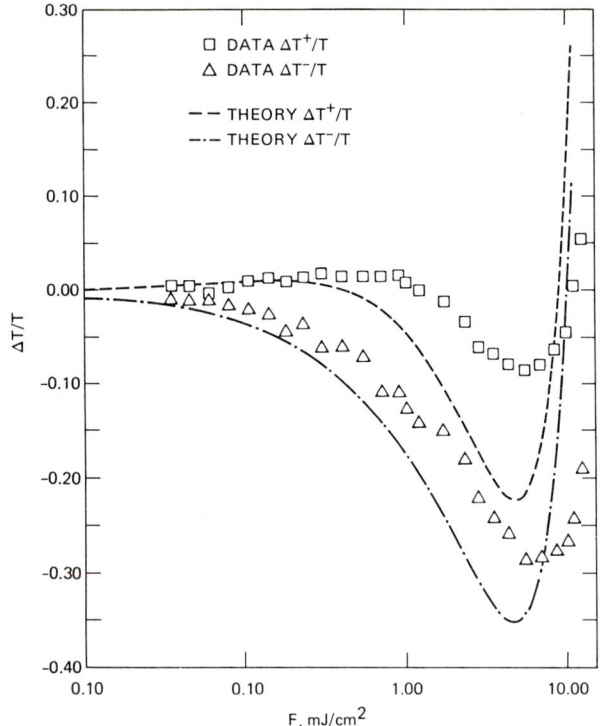

Fig. 1. The normalized change in the probe transmission for the sample oriented for photorefractive energy transfer from strong to weak beam, $\Delta T^+/T$ (squares), and the normalized change in probe transmission for the sample oriented for photorefractive energy transfer from weak to strong, $\Delta T^-/T$ (triangles), as a function of the pump fluence for a grating spacing of 1.7 μm and a constant pump-to-probe ratio of 20:1. The dashed and dashed-dot curves are the results of theoretical calculations for the two orientations, respectively.

is believed to be an order of magnitude larger than the latter. For higher fluences and these pulsewidths, investigations of the photorefractive gratings are complicated by the onset of two-photon absorption and the accompanying absorptive and refractive index changes. In fact, at the highest fluences, the photorefractive energy transfer is dominated by a space-charge field that originates from the separation between mobile free electrons and mobile free holes generated by two-photon absorption.

An example of such measurements is shown in Fig. 1. Here, the normalized change in probe transmission $\Delta T/T$ is shown as a function of pump fluence for a pump-to-probe ratio of 20:1. The quantity $\Delta T/T$ is defined as the probe energy transmission with the pump present minus the probe transmission without the pump, both divided by the probe energy transmission without the pump:

$$\Delta T/T = [T \text{ (with pump)} - T \text{ (without pump)}]/T \text{ (without pump)}. \qquad (1)$$

The squares ($\Delta T^+/T$) represent data acquired with the crystal oriented such that photorefractive energy transfer is from the strong to the weak beam; the triangles ($\Delta T^-/T$) represent data taken after the crystal was rotated 180° about the surface normal that bisects the angle between the two beams. We have performed theoretical calculations based on a simple set of materials equations combined with Maxwell's equations. The materials equations used here differ from those used previously [12,13,14] in that they were modified to include two-photon absorption and neglect recombination. The results of such a numerical calculation for the two crystal orientations are shown by the dashed and dot-dashed curves in Fig. 1. We wish to emphasize that these curves are the result of a zero-fit-parameter calculation. That is, all material parameters were either taken from the literature or independently measured for this sample. Quantitative agreement between experiment and theory can be easily achieved by varying these material parameters within reasonable uncertainties, but without additional independent measurements there is no justification for such a forced fit. In any event, qualitative agreement between calculated and measured data is quite satisfactory.

The qualitative interpretation of the data of Fig. 1 that emerges from such studies is as follows. For the top curve, photorefractive gain initially competes with loss from two photon absorption. For the lowest fluences, photorefractive transfer is larger than the two-photon loss, and a net gain is observed. Eventually, as the fluence is increased, all of the EL2 donors are ionized, and two-photon absorption begins to dominate, with a zero crossing near 1 mJ/cm$^2$. Finally, however, we note that two-photon absorption is accompanied by the generation of electron-hole pairs. That is, a free-carrier grating is produced. Because the real part of the Drude index of refraction is much larger than the imaginary part, an index grating is created that is in phase with the modulated intensity profile. It is well known that such a grating cannot transfer energy in steady state; however, in the transient regime, there is a higher-order transfer that is always from strong to weak beam [15]. This transient transfer was found to dominate at the higher fluences, resulting in probe gain for fluences larger than approximately 10 mJ/cm$^2$, as shown. For the lower curve, the photorefractive energy transfer is from probe to pump, and photorefractive effects as well as two photon absorption contribute to increasing probe loss with increasing fluence. Again, however, at the highest fluences, transient energy transfer begins to dominate and to reverse the sign of the slope.

Finally, we can isolate and separate the photorefractive contributions from those of two-photon absorption and free-carrier transient energy transfer by recognizing that (in the small signal limit) only photorefrac-

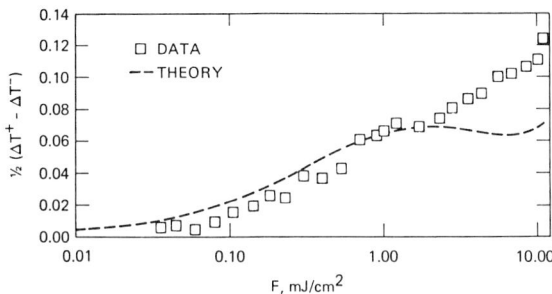

Fig. 2. Half the difference between the normalized change in the probe transmission for the sample oriented for photorefractive probe gain ($\Delta T^+/T$) and the change in the probe transmission for the sample oriented for photorefractive probe loss ($\Delta T^-/T$) as a function of pump fluence for a grating spacing of 1.7 μm and a constant pump-to-probe ratio of 20:1 (i.e., half the difference between the squares and the triangles in Fig. 1). The dashed curve is one half the corresponding difference between the two theoretical curves of Fig. 1.

tive energy transfer explicitly depends on crystal orientation. When the crystal is oriented for weak to strong beam transfer, the photorefractive grating contributes to probe loss; when the crystal is rotated 180° about the surface normal, the photorefractive energy transfer is reversed and is from strong to weak, while the other two contributions maintain their sign. Consequently, twice the photorefractive energy transfer is obtained by subtracting the two curves shown in Fig. 1. The photorefractive energy transfer obtained in this way is shown in Fig. 2. We emphasize that there are two distinct contributions to the photorefractive transfer. At low fluences, the space charge field is primarily between one-photon-generated mobile free electrons and immobile ionized $EL2^+$ donors. At higher fluences, the space charge field (i.e. Dember field) between two-photon-generated free electrons and free holes also contributes.

## References

1. P. Gunter, Phys. Repts. **93**, 200 (1983).
2. C.-T. Chen, D.M. Kim, and D.von der Linde, Appl. Phys. Lett. **34**, 321 (1979).
3. L.K. Lam, T.Y. Chang, J.Feinberg, and R.W. Hellwarth, Opt. Lett. **6**, (1981).
4. J.P. Hermann, J.P. Herriau, and J.P. Huignard, Appl. Opt. **20**, 2173 (1981).
5. D. Von Der Linde, A.M. Glass, and K.F. Rodgers, Appl. Phys. Lett. **25**, 155 (1974).
6. D. von der Linde, O.F. Schirmer, and H. Kurz, Appl. Phys. **15**, 167 (1978).
7. D. von der Linde, A.M. Glass, and K.F. Rodgers, Appl. Phys. Lett. **26**, 22 (1975).
8. A.M. Glass, A.M. Johnson, D.H. Olson, W. Simpson and A.A. Ballman, Appl. Phys. Lett. **44**, 948 (1984).
9. M.B. Klein, Opt. Lett. **9**, 350 (1984).
10. J. Strait and A.M. Glass, J. Opt. Soc. Am.B **3**, 342 (1986); A.M. Glass, M.B. Klein, and G.C. Valley, Electron. Lettr. **211**, 220 (1985).
11. J. Strait and A.M. Glass, Appl. Opt. 255 338 (1986).
12. V.L. Vinetskii and N.V. Kukhtarev, Sov. Phys. Solid State, **166**, 2414 (1975).
13. N.V. Kukhtarev, Pis'ma Zh. Tekh. Fiz. 2, 1114 (1976) [Sov. Tech. Phys. Lett 2, 438 (1976)]; G.C. Valley, IEEE J. Quantum Electron., QE-19, 1637 (1983).
14. G.C. Valley, J. Appl. Phys. 59, 3363 (1986); F.P. Strohkendl, J.M.C. Jonathan, and R.W. Hellwarth, Opt. Lett. **11**, 312 (1986).
15. V.L. Vinetskii, N.V. Kukhtarev, and M.S. Soskin, Sov. J. Quantum Electron. **7**, 230 (1977).

# Time-Resolved Photoluminescence Measurements in $Al_xGa_{1-x}As$ Under Intense Picosecond Excitation

K. Bohnert[1], H. Kalt[1], D.P. Norwood[1], T.F. Boggess[1], A.L. Smirl[2], and R.Y. Loo[2]

[1]Center for Applied Quantum Electronics, Department of Physics, North Texas State University, Denton, TX 76203, USA
[2]Hughes Research Laboratories, 3011 Malibu Canyon Road, Malibu, CA 90265, USA

The ternary semiconductor alloy $Al_xGa_{1-x}As$ exhibits a direct-indirect gap crossover with increasing aluminum concentration x at $x_c=0.435$ [1]. By varying the alloy composition, the relative energy separation between direct and indirect gaps in the vicinity of the crossover composition can be varied, and the distribution of photo-excited electrons among direct and indirect conduction band valleys can be externally controlled. In the following, we report preliminary results that illustrate the influence of the variation of the band structure with composition on the nature and dynamics of radiative carrier recombination following intense picosecond excitation. We further analyzed the photoluminescence with respect to bandgap renormalization which influences the spectral position of the emission bands. For appropriate energy separations between direct and indirect gaps (at compositions near $x_c$), the number of photo-excited electrons in the energetically higher lying conduction band valley(s) can be arranged to be significantly smaller than in the lower valley(s). The electron exchange contribution to the renormalization of the larger gap then is strongly reduced, whereas the lower gap is renormalized by full exchange and correlation effects. This, in principle, allows one to experimentally separate correlation and exchange contributions to band gap renormalization.

The investigated $Al_xGa_{1-x}As$ layers were grown on GaAs substrates by liquid phase or molecular beam epitaxy. Their thicknesses ranged between 2 and 10 $\mu$m. The samples were mounted in a closed-cycle refrigerator for temperature-dependent measurements. The excitation source was a frequency doubled, actively/passively mode-locked Nd:YAG laser, providing single pulses with a temporal width of 32 ps (FWHM) at 532 nm. The photoluminescence from the center of the 700 $\mu$m (FWHM) excitation spot was dispersed in a 0.25 m spectrometer and time-resolved with a highly sensitive streak camera. The temporal resolution was determined by the width of the selected time window (typically between 30 and 70 ps).

Initially, an alloy composition well within the direct gap regime (x=0.23) was investigated where a significant interference by indirect band extrema was excluded. Plasma luminescence typical for a direct-gap semiconductor was observed. The temporal evolution of the spectrally-resolved emission was measured on a picosecond time scale, and carrier density and temperature were extracted as a function of time. The temporal evolution of the carrier temperature in room temperature $Al_{0.23}Ga_{0.77}As$ is shown in Fig. 1a. The exciting fluence was 2 mJ/cm$^2$. The initial excess energy of the photo-excited electron-hole pairs was 630 meV. The temperature during the pulse maximum (t=0 ps) is around 450 K. This

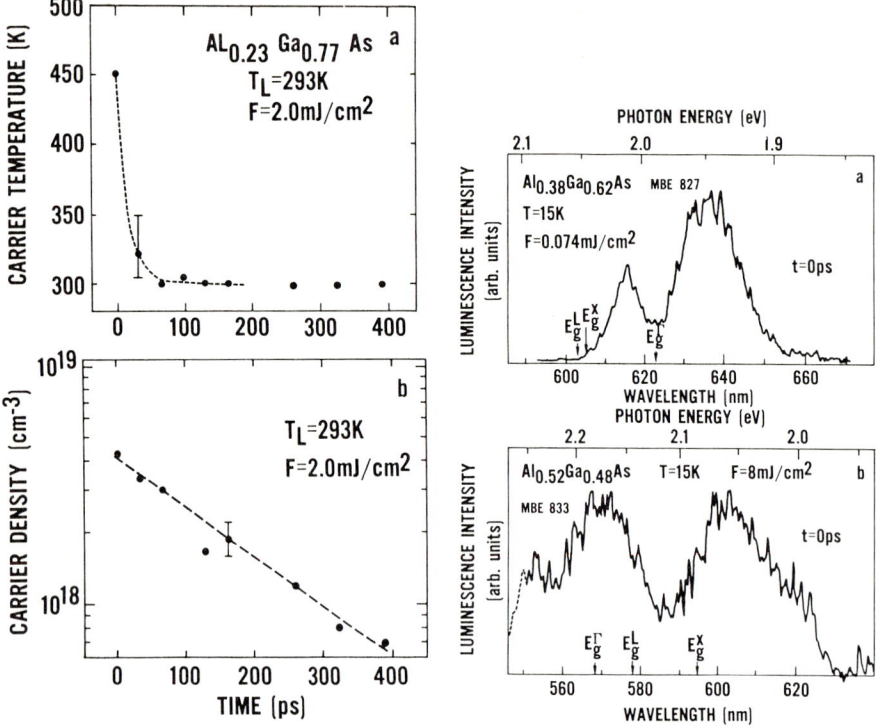

Figure 1. Temperature (a) and density (b) of the photoexcited carriers for room temperature $Al_{0.23}Ga_{0.77}As$ as a function of time

Figure 2. Low temperature photoluminescence spectra at t=0 ps for $Al_{0.38}Ga_{0.62}As$ (a) and $Al_{0.52}Ga_{0.48}As$ (b)

indicates that the carriers lost most of their initial excess energy on a time scale short compared to the pulsewidth. The carrier temperature reaches lattice temperature essentially as soon as excitation ends. On time scales given by our temporal resolution, no definite evidence for a reduced carrier cooling rate due to screening of the carrier-phonon interaction [2] and/or the build-up of a nonequilibrium phonon distribution [3] at high densities of photo-excited carriers was found. The carrier density (Fig. 1b) shows essentially an exponential decay with a time constant of ~200 ps.

Figure 2a shows a low temperature photoluminescence spectrum for $Al_{0.38}Ga_{0.62}As$ at t=0 ps. For this alloy composition the direct conduction band valley is only approximately 60 meV below the indirect X-valleys. The nonrenormalized gaps are indicated by arrows. The low and high energy peaks in this spectrum are attributed to recombination from direct and indirect valleys, respectively. The decay times for the direct and indirect emission bands are 105 ps and 60 ps, respectively. Emission from indirect valleys for alloy compositions near the direct-indirect crossover occurs predominantly without participation of phonons due to an electron scattering by random potential fluctuations [4]. The relative intensity of the indirect emission band indicates a high efficiency for this recombination channel. The shorter decay time for the indirect emission is consistent with electron transitions from the indirect valleys to the lower $\Gamma$-valley which help to adjust an intervalley equilibrium distribution for the photo-excited electrons.

Figure 2b shows a luminescence spectrum at t=0 ps for x=0.52. For this composition, the lowest conduction band edge occurs at the X-point. It is separated from the direct $\Gamma$-minimum by approximately 100 meV. The non-renormalized band gaps are indicated by arrows. Again two distinct emission bands are observed. The direct emission (high energy band) is observed only during excitation, i.e., during the relaxation of the photo-excited electrons to the bottom of the bands. The decay time for the indirect emission now is increased to 640 ps. This is consistent with a decreasing probability for no-phonon transitions with increasing direct-indirect gap separation in the indirect gap regime [5].

An examination of band gap renormalization shows an enhanced renormalization of the $\Gamma$-gap for compositions below but close to $x_c$. This is explained by an increase of the effective electron mass due to multivalley effects [6]. The renormalization of gaps above the lowest gap appears to be reduced due to a reduced exchange contribution.

In conclusion, for alloy compositions near the direct-indirect gap crossover, emission from direct and indirect conduction band valleys is observed simultaneously. The relative intensity of the indirect (no-phonon) emission band indicates a high radiative recombination rate due to random potential fluctuations. Bandgap renormalization near $x_c$ is influenced by disorder effects and by the distribution of the photo-excited electrons among several conduction band valleys.

References

1. R. Dingle, R. A. Logan, and J. R. Arthur: In GaAs and Related Compounds, ed. by C. Hilsum (Institute of Physics, London, 1977), p. 210
2. H. J. Zarrabi and R. R. Alfano: Phys. Rev. B 32, 3947 (1985)
3. H. M. van Driel: Phys. Rev. B 19, 5928 (1979)
4. E. Cohen, M. D. Sturge, M. A. Olmstead, and R. A. Logan: Phys. Rev. B 22, 771 (1980)
5. A. W. Pikhtin: Fiz. Tekh. Poluprovodn. 11, 425 (1977) [Sov. Phys. Semicond. 11, 245 (1977)]
6. P. J. Pearah, W. T. Masselink, J. Klem, T. Henderson, H. Morkoc, C. W. Litton, D. C. Reynolds: Phys. Rev. B 32, 3857 (1985)

# Picosecond Excite-Probe and Transient Grating Studies of $Ga_xIn_{1-x}As_yP_{1-y}$

R.J. Manning[1], A. Miller[1], A.M. Fox[1], and J.H. Marsh[2]

[1]Royal Signals and Radar Establishment, Gt Malvern,
 Worcs WR14 3PS, UK
[2]Department of Electronic and Electrical Engineering,
 University of Sheffield, S1 3JD, UK

The bandgap of the quaternary alloy semiconductor $Ga_xIn_{1-x}As_yP_{1-y}$, lattice matched to InP, is adjustable by composition between $0.92 \mu m \leqslant \lambda_g \leqslant 1.65 \mu m$. This property makes it especially attractive for use with fibre optic systems which have minimum loss and dispersion at $1.55 \mu m$. An understanding of the nonlinear optical properties of these materials is important for their assessment in optical logic and signal processing applications. We have previously reported the observation of ultrafast absorption saturation (~ 10 psec) in excite-probe studies of GaInAsP having a 0.86 eV bandgap ($\lambda_g = 1.45 \mu m$) using 5-psec pulses at $1.054 \mu m$ [1]. This paper describes excite-probe studies on a second, 1.03 eV bandgap sample ($\lambda_g = 1.21 \mu m$), and three pulse transient grating measurements on both samples. We observe significantly different dynamical behaviour for the two samples in the excite-probe configuration, whilst in transient grating measurements both samples exhibit decays of a similar magnitude (150-250 psec).

All experiments used single 5-psec, $TEM_{00}$ mode pulses from a Nd: Phosphate glass laser. These were divided into two equal "excite" beams which were focussed to a spot size of $100 \mu m$ (FWHM) on the sample, coincident in space and time, at an angle $\theta = 10°$ to each other. An attenuated "probe" pulse was sent through a variable delay and focussed to sample only the central part of the excited region. This probe beam bisected the angle between the two excite arms, and was orthogonally polarised. For excite-probe studies, one of the excite arms was blocked to form the conventional arrangement. In the transient grating studies, a detector monitored the first-order diffracted probe from the population grating created by the interference of the overlapping excite pulses. The grating spacing (= $\lambda/2 \sin \theta/2$) was $6 \mu m$. Both samples used in these experiments were grown by liquid phase epitaxy, matched on a semi-insulating Fe-doped InP substrate. Absorption edges were consistent with the predicted bandgap energies from X-ray analysis (sample 1, 0.86 eV, x = 0.35, y = 0.78, thickness $2.5 \mu m$; sample 2, x = 0.23, y = 0.49, thickness = $1.4 \mu m$). The low power absorption coefficients were $1.8 \times 10^4 cm^{-1}$ and $1.5 \times 10^4 cm^{-1}$ respectively at $\lambda = 1.054 \mu m$ (photon energy = 1.176 eV). The background carrier concentrations of $10^{16} cm^{-3}$ were well below the densities created in these experiments.

Figure 1a shows the transmission of the probe pulse as a function of delay after excitation for sample 1 as previously reported [1]. This is to be compared with Fig. 1b, the probe transmission as a function of delay for the 1.03 eV sample. The transmission of both samples is bleached at high intensities due to band-filling by the optically created carriers up to and beyond the optically coupled states. For the 0.86 eV sample, the probe transmission rose from 0.5% to 10% for incident excite pulse energy densities between 1 and 15 $mJ/cm^2$. The high speed recovery (~10 psec) is attributed to Auger recombination. For the 1.03 eV sample, the

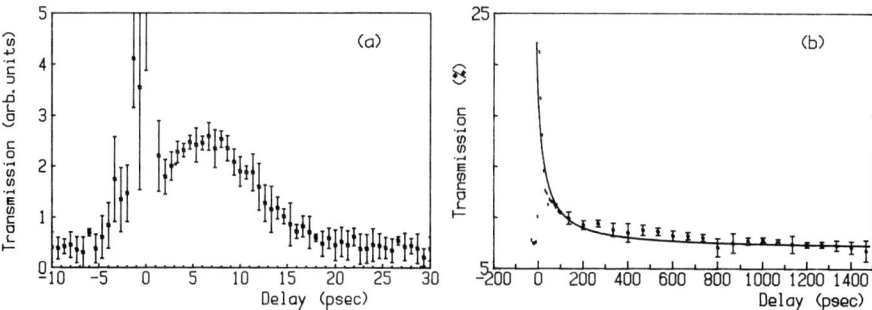

Figure 1. Probe pulse transmission versus delay for GaInAsP
a) 0.86 eV bandgap, ~ 15 mJ/cm$^2$, b) 1.03 eV, ~ 8 mJ/cm$^2$.

transmission rose from ~7% to 25% for incident energy densities between 0.4 and 8 mJ/cm$^2$. This was consistent with the lower bleaching energies expected for the second sample since the optically coupled states lie closer to the band edge and fewer carriers are required for band filling. Intervalence band absorption [1] is indicated by limting of the maximum transmission in both samples to well below 50%.

The observed saturation recovery is much longer for the second sample. High resolution scans near zero delay reveal a fast component (15-30 psec) which we again attribute to Auger recombination, but this is then followed by a much longer recovery, ~ 500 psec (1/e), which we attribute primarily to bimolecular (radiative) recombination. We assume that this bimolecular component is not observed in the first sample because Auger recombination removes carriers from the optically coupled states sufficiently quickly for the sample to become highly absorbing before bimolecular recombination becomes dominant. The lower carrier densities achieved in the second case mean that Auger recombination is less significant and the bimolecular component is observed.

A numerical calculation evaluated the absorption coefficients as a function of carrier density, N, and temperature, $T_c$. In the 1.03 eV sample, saturation should occur at $N \sim 6.5 \times 10^{18}$cm$^{-3}$ ($T_c$ = 300 K). The fit (solid line) shown in Fig 1(b), uses an Auger coefficient, $\gamma$, of $5 \times 10^{-28}$cm$^6$s$^{-1}$, and a bimolecular coefficient, $\beta$, of $1.2 \times 10^{-10}$cm$^3$s$^{-1}$. On the very short timescales encountered for the 0.86 eV sample, thermal equilibrium with the lattice has not been achieved. However, saturation should occur at $N \sim 2 \times 10^{19}$cm$^{-3}$ for $T_c$ = 1200 K implying $\gamma \sim 2 \times 10^{-28}$cm$^6$s$^{-1}$. These values for $\gamma$ are somewhat larger than previously reported [1-3] but agree better with calculations [4].

Transient grating measurements using the two samples are shown in Fig. 2a and 2b, where the first-order diffracted probe signal is plotted as a function of delay after excitation. The 0.86 eV bandgap sample (Fig. 2a) again shows the fast decay observed in excite-probe studies, but now a subsequent much longer "tail" is also observed, of duration ~ 250 psec (1/e). This persistence of the grating after the saturation relaxation of the optically coupled states suggests that the grating is primarily refractive. The 1.03 eV bandgap (Fig. 2b) shows a decay of duration ~ 160 psec (1/e), of similar magnitude to the longer decay component of Fig. 2a. A measurement with higher temporal resolution revealed the fast

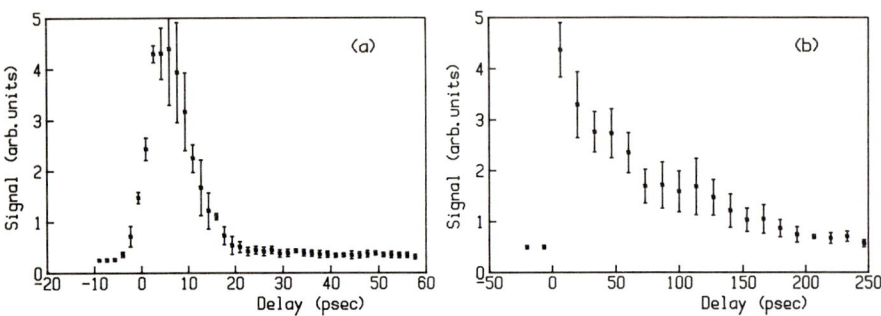

Figure 2. First-order diffracted probe signal versus delay for GaInAsP a) 0.86 eV b) 1.03 eV bandgap.

Auger component at the beginning of the grating decay, similar to the excite-probe measurements shown in Fig. 1b. The slower component of the second sample is shorter than expected from the decay obtained in excite-probe measurements. This is almost certainly due to ambipolar diffusion "washing out" the population grating. Values [5] for the ambipolar diffusion coefficients $D_a (\simeq 2D_h)$ of 0.86 and 1.03 eV GaInAsP imply grating diffusion times $(= \Lambda^2 / 4 \pi^2 D_a)$ of 2.7 nsec and 2.9 nsec respectively. This would imply diffraction efficiency decay times of ~200 psec for both samples (assuming a bimolecular dominated decay of ~500 psec).
In conclusion, we have demonstrated that excite-probe studies of GaInAsP for two different band-gaps give radically different observed recovery times. This has been shown to be due to the density dependent recombination processes occurring in the material, and the different saturation characteristics. Conversely, the slow components in transient grating measurements have similar magnitudes, consistent with bimolecular recombination and diffusion.

References

[1] A Miller, R J Manning, A M Fox, J H Marsh: Electron Lett 20, 601 (1984).
[2] B Sermage, H J Eichler, J P Heritage, R J Nelson, N K Dutta: Appl Phys Lett 42, 259 (1983).
[3] E Wintner, E P Ippen: Appl Phys Lett 44, 999 (1984).
[4] N K Dutta, R J Nelson: J Appl Phys, 53, 74 (1982).
[5] In GaInAsP Alloy Semiconductors, ed by T P Pearsall (Wiley - Interscience, New York, 1982).

Controller, HMSO, London, 1986.

# Ultrafast Dynamics in GaAlAs Diode Laser Amplifiers

*M.S. Stix, M.P. Kesler, and E.P. Ippen*

Research Laboratory of Electronics, Massachusetts Institute of
Technology, Cambridge, MA 02139, USA

In this paper we report the results of our experimental and theoretical
investigations of subpicosecond dynamics in GaAlAs laser diodes. The experiments, utilizing 0.4-0.5 ps fiber-compressed pulses in the near
infrared, indicate the presence of at least one subpicosecond process
occurring in the devices. In this paper we detail our experimental technique and describe the theoretical framework in which we have interpreted
the results[1]. We also discuss how our experimental findings relate to
the macroscopic dynamics of the laser devices by outlining the connection
between our work and the gain compression factor found in the rate equation
description[2].

To acquire subpicosecond resolution for the measurements we have employed
the fiber compression technique[3] to reduce the 8-10 ps pulses from our
synchronously pumped dye laser to 0.4-0.5 ps. Our laser system is composed of a cavity dumped 0X750 dye laser pumped by a Krypton laser, and
has output pulses tunable from 780 nm to 860 nm - wavelengths around
the GaAs bandgap. In addition to a 3-plate birefringent filter set, a
fourth bandwidth-limiting plate was used to facilitate the generation of
transform limited pulses. The compressor, using 35 m of single-mode polarization-preserving fiber, uses a four-pass grating delay line[4] to ensure a circular output beam and protect against unintentional spectral filtering of the pulses when coupling into the laser diode mode.

The pump-probe experiments were performed by coupling the pump and probe
pulses into an AR coated GaAlAs laser diode biased above the uncoated threshold by a constant DC current source. Orthogonal polarizations were used
for the pump and probe beams which excited, respectively, the TE and TM
modes. At the output of the diode a polarizer selects only the probe pulses
for detection.

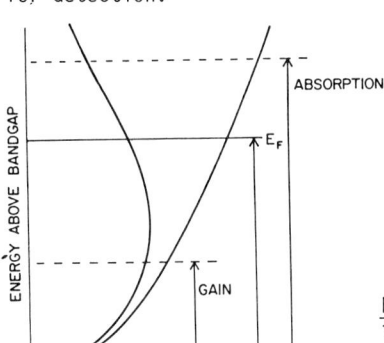

Fig. 1. Schematic of carrier density in conduction band illustrating the three regions in which the experiments were performed.

In GaAlAs biased for lasing, a high density (approximately $2 \times 10^{18}/cm^3$) electron-hole gas exists in the active layer. Figure 1 shows the conduction band and highlights, schematically, the three wavelength regions at which it is particularly interesting to perform pump-probe experiments. One wavelength region is that below the Fermi level where there is gain, one region is that above the Fermi level where there is absorption, and the third region is right at the Fermi level where the medium is transparent (neither saturable gain nor saturable absorption). To avoid confusion concerning the transparency wavelength an aside is worthwhile: at this wavelength the pump pulse will not tend to remove carriers from the system (as happens at gain wavelengths), nor will the pump photons tend to add electrons to the system (as happens in absorption). At this wavelength carriers are neither added to nor subtracted from the conduction band.

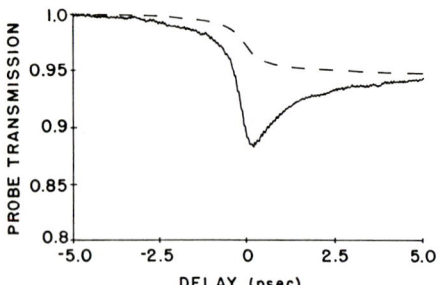
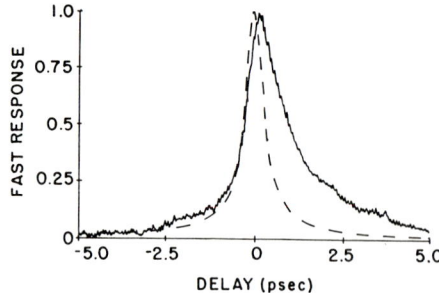

Fig. 2. Pump-probe data for gain region along with the integral of the intensity autocorrelation (dashed line).

Fig. 3. Isolation of time response for the gain data. Dashed line is the intensity autocorrelation.

Figure 2 presents data obtained for a diode laser biased for X5 single pass gain. Two features are noticeable: first, as a result of stimulated emission the gain of the device is partially saturated and the transmission at +5 ps is less than that at -5 ps. Second, a significant overshoot, or fast response, is present at $\tau = 0$. To analyze the data it is helpful to isolate the fast response from the slower component of the signal. These slower components, due to changes in carrier population, recover on much longer timescales (ns), and so appear as step responses in the signal. The isolation simply amounts to subtracting the integral of the intensity autocorrelation (step response) from the data[5]. In Fig. 2 this integral is given by the dashed line, and the desired signal is the difference between the dashed and solid curves. In Fig. 3 the fast component of the response is shown along with the intensity autocorrelation, and it is seen that the experimental traces clearly reveal the presence of a fast relaxation process occurring in the device.

Consider now the data obtained in the absorption wavelength region, Fig. 4. Here the data deviates again from the step response, but the fast component is now opposite in sign to that of the longer time absorption saturation. This fast response is isolated and shown in Fig. 5. Next, examine data obtained at the transparency wavelength. At this wavelength, the pump pulse induces no net change in the carrier density and thus we expect no saturation component to the trace. The data we obtain at this wavelength, Fig. 6, does show quite a large transient near $\tau = 0$, however. Inverting

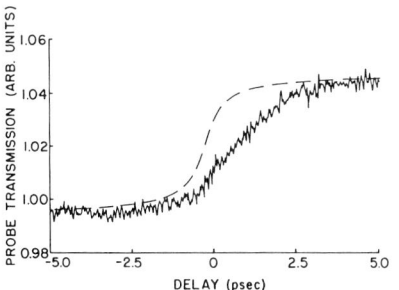

Fig. 4. Probe transmission versus delay in the absorption region. The integral of the intensity autocorrelation is also shown (dashed line).

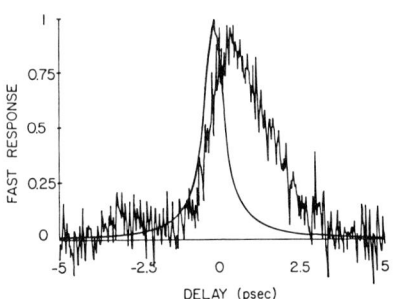

Fig. 5. Fast response for absorption data along with intensity autocorrelation.

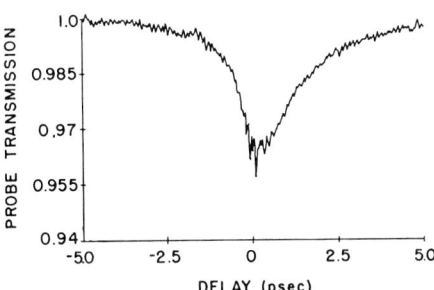

Fig. 6. Pump-probe data for transparency region. The levels of probe transmission at $\tau = -5$ ps and $\tau = +5$ ps are equal since there is no change in carrier density.

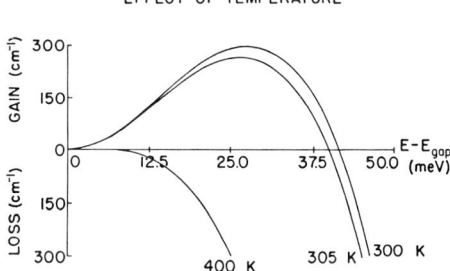

Fig. 7. Calculated gain versus energy for different carrier temperatures. Note that the gain is always reduced by increasing the carrier temperature.

this data (see Fig. 8) gives a signal which is very similar to the other fast responses obtained.

To explain the data we propose a dynamic temperature model in which the effect of the pump pulse is to change the temperature of the carrier distributions in the active layer. In this model, it is assumed that carrier-carrier interactions are occurring sufficiently fast that Fermi-Dirac carrier distributions are maintained, although not at an equilibrium (lattice) temperature. One mechanism for this temperature change is free carrier absorption, which would cause a heating of the distributions. Another mechanism is that of stimulated emission which removes carriers close to the band edge which are "cool" compared to the rest of the carriers in the conduction band (and thus indirectly heats the gas). The signal we see in our data is a result of the carrier temperature relaxing back to the lattice temperature. With this interpretation we are able to explain our data.

Consider first the data observed for the transparency region. At this wavelength there is no net change in carrier population, but since the pump pulse causes free carrier absorption, the temperature of the carriers is raised. This increase in temperature results in a shift of the Fermi level to lower energies, thus our probe wavelength is now above the Fermi level and it experiences absorption. As the carriers cool to the lattice temperature by emitting optical phonons, the Fermi level increases and the probe absorption is decreased. Considering the probe transmission, we expect it to decrease as the pump pulse heats up the carrier distribution, and then to increase as the carriers cool to the lattice temperature. Indeed, this is what is observed experimentally in this situation.

A similiar explanation can be given for the data obtained in the gain and absorption regions, also. The pump pulse causes a partial saturation due to the change in carrier density, but also heats the carrier distributions via free carrier absorption. This heating has the effect of reducing the gain (or increasing the absorption) seen by the probe pulse. Thus the probe samples a dynamically changing gain as the carriers relax to the lattice temperature. The dynamic carrier temperature model gives good agreement between expected results and data.

We would like to point out how sensitive the gain is to very small changes in the carrier temperature. The changes in temperature we are observing in these experiments are very modest, typically 1K. These small changes, however, have surprisingly significant effects on the gain. We show in Fig. 7 a calculation[1] of gain versus photon energy (above bandgap) for a GaAs sample having $n = p = 2 \times 10^{18}/cm^3$ for 3 different carrier temperatures. A 5K rise in temperature (1.5% increase) causes a substantial drop in gain (12.5%). A 100K rise in temperature causes the sample to become highly absorbing. Clearly, then, very small changes in carrier temperature are detectable. Hence, the changes in transmission we observe in our experiments are of the proper order of magnitude one would expect for the typical increases in carrier-temperature brought about by free carrier absorption and stimulated emission.

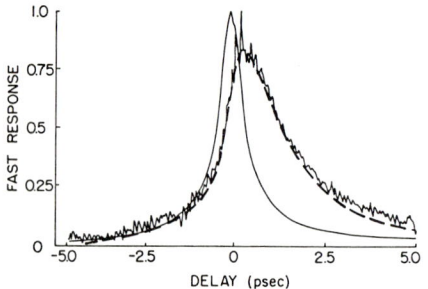

Fig. 8. Theoretical fit for transparency data (dashed line) obtained by convolving a 0.9-ps exponential with the intensity autocorrelation (smooth curve).

The data would therefore tend to strongly support this dynamic temperature model. However, in fitting our data we have found evidence of a much faster response. A temperature relaxation time of 0.9 ps provides an adequate fit to the main portion of all of our data, but the addition of a small component of much faster time (< 0.1 ps) gives a much better fit for the gain and absorption data. No additional response is necessary for the transparency data. The origin of this faster response could be spectral hole burning or it might result from the relaxation of an oriented distribution of carriers. We present in Fig. 8 a fit to the transparency data which is simply a 0.9 ps exponential convolved with the intensity autocorrelation[5].

Finally, we would like to conclude our paper with a discussion of how the effects we have observed in our experiments can be related to the gain compression factor found in the laser diode rate equations. This factor plays an important role in the damping of relaxation oscillations, as well as in the current modulation response, and arises from the small decrease in laser gain with photon density in the cavity (attributed in part to spectral and spatial hole burning). From the considerations put forward so far it is very probable that direct heating of the carriers in the active layer by the lasing photons results in a tendency to raise the carrier temperature. This tendency will, in turn, be offset by the carriers trying to relax back to the lattice temperature with a 0.9-ps characteristic time. A balance will be struck between the heating and cooling processes resulting in a slight steady-state heating of the carrier distribution, and thus a slight decrease in gain. Preliminary calculations show that this contribution to the gain compression factor can be substantial, again resulting from the extreme sensitivity of the gain to carrier temperature.

This work was supported in part by Grant No. ECS-8406290 from the National Science Foundation.

REFERENCES

1. M. S. Stix, M. P. Kesler, E. P. Ippen: Appl. Phys. Lett. $\underline{48}$, 1722 (1986)
2. R. S. Tucker: J. Lightwave Tech. $\underline{\text{LT-3}}$, 1180 (1985)
3. B. Nikolaus, D. Grischkowsky: Appl. Phys. Lett. $\underline{42}$, 1 (1983)
4. A. M. Johnson, R. H. Stolen, W. M. Simpson: Appl. Phys. Lett. $\underline{44}$, 729 (1984)
5. E. P. Ippen, C. V. Shank: in Ultrashort Light Pulses: Picosecond Techniques and Applications, S. L. Shapiro, ed. (Berlin: Springer Verlag, 1977)

# Electronic Energy Relaxation and Localization in Two II-VI Compound Semiconductor Quantum Well Structures

Y. Hefetz, W.C. Goltsos, D. Lee, and A.V. Nurmikko

Division of Engineering and Department of Physics, Brown University, Providence, RI 02912, USA

## 1. Introduction

With emerging new semiconductor superlattice structures it is important to evaluate their electronic characteristics, particularly in the limit of very thin constituent layers (<100 Å). We report here on the application of transient luminescence techniques to new II-VI compound semiconductor quantum well structures with two examples, based on the ZnSe/(Zn,Mn)Se [1] and CdTe/ZnTe heterostructures whose growth by molecular beam epitaxy (MBE) has been pioneered at Purdue University and University of Illinois (Chicago), respectively [2]. In this work we have employed a tunable, high repetition rate picosecond source in the blue and red regions of the spectrum (Fig. 1). An optimum compromise for temporal broadening in the spectroscopic measurements was achieved by using a spectrometer with a pair of gratings in the subtractive mode, joined together with a Hamamatsu synchroscan streak camera. In this way a temporal/spectral resolution of approximately 10 psec/5 Å was achieved. As a consequence, time-resolved spectra from the quantum well structures described below have yielded direct information of key energy relaxation and localization processes of excitons.

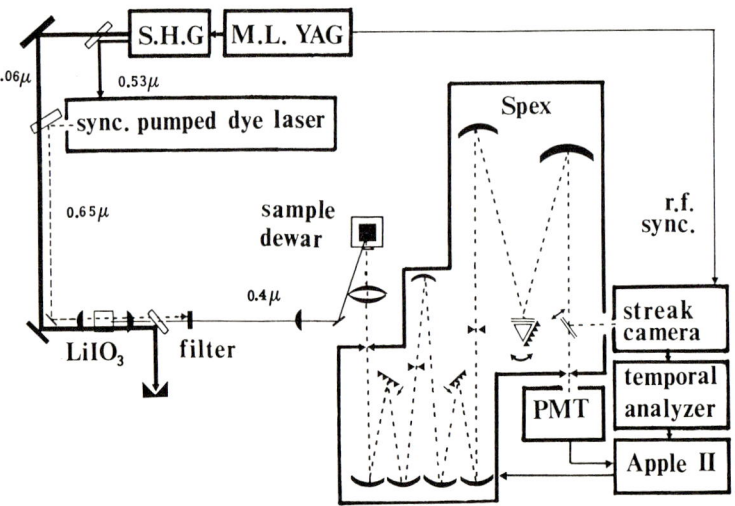

Fig. 1: The experimental arrangement showing the generation of picosecond pulses in the 370-450 nm region by nonlinear mixing in $LiIO_3$. Average power of 0.5 mW in short (<5 ps) pulses at 100 MHz repetition rate was achieved. The spectrometer is a SPEX Triple Spectrometer in which the third grating is replaced by a flat mirror.

The two strained layer quantum well structures discussed here are characterized by small valence band offsets but yet offer strong excitonic effects (typical in II-VI compounds), with Coulomb-correlations which vary according to the degree of conduction electron confinement.

## 2. Carrier Dynamics and Exciton Formation in ZnSe/(ZnMn)Se MQW's

The new ZnSe/(ZnMn)Se multiquantum well structures (MQW) are of particular interest due to their large direct band gap ($E_g \approx 2.8$ eV) and giant Zeeman splitting in the Dilute Magnetic Semiconductor $Zn_{1-x}Mn_x Se$ [3].

An important dynamical question of initially photoexcited barrier layers involves the competition between the two major energy relaxation paths: (i) transfer of photoexcited electron-hole pairs into the ZnSe quantum wells and (ii) conversion of the electronic energy into the internal excitation of the Mn-ion within the barrier, (the latter is the source of strong yellow luminescence in bulk $Zn_{1-x}Mn_x Se$ for $x \geq 0.03$). This is illustrated in Fig. 2 by the rates $1/\tau_d$ and $1/\tau_c$.

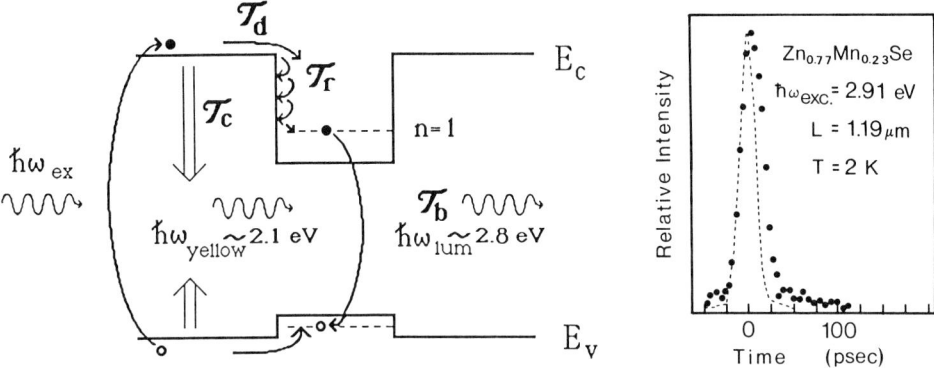

Fig. 2: Illustration of the carrier dynamics in the ZnSe/(ZnMn)Se MQW. Following photo excitation above the barrier, some of the carriers are captured by the well, relax by phonon assisted cooling, and form excitons. Other electron-hole pairs convert their energy into Mn-ion d-electron excitations which emit yellow photons.

Fig. 3: Transient exciton luminescence from a $Zn_{0.77}Mn_{0.23}Se$ single crystal film at T=2K (black dots), excited by picosecond pulses across the bandgap at 2.91 eV. The dashed line indicates system time resolution.

In Fig. 3 we show that the (blue) exciton decay from a single crystal film of (Zn,Mn)Se (x=.23) is strikingly rapid ($\tau_c \approx 15$ps). This occurs as a result of efficient energy transfer from the Bloch-like states to the Mn-ion d-electron state. In a quantum well structure, such a transfer can be bypassed by the even more efficient collection of the excess electron-hole pairs into the ZnSe wells. The efficiency of collection can be inferred by examining the relative intensity of the yellow versus blue emission in time integrated spectra [4]. For example, in a ZnSe/$Zn_{.77}Mn_{.23}$Se MQW sample of well thickness $L_w=67$Å, the ratio of these emissions after spectral integration is $W_{blue}/W_{yellow}=16.8$. In sharp contrast, the exciton emission is dominated by the Mn-ion emission in the single crystal

(Zn,Mn)Se film: $W_{blue}/W_{yellow} = 4.31 \times 10^{-3}$. Direct measurement of the yellow emission rise time was well beyond our experimental sensitivity; this follows from the fact that the corresponding lifetime is very long ($\tau_y > 1 \mu s$) so that the number of photons emitted per picosecond is low (<<1 psec for our conditions).

Additional information is obtained by examining the time-resolved emission by excitons in the ZnSe quantum well [4]. Figure 4 shows the growth and decay of the ground state (light-hole) exciton luminescence in a structure with well thickness of 67 Å and x(Mn)=0.23, excited by picosecond pulses at photon energy well in excess of the (Zn,Mn)Se 'barrier' bandgap. (For reference, the bulk exciton Bohr radius in ZnSe is about 28 Å.) The exponential decay time is $\tau_b \approx 200 ps$. The rise time of $\approx 90 ps$ is a combination of the transfer time $\tau_d$, the optical phonon assisted relaxation $\tau_r$, and the final stage of acoustical phonon assisted thermalization, exciton formation, and localization time $\tau_f$ (excitons at ground state are weakly localized by quantum well width fluctuations).

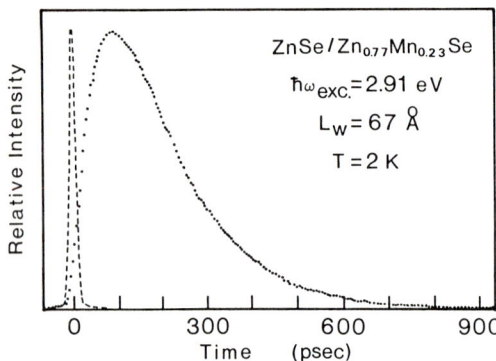

Fig. 4: Transient exciton luminescence from a MQW structure, light-hole ground state energy at 2.795 eV at T=2K, excited by picosecond pulses at 2.91 eV. The dashed line shows the time response of the monochromator/streak camera system.

To further investigate the contributions of these processes, the photon energy of excitation was varied over a wide range. When exciting directly into the well, the rise time of the PL (Fig. 4) did not change until within ~3 meV from the luminescence peak, where the rise time abruptly shortened. This shows that the actual capture of hot photoexcited e-h pairs into the ZnSe wells and the optical phonon emission steps occur on a short timescale (certainly less than 10 psec), and that the finite rise time is mainly due to the last stages of formation of the exciton in the ZnSe layers (cooling of e-h pairs by acoustic phonon emission).

### 3. Strong 2-D Localization Effects in CdTe/ZnTe MQW

The CdTe/ZnTe superlattice is a new II-VI compound semiconductor structure with very large lattice mismatch (>0.06), which nevertheless can be elastically accommodated in sufficiently thin layered samples. Initial X-ray diffraction [2] and photoluminescence [5] data on MBE-grown structures indicate that they are of high structural quality if proper care is taken in their 'strain design' to minimize misfit dislocations. The typical structure is composed of CdTe 'wells' and ZnTe 'barriers' which are approximately 20 to 50 Å thick (the exciton Bohr radius in bulk CdTe is some 60 Å). Calculation of the mismatch-induced stress in the structure shows that it significantly modifies the electronic energies (on the order of 100 meV) [5]. We have specifically considered the additional effects of interfacial roughness under these unusual conditions of high stress and small valence band offset. Our estimates show that irregularities on the order of one monolayer in the well width will result in a 20-40 meV fluctuation in the heavy-hole exciton ground state,

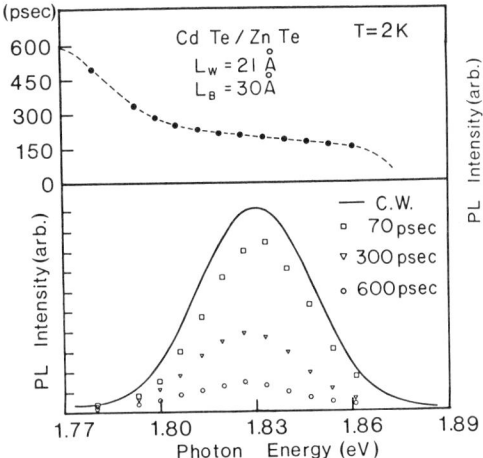
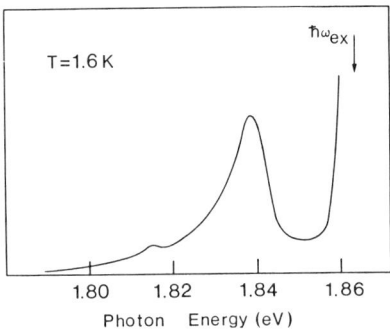

Fig. 5: Continuous wave and time-resolved photoluminescence spectra from a CdTe/ZnTe superlattice (lower panel), when excited above the mobility edge. The upper panel shows the exciton decay time in this range of localized states.

Fig. 6: The appearance of two strong LO-phonon sidebands in luminescence when photoexcitation is directly into the localized exciton states. The observed phonon energies (24 meV) suggest coupling to CdTe/ZnTe strain modified phonons. Transient spectra shows that this is not a Raman line but is likely to be associated with optical phonon intermediated tunneling in the 2-D localized states.

mainly through the high local stress around isolated islands of different local layer thickness. Under these conditions we expect to see strong 2-dimensional localization effects with mean energy of localization larger than kT at low and moderate temperature and also in excess of the exciton binding energy ($E_x \approx 10$ meV in 3-D CdTe).

Figure 5 (bottom part) shows time-resolved luminescence spectra for a CdTe/ZnTe MQW sample ($L_w$=30 Å, $L_b$=30Å) following excitation above the exciton ground state at t=0 psec (the solid line shows the cw spectra for comparison). The spectra are rather broad as expected from the local fluctuation discussed above. The upper portion of the figure summarizes the lifetime measured across the spectrum. The relevant observations are that (i) there is little spectral shift with time and (ii) the exciton lifetime (at T=2K) is almost constant ($\approx$170ps) except on the lower and higher energy sides of the spectrum. Increasing the temperature to 40 K shows little change in the spectral or temporal characteristics. A fast rise time (<15ps) is also observed to be constant across this energy range. These features are consistent with our interpretation based on exciton localization by the (strain driven) quantum well potential fluctuations. This manifests itself in a surprisingly large degree of spatial immobility of the excitons and only a limited amount of energy diffusion within the exciton lifetime (probably dominated by nonradiative recombination).

To substantiate this interpretation we looked for luminescence line narrowing which is expected for a system of localized resonances under resonant excitation [6]. Here we took advantage of the double monochromator that allowed us to probe as close as 3 meV to the excitation line. There was no change in the spectral or temporal signature of the spectra until the excitation energy was lowered to within the luminescence line where the signal weakened, indicating low absorbance by the low density of localized states [8].

To further test this interpretation, we performed time-resolved experiments under conditions of resonant excitation into the localized exciton states. The use of the double monochromator allowed us to probe to within ≤3 meV of the excitation line. Upon lowering the photon energy of excitation into the localized states, we verified the presence of narrow resonantly excited luminescence peaks with associated phonon sidebands (Fig. 6). The time dependence of these features was that of the luminescence data obtained under non-resonant pumping (Fig. 5), showing the absence of spectral diffusion. Further, the phonon sidebands could thus be identified as phonon assisted recombination (as opposed to a Raman process). In identifying the phonons involved we note that $\hbar\omega_{LO}(ZnTe)=26meV$ and $\hbar\omega_{LO}(CdTe)=21meV$ for unstrained crystal; however, when we calculate their energy under the stress in the structure, both phonons will have energy of about 24 meV [9].

## 4. Acknowledgements

We would very much like to thank and acknowledge the two groups which developed the MQW structures used in this work, headed by Professors R. L. Gunshor and L. Kolodziejski at Purdue University, and Professor J. P. Faurie at the University of Illinois (Chicago). This work was supported by NSF/ECS and DARPA.

## 5. References

1. R. L. Gunshor and L. A. Kolodziejski, R. L. Gunshor, T. C. Bonsett, R. Venkatasubramanian, S. Datta, R. B. Bylsma, W. M. Becker, and N. Otsuka: *Appl. Phys. Lett.* 47, 169 (1985).

2. S. Sivanathan, X. Chu, and J.-P. Faurie: *Appl. Phys. Lett. (in press)*.

3. Y. Hefetz, J. Nakahara, A. V. Nurmikko, L. A. Kolodziejski, R. L. Gunshor and S. Datta: *Appl. Phys. Lett.* 47, 989 (1985).

4. Y. Hefetz, W. C. Goltsos, A. V. Nurmikko, K. A. Kolodziejski and R. L. Gunshor: *Appl. Phys. Lett.* 48, 372 (1986).

5. R. H. Miles, G. Wu, M. Johnson, T. C. McGill, J.-P. Faurie, and S. Sivanathan: *Appl. Phys. Lett.* 48, 1383 (1986).

6. E. Cohen, and M. D. Sturge : *Phys. Rev.* B25, 3828 (1982); Y. Saski, H. Serizawa and Y. Nishina: *Jour. of Non-Crystalline Solids* 59, 1003 (1983).

8. Y. Hefetz, D. Lee, A. V. Nurmikko and J.-P. Faurie: to be published.

9. B. A. Weinstein and R. Zallen: Topics in Applied Physics Vol. 54 (Springer-Verlag, 1984) pp.463-525

# Transient Raman Scattering in Multiple Quantum Well Structures

*D.Y. Oberli, D.R. Wake, M.V. Klein, J. Klem, and H. Morkoç*

University of Illinois at Urbana-Champaign, Materials Research Laboratory and Coordinated Science Laboratory, 104 South Goodwin Avenue, Urbana, IL 61801, USA

1. Introduction

Over the past few years, hot electrons in multiple quantum well structures of GaAs/AlGaAs have been studied extensively [1,2,3]. The dynamics of free carriers and their interaction with phonons play a major role in the physics of high field devices. It is therefore important from a fundamental point of view, to study the relaxation of electrons created by the absorption of a picosecond light pulse. Time-resolved photoluminescence experiments have been used to measure the temperature changes of a photo-excited plasma under different excitation conditions [3,4]. However, they do not allow a direct determination of the electronic densities in the various subbands.

We have developed an experimental technique ideally suited to investigating the carrier dynamics within a GaAs well. In a picosecond time regime, the variation of the density of a photoexcited plasma is monitored by Raman scattering. Inelastic light scattering from electronic intersubband transitions in a quantum well gives direct access to the relative carrier density of each subband [5].

2. Experimental description

Our experimental set-up includes two dye lasers operating at different wavelengths synchronously pumped by a mode-locked Ar ion laser at a repetition rate of 76 MHz. The scattered light is analyzed and dispersed in a Triplemate spectrograph and the entire spectrum is recorded at once by a charge coupled device. The sample is kept in a cryostat at a temperature of 5 K by a flow of cold helium gas. The GaAs-$Al_{.3}Ga_{.7}As$ multiple quantum well structure was grown by MBE on a <100> GaAs substrate and consists of 30 periods of 100Å of $Al_{.3}Ga_{.7}As$ and 220Å of GaAs.

The temporal information is obtained in the following manner: an I.R. pulse first photoexcites electrons directly into the GasAs layers and a second pulse probes the electronic distributions on separate subbands after a given delay time. Raman scattering measurements are performed in the pseudo-backscattering geometry. Polarization selection rules of electronic Raman scattering [5] separate single-particle intersubband excitation in $z(x',y')\bar{z}$ (depolarized spectrum) from collective intersubband excitations in $z(x',x')\bar{z}$ (polarized spectrum).

In figure 1, we show Raman spectra taken under picosecond excitation after a time delay of 650 ps. The peak in the depolarized spectrum is due to single-particle intersubband excitation. Its energy shift is equal to the energy separation between the first and second electronic subbands. In

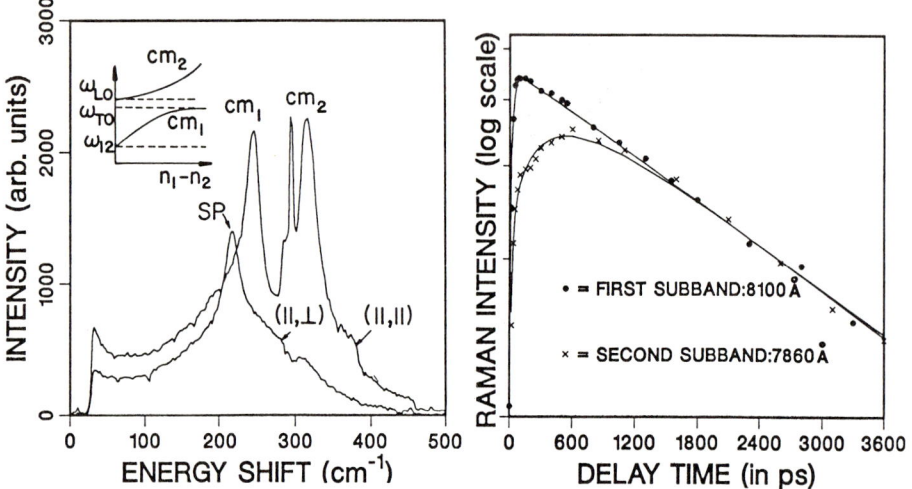

Fig. 1. Raman scattering spectra after a 650 ps delay for both polarizations. The inset illustrates the energy shifts of the collective modes with the relative electron density.

Fig. 2. Intensity of the single-particle intersubband excitation at various delays for two pump wavelengths. Solid lines are the theoretical fits described in the text.

the polarized spectrum, we observe two peaks resulting from the coupling of collective intersubband excitation with LO phonons [6]. In the inset, the energy shifts of these coupled modes are plotted schematically as a function of the relative carrier density in the lowest subbands. The lower energy mode is shifted to higher energy by the depolarization field associated with the collective excitation.

In figure 2, the intensity of the depolarized spectrum integrated over the whole spectral peak is plotted as a function of time delay. Solid points correspond to the excitation of the first subband alone. After an initial rise the intensity decreases exponentially with a time constant of 3200 ps for the 200 Å MQW structure. This decay time is the result of radiative and non-radiative recombination of the electron plasma. When electrons are initially excited in both the first and second subbands, the overall intensity is significantly reduced during the first 300 ps due to the presence of carriers in the second subband. In the next section, we describe the model used to fit the data. An intersubband scattering time equal to 300 ps corresponds to the best fit to the cross data points. The energy shift of the collective modes with delay time is consistent with the variation of intensity of the single-particle peak.

## 3. Discussion of experimental results

The temporal resolution of the experiment, mainly due to the relative dye laser time jitter, is estimated from the cross correlation to be 8 ps. Within this time scale and for the large electron densities studied here ($10^{11} cm^{-2}$) electron-electron collisions constitute the dominant scattering mechanism[8]. They effectively randomize the electron momentum and redistribute the energy of the electrons to establish a quasi-equilibrium distribution characterized by a temperature $T_e$ above the lattice temperature[7]. When different energy subbands are populated, the carrier distributions are held at the same temperature by these collisions and characterized by individual quasi-Fermi energies.

In a time regime much slower than the above mechanism, emission of acoustic phonons cools the electrons in an intrasubband scattering process and changes the relative subband populations in an intersubband scattering process. The single-particle peak in our Raman spectra is due to a photon from the probe pulse scattering an electron inelastically from the first to the second subband. The Raman cross section is proportional to the transition probability of an electron between the two subbands:

$$\frac{d^2\sigma}{d\omega\,d\Omega} \propto \int_0^\infty f_{FD_1}(E)\,(1 - f_{FD_2}(E))\,\frac{(\Gamma/2)^2}{(E - E_r)^2 + (\Gamma/2)^2}\,dE \quad,$$

where $f_{FD}$ is the Fermi-Dirac distribution, $\Gamma$ is a resonance width and $E_r$ is the resonant energy of the electrons. The last term arises from the Raman inelastic scattering mechanism [6] which exhibits a strong resonance enhancement at the $E_0 + \Delta_0$ optical gap. If the two electronic distributions are degenerate and the linewidth of the Lorentzian is much larger than the quasi-Fermi energies, the Raman cross section varies simply as the difference of the two electronic densities. We describe the population change of each subband by a system of rate equations:

$$\frac{dn_1}{dt} = -\frac{n_1}{\tau_r} + \frac{n_2}{\tau_{12}} \quad \text{and} \quad \frac{dn_2}{dt} = -\frac{n_2}{\tau_r} - \frac{n_2}{\tau_{12}} \quad.$$

$n_1(n_2)$ is the electronic density on the first (second) subband, $\tau_r$ and $\tau_{12}$ are respectively the electronic lifetime and the intersubband scattering time. We find two important contributions to the initial fast rise of the Raman intensity. The first one originates from the finite pulse width and time jitter, the second one is caused by hot carriers which have enough kinetic energy to lie outside the range of the resonant-Raman probe. After an estimated time of 50 ps these two effects have subsided. Indeed, a fit to the solid data points shows an exponential decay of the Raman intensity with a time constant of 3200 ps corresponding to $\tau_r$. This model also successfully describes the relaxation of electrons excited into the first and second subbands. Fitting the model to the set of crosses in figure 2 results in an intersubband scattering time of 300 ps for this MQW[9]. Presently, our fit requires an initial ratio of 3 to 1 of the electronic densities in the first and second subbands to explain the amount the signal is decreased with respect to pumping the first subband alone, although one would expect a ratio closer to unity. Since the ratio only arises as a fitting parameter, we are presently seeking to independently measure this subband population ratio by other means. Our results should be regarded as preliminary since the determination of the intersubband scattering time depends somewhat on this ratio.

This determination of the intersubband scattering time is unique since it cannot be obtained from transport experiments. Mobility measurements are dominated by impurity scattering at low temperature and LO phonon scattering at room temperature. Thus, the intensity of the single-particle intersubband scattering is a sensitive probe of the relative carrier distributions.

4. Conclusion

We have studied the dynamics of electrons photoexcited in a GaAs MQW at 5K. Using a novel Raman scattering version of the pump and probe technique, our measurements reveal the band filling of the first subband from the second subband. This leads to a determination of an intersubband scattering time of 300 ps for the 220 Å GaAs well.

This work was supported by NSF under DMR-82-03523 and 83-16981, by AFOSR and JSEP.

References

1. J. Shah, A. Pinczuk, A. C. Gossard and W. Wiegmann; Phys. Rev. Lett. 54, 2045 (1985).
2. C. H. Yang, J. M. Carlson-Swindle, S. A. Lyon and J. M. Worlock; Phys. Rev. Lett. 55, 2359 (1985).
3. J. F. Ryan, R. A. Taylor, A. J. Turberfield, Angela Maciel, J. M. Worlock, A. C. Gossard and W. Wiegmann; Phys. Rev. Lett. 53, 1841 (1984).
4. Kathleen Kash and Jagdeep Shah; Appl. Phys. Lett. 45, 401 (1984).
5. E. Burstein, A. Pinczuk and D. L. Mills; Surf. Sci. 98, 451 (1980).
6. A. Pinczuk, J. Shah, A. C. Gossard and W. Wiegmann; Phys. Rev. Lett. 46, 1341 (1981).
7. J. Shah; J. Phys. C7, 42, 445 (1981).
8. J. L. Oudar, A. Migus, D. Hulin, G. Grillon, J. Etchepare and A. Antonetti; Phys. Rev. Lett. 53, 384 (1984).
9. Independent theoretical estimate of a 300 ps intersubband scattering time was given by J. P. Leburton, private communication.

# Fast Energy Relaxation of Hot Electrons in Bulk GaAs and Multi-Quantum Wells

## C.H. Yang and S.A. Lyon

Department of Electrical Engineering, Princeton Universiy, Princeton, NJ 08544, USA

When a small density of carriers with large excess kinetic energy is injected into GaAs the relaxation can be broken into several stages [1]. First, the hot carriers relax to energies near the band edges through rapid emission of longitudinal optical (LO) phonons. These carriers typically then equilibrate among themselves, but with a temperature well above that of the lattice. If the sample initially contains a significant density of cold carriers, the first stage of the relaxation proceeds even more rapidly because electron-electron scattering as well as LO phonons contribute to the energy loss by the injected carriers.

To date, most experiments have probed hot carrier relaxation after the thermalized distribution has been established. The first stage of the relaxation is of both fundamental and practical interest, but very difficult to study. Scattering times are of the order of 100fs or less [2] for electrons injected with a kinetic energy larger than the LO phonon energy (36.7meV [3]). Direct time-resolved optical measurements are difficult in this temporal regime, especially with the constraint that the injected carrier density be kept low.

The steady-state measurements of MIRLIN et al. [2] have provided clear evidence for the dominance of LO phonon emission at low carrier densities, and a rather clean measurement of the emission rate. These authors measured the hot luminescence arising from electrons recombining with neutral acceptors. Unfortunately, their technique cannot be used in the presence of a large density of cold electrons, which are needed to study electron-electron scattering.

Here we present the results of measurements of hot electron distributions and relaxation rates obtained with a new quasi-steady-state hot luminescence technique [4]. We have studied the initial relaxation of hot electrons in both modulation-doped quantum wells (QW) and bulk doped GaAs. In the QW we find that the rate is relatively constant over the range ($\sim$100 meV) we have studied. In bulk GaAs, on the other hand, the relaxation rate appears to decrease for higher energy electrons.

The quasi-steady-state technique has been described previously [4]; here we will give only a brief explanation. The technique uses the ratio of two luminescence intensities to obtain the hot electron density (n(E), in units of [e$^-$/cm$^3$·eV] for bulk material and [e$^-$/cm$^2$·eV] per well for QW) in terms of the density of "cold" (doped-in) electrons ($n_0$). The hot luminescence is proportional to the product of the hot electron and hot hole ($p(E_h)$) densities. However, since a heavy hole in GaAs is much more massive than an electron, the hole is within a few kT of the band edge and in thermal equilibrium with the "cold" electrons. The ratio of the density of hot to cold holes is then given by a Boltzmann factor

$$\frac{p(E_h)}{p_0} = e^{-E_h/kT_c}, \qquad (1)$$

where $E_h$ is the energy of the hot hole (relative to the valence band edge), $p_0$ is the density of holes at the band edge, k is Boltzmann's constant, and $T_c$ is the carrier temperature. The intensity of the band-edge luminescence is proportional to the density of the cold electrons and holes. Then the density of electrons at an energy E is given by

$$n(E) = C \left(\frac{I(h\nu)}{I_o}\right) e^{E_h/kT_c} \quad , \qquad (2)$$

where $I_o$ ($I(h\nu)$) is the band-edge (hot) luminescence intensity, $h\nu = E + E_h + E_g$, and C is a constant which depends upon the ratio of the matrix elements for the hot and band-edge luminescence and upon the band-edge density of states.

A complication arises in the measurement of the hot luminescence intensity. One contribution is from processes like those described above. Another contribution comes from geminate recombination in which the electron recombines with the hole it was originally created with. The above analysis breaks down because this hole is not in thermal equilibrium with the other carriers. We have found that the contribution from geminate processes varies widely from sample to sample and may be related to defects or surface contamination [5]. We use two lasers, one with a wavelength of 6721Å and the other at 7993Å, and measure the hot luminescence at an energy between them in order to avoid the geminate recombination. By chopping the IR laser and taking the difference spectra between having it on and off, the signal due to geminate processes cancels out. A Charge-Coupled-Device (CCD) array was used to detect the weak hot luminescence.

The average relaxation rate of electrons with energy E can be obtained from the measured density and generation rate (G) assuming conservation of particles:

$$R(E) = \frac{G}{n(E)} \quad , \qquad (3)$$

where R has units of eV/s. For a single electron Eq. 3 makes the assumption that it "slides" down the bands in the sense that the electron loses energy in increments that are small compared to the energy window (set by the spectrometer slits) used to measure the luminescence. In fact, the electrons probably lose energy in larger increments. However, in the present experiments we measure an ensemble averaged rate. Only if both the size of the increments and the initial energy are the same for each electron will Eq. 3 become invalid. This situation would occur at low carrier densities near the excitation energy, in which case it is known that the electrons lose energy in increments of the LO phonon energy, and peaks in the hot luminescence spectra are clearly observed [2]. We do not find peaks in the hot electron density for the present experiments, and thus we expect Eq. 3 to remain valid.

The QW samples used in these experiments consisted of 15 periods of alternating GaAs wells and $Al_xGa_{1-x}As$ (x = 0.28) barriers. The samples were modulation doped and had an electron density of $2.5 \times 10^{11} e^-/cm^2$ per well, as measured in a Shubnikov - de Haas experiment. From excitation spectra obtained on this sample, we have determined that the transition from the top of the third heavy-hole subband to the bottom of the third electron subband occurs at 7425Å. Another transition, probably from the second light hole to the second electron subband occurs at approximately 7150Å. The energy-loss rate obtained for electrons in the QW is shown in Fig. 1. These data were taken with a sample temperature of 83K, which was chosen to optimize the hot hole density. The density of hot electrons is proportional to the reciprocal of the relaxation rate. The fall in the rate for wavelengths below 7100Å comes in a region of the spectrum where the signal was weak. Further experiments will be necessary in order to determine if there is a real reduction in the relaxation rate at these energies. A preliminary report [4], based upon data obtained point-by-point with a photomultiplier, that there may be a reduction in the energy-loss rate for electrons at the bottom of the third subband is not substantiated by the present experiments.

In Fig. 2 we show the energy-loss rate for hot electrons in bulk n-type (Si-doped, n = $5 \times 10^{17}/cm^3$) GaAs. Here we see a trend that electrons at lower energies relax more rapidly than those at higher energies. Over the range of energies shown in Fig. 2, the rate varies by about a factor of 3. Further experiments will be necessary in order to determine the cause of this variation in relaxation rate for bulk GaAs.

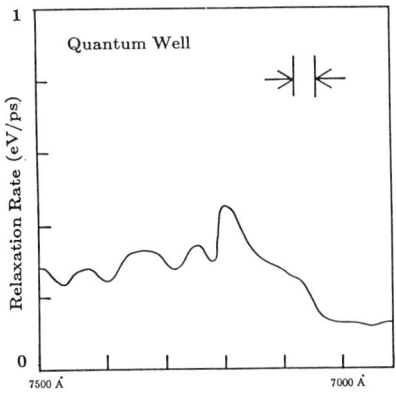
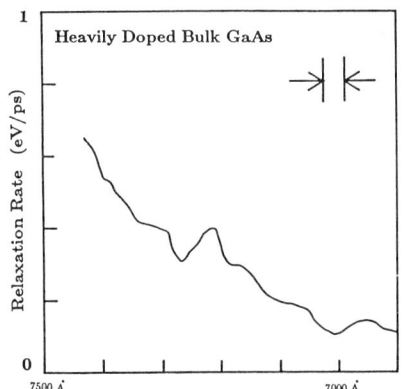

FIG. 1: Energy-loss rate for hot electrons in 150Å, modulation-doped quantum wells at a carrier temperature of 85K

FIG. 2: Energy-loss rate for hot electrons in n-type bulk GaAs at a carrier temperature of 85K

In the bulk GaAs sample we also observed hot luminescence in the energy range of Fig. 2 with the IR laser alone. The observation energy is more than 100meV above the laser energy. The origin of these hot electrons is not clear; Auger recombination is a possible source but it seems unlikely to be able to produce a large enough density. Their density was comparable to that produced by the red laser. The presence of these extra hot electrons could produce an error in our measurement of the energy-loss rate for this sample. The correction could raise the relaxation rates by a factor of up to 2. In contrast, for the QW sample the hot electron density in the absence of the red laser was negligible.

In conclusion we have presented the first energy-resolved measurements of energy-loss rates for hot electrons in GaAs at energies well above the LO phonon and in the presence of a high density of cold carriers. We have studied relaxation in both quantum wells and doped bulk GaAs. The energy-loss rates are of the same order of magnitude in the two samples, and in the range observed in previous time-resolved (but not energy-resolved) [6,7] and steady-state [2,4] experiments. The relaxation rate in the QW is approximately constant over the energy range we studied with a magnitude of about 0.3eV/ps. The energy-loss rate in the bulk GaAs, on the other hand, appears to decrease with increasing electron energy, varying from 0.6eV/ps to 0.2eV/ps. In addition we have observed hot luminescence from bulk GaAs, even when exciting with photons of considerably lower energy.

The CCD system was jointly developed with J.M. Worlock and the authors would also like to thank A.C. Gossard and W. Wiegmann for the multi-quantum well sample. This work was supported in part by NSF, Ford, RCA, GE and Varian through the Presidential Young Investigator Program and by DARPA through ONR under contract No. N00014-83-K-0739.

### References

1. Jagdeep Shah, Solid-State Electronics **21**, 43 (1978).
2. D. N. Mirlin, I. Ya. Karlik, L. P. Nikitin, I. I. Reshina, and V. F. Sapega, Solid State Commun. **37**, 757 (1980).
3. A. Mooradian and G. B. Wright, Solid State Commun. **4**, 431 (1966).
4. C. H. Yang and S. A. Lyon, Physica **134B**, 305 (1985).

5. C. A. Murray and T. Greytak, Phys. Rev. **B20**, 3368 (1979).
6. D. H. Auston, S. McAfee, C. V. Shank, E. P. Ippen, and O. Teschke, Solid-State Electronics **21**, 147 (1978); C. V. Shank, D. H. Auston, E. P. Ippen, and O. Teschke, Solid State Commun. **26**, 567 (1978).
7. C. L. Tang and D. J. Erskine, Phys. Rev. Lett. **51**, 840 (1983); D. J. Erskine, A. J. Taylor, and C. L. Tang, Appl. Phys. Lett. **45**, 54 (1984).

# Picosecond Photoluminescence and Energy-Loss Rates in GaAs Quantum Wells Under High-Density Excitation

*T. Kobayashi[1], H. Uchiki[1], Y. Arakawa[2], and H. Sakaki[2]*

[1] Department of Physics, Faculty of Science, University of Tokyo, Tokyo 113, Japan
[2] Institute of Industrial Science, University of Tokyo, Tokyo 106, Japan

Among the features of the luminescence from GaAs/AlGaAs semiconductors with quantum well structures studied, carrier dynamics in quantum wells such as relaxation processes of electrons [1-7] are not fully understood at the present stage. We report in the present paper the experimental observations of the picosecond photoluminescence properties of GaAs/$Al_xGa_{1-x}As$ quantum well structures at 77 [K] under high-density excitations up to $10^{13}$ [/cm$^2$]. The screening effects on the relaxation of excitation and on the decay kinetics of carriers are taken into consideration theoretically to explain the experimental results under the high-density excitations.

The well width ($L_w$) and barrier width ($L_b$) of the three samples are given in Table 1. The samples were photoexcited by a second harmonic (532 [nm]) pulse from a mode-locked Nd:YAG laser of about 25 [ps] at FWHM.

Table 1. GaAs/$Al_xGa_{1-x}As$ MQW samples used in the present study

| Sample | x | $L_w$[A] | $L_b$[A] | Number of periods |
|---|---|---|---|---|
| A | 0.3 | 78 | 97 | 50 |
| B | 1.0 | 103 | 72 | 75 |
| C | 0.3 | 200 | 200 | 25 |

The observed time-integrated photoluminescence spectra of samples listed in Table 1 are shown in Fig. 1. The transition between n=2 e-h subbands appears under high-density excitation. The transition due to the recombination of electrons and light holes could not be discriminated from that of electrons and heavy holes because of the limited spectral resolution of 10 [meV] in our experiment.

The time dependence of the energy-resolved luminescence intensity was measured with a streak camera operated in a two-dimensional mode. Figure 2 shows the dependence of the decay time on the luminescence photon energy. The time-resolved luminescence spectrum, $I(\omega, t)$, can be constructed from the data of both the time-integrated spectrum, $I_0(\omega)$, and the energy-resolved decay time, $\tau(\omega)$: $I(\omega, t) = \exp(-t/\tau(\omega))I_0(\omega)/\tau(\omega)$. The time-resolved luminescence spectra thus reorganized for samples A and B at 50, 100, 200, 300, and 400 [ps] after excitation are shown in Fig. 3. We fitted these spectra with the spontaneous emission spectra calculated without k-selection rule and obtained Fermi energies ($E_{Fe}$ and $E_{Fc}$) and carrier temperatures (T's) as fitting parameters.

The dependence of the carrier energy and the energy-loss rate on the delay time is shown in Fig. 4 for sample B. We calculated energy-loss rates of electrons due to the interaction with LO-phonons in a GaAs single

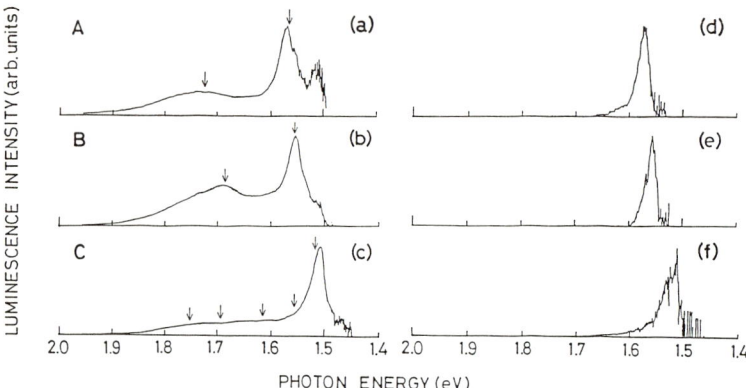

Fig. 1. Time-integrated luminescence spectra of samples A ((a), and (d)), B ((b), and (e)), and C ((c), and (f)). The arrows denote the interband energy separation between electron and heavy-hole subbands calculated by the Kronig-Penny model. (a), (b), and (c): high-density excitation (50±5 [MW/cm$^2$]). (d), (e), and (f): low-density excitation (0.48±0.05 [MW/cm$^2$]).

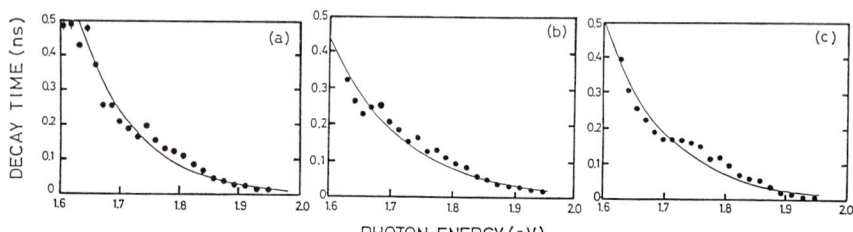

Fig. 2. Energy dependence of the decay time of the luminescence of samples A (a), B (b), and C (c). Excitation intensity is 50±5 [MW/cm$^2$]. The solid curve shows the decay time used in the calculation of time-resolved luminescence spectra.

QW. The energy-loss rate was calculated with the use of the experimental values of the carrier density and the temperature. The results with and without the screening effect being taken into account are shown by the open and closed circles in Fig. 4, respectively. The calculated energy-loss rates with the screening effect are in fair agreement with the observed values up to the delay time of about 0.4 [ns].

In conclusion, the time-resolved luminescence spectra with emphasis on excited electron and hole subbands from three samples of GaAs/Al$_x$Ga$_{1-x}$As (x=0.3 and 1) semiconductors with MQW structures were obtained under high-density excitations. Screening of the electron-phonon interaction by the photogenerated carriers gives the energy-loss rates of electrons which agree with the observed values.

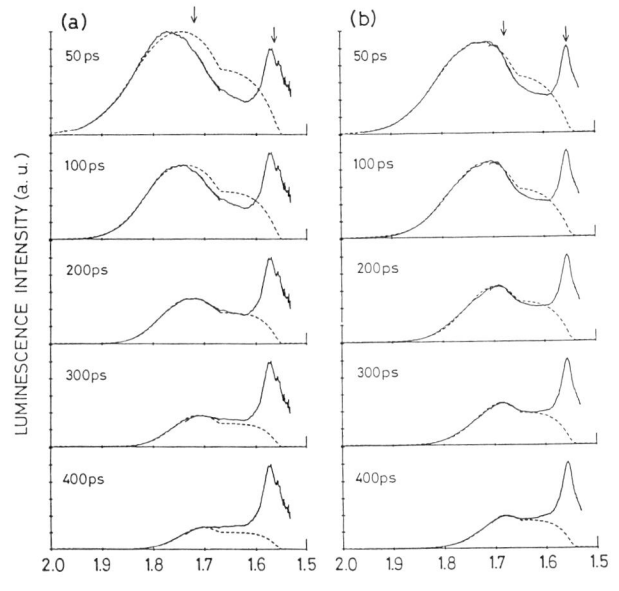

Fig. 3. Time-resolved luminescence spectra of samples A (a) and B (b) at various delay times after excitation. The broken curves show the bestfitted spectra.

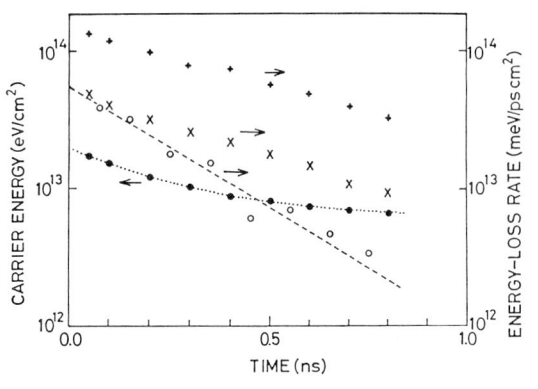

Fig. 4. Temporal behavior of the carrier energy (●) and the energy-loss rate (○) for sample B. The cross and plus marks show the energy-loss rates theoretically calculated with and without the screening effect being taken into account, respectively.

## References

1. C.V. Shank, R.L. Fork, R. Yen, J. Shah, B.I. Greene, A.C. Gossard, C. Weisbuch: Solid State Commun. 47, 981 (1983)
2. E.O. Gobel, H. Jung, J. Kuhl, K. Ploog: Phys. Rev. Lett. 51, 1588 (1983)
3. Z.Y. Xu and C.L. Tang: Appl. Phys. Lett. 44, 692 (1984)
4. J.F. Ryan, R.A. Taylor, A.J. Turberfield, A. Maciel, J.M. Worlock, A.C. Gossard, W. Wiegmann: Phys. Rev. Lett. 53, 1841 (1984)
5. J. Shah, A. Pinczuk, A.C. Gossard, W. Wiegmann: Phys. Rev. Lett. 54, 2045 (1985)
6. H. Uchiki, Y. Arakawa, H. Sakaki, T. Kobayashi: Solid State Commun. 55, 311 (1985)
7. H. Uchiki, T. Kobayashi, H. Sakaki: submitted to Phys. Rev. B

# Broad Tuning of the Photoluminescence Energy and Lifetime by the Quantum-Confined Stark Effect

*H.-J. Polland*[1], *L. Schultheis*[1], *J. Kuhl*[1], *E.O. Göbel*[2], *and C.W. Tu*[3]

[1]Max-Planck-Institut für Festkörperforschung,
 D-7000 Stuttgart 80, Fed. Rep. of Germany
[2]Philipps-Universität, FB Physik, D-3550 Marburg, Fed. Rep. of Germany
[3]AT & T Bell Laboratories, Murray Hill, NJ 07974, USA

## 1. Introduction

Quantum wells (QWs) exposed to electric fields perpendicular to the layers may significantly change their optical properties.[1-4] Low-energy shifts of the absorption band edge exceeding the exciton binding energy ("quantum-confined Stark effect", QCSE) have been reported[1] and applied for e.g. high-speed optical modulators[5] and electro-optical bistable devices.[6] The recent detection of Stark shifts of the photoluminescence energy[2,3] as large as 100meV suggests application of the QCSE for high-speed electrically tunable lasers.

In the first part of this paper we report on low temperature picosecond photoluminescence studies of GaAs/Al$_{0.3}$Ga$_{0.7}$As QWs of different thicknesses. For thick wells a large Stark shift with the electric field is found which is, however, accompanied by a drastic increase of the lifetime and implies an undesired strong decrease of the oscillator strength. This decrease with the electric field severely restricts the practical utilization of the QCSE for the development of electro-optical devices, e.g. wavelength tunable lasers or detectors with tunable spectral sensitivity. In the second part we propose a novel structure, the graded quantum well, where the bandgap varies within the well. For this structure the theory predicts similar strong energy shifts, whereas the field-dependence on the oscillator strength becomes much weaker.

## 2. Picosecond Electroluminescence Studies on GaAs/AlGaAs Quantum Wells

We have investigated the photoluminescence properties at 7K for four different QW thicknesses of 5nm, 10nm, 20nm and 28nm as a function of the electric field strength. For the time-resolved measurements of the photoluminescence signal we have used a synchroscan streak camera and a synchronously mode-locked dye laser (photon energy 1.675eV, intensity 5x10$^{12}$photons per cm$^2$ and pulse, pulse duration ~3ps). Details of the sample preparation and experimental parameters are given in Ref.2.

Time-integrated photoluminescence spectra exhibit a significant low-energy shift of the photoluminescence peak with increasing electric field (Fig.1). Numerical calculations of the Stark-shift in the single particle picture for the finite quantum well (solid lines) give excellent agreement with the experimental points, if slightly smaller values of the QW thicknesses (4nm, 8nm, 16nm and 20nm) than the nominal values as derived from the growth conditions are assumed. For the calculation the effective electron and hole masses are taken from Ref.1 for a conduction band discontinuity of 0.57. Fig.1 (r.h.s.) displays the corresponding photoluminescence lifetimes as deduced from our measured streak curves. The data are normalized to the values at zero electric field, which are 180ps, 290ps, 660ps and 950ps for the 5nm, 10nm, 20nm and 28nm quantum wells, respectively. The strong field-induced lifetime enhancement up to a factor of 100 demonstrates efficient charge separation in the wider quantum wells and a corresponding decrease in electron-hole wavefunction overlap. The experimental results are excellently described by the solid lines which are calculated with the same theoretical model as used in Fig.1 (l.h.s.) for the energy shift. A sharp decrease of the photoluminescence lifetime is found for the 5nm quantum well at fields above 130kV/cm and is attributed to quantum-mechanical tunneling. This tunneling is most important for the thinner QW as expected owing to the higher subband energy.

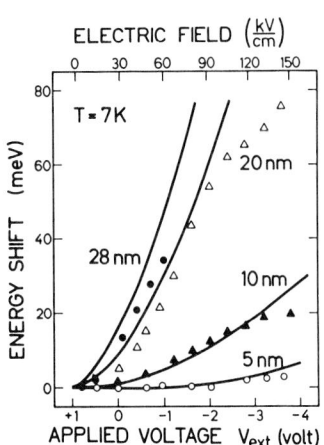
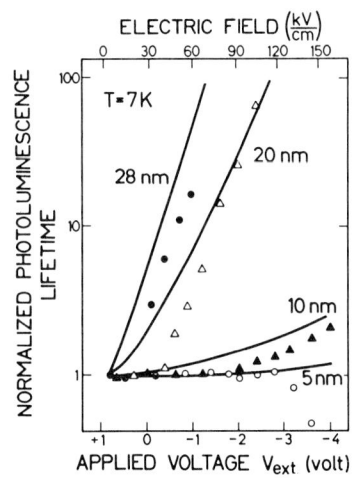

Fig.1: Low-energy shift of the photoluminescence peak (l.h.s.) and normalized photoluminescence lifetime (r.h.s.) versus external voltage for quantum wells with nominal thicknesses of 5nm 10nm, 20nm and 28nm. Solid lines are calculated for the finite quantum well model using thicknesses of 4nm, 8nm, 16nm and 20nm, respectively

Our results demonstrate that especially for the wider wells the energy shift involves a strong decrease of the oscillator strength. These features originate in the strong separation of electrons and holes in the wider wells.

## 3. The Graded Quantum Well in the Regime of the QCSE

The structure which we propose to overcome the undesired decrease of the oscillator strength consists of a quantum well with a

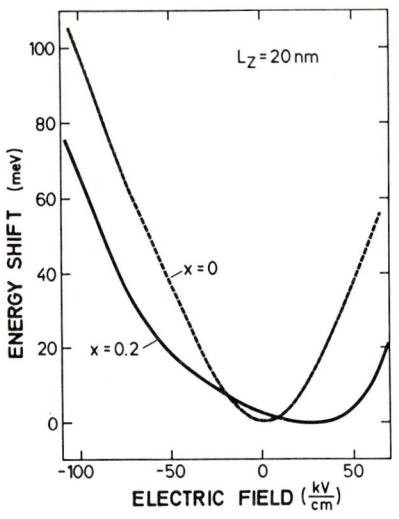 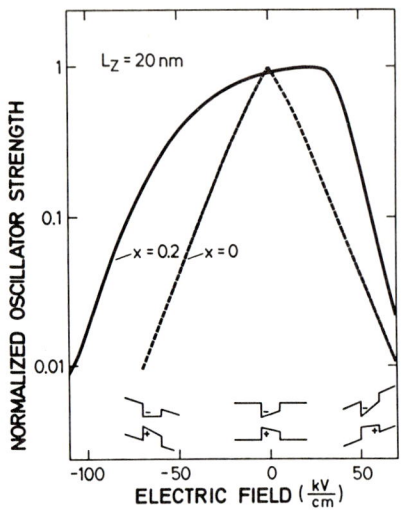

Fig. 2: Calculated low-energy shift (l.h.s.) and oscillator strength (r.h.s.) of the lowest transition in dependence on the electric field for a linearly graded quantum well ($Al_{0.3}Ga_{0.7}As/Al_xGa_{1-x}As/Al_{0.3}Ga_{0.7}As$ with x between 0 and 0.2; solid line) and an ungraded quantum well (dashed line). The well width is 20nm. Band schemes for three field values are schematically shown in the right figure (insert)

linearly increasing or decreasing bandgap across the well width (insert of Fig. 2, r.h.s. at zero electric field). The graded QW is e.g. obtained by the growth of an $Al_xGa_{1-x}As$ layer with continuously varying Al-content which is cladded by layers with a still higher portion of Al. The opposite bending of the conduction and valence band edge in this well leads to strong localization of carriers near the left-side interface. Spatial charge separation, i.e. strong decrease in oscillator strength, by an electric field is thus weaker in the field regime, where the gradients of the conduction and valence band edges are opposite.

Figure 2 displays the energy shift (l.h.s.) and normalized oscillator strength (r.h.s.) of a 20nm thick quantum well with linearly increasing Al-content from x=0 at the left to x=0.2 at the right interface. For the calculation of these curves the wavefunctions are numerically computed in the single particle picture for the finite quantum well (30 percent Al-content in the cladding layers). Also shown are theoretical results for the ungraded quantum well (dashed line) The energy shift (l.h.s.) of up to 70meV for electric fields of ∼-100kV/cm is comparable to that of the ungraded structure (x=0) and predicts strong tunability of the photoemission for the novel QW structure in the regime of the QCSE. In addition, carriers remain essentially localized at the left interface of the well for electric fields between ∼ -100kV/cm and ∼ 30kV/cm which results in a weaker field dependence of the oscillator strength (r.h.s.).

It should be noted that the structure of the linearly graded quantum well represents one special example where tuning of the energy can be achieved without significant charge separation in the well. More sophisticated structures with opposite gradients of conduction and valence band edges are imaginable for which comparable field-induced Stark shifts but negligible field influence on the oscillator strength can be expected.

## 4. Conclusions

In summary, we have demonstrated that the quantum-confined Stark effect (QCSE) leads to a drastic decrease of the oscillator strength of the $n=1$ transition accompanied by a red shift of the photoluminescence. These features are consistent with the theoretical predictions of the quantum—confined Stark effect. The polarization of the confined electron-hole pairs which is strongest for the wider wells, is significantly weaker in more sophisticated structures such as linearly graded quantum wells.

## 5. Literature

[1] D.A.B.Miller, D.S.Chemla, T.C.Damen, A.C.Gossard, W.Wiegmann T.H.Wood, and C.A.Burrus, Phys.Rev.B$\underline{32}$, 1043 (1985).
[2] H.-J.Polland, L.Schultheis, J.Kuhl, E.O.Göbel, and C.W.Tu, Phys.Rev.Lett.$\underline{55}$, 2610 (1985).
[3] L.Vina, R.T.Collins, E.E.Mendez, and W.I.Wang, Phys.Rev.B$\underline{33}$, 5939 (1986).
[4] M.Yamanishi, Y.Usami, Y.Kan, and I.Suemunc, Jpn.J.Appl.Phys. $\underline{24}$, L586 (1985).
[5] T.H.Wood, C.A.Burrus, D.A.B.Miller, D.S.Chemla, T.C.Damen, A.C.Gossard, and W.Wiegmann, IEEE J.Quantum Electron.$\underline{21}$, 117 (1985).
[6] D.A.B.Miller, D.S.Chemla, T.C.Damen, A.C.Gossard, W.Wiegmann, T.H.Wood, and C.A.Burrus, Appl.Phys.Lett.$\underline{45}$, 13 (1984).

## 6. Acknowledgement

The authors gratefully acknowledge the expert technical assistance of K.Rother and H.Klann.

# Auger Heating of Silicon-on-Sapphire by Femtosecond Optical Pulses

*M.C. Downer\* and C.V. Shank*

AT & T Bell Laboraories, Holmdel, NJ 07733, USA

Previous experiments have shown that a silicon surface becomes highly reflective [1,2] and microscopically disordered [3] in less than one picosecond following femtosecond pulsed excitation above a threshold fluence of approximately 0.1 J/cm$^2$. The present experiment investigates the temperature rise of crystalline silicon excited below the threshold for melting through time-resolved measurements of refractive index changes in a submicron silicon film [4]. This technique [5,6] relies upon the well characterized temperature dependence [7] of the refractive index n + ik of silicon in the wavelength range between the indirect (1.15 eV) and direct (3.0 eV) band gaps. The temperature dependence of the index results from a downward shift of the direct band edge of silicon caused by renormalization of the band energy by the electron-phonon interaction, with a small contribution from lattice thermal expansion [8]. Our data resolves a delay of tens of picoseconds in lattice heating.

## 1. Experiment

A 0.5 micron thick silicon layer on a sapphire substrate was photo-excited with 100 fsec pulses at a fluence of 0.02 J/cm$^2$, and probed with normally incident white light continuum pulses. Use of a thin film permits sensitive measurement of small heat-induced index changes, permits simultaneous measurement of reflected and transmitted probe light, and suppresses carrier diffusion on the time scale of the experiment. Probe light from a 10 micron diameter circular area near the center of the photoexcited region was selected by imaging transmitted and reflected probe beams onto irises placed in the image planes. These spatially filtered beams, along with a reference probe beam were collected and analyzed with an optical multichannel analyzer.

## 2. Results

Fig. 1 shows a) reflectivity and b) transmission spectra at time delays $\Delta t$=-1ps, 0.5 ps, and 200 ps following photo-excitation. In Fig. 1a, the initial blue shift of the Fabry-Perot fringes at $\Delta t$=0.5 ps results from a decrease in the real index n caused by the presence of a dense electron-hole plasma ($N_0 = 5 \times 10^{20}$ cm$^{-3}$). At later time delays the fringes drift back towards the red, passing the original positions in a few tens of picoseconds, before reaching a final red-shifted position evident in the curve at $\Delta t$=200 ps. This red shift results from lattice heating [7,8] which accompanies the decay of the electron-hole plasma through Auger recombination. The corresponding transmission spectra in Figure 1b show qualitatively similar shifts in fringe positions. The strong initial drop in transimission evident in the curve at $\Delta t$=0.5 ps, followed by a slow partial recovery suggests that plasma-induced absorption, along with heating, contributed to the changes in k.

---

\* Present Address: Physics Department, University of Texas at Austin

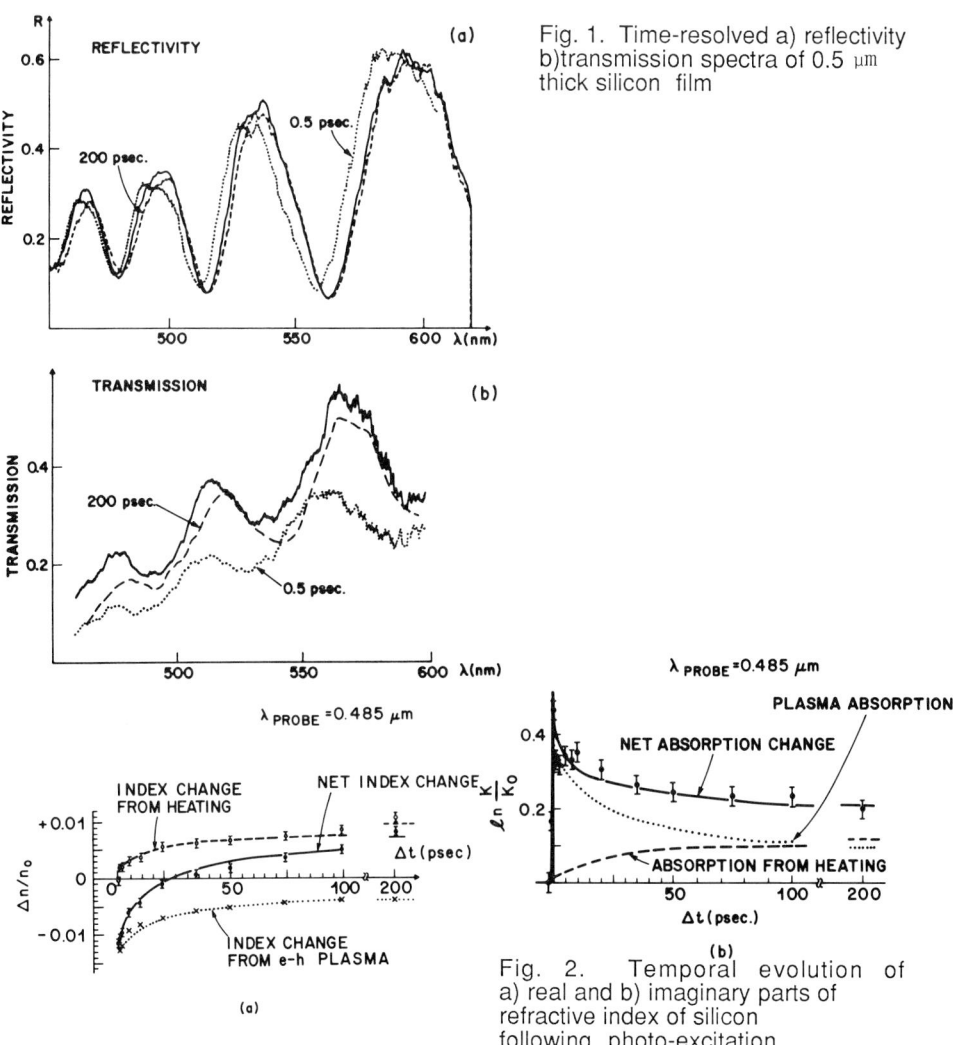

Fig. 1. Time-resolved a) reflectivity b) transmission spectra of 0.5 μm thick silicon film

Fig. 2. Temporal evolution of a) real and b) imaginary parts of refractive index of silicon following photo-excitation

## 3. Analysis

We analyzed the data in Fig. 1 by extracting values of n and k using the thin film optics equations [9]. Fig. 2 shows the results of this analysis at probe wavelength 485 nm, where the heating effect is large [7] and the plasma effect as small as possible. The real index n drops immediately by 1.2 percent upon photoexcitation (plasma effect), then rises toward a value 1 percent higher than its initial value as the lattice heats and plasma density decreases, primarily through Auger recombination. In order to deconvolve the plasma effect from this data, we took concurrent data at probe wavelength 1.0 micron, where the heat-induced index change is three times smaller [7] and the plasma-induced index change four times larger than at 485 nm. Thus, within our experimental error, the infrared data represent a pure plasma-induced index change and, when scaled by the Drude model factor $\lambda^2$ (as are x's in Fig. 1a), model the plasma-induced index change at 485 nm. Subtraction of these two sets of data yields the pure heat-induced index change, shown by the open circles. Note the delay of tens of picoseconds in lattice heating.

The curves in Fig. 2 are based on a quantitative model of plasma decay and lattice heating caused by Auger recombination. According to our model, the initially photoexcited plasma loses 10 to 15% of its energy by direct phonon emission within a picosecond. Further energy transfer to the lattice, however, is blocked by the Pauli exclusion principle, since hot carriers can relax only to the quasi-Fermi levels. Further lattice heating therefore awaits Auger recombination of the degenerate plasma, which re-excites carriers, which then relax again by phonon emission to the new quasi-Fermi level. The e-h pair density N thus evolves in time according to the Auger formula:

$$dN/dt = - C N^3 . \qquad (1)$$

The fractional index change induced by the evolving e-h plasma is then:

$$\Delta n/n_0 = - (2\pi e^2 / \varepsilon_0 m^* \omega^2) N, \qquad (2)$$

where $\varepsilon_0 = 19$ at 485 nm, and $m^* = 0.2 m_e$ is the reduced e-h effective mass. The simultaneous solution of (1) and (2) yields an expression for the fractional plasma-induced index change at 485 nm. Using $N_0 = 5 \times 10^{20}$ cm$^{-3}$, the curve through the scaled infrared data in Fig. 2a was obtained. Using N(t) thus obtained, Auger heating is then described by the energy balance equation:

$$C_p dT/dt = C N^3 (E_g + kTH), \qquad (3)$$

where H is a degeneracy factor equal to 3 for a nondegenerate plasma and increasing with degeneracy and carrier density. The positive fractional index change caused by Auger heating is related to the temperature evolution by:

$$\Delta n/n_0 = \beta \Delta T(t) / n_0 , \qquad (4)$$

where $\beta = 8.0 \times 10^{-4}$ K$^{-1}$ at 485 nm. [7]. The upper theoretical curve in Fig. 2a was obtained from simultaneous solution of (3) and (4), using the same values of $N_0$ and C as for the upper curve. The middle curve, which fits the original data at 485 nm, is the sum of the upper and lower curves. Clearly this simple model satisfactorily explains the magnitude as well as the temporal evolution of the index changes shown in Fig. 2a. The temporal evolution of the fractional change in k shown in Fig. 2b also appears to reflect the combined influence of plasma and lattice heating effects. An analysis of the heat-induced and plasma induced absorption, again using the same $N_0$ and C values as before, yields the curves shown if a plasma induced absorption cross section of $1.6 \times 10^{-17}$ cm$^2$ is assumed.

4. Conclusions

Extrapolation of the present results to fluences required for melting (>0.1 J/cm$^2$) indicates that energy transfer to the silicon lattice through Auger heating would occur within a picosecond because of the $N^3$ dependence of the energy transfer rate, consistent with melting times inferred from surface reflectivity [1,2] and surface second harmonic generation [3] experiments. In metallic samples, only rapid direct heating would be expected.

Acknowledgment: We are grateful to F.A. Beisser and D. W. Taylor for expert technical assistance.

References

1. C. V. Shank, R. Yen, and C. Hirlimann, Phys. Rev. Lett. 50, 454 (1983).
2. M. C. Downer, R. L. Fork, and C. V. Shank, J. Opt. Soc. Am. B 2, 595 (1985).
3. C. V. Shank, R. Yen, and C. Hirlimann, Phys. Rev. Lett. 51, 900 (1983).
4. M. C. Downer and C. V. Shank, Phys. Rev. Lett. 56, 761 (1986).
5. K. Murakami. H. Itoh, K Takita, and K. Masuda, Physica 117B, 1024 (1983).
6. L.A. Lompre, J.M. Liu, H. Kurz, and N. Bloembergen, Appl. Phys. Lett. 43, 168 (1983).
7. G. E. Jellison, Jr., and F. A. Modine, Appl. Phys. Lett. 41, 180 (1982).
8. P. B. Allen and M. Cardona, Phys. Rev. B23, 1495 (1981).
9. O. S. Heavens, Optical Properties of Thin Solid Films (Dover, 1985).

# The Origin of Picosecond Photoinduced Absorption Decays in Hydrogenated Amorphous Silicon

*W.B. Jackson, C. Doland, and C.C. Tsai*

Xerox Palo Alto Research Center, 3333 Coyote Hill Road,
Palo Alto, CA 94304, USA

A systematic study of the dependence of picosecond photoinduced absorption (PA) in undoped hydrogenated amorphous silicon (a-Si:H) on sample thickness, pump wavelength, and trap density has been undertaken. As for the case of doped and compensated films, the results rule out bimolecular recombination, surface recombination, and deep trapping as mechanism for explaining the decays. The decays are consistent with thermalization within the band tails.

## 1. INTRODUCTION

The study of ultrafast relaxation of carriers in amorphous materials such as a-Si:H has been extensively studied using photoinduced absorption (PA).[1-6] The decay of the absorption of a weak probe due to carriers excited by a stronger pump beam has been interpreted using a number of conflicting models. The primary candidates are direct bimolecular recombination between band tail electrons and holes,[1] diffusion to and recombination at the sample surface,[2] thermalization and subsequent deep trapping of the excited carriers in deep traps,[3,4] and thermalization of carriers within the band tails.[5,6] Because the above models yield reasonable fits to various sets of data, it can be concluded that a fit to a given model is not sufficient evidence for the model applicability. Consequently, in this paper, we compare the predictions of various models for a wide range of sample conditions in order to exclude some of the proposed models. Recent results have indicated that relaxation in the band tails accounts for the decays in all materials both doped and compensated.[6] This paper provides further evidence for this conclusion.

## 2. EXPERIMENTAL SETUP

The characteristics of the PA experimental setup are very similar to those found in Refs. [3,4] and described more fully in Ref. [6]. The 614 nm laser pulses, focussed to a 30$\mu$m diameter spot, were produced by a synchronously pumped cavity dumped dye laser system with an energy up to 8 nJ, a 12 psec pulse width, and a 800 kHz repetition rate. The excited carrier densities ranged roughly from $10^{18}$ to $10^{19}$ carriers per cm$^3$. Delays up to 8 nsecs could be obtained using a feedback stablized delay line which insured overlap of the pump and probe spots over the entire delay.[7] The series of undoped films were produced by rf plasma decomposition of silane at low r.f. power onto glass or quartz substrates. Further characteristics will be described below and in Ref. [6].

## 3. RESULTS AND DISCUSSION

### 3.1 Surface Recombination

Consider first the possiblility that the decay is due to diffusion of carriers to the surface where they recombine. The time to reach the surface is given approximately by the relation $\tau \simeq L^2/D$ where L is the sample thickness and D is the diffusion coefficient. Consequently, the decay time should exhibit a strong dependence on thickness. Despite a change of a factor of 10 in sample thickness in Fig. 1, there is no discernible change in the decay rate. The observed changes due to etalon effects are much smaller than the factor of 100 change expected.[8] Another prediction of the diffusion model is that the decay rate should decrease for more strongly absorbed light since the carriers are generated closer to the surface. The observed decays presented in Fig. 2, however, again do not exhibit any discernible changes despite a change in the absorption depth by a factor of 3 demonstrating that for thicker films, unlike the case for thinner films, the decays are not controlled by carrier diffusion.[2]

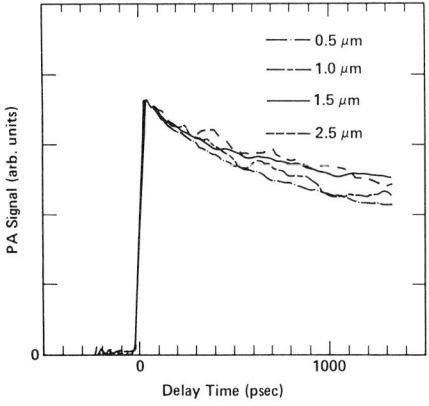

Fig. 1. PA decays versus delay time for various thickness films.

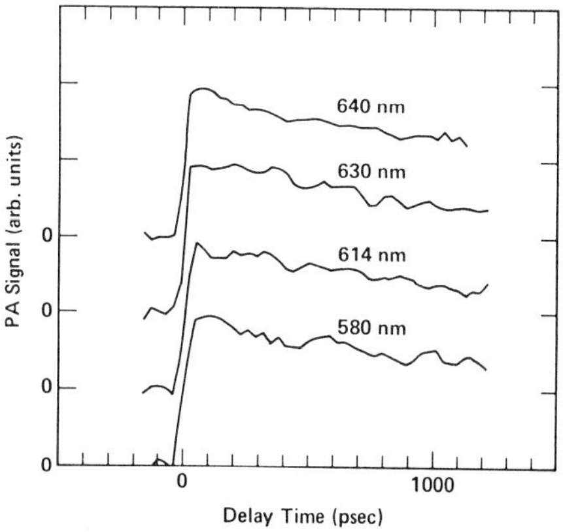

Fig. 2. PA decays versus delay time for different excitation and probe wavelengths.

### 3.2 Bimolecular Recombination

The second possibility to consider is that the decays are due to bimolecular recombination of band tail electrons and holes. The results of Fig. 2 are inconsistent with this possibility since the carrier density is changing by a factor of 3. One would expect that the decay rate would also change by a similar factor. Further evidence that the decay does not depend on power density is presented in Ref. [6]. Hence, bimolecular recombination does not explain the observed decays.

### 3.3 Deep Trapping

Next, we consider the possibility that the PA decays are due to thermal excitation of carriers from the band tails into deep traps which have a lower optical cross section. This hypothesis, using multiple trapping, predicts that the deep trapping time and consequently the PA decay, should depend on the number of traps, $N_s$, according to the relation $\tau = (N_s)^{-\alpha}$ where $0.5 < \alpha < 1$. Despite an increase in the defect densities of 2 orders of magnitude for a series of samples with different spin densities, the PA decay times actually increase slightly—a result not at all consistent with deep trapping. Other data in conflict with the deep trapping model, presented in Ref. [6], found that compensated and phosphorus doped material exhibited nearly the same decays despite a factor of 1000 difference in defect densities. Furthermore, the number of traps is several orders of magnitude less than the number of excited carriers in undoped and compensated specimens so the decays cannot be due to carriers thermalizing into deep traps. Consequently, the deep trapping model fails several important tests and hence, does not play a major role in PA decays on this time scale.

## *3.4 Band Tail Thermalization*

Finally, we consider the fourth and most likely explanation for the observed PA decays; namely, the decay is due to a decrease in the optical cross section as the carriers thermalize within the band tails. This model, recently proposed for PA decays in compensated material,[5] is consistent with the observed PA decays in all samples.[6] For example, the model is consistent with the doping and compensation studies presented elsewhere. The model requires that the optical cross section is fairly sensitive to the precise character of the band tail states. Consequently, alteration of the band tail states through doping should give rise to observable changes in the decays. At the same doping density of $\sim 10^{-4}$ for boron, phosphorus, and compensated material, the decays are dramatically altered making a transition from photoinduced absorption to photoinduced transparency. This doping level is the same level at which the hyperfine measurements indicate that the donor and acceptor levels dominate the band tails. The wavefunctions of the donor and acceptor levels, possessing significantly different character than band tail states, give rise to major changes in the PA decays.

## *4. ACKNOWLEDGEMENTS*

We would like to thank R. Street, M. Stutzmann, and J. Tauc for helpful discussions. This work was supported in part by Solar Energy Research Institute Contract No. XB−3−03112−1.

## *REFERENCES*

1. Z. Vardeny, P. O'Connor, S. Ray, and J. Tauc, Phys. Rev. Lett. 44, 1267 (1980).
2. V. J. Newell, T. S. Rose, and M. D. Fayer, Phys. Rev. B 32, 8035 (1985).
3. J. Strait. Ph.D. Thesis. (Brown University. May, 1985).
4. J. Strait and J. Tauc, Appl. Phys. Lett., 47, 589 (1985).
5. C. Thomsen, H. Stoddart, T. Zhou, and J. Tauc, Phys. Rev. B, 33, 4396 (1986).
6. W. B. Jackson, C. Doland, and C. C. Tsai (to be published).
7. C. Doland, W. B. Jackson, and A. Anderson (to be published) and in this proceedings.
8. D. M. Roberts and T. L. Gustafson, J. Non-Cryst. Solids 77 &78, 551 (1985) and D. M. Roberts, J. F. Palmer, and T. L. Gustafson (submitted for publication).

# Picosecond Decay of Photoinduced Absorption in Hydrogenated Amorphous Silicon

*D.M. Roberts and T.L. Gustafson*

Standard Oil Research & Development, 4440 Warrensville Center Road, Warrensville Heights, OH 44128, USA

The study of the transient behavior of photogenerated carriers is helping to elucidate the mechanisms for carrier relaxation in amorphous silicon (a-Si) materials. The dynamics of hot carrier thermalization, localization, trapping, and recombination can all be probed using time resolved optical techniques. There are three primary methods for obtaining this information: photoconductivity (PC) probes carriers above the mobility edge;[1] photoluminescence (PL) measures radiative and non-radiative recombination rates;[2] and photoinduced absorption (PA) probes changes in the optical absorption due to the presence of excited carriers.[3]

The recent picosecond PA experiments have used the same wavelength for both the pump and the probe.[3] In this work we present picosecond PA results on intrinsic a-Si:H that were obtained using independently tunable pump and probe lasers. We have shown that the observed PA decay, as monitored using both transmittance and reflectance, is different at different sample thicknesses.[4,5] This distortion is caused by an etalon effect in the sample and will be present for all thin film samples if the steady-state optical absorption spectrum shows interference fringes. We interpret the observed decays using the model for dispersive transport. We also show that the PA decay depends on carrier density and source repetition rate. We present preliminary results on the temperature dependence of the PA decays from 200 to 400 K.

We have presented the details of the experimental apparatus elsewhere.[6] Briefly, a mode locked argon ion laser pumped two synchronously pumped cavity dumped dye lasers to produce independently tunable pump and probe pulses. The pump dye laser was R590, tunable from 1.94-2.18 eV; the probe dye laser was either R590 or DCM, providing tunability from 1.70-2.18 eV. The cross correlation between the pump and probe pulses gave a time resolution of ~8 ps. The pulse repetition rate could be varied and was typically 500 kHz. The pulses were focused through a 100 mm focal length lens to a spot size of ~50 microns, with the pump being slightly larger in diameter at the focus than the probe. We detected the increase in absorption using a time modulation technique in order to eliminate the thermal background that is present with mechanical chopping. In order to obtain undistorted decays of $\Delta\alpha$ we either summed the $\Delta T$ and $\Delta R$ decays or we collected the decay where $\partial T/\partial n = 0$.[5,7] Data collection times were typically 300 or 600 seconds. The sample used in this work was deposited in a capacitively coupled rf glow discharge system. The sample was assessed as having a low defect density on the basis of the steady state photoluminescence intensity.

We model the data by assuming that there are three levels contributing to the induced absorption, each with a different absorption cross section, and that the carriers in the band tail states relax according to the multiple trapping model (i.e.

dispersive transport).[8] The resulting equation for the decay of the induced absorption is as follows:

$$\Delta\alpha(t) = \frac{n_0(\sigma_1-\sigma_2)}{1 + (t/\tau)^\beta} + n_0(\sigma_2-\sigma_0),  \quad (1)$$

where $\Delta\alpha$ is the induced absorption; $n_0$ is the initial number of excited carriers; $\sigma_0$, $\sigma_1$, and $\sigma_2$ are the cross sections for the transition between the valence band and conduction band, for the absorption of excited carriers in the band tail states, and for the absorption of carriers in a deeper state, respectively; $\tau$ is the mean trapping time; and $\beta$ is the dispersion parameter.

We have found that the decay curves at all repetition rates, pump intensities, and temperatures can be fit using (1). At a given temperature, the dispersion parameter appears to be constant. As the peak carrier density increases the $n_0(\sigma_2-\sigma_0)$ term increases linearly, as expected, for a given repetition rate. The slope of the $n_0(\sigma_2-\sigma_0)$ term versus $n_0$ increases as the repetition rate increases. Since $\sigma_2$ is time invariant on the picosecond time scale, the change in slope represents a "filling up" of state 2 as the repetition rate increases; the effective $\sigma_2$ is modified by the change in the population in state 2 at the different repetition rates. These data suggest that state 2 has a lifetime of ~1 $\mu$s. In Fig. 1 we plot the mean trapping time as a function of peak carrier density at repetition rates of 250 kHz, 500 kHz, and 1 MHz. The mean trapping time, $\tau$, increases as the peak carrier density decreases until the peak carrier density approaches ~$10^{18}$/cm$^3$, at which point $\tau$ remains constant within our signal-to-noise. The presence of a large residual carrier concentration appears to shorten the mean trapping time significantly.

From the intensity and repetition rate study we suggest that it is important to work at carrier densities less than $10^{18}$/cm$^3$ when studying intrinsic a-Si:H. For practical reasons 500 kHz is a suitable choice for the repetition rate; data acquisition times become unreasonably long at lower repetition rates.

We also obtained the temperature dependence of the PA decay. Over the range from 200 to 400 K the dispersion parameter increased linearly from 0.5 to

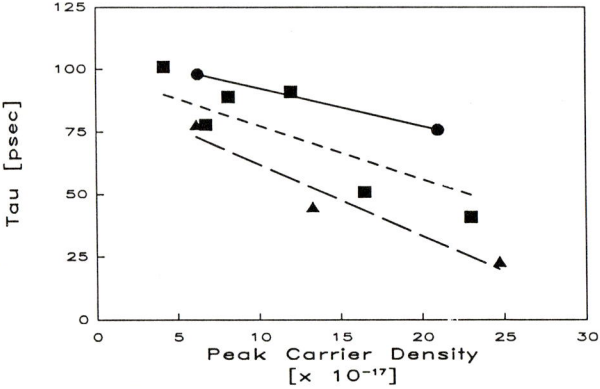

Figure 1. Mean trapping time, $\tau$, as a function of peak carrier density for low defect density, intrinsic a-Si:H at repetition rates of 250 kHz (circles), 500 kHz (squares), and 1 MHz (triangles); lines are a guide to the eye

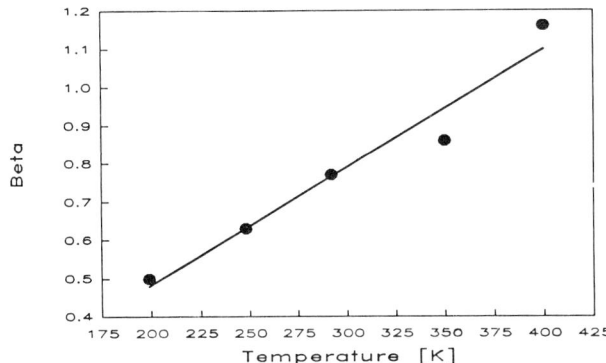

Figure 2. The dispersion parameter, $\beta$, as a function of temperature for low defect density, intrinsic a-Si:H; the line represents the linear least squares fit with a slope of 320 K

~1.1. These data are shown in Fig. 2. The slope of $\beta$ as a function of temperature suggests that electrons are the carriers we are observing in the PA experiment for the intrinsic materials. The temperature dependence of $\tau$ was very interesting. At temperatures below 300 K, $\tau$ increased as the temperature decreased, as expected. However, at temperatures above 300 K, $\tau$ remained constant and then increased. This behavior is quite unexpected; additional work is in progress in order to establish the reason for this anomalous variation in $\tau$.

References

1. A. M. Johnson, D. H. Auston, P. R. Smith, J. C. Bean, J. P. Harbison, and A. C. Adams, Phys. Rev. B 23, 6816(1981).
2. B. A. Wilson, T. P. Kerwin, and J. P. Harbison, Phys. Rev. B 31, 7953(1985).
3. J. Tauc, Hydrogenated Amorphous Silicon, Part B, J. Pankove, Ed., (Academic Press, New York, 1984), pp. 299, and references therein.
4. D. M. Roberts and T. L. Gustafson, J. Non-Cryst. Solids 77 & 78, 551(1985).
5. D. M. Roberts, J. F. Palmer, and T. L. Gustafson, J. Appl. Phys., in press.
6. D. M. Roberts and T. L. Gustafson, Opt. Commun. 56, 334(1986).
7. H. T. Grahn, C. Thomsen, and J. Tauc, Opt. Commun., in press.
8. H. Scher, J. Phys., Paris 42, 547(1981).

# Femtosecond Spectroscopy of Hot Carriers in Germanium

*P.M. Fauchet*[1], *D. Hulin*[2], *G. Hamoniaux*[2], *A. Orszag*[2], *J. Kolodzey*[1], and *S. Wagner*[1]

[1]Department of Electrical Engineering, Princeton University,
Princeton, NJ 08544, USA
[2]Laboratoire d'Optique Appliquée, ENSTA, Ecole Polytechnique,
F-91120 Palaiseau, France

## 1. Introduction

The availability of femtosecond laser sources has made possible the study of electronic processes under extreme non-equilibrium conditions in semiconductors. Such studies are important from a basic physics viewpoint and for electronic or optical devices. We report here some results of a femtosecond spectroscopic study of very hot carriers in germanium. The laser source is a CPM dye laser amplified to 1 mJ at 10 Hz. Pump and probe reflectivity and transmission measurements are performed on a 250 nm-thick polycrystalline Ge film deposited on a transparent substrate. The pump is usually at 620 nm although experiments have been performed at 310 nm. The probe is the spectrally filtered output of a white light continuum generation cell and is nearly normal-incident on the sample. The sample is held at room temperature and raster-scanned to avoid cumulative heating and damage.

## 2. Results

First, we focus our attention on a 10 ps window around zero time delay. Figure 1 shows typical results for transmission T and reflectivity R at 850 nm. The very intense 620 nm pump pulse is coincident with the decrease in both T and R. The transmission drops abruptly, then recovers in part, and finally starts a gentle decrease. The reflectivity also drops abruptly and then recovers to essentially its initial value in a few ps. Figure 2 shows the behavior of T on a longer time scale. The magnitude of the initial drop is a strong function of pump intensity, whereas the partial recovery appears independent of pump intensity. The transmission curves at 771 nm, the other probe wavelength that was thoroughly investigated, are qualitatively similar. Although R displays a more monotonic behavior, the sign of the variation is frequency dependent: R drops at 850 nm and increases at 771 nm. At both wavelengths, R recovers in less than 10 ps and shows very little variation thereafter. Finally, pumping at 310 nm does not lead to quantitatively different results. In particular, on a 10 ps time scale, the transmission still displays a weak initial drop followed by a net increase reminiscent of the low pump intensity curve of Fig. 2.

## 3. Discussion

From T and R, we calculate the time evolution of the absorption coefficient $\alpha$ and the refractive index n. At the wavelengths of interest, the optical density of the film is such that interference effects can be neglected, which simplifies the analysis [1]. Figure 3 shows the variations of $\alpha$ as a function of time for different pump intensi-

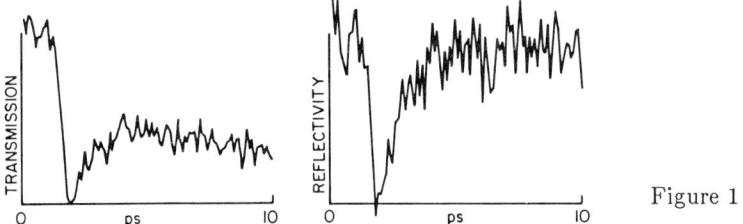

Figure 1

Transmission and reflectivity at 850 nm after high intensity excitation at 620 nm. Zero time delay is coincident with the sharp decrease in each curve.

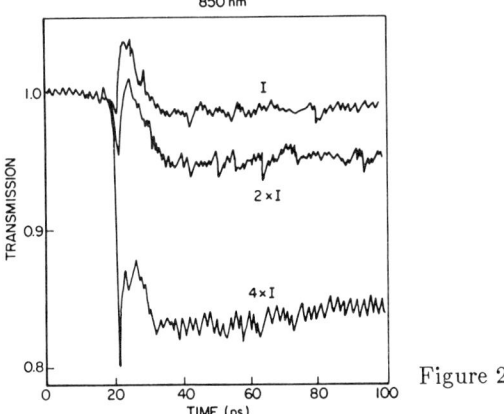

Figure 2

Transmission at 850 nm after excitation at 620 nm with three different pump intensities. Zero time delay is coincident with the sharp decrease in each curve.

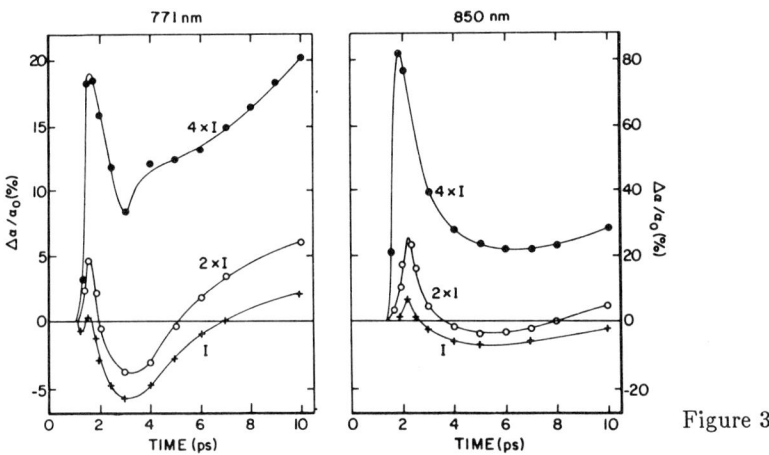

Figure 3

Time evolution of the absorption coefficient obtained from the transmission and reflectivity curves at two probe wavelengths and three pump intensities. Note the different vertical scales. Zero time delay is coincident with the increase in absorption.

ties and probe wavelengths. Several remarkable facts deserve notice. First, $\alpha$ increases sharply for 0.5 to 1 ps. At low pump intensity, $\alpha$ then undergoes a net decrease: the sample becomes less opaque. At high pump intensity, $\Delta\alpha$ remains positive. The initial increase is four times larger at 850 nm than at 771 nm and the "transparency" period is longer and tends to start later. The variations in $\alpha$ follow closely those of T. Similarly, the variations of n follow closely those of R, and thus do not display such strong oscillations.

We propose the following model to explain our data up to 5 ps after excitation. The pump pulse injects between $10^{20}$ and $10^{21}$ carriers /cm$^3$. If we assume that carrier-carrier scattering is very fast on that time scale, the plasma can be characterized by a two quasi-Fermi levels and one electronic temperature $T_e$. If we further assume that little or no energy is lost from the plasma to the lattice on that time scale by emission of phonons, the initial $T_e$ is in excess of 5000K. The plasma cools down in approximately 1 ps after which the lattice temperature increases somewhat and the plasma density starts to drop by Auger recombination [2]. The bandgap is reduced because of the lattice temperature increase (a small and gradual effect) and because of the renormalization that takes place in the presence of the plasma (a larger and abrupt effect, which tappers off during recombination).

We have calculated the valence to conduction band and the intravalence band [3] absorption coefficients under these conditions. Only direct transitions are considered although the influence of L and X valleys is included to calculate the distribution of electrons in the conduction band. The dependence of $\alpha_{v \to c}$ is a rather complicated function of the carrier density and distribution. In general, the intravalence absorption increases with increasing density and decreasing $T_e$ at the wavelengths of interest. The net result of these calculations is to predict a large positive $\Delta\alpha$ at t=0, a negative $\Delta\alpha$ at low pump intensity as $T_e$ decreases, and a longer and slightly delayed "transparency" period at 850 nm. The amplitude of the large initial increase is a very sensitive function of the exact initial density and temperature. These calculations and a more complete description of our experimental results will be presented elsewhere [4].

In conclusion, we have shown that the large variations in transmission and in reflectivity recorded after intense femtosecond excitation of Ge can be attributed to cooling of hot carriers. This is in contrast to the interpretation proposed by others for similar experiments performed in silicon [5]. It was suggested that two-photon absorption (one probe and one pump) at zero time delay could be responsible for the initial $\Delta\alpha$. The scaling of the initial increase with pump intensity and probe wavelengths, its duration that exceeds the cross-correlation of the pump and the probe pulses, and the subsequent "transparency" all concur to rule out two-photon absorption in our experiment.

P.M.F. acknowledges an IBM Faculty Development Award, and a visiting professorship at Université de Paris VI when the experiments were performed. He also thanks N. Bambha for computer simulations.

1. H.T. Grahn, C. Thomsen and J. Tauc, Optics Commun., in press.
2. D.H. Auston, C.V. Shank and P. Lefur, Phys. Rev. Lett. **35,** 1022 (1975).
3. R.B. James, IEEE J. Quantum Electron. **QE-19,** 701 (1983).
4. P.M. Fauchet et al., to be published.
5. M.C. Downer and C.V. Shank, Phys. Rev. Lett. **56,** 761 (1986).

# Spin Dephasing Kinetics of Free Carriers in Alloy Semimagnetic Semiconductors $Cd_{1-x}Mn_xSe$ by One and Two Photon Excitation

*M.R. Junnarkar and R.R. Alfano*

Institute for Ultrafast Spectroscopy and Lasers,
The City College of New York, NY 10031, USA


1. Introduction

Spin-oriented carriers generated in the conduction band by "polarized optical pumping" from heavy hole and light hole valence bands has been known [1,2,3] for over two decades. In zinc-blende structures the optical transition from the valence bands $\Gamma_8$ to the conduction band $\Gamma_6$, shows that there are three times as many electrons excited to a state with spin antiparallel to the excitation photon angular momentum as compared with the spin parallel to it. A similar situation arises for the semimagnetic semiconductor $Cd_{1-x}Mn_xSe$ ($0 \leq x < 0.5$) having a wurtzite lattice. The spin polarization of the conduction electrons depends not only on the selection rules but also on the relaxation mechanisms.

The spin relaxation time $T_S$ can be measured from the time dependence of the luminescence polarization factor[1] which is defined by

$$L(t) = \frac{o^+(t) - o^-(t)}{o^+(t) + o^-(t)}, \qquad (1)$$

where $o^+(t)$ and $o^-(t)$ are luminescence intensities for right and left circular polarization at time t. For the case of crystal field split valence bands $\Gamma_9$ and $\Gamma_7$ in a wurtzite crystal[4] and if one assumes that the holes depolarize very fast ($<10^{-13}$ sec), the luminescence polarization factor $L(t)$ is equal to spin polarization factor[1] $\rho(t)$ for free electrons.

2. Experimental

Photoexcited carriers were produced using a mode-locked Nd:glass laser system generating 6-psec pulses. Two photon excitation was obtained with laser fundamental wavelength ($\lambda$=1060 nm) and single photon excitation ($\lambda$=530 nm) was obtained using second harmonic generation in KDP. Stokes-shifted stimulated Raman photon excitation at 623 nm was achieved using a frequency-doubled pulse (530 nm) passing through ethanol. The details of the time resolved spin polarized photoluminescence setup and picosecond laser system have been described elsewhere[2]. The linearly polarized output of the laser system was circularly polarized (left or right) before excitation and the circularly polarized (left and right) luminescence from the sample was analyzed using a broad band quarterwave plate. A Wollaston prism was placed in front of the entrance slit of a streak camera to spatially resolve the left and right circular polarization of the luminescence. A prepulse was used as a marker for overlapping various data files for averaging, adding, and subtracting using PDP-11 computer. Single shot kinetics for left and right circular polarization of the luminescence was investigated simultaneously to measure $\rho(t)$ and $T_S$.

## 3. Discussion

Using Kane's model for band structure in semiconductors, the spin polarization factor $\rho_1(0)$ for one photon and two photon absorption ($\rho_2(0)$) can be calculated in terms of the band gap and excitation photon energy. For one photon excitations at 530 nm and 623 nm in the case of CdSe with band gap of 1.81 eV (684 nm), the initial polarization factor is calculated to be 48% and 50%, respectively. This is in good agreement with the experimental values. The measured values for $\rho_1(0)$ are 44% and 48%, respectively. The band gap of $Cd_{1-x}Mn_xSe$ for x=0.05 is 1.89 eV. The calculated value of $\rho_2(0)$ is 63%, which compares poorly with the experimental value of 18±5%. Similarly for $Cd_{0.9}Mn_{0.1}Se$ with a band gap of 1.97 eV, the calculated and experimental values of $\rho_2(0)$ are 64% and 28±5%, respectively (Fig.1). This is believed to be due to the existence of $Mn^{2+}$ excited states ($^4G$) in the conduction band. Angle-resolved ultraviolet photoemission spectroscopy studies had clearly indicated strong hybridization of $Mn^{2+}$ d levels with p-like states in the valence band. In such a case, the optical transition is from a mixed ($p-_6A^1$) valence band to a mixed (s-$^4A_1$, $^4E$, $^4T_1$, and $^4T_2$) conduction band. These mixed levels may play an important role in reducing the polarization factor $\rho_2(0)$ by modifying the selection rules.

The spin relaxation times measured in CdSe, $Cd_{0.95}Mn_{0.05}Se$ and $Cd_{0.9}Mn_{0.1}Se$ are ~26 psec, ~16 psec and ~20 psec, respectively (Fig.1). The observed fast spin relaxation of free carriers in semiconductors with the wurtzite structure will be discussed within the framework of the theory proposed by D'yakonov and Perel'[3,5] and extended by Margulis et al.[6]. This theory is applicable to both semiconductors, i.e. CdSe and $Cd_{1-x}Mn_xSe$. A semiconductor without an inversion center has a spin relaxation mechanism whose role rapidly increases in effectiveness with increasing electron energy. This mechanism involves spin splitting of the conduction band proportional to the quasimomentum. This splitting is equivalent to an effective magnetic field acting on the spins, whose direction depends on the direction of this momentum. For a strongly degenerate carrier distribution the spin relaxation rate is given by[6]

$$\frac{1}{T_s} = \frac{16}{3} \frac{\gamma^2 m_e \xi \langle \tau \rangle}{\hbar^4} . \qquad (2)$$

Using $m_e = 0.13 m_0$, $\gamma = 2.56 \times 10^{-31}$ joules-m, and a collision time $\langle \tau \rangle$ of $1 \times 10^{-14}$ sec, the expression for the spin relaxation rate reduces to

$$\frac{1}{T_s} = 6.67 \times 10^{11} \xi, \qquad (3)$$

where the Fermi energy $\xi$ is in eV. This implies that the spin relaxation time in CdSe is 21 psec, for a photogenerated carrier concentration of $5 \times 10^{18}/cm^3$ ($\xi=0.072$ eV), which is quite close to the measured values of ~ 26 psec.

Ultrafast spin relaxation and the initial small spin alignment in $Cd_{1-x}Mn_xSe$ is an indication of rapid spin exchange[7,8] of carrier spins with $Mn^{2+}$ localized spins of S=5/2. For this case, the spin relaxation rate is given by

$$\frac{1}{T_s} = (\frac{1}{9}) \omega_c(x)^2 \tau_c(x), \qquad (4)$$

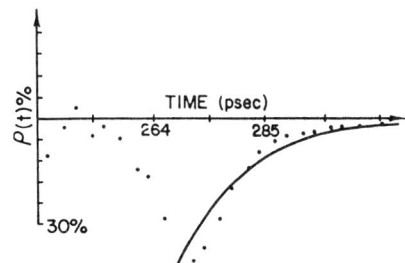

Figure 1. ρ(t) (...) is plotted vs.t for CdSe. The solid line is $\rho(t) = \rho(0)\exp(-2t/T_s)$.

where $\tau_c(x)$ (~$1/x$) is the collision time for electron-$Mn^{2+}$ ion small angle scattering[8] and $\omega_c(x)$ is the precession frequency corresponding to the mean effective magnetic field[7] (~ $x^{3/2}$).

Taking into account the spin relaxation mechanisms of the host crystal (CdSe) and the localized $Mn^{2+}$ ions, the theoretical total spin relaxation rate (for $\xi$=72 meV) is given by

$$\frac{1}{T_s} = 1.08 \times 10^{12}(x^2) + 4.75 \times 10^{10} \text{ in sec}^{-1}. \tag{5}$$

The total spin relaxation times calculated are 17 and 20 psec for x=0.1 and 0.05, respectively. These values are in reasonable agreement with our measurements.

## 4. Summary

The fast spin relaxation observed in CdSe results from a mechanism associated with the noncentrosymmetric character of the band structure for this material. This process is similar to the one proposed by D'yakonov and Perel' for the zinc blende crystal structures. The observed spin polarization factor for carriers in CdSe is in good agreement with theory. The spin relaxation times are < 20 psec in $Cd_{1-x}Mn_xSe$ and are consistent with spin flip Raman scattering measurements. The increase in spin relaxation rate relative to CdSe is explained in terms of the carrier spin exchange between the carriers and the magnetic spin sites.

This research is supported by NSF and AFOSR.

References
1. 'Semiconductors probed by Ultrafast Spectroscopy Vol.2' edited by, R.R. Alfano, Academic Press 1985,p.199.
2. R.J. Seymour and R.R. Alfano,Appl.Phys.Lett. 37, 231 (1980).
3. R.J. Seymour, M.R. Junnarkar and R.R. Alfano, Phys.Rev. B24, 3623 (1981).
4. Donald Long,'Energy bands in semiconductors',chap.7,1968.
5. M.I. D'yakonov and V.I. Perel', Sov. Phys. JETP 33,1053 (1971).
6. A.D. Margulis and V.I.A. Margulis, Sov. Phys. Semicond. 18, 305, (1984)
7. M.I. D'yakonov and V.I. Perel', Sov. Phys. JETP 38, 177 (1974).
8. D. Heiman, P.A. Wolff and J. Warnock, Phys. Rev. B27, 4848 (1983).

# Detection of Higher Order Fourier Components of Index Gratings in Picosecond Transient Grating Experiments

E.O. Göbel[1] and H. Saito[2]

[1] Philipps-Universität, Fachbereich Physik, Renthof 5,
D-3550 Marburg, Fed. Rep. of Germany
[2] Okayama University of Science, Department of Applied Physics,
Okayama 700, Japan

In transient induced grating experiments the decay of an intensity or phase grating produced by interference of two coherent light pulses in e.g. solids or liquids is monitored by diffraction of a third probe pulse. The intensity of the first order diffracted beam is generally detected as a function of time delay between the excitation pulses and the probe pulse to provide information on the dynamics of diffusion and recombination of the photoexcited species /1/. Phase relaxation processes can also be studied if the two interfering excitation pulses are additionally delayed with respect to each other /2/. We report a further extension of this technique, which allows the study of the temporal variation of the spatial shape of an originally sinusoidal index grating by simultaneous detection of the decay of different diffraction orders.

The experiments are performed on thin (10 to 50 μm thick) single crystal platelets of CdS. The excitation and probe pulses were provided by an active-passive mode-locked Nd-YAG laser (25 ps pulse width). The third harmonic pulses ($\lambda$ = 355 nm) with photon energy larger than the band gap of CdS have been used for excitation. The decay of the respective index grating is monitored by measuring the first and second order diffracted light intensity of the second harmonic pulses ($\overline{\lambda}$ = 532 nm), which are transparent for CdS.

The diffraction of the probe pulse is due to the spatially periodic change of the refractive index of the sample by the free carriers and/or excitons. If the j-th Fourier component of the index change is represented by $(\Delta n)_j$, the respective phase modulation of the probe pulse, $\varphi_j$, at wavelength $\lambda$ is given by /1/:

$$\varphi_j = \frac{2\pi d}{\lambda} (\Delta n)_j \quad . \qquad (d = \text{thickness of the grating}) \tag{1}$$

The amplitude of the n-th order diffracted beam, $A_n$, for normal incidence in the Raman-Nath regime is determined by /3/:

$$\frac{\partial A_n}{\partial z} + \frac{1}{2d} \sum_{j=1}^{\infty} \varphi_j [A_{n-j} \exp(i\delta_j) - A_{n+j} \exp(-i\delta_j)] = 0 \quad ; \tag{2}$$

$\delta_1$ is the phase factor. For a purely sinusoidal grating (j=1) the solution is given by the Bessel function of n-th order, i.e. $A_n = J_n(\varphi_1)$. Because $|\varphi_1| \ll 1$, the n-th order Bessel function can be approximated by $(\varphi_1)^n/2n!$ The intensity of the n-th diffraction order is then given by:

$$I_{dif}^{(n)} = A_n^2 \propto \varphi_1^{2n} \propto (\Delta n)^{2n} \quad . \tag{3}$$

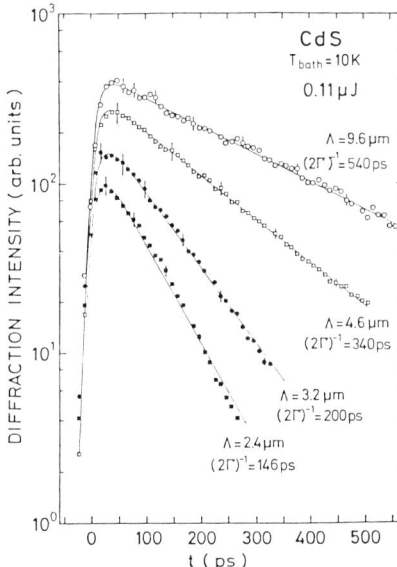

Fig. 1: First order diffraction intensity vs. time for different grating periods at T = 10 K and 0.11 µJ pulse energy

The index change is proportional to the difference of the carrier (or exciton) density at the maximum and the minimum of the grating

$$\Delta n \propto N(0,t) - N(\Lambda/2,t) \equiv \Delta N \quad , \quad (4)$$

where $\Lambda$ is the grating period. The solution of the recombination diffusion equation for a linear recombination process yields:

$$I_{dif}^{(n)}(\Lambda,t) \propto \exp(-2n\Gamma t) \; ;$$
$$\Gamma = \left( \frac{4\pi^2 \cdot D}{\Lambda^2} + \frac{1}{\tau} \right) \; ; \quad (5)$$

D = diffusion coefficient, $\tau$ = recombination lifetime.

We thus obtain an exponential decay for all the different diffraction orders with a time constant which decreases linearly with the order n. Nonlinear processes, however, will result in a distortion of the originally sinusoidal shape of the index grating and thus higher Fourier components will appear. Deviations from eq. (5) are therefore expected. Simultaneous detection of the different diffraction orders then provides a Fourier analysis of the instantaneous index profile.

The low temperature (T~5K) experiments on CdS revealed an exponential decay of the first order diffracted beam at low excitation intensities (0.11 µJ pulse energy corresponds to about 1 mW/cm$^2$) as depicted in Fig. 1. The index changes at these excitation intensity can be attributed to excitons /4/. The time constant $(2\Gamma)^{-1}$ becomes faster with decreasing grating period due to diffusion. The full lines represent calculated curves according to eq. (5) with $\tau$ = 600 ps and D = 21 cm$^2$/s.

At higher excitation intensities the decay of the first order diffraction be-

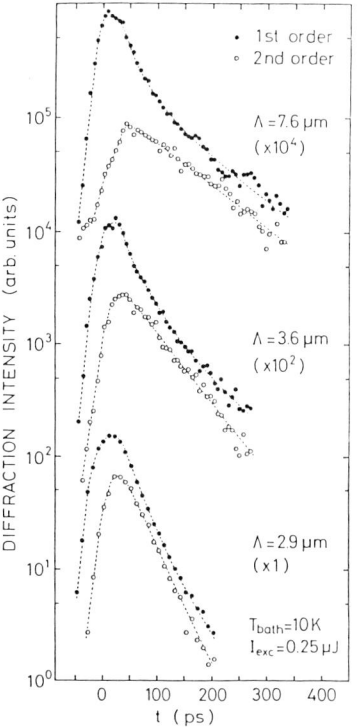

Fig. 2: First (full points) and second order diffraction intensity (open circles) vs. time for different grating periods at T=10K and 0.25 µJ pulse energy

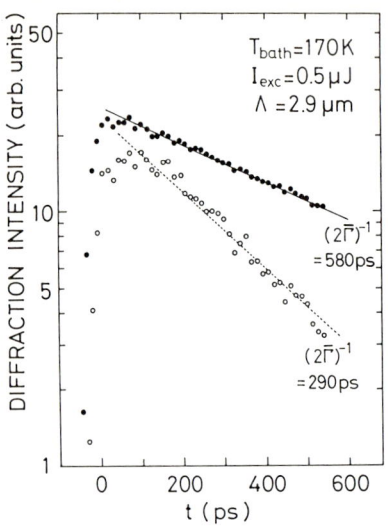

Fig. 3: First (full points) and second order diffraction intensity vs. time at T=170K and 0.5 µJ pulse energy

comes non-exponential and simultaneously the time behaviour of the higher orders deviates from the predictions of eq. (5). Figure 2 depicts the result for an excitation pulse energy of 0.25 µJ. The first order diffraction (full points) shows a fast initial decay followed by a slower decrease as most obvious for $\Lambda = 7.6$ µm. The second order diffraction decays exponentially, however, the position of the maximum is delayed with respect to the first order. In any case, the time constant of the second order is not half of the first order decay constant as expected for a purely sinusoidal grating. We therefore have to conclude that a distortion of the index grating occurs with a pronounced second Fourier component (j=2), which contributes to the second order diffraction (n=2). The deformation of the originally sinusoidal index grating can be attributed to an exciton-free carrier transition occurring at the intensity maxima of the interference pattern due to screening of the Coulomb interaction at high free carrier densities /5/. Excitons thus will be formed only at the wings of the interference pattern. A distortion of the sinusoidal shape of the index profile can be accounted to the different contribution of excitons and free carriers to the refractive index /4/.

Excitonic effects can be neglected at appreciably higher temperatures because of rapid thermal dissociation of the excitons and the index change can be attributed to free carriers only. Figure 3 shows the experimental result for T = 170 K. Both the first and second order decay exponentially with a time constant of 580 ps and 290 ps, respectively, in perfect agreement with eq. (5). This demonstrates that at least no second Fourier component (j = 2) occurs, opposite to the low temperature result.

In conclusion, we have demonstrated that simultaneous detection of different diffraction orders in transient grating experiments provides information on the instantaneous spatial shape of the index (or absorption) grating. Spatially and temporally coupled processes thus can be studied with a resolution determined by the wavelength and the pulse width of the laser pulses, respectively.

1. see e.g. A.L. Smirl: In Semiconductors Probed by Ultrafast Laser Spectroscopy, Vol. 1 (R.A. Alfano, ed.; Acad. Press, Orlando, 1984) p. 197
2. A.M. Weiner, E.P. Ippen: Optics Lett. 9, 53 (1984)
3. W.R. Klein, B.D. Cook: IEEE Trans. Sonics & Ultrasonics 14, 123 (1967)
4. H. Saito: Journ. Luminesc. 30, 303 (1985);
   H. Saito, E.O. Göbel: Phys.Rev. B 31, 2360 (1981)
5. H. Saito, E.O. Göbel: Opt. Lett., June 1986

# Transient Thermoreflectance Studies of Thermal Transport in Compositionally Modulated Metal Films

*G.L. Eesley, C.A. Paddock, and B.M. Clemens*

Physics Department, General Motors Research Laboratories,
Warren, MI 48090, USA

Transient thermoreflectance (TTR) measurements of thermal diffusion use two synchronous picosecond laser pulses. The first pulse produces ultra-fast heating. The second pulse has a variable delay with respect to the heating pulse, and it is used to measure the thermally induced change in surface reflectivity. For small temperature deviations the reflectivity change is linear in temperature, and a temperature profile over several hundred picoseconds can be measured with this technique[1,2]. The penetration depth of visible light in a metal is approximately 20 nm, and thermal diffusion out of this region occurs in a few hundred picoseconds. Therefore, for film thicknesses of 100 nm or greater, the TTR measurement can be completed before substrate effects become important.

Since the heating depth is small compared to the diameter of the illuminated surface, a one-dimensional heat flow model can be fitted to our measurements. Our fitting routine solves the one-dimensional heat flow equation by the method of finite differences. The routine involves a two parameter fit with the thermal diffusivity and a constant scaling factor as the free parameters[2]. The accuracy of the fit is sensitive to the value of the optical absorption coefficient of the material, since we are actually monitoring the flow of heat out of the optical heating depth. We determine the complex refractive index of our samples by measuring the ratio of s-polarized to p-polarized reflectivity for both 30° and 70° angles of incidence. We fit these measurements to the Fresnel reflection formula, and determine the real(n) and imaginary(k) parts of the metal refractive index. The imaginary part is then used to calculate the absorptivity from the relation $\alpha = 4\pi k/\lambda$, where $\lambda$ is the heating laser wavelength.

We have used TTR to measure the thermal diffusivity of single crystals, deposited single element metal films, and compositionally modulated Ni-Zr and Ni-Ti films[2]. We find the thermal diffusivity of the modulated metal films is substantially smaller than that measured for the constituent single element films. This indicates that the interface between two different metallic layers can alter the thermal transport in a direction perpendicular to the film plane. Several analytical models of heat flow in multilayer systems have been developed to include the thermal impedance of boundaries where there is an abrupt change in thermal properties[3,4]. Our eventual goal is to use TTR to measure the thermal impedance of interfaces, by fitting our results to a one-dimensional heat flow model which includes the boundary conditions appropriate to thermal transport across an interface. In this work, we report on a series of measurements of thermal transport across single metal-metal interfaces which have different lattice size mismatch.

The TTR apparatus consists of two synchronously pumped dye lasers which generate 4-psec pulses at a wavelength of 633 nm for the heating pulse, and 595 nm for the probing pulse. The probing pulse can be time delayed relative to the heating pulse, and zero time delay is determined by performing a cross-correlation measurement of the two pulses in a nonlinear crystal placed in the sample position. Lock-in detection techniques are used to measure the thermally induced change in sample reflectivity as a function of time delay between the modulated heating pulse train and the probing pulse train[1,2].

A series of bilayer samples were fabricated to consist of a 30 nm Ni cap over a 300 nm thick metal underlayer. The underlayer metals used were Cu, Mo, Ti and Zr. Multilayer samples of the same metal pairs were also fabricated, in addition to 300 nm thick single element films. The samples were prepared in a dual source magnetron sputter deposition system with a base pressure of $10^{-7}$ Torr. Silicon wafers and Kapton films (for transmission x-ray diffraction) were used as substrates under the simultaneously operating, shielded sputter sources. Deposition rates were held constant at 0.5 nm/sec and the sputtering atmosphere was 2 mT of argon.

The sample atomic structure is currently being investigated by x-ray diffraction and details will be presented in a later publication. However, some initial observations are relevant to the present investigation. The structural order of metal-metal interfaces is very sensitive to the atomic lattice mismatch between the constituent elements. For Ni-metal interfaces, the size mismatch increases in the order Cu, Mo, Ti and Zr. In an isostructural system with small mismatch, such as Ni-Cu, coherent interfaces are produced and some interdiffusion is expected. Despite the structural difference and larger mismatch in the Ni-Mo system, evidence for interface coherence is also observed. On the other hand, there is evidence for a strained or amorphous interfacial region in Ni-Ti and Ni-Zr samples, with the interfacial region being larger in the Ni-Zr sample.

A graphic example of the degradation in thermal transport produced by the presence of a single interface is shown in Fig.1. Transient thermoreflectance measurements are shown for both single element Ni and Ti films, and for a Ni-Ti bilayer film. The decrease in thermal transport due to the metal-metal interface is clearly demonstrated and this effect is also observed in the other bilayer films.

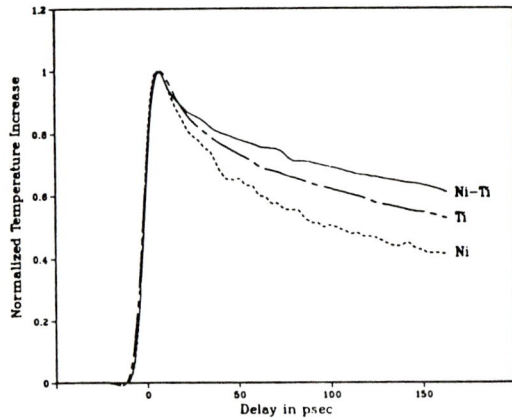

Fig. 1 TTR measurements of Ni, Ti and the Ni-Ti bilayer films

The results of fitting the TTR measurements to our heat flow model are tabulated in Table 1. We have included measurements of the film optical properties, as well as the fitted effective thermal diffusivity of the bilayer and single element films. Figure 2 shows the TTR measurements for all four bilayer films investigated. The thermal diffusion is faster in the Ni-Cu film, followed by the Ni-Mo bilayer film. This ordering would be expected since the thermal diffusivity of Cu and Mo is larger than Ni, Ti or Zr. Also the degree of interfacial order is expected to be better for Ni-Cu and Ni-Mo metal pairs.

A somewhat unexpected result can be found by comparing the thermal diffusion in the Ni-Ti and Ni-Zr bilayer films. In this case we find that the diffusion in the Ni-Ti film is faster than in the Ni-Zr film, even though the thermal diffusivity of Zr is larger than that of Ti (see Table 1). We believe this reversal in trend may be attributed to the fact that the thermal impedance of the Ni-Zr interface is larger as a result of a larger lattice mismatch and the higher degree of interfacial disorder. The relative lattice mismatch for the bilayer samples is calculated from literature values of the nearest-neighbor lattice constants[6], and the results are contained in Table 1. We find that the trend in degradation of thermal diffusion due to a metal-metal interface is correlated with the relative lattice mismatch of the metal

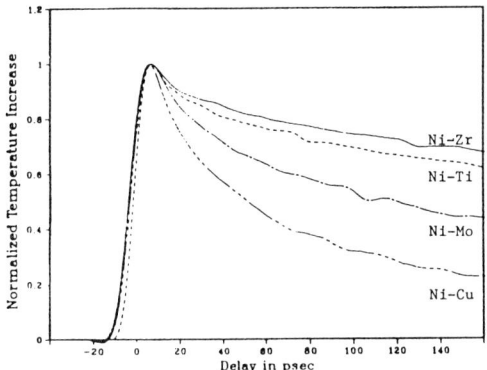

Fig. 2 TTR measurements of the single interface bilayer films

Table 1  Measured values of the metal film refractive index (n,k), the calculated optical absorptivity ($\alpha$) and the fitted thermal diffusivity (D). The lattice mismatch is relative to Ni and is calculated using literature values for the metal nearest-neighbor distance[6]

| Sample | | n | k | $\alpha$ ($\times 10^7$ m$^{-1}$) | D ($\times 10^{-6}$ m$^2$/sec) | Lattice Mismatch (%) |
|---|---|---|---|---|---|---|
| Ni | single element | 1.80 | 3.71 | 7.4 | 4.4 | – |
| Mo | | 3.07 | 3.84 | 7.6 | 8.7 | – |
| Ti | | 2.34 | 3.28 | 6.5 | 1.5 | – |
| Zr | | 2.13 | 3.31 | 6.6 | 2.5 | – |
| Ni–Cu | bilayer film | 1.73 | 3.70 | 7.3 | 32.0 | 2.8 |
| Ni–Mo | | 1.80 | 3.85 | 7.6 | 3.1 | 9.2 |
| Ni–Ti | | 1.63 | 3.54 | 7.0 | 0.51 | 16.1 |
| Ni–Zr | | 1.81 | 3.80 | 7.5 | 0.27 | 27.3 |

constituents. The results of our x-ray diffraction analysis will provide more accurate lattice mismatch values and possibly a quantitative correlation with the measured thermal diffusivity.

We are currently developing a numerical model of interfacial heat flow which can be fit to our TTR measurements on the bilayer systems. We will obtain values for the interfacial thermal impedance and search for a basic correlation with the lattice mismatch/interfacial disorder.

1. C. A. Paddock and G. L. Eesley, Optics Letters 11, 273(1986).
2. C. A. Paddock and G. L. Eesley, to be published in J. Appl. Physics(July 1986).
3. D. L. Balageas, J. C. Krapez and P. Cielo, J. Appl. Physics 59, 348(1986).
4. J. Opsal and A. Rosencwaig, J. Appl. Physics 53, 4240(1982).
5. A measurement of the Cu diffusivity is complicated by nonequilibrium heating effects. See: G. L. Eesley, Phys. Rev. B33, 2144(1986).
6. C. Kittel, Introduction to Solid State Physics, 4th Edition, Ch.1, Table 5 (John Wiley and Sons, Inc., New York, 1971).

# Femtosecond Studies of Nonequilibrium Electronic Processes in Metals

*R.W. Schoenlein[1], W.Z. Lin[1], J.G. Fujimoto[1], and G.L. Eesley[2]*

[1] Department of Electrical Engineering and Computer Science and Research Laboratory of Electronics, Massachusetts Institute of Technology, Cambridge, MA 02139, USA
[2] Physics Department, General Motors Research Laboratories, Warren, MI 48090, USA

In this paper we investigate nonequilibrium electron heating in gold using femtosecond pump and continuum probe techniques. Time resolved measurements of reflectivity allow a characterization of both the excited electron thermal distribution and the dynamics of its cooling to the lattice temperature.

Although this process was predicted over a decade ago, it was not until the development of picosecond and femtosecond laser sources that experimental observations of this phenomena were possible[1-4]. Evidence for nonequilibrium electron heating has been obtained from time resolved reflectivity, photoemission, and transmissivity experiments[2-4]. Recent advances in the generation of femtosecond pulses along with higher resolution measurement techniques now enable a more comprehensive investigation of nonequilibrium electron heating and provide the basis for the examination of a wide variety of nonequilibrium processes in metallic systems.

An ultrashort laser pulse incident on a metal interacts first with the electrons which thermalize rapidly via electron-electron scattering. If the pulses are sufficiently short, electron temperatures in excess of the lattice temperature are generated since the electronic heat capacity is much less than that of the lattice. Thermal relaxation of the electrons occurs primarily through electron-phonon interaction. The dynamics of this process are nonlinear, since the electronic heat capacity is temperature dependent, and may be modelled by coupled nonlinear heat flow equations[1].

Changes in the electronic temperature are indicated by changes in the occupancy of states about the Fermi level $\Delta \rho = \rho(T_1) - \rho(T_0)$ where $\rho(T)$ is the Fermi function. Figure 1 shows this smearing of the occupancy at various temperatures. The line shape displays an inflection point about the Fermi energy, indicating increased occupancy at higher energies, and decreased occupancy at lower energies. At lower temperatures, the line shape narrows and the peaks shift slightly toward the Fermi energy. These changes are observed experimentally through transient reflectivity measurements which monitor electronic transitions from the d-bands to conduction band regions near the Fermi energy. The threshold for this transition in gold is 2.38 eV (522 nm).

Experimental measurements are performed in thin gold films using 65-fs pump pulses at 2.0 eV with a continuum reflectivity probe. The laser system consists of a CPM dye laser with an 8 kHz copper vapor laser amplifier[5]. Differential measurements of transient reflectivity are performed by monitoring the reflected signal from the sample along with a reference signal from the continuum. Signal and reference are simultaneously filter by a monochromator with ~3-nm resolution and transient measurements are performed for different probe photon energies.

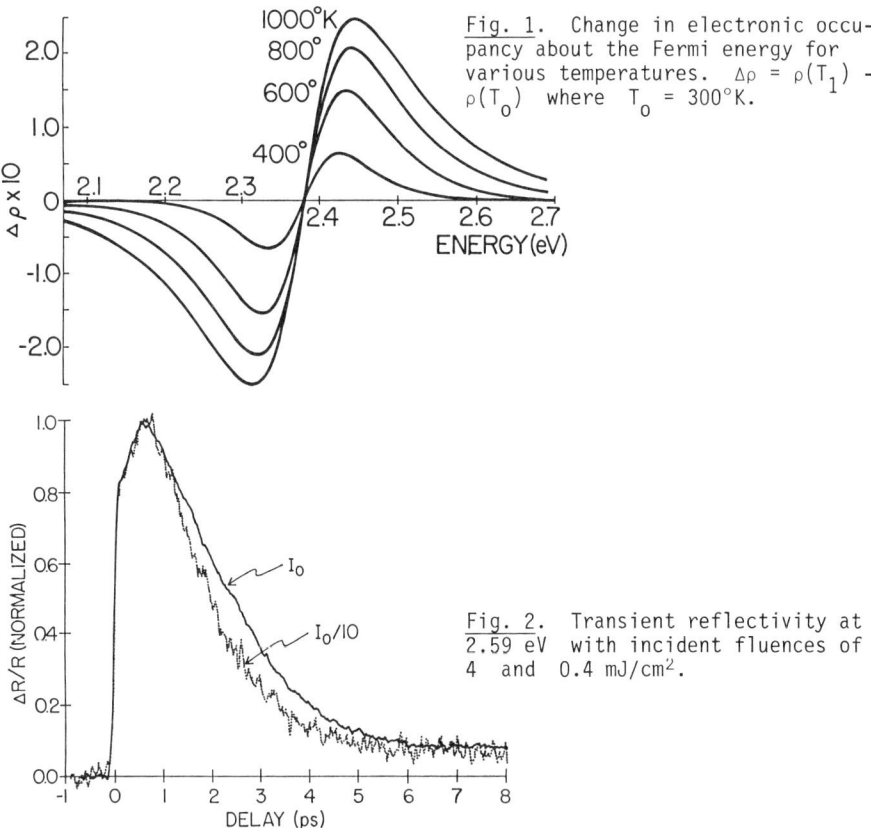

Fig. 1. Change in electronic occupancy about the Fermi energy for various temperatures. $\Delta\rho = \rho(T_1) - \rho(T_0)$ where $T_0 = 300°K$.

Fig. 2. Transient reflectivity at 2.59 eV with incident fluences of 4 and 0.4 mJ/cm².

Figure 2 shows transient reflectivity measurements at 2.59 eV for pump fluences of 4 and 0.4 mJ/cm. We observe an increase in reflectivity $\Delta R/R > 0$ and relaxation times of ~3 and ~2 ps respectively. At higher pump fluence $I_0$ the peak change in reflectivity $\Delta R/R$ is ~$10^{-2}$ and the estimated electron temperature is ~1000°K while the change in lattice temperature is several tens of degrees. At the lower pump fluence the peak reflectivity change is $10^{-3}$. The long-time component of the reflectivity signal results from residual lattice heating which recovers in tens of picoseconds according to diffusion cooling. Probing below the Fermi energy we observe a decrease in reflectivity $\Delta R/R < 0$ which displays similar relaxation behavior. These results correspond to increased electronic occupancy above the Fermi energy and decreased occupancy below the Fermi energy resulting from electron heating. The increase in relaxation time at higher pump fluence is consistent with the temperature dependence of the electronic heat capacity.

Figure 3 shows transient reflectivity measurements at probe energies near the largest change in electronic occupancy 2.59 eV and in the tails of the distribution 2.69 eV. The decreased relaxation time observed at 2.59 eV arises from the fact that the change in the distribution $\Delta\rho$ narrows with decreasing temperature. This is consistent with uniform electron cooling. Complementary behavior is observed at comparable probe photon energies below the Fermi energy.

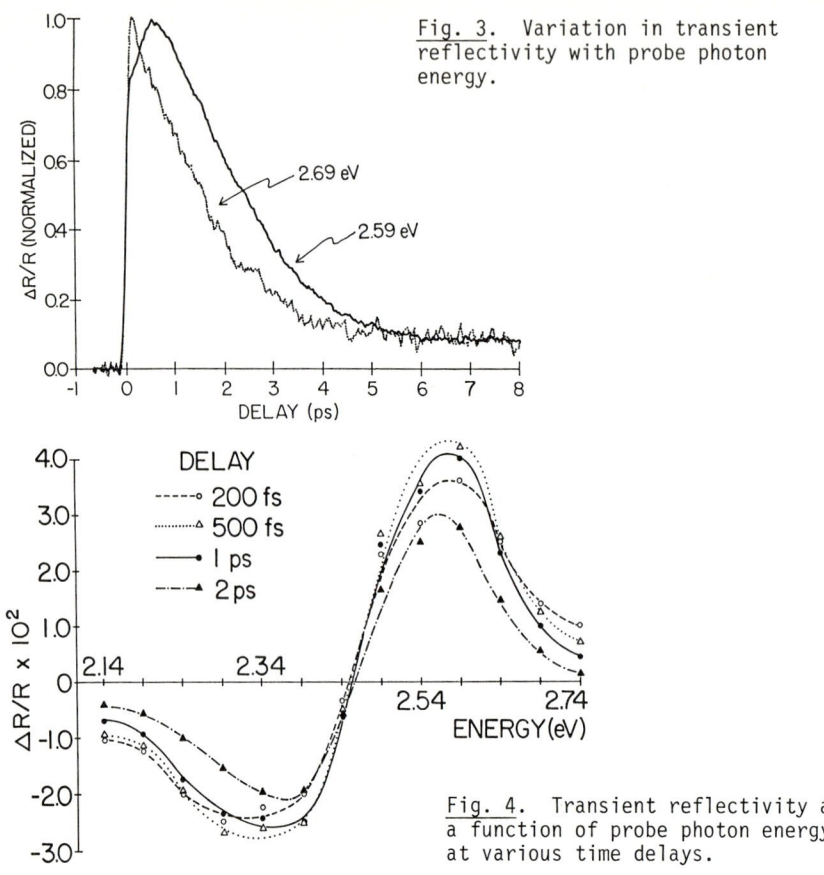

Fig. 3. Variation in transient reflectivity with probe photon energy.

Fig. 4. Transient reflectivity as a function of probe photon energy at various time delays.

Figure 4 shows measured transient reflectivity as a function of energy at various time delays. This is a compilation of time-resolved traces obtained for different probe photon energies. The shape follows the change in electronic occupancy (Fig. 1), displaying an inflection point near the Fermi energy and antisymmetric peaks. The narrowing of these peaks, and their shift toward the Fermi energy with increasing time delay indicates cooling of the electronic distribution and is consistent with theoretical predictions.

An interesting feature observed in the data of Figs. 2-4 is the transient increase in $\Delta R$ for time delays less than 1 ps. This is observed for probe photon energies near peak changes in occupancy above and below the Fermi energy. At present, the origin of this behavior is not completely understood. However, we believe that it is not a delayed heating process since the reflectivity line shape suggests a uniform cooling immediately after the pump pulse. More detailed continuum studies as well as consideration of the connection between transient reflectivity and electronic occupancy will yield a better understanding of this process.

In summary, femtosecond continuum pump-probe techniques have been applied to investigate nonequilibrium electron dynamics in metals. Relaxation times of the transient reflectivity were measured and found to depend on the conduction band energy being probed, and on the incident fluence. These results are consistent with the dynamic cooling of the electronic distribution. The measurement technique allows the separation of electronic processes from phonon or lattice heating effects. Future extensions of these studies may allow the investigation of electron transport in metals and nonequilibrium phenomena in semiconductor devices.

We wish to thank E. P. Ippen and S. D. Brorson for helpful discussions and technical assistance. W. Z. Lin is visiting from the Zhongshan University, Guangzhou, People's Republic of China. This work was supported in part at MIT by the Joint Services Electronic Program Contract DAAL03-86-K-0002.

REFERENCES

1. S. Anisimov, B. Kapeliovich, T. Perel'man: Sov. Phys. JETP 39, 375 (1974)
2. G. Eesley: Phys. Rev. Lett. 51, 2140 (1983); G. Eesley: Phys. Rev. B 33 (1986)
3. J. Fujimoto, J. Liu, E. Ippen, N. Bloembergen: Phys. Rev. Lett. 53, 1837 (1984)
4. H. Elsayed-Ali, M. Pessot, T. Norris, G. Mourou: Ultrafast Phenomena V, Snowmass, CO (1986)
5. W. Knox, M. Downer, R. Fork, C. Shank: Opt. Lett. 9, 552 (1984)

# Time-Resolved Observation of Electron-Phonon Relaxation During Femtosecond Laser Heating of Copper

*H. Elsayed-Ali, M. Pessot, T. Norris, and G. Mourou*

University of Rochester, Laboratory for Laser Energetics,
250 East River Road, Rochester, NY 14623, USA

Thermal modulation of the optical properties of metals is a widely used technique in studying critical points in band structure. Recently the modulation of reflectivity of copper has been used to observe nonequilibrium electron-lattice temperatures during picosecond (~5 ps FWHM) laser heating of up to a few degrees, EESLEY [1]. Although nonequilibrium heating was demonstrated in these experiments, the time resolution was insufficient to resolve electron-phonon relaxation. In a subsequent report, the phenomenon of thermally enhanced multiphoton photoemission was used to time-resolve electron-phonon relaxation in tungsten, FUJIMOTO et al. [2]. Results indicated that such relaxation is accomplished in a few hundred femtoseconds. We report results obtained using amplified 150-300 fs laser pulses to time-resolve electron-phonon relaxation by monitoring the laser-heating-induced modulation of the transmissivity of 200 Å copper films.

A 1 KHz synchronously amplified colliding pulse mode-locked laser, DULING et al. [3], ($\lambda \simeq 620$ nm) was used for the pump-probe experiments. The sample was heated using the 620 nm fundamental. Probing was accomplished at 620 nm or using a 10 nm (FWHM) band from white light generated by focussing the probe beam on an ethylene glycol cell. The pump and probe were incident collinearly normal to the copper film (polarized perpendicular to each other) and focussed to ~27 and ~14 µm diameter spots respectively, such that the probe was near the center of the pump.

The transmissivity of the thin copper films at $\lambda = 620$ during laser heating (~300 fs FWHM) for a pump laser fluence of 15 nJ (a) and 65 nJ (b) are shown (Fig. 1). The initial response of the transmissivity appears to integrate the heating pulse. The decay of the fast transient was found to be 1-4 ps and increases with the heating pulse fluence. This effect is due to larger differences between electron and lattice temperatures for higher fluences, thus, more electron-phonon collisions are required for thermalization.

A simplified numerical model of nonequilibrium heating of copper was constructed and applied to conditions in Fig. 1. This model is based on a solution of two coupled nonlinear differential equations in the form, EESLEY [1] and FUJIMOTO et al. [2]:

$$C_e(T_e) \frac{dT_e}{dt} = P_o(t) - G(T_e - T_l), \qquad (1)$$

$$C_l \frac{dT_e}{dt} = G(T_e - T_l), \qquad (2)$$

where $C_e(T_e)$ is the electronic heat capacity which is directly proportional to the electron temperature, $P_o(t)$ represents the laser heating pulse, and G is the

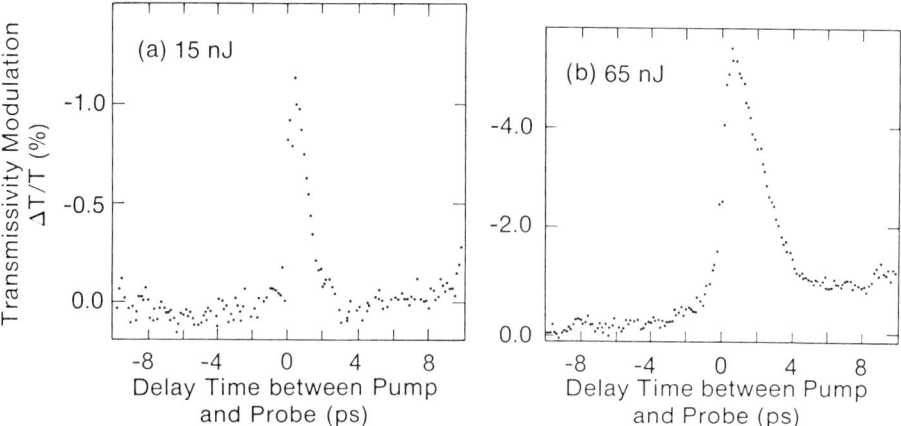

Fig. 1  Time-resolved transmissivity of 200 Å copper film during laser heating (~300 fs FWHM at $\lambda \simeq 620$ nm).

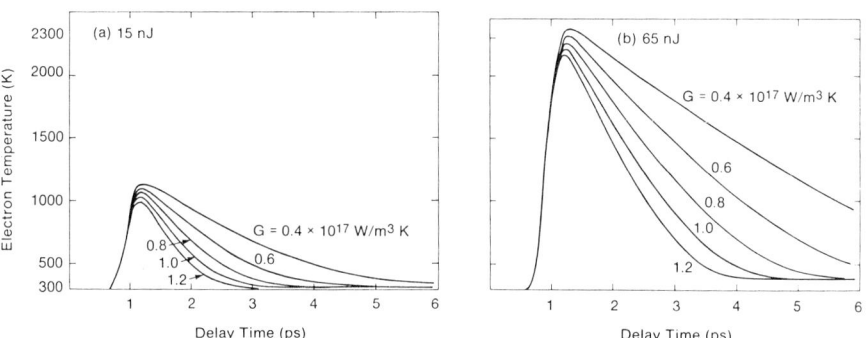

Fig. 2  Numerical modeling of the time evolution of electron temperature for experimental condition of Fig. 1.

electron-phonon coupling constant. Thermal conductivity losses were ignored due to the thin film geometry used in the present work. Simulations of conditions used in Fig. 1 for different values of electron-phonon coupling constant, G, are shown in Fig. 2. Results indicate that G has a value of $\sim 1 \times 10^{17}$ W/m$^3$·K. For a pulse energy of 65 nJ (peak fluence $\simeq 3.8 \times 10^{10}$ W/cm$^2$), the model predicts a peak electron temperature of 2200 K and an equilibrium electron-phonon temperature of 385 K.

Using white light in 10 nm steps from $\lambda = 560$ to 640 nm ($\lambda = 590$ nm corresponds to an electron transition from the top of the d-band to the Fermi level) showed similar behavior as when probing at $\lambda = 620$ nm.

We have directly measured the electron-phonon relaxation time in copper as a function of pump laser fluence and probe photon energy for $\lambda = 560$ to 640 nm. We have demonstrated nonequilibrium heating with a large (few thousand K) difference between electron and lattice temperatures. Electronic and lattice effects on the optical properties of copper were separated in time. Extension of probe measurements to the near IR and UV parts of the spectrum would separate effects of bound and free electrons on the optical properties and provide considerable information on the band structure.

## Acknowledgement

This work was supported by the Laser Fusion Feasibility Project at the Laboratory for Laser Energetics which has the following sponsors: Empire State Electric Energy Research Corporation, General Electric Company, New York State Energy Research and Development Authority, Ontario Hydro, Southern California Edison Company, and the University of Rochester. Such support does not imply endorsement of the content by any of the above parties.

## References

1. G.L. Eesley: In Phys. Rev. Lett. 51, 2140 (1983).
2. J.G. Fujimoto, J.M. Liu, E.P. Ippen, and N. Bloembergen: In Phys. Rev. Lett. 53, 1837 (1984).
3. I.N. Duling III, T. Norris, T. Sizer II, P. Bado, and G.A. Mourou: In J. Opt. Soc. Am. B 2, 616 (1985).

# Femtosecond Carrier Relaxation in Semiconductor-Doped Glasses

## M.C. Nuss*, W. Zinth, and W. Kaiser

Physik Department der Technischen Universität München, Arcisstr. 21, D-8000 München 2, Fed. Rep. of Germany

Semiconductor-doped glasses incorporating small $CdS_xSe_{1-x}$ or $CdSe_xTe_{1-x}$ crystallites in a glass matrix, show strong nonlinear absorption /1/ and large values of the nonlinear optical susceptibility $\chi_3$ /2/. The semiconductor crystallite system currently attracts increasing attention when the semiconductor inclusions are small enough to expect quantum effects.

The semiconductor-doped glassfilters studied (Schott RG 830, RG 715, and RG 645) are characterized by an exponential absorption edge $\sim$ three times less steep than the corresponding bulk semiconductors. The frequency position of the absorption edge varies with composition, but it may as well be influenced by the size of the semiconductor inclusions. No confinement effects are observed in the absorption spectra at 300 K, most probably due to a certain variation of crystallite size.

Femtosecond absorption recovery of all three Schott filter glasses was studied. In the experiment, femtosecond light pulses of $\simeq$ 60 fs duration and $\lambda$ = 620 nm from a colliding-pulse modelocking dye laser generate an electron-hole plasma with a carrier density $N \simeq 3 \times 10^{17}$ cm$^{-3}$ in the semiconductor inclusions. A weaker probing pulse samples the absorption changes as a function of the delay time between exciting and probing pulses. All data were recorded at room temperature.

Fig.1a,b shows the time-resolved absorption changes for the semiconductor-doped glass filters RG 645 and RG 830. The absorption change $\Delta A$ is plotted versus delay time between exciting and probing pulses. The dashed curves are the cross-correlation traces determining time zero. In the RG 830 semiconductor-doped glass (Fig.1b) the energy of the femtosecond laser pulses is well above the bandgap, creating carriers with an excess energy $\Delta E \simeq 500$ meV. A fast recovery of the initial bleaching is observed with a time constant of 230 fs. This process can be identified with cooling of the initially hot electron gas to the lattice temperature. A comparison with the expected energy loss rate of the electron gas due to the emission of LO-phonons in crystalline semiconductors /3/ yields similar relaxation times for the absorption recovery. In RG 645 (Fig.1a) the opposite extreme is realized. The excess energy is only slightly larger than the bandgap. The absorption decreases with the integrated pulse intensity. No fast carrier relaxation process is observed since the carriers are injected with low excess energy. Consequently, the temperature of the electron gas will not change and no fast absorption recovery is observed. The bleaching is due to the filling of states in the conduction and valence bands and remains essentially constant over the plotted time.

---

*Present address: AT&T Bell Laboratories, 600 Mountain Ave., Murray Hill, N.J. 07974, USA

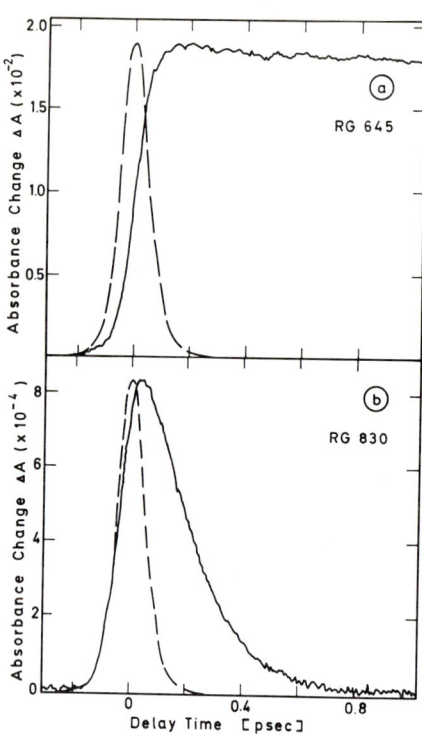

Fig.1  Femtosecond absorption recovery of semiconductor-doped glass filters RG 645 (Fig.1a) and RG 830 (Fig.1b).

Actually, $\Delta A$ decays with a long time constant of $\sim$ 100 ps (not shown in the figure). We attribute the 100 ps time constant to electron-hole recombination. This short recombination time is not unreasonable in the light of the large number of surface states in the semiconductor-doped glasses, reacting as efficient recombination centers.

The most interesting case is the semiconductor-doped glass RG 715 (Fig.2), where the initial carrier excess energy $\Delta E$ is $\sim$ 300 meV. Here, both the femtosecond decay process as well as the bandfilling can be traced. The three solid curves correspond to different excitation densities, where the lower curves are for two times (curve b) and four times (curve c) attenuated laser pulses, respectively. A semilog plot of the data reveals two time constants; 250 fs for the fast initial absorption recovery and 85 ps for the carrier recombination process. Significant is the superlinear decrease with carrier density of $\Delta A$ for times > 1 ps: $\Delta A$ decreases four times faster if the carrier density is reduced by a factor of 2.

The data can be explained assuming a carrier density dependent bandfilling process. In particular, we think of energy levels or traps below the conduction band, acting as a sink for the conduction band electrons. If the density of these levels is about the same as the carrier density ($\sim 10^{17}$ $cm^{-3}$) and the trapping time is several hundfred femtoseconds, the capture of the carriers into these levels becomes saturated, leading to the nonlinear dependence of $\Delta A$ on carrier density observed in Fig.2. The fast 250 fs relaxation process is a combination of cooling of the electron gas due to LO-phonon emission and trapping of charge carriers. From there, electron-hole recombination takes place via surface states in the middle of the bandgap with a time constant of 85-100 ps.

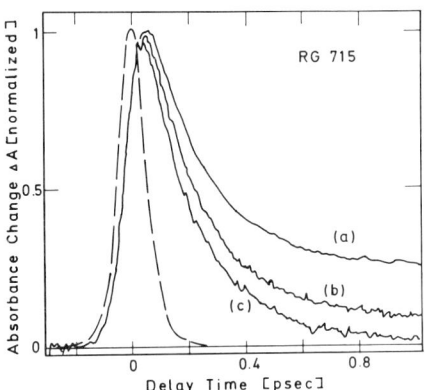

Fig.2 Absorption recovery of the semiconductor-doped glass RG 715

In conclusion, we point out that the absorption changes observed for the three semiconductor-doped glasses investigated here can be well explained within the framework of bulk semiconductor physics. The observed processes are cooling of a hot electron, electron trapping, and recombination. There is no indication of quantum size effects in the currently used glasses. We expect that excitonic effects will only be apparent for semiconductor-doped systems with smaller crystallites and with homogeneous size distribution.

1  G. Bret, F. Gires, Appl. Phys. Lett. 4 (1964) 175
   B. Danielzik, K. Nattermann, D. von der Linde, Appl. Phys. B38 (1985), 31
2  R.K. Jain, R.C. Lind, J. Opt. Soc. Am. 73 (1983) 647
   P. Roussignol, D. Ricard, K.C. Rustagi, C. Flytzanis, Optics Commun. 55 (1985) 143
3  J. Shah, Solid State Electron. 21 (1978) 43

# Femtosecond Dynamics of Electron-Hole Plasma in Semiconductor Microcrystallite Doped Glass

G.R. Olbright, B.D. Fluegel, S.W. Koch*, and N. Peyghambarian

Optical Sciences Center, University of Arizona, Tucson, AZ 85721, USA

Using femtosecond laser excitation, electrons in $CdS_xSe_{1-x}$ semiconductor microcrystallites are excited a few LO phonon energies above the bandgap at room temperature. The relaxation of this initially nonthermalized carrier distribution to quasi-thermal-equilibrium is time resolved and its subsequent cooling to the lattice temperature is studied. The dynamics of photo-excited electron-hole plasma in semiconductor microcrystallites reveals time-dependent participation of the many-body effects of plasma screening, bandgap renormalization, broadening of tail states and bandfilling. Renormalization of the bandgap, broadening of the resonances and screening of the continuum states is especially important for t <100 [fs]. Bandfilling due to carrier relaxation dominates after 200 [fs], with the onset of electron-hole recombination in a few [ps].

A high-repetition-rate femtosecond pump-probe technique is used to measure the optically induced changes in transmittance of the sample. The output of a balanced colliding-pulse mode-locked ring dye laser [1] is amplified with a copper vapor laser [2] at 7.3 [kHz] to 2 [μJ] energy and 200 [fs] duration. A fraction of the amplified pulse at 620 [nm] is used as a pump pulse while the remainder is focused to $10^{13}$ [W/cm²] on an ethylene glycol jet producing a broad-bandwidth probe pulse. The transmission of the broad-bandwidth probe was measured using a spectrometer and an optical multichannel analyzer (OMA III). The sample consisted of a 750-μm-thick optically polished $CdS_{0.24}Se_{0.76}$ Corning CS 2-64 sharp cut color filter consisting of nominally ≈120 Å microcrystallites embedded in a borosilicate glass matrix. The large size distribution of the microcrystallites in commercially available sharp cut color filters results in the observation of a featureless absorption edge at liquid helium temperatures, evidence of extreme inhomogeneous broadening of excitonic resonances and the absence of quantum size effects.

Fig. 1a and 1b show the absorption of the probe pulse as a function of incident photon energy for various time delays between the pump and probe pulses. The spectrum labeled -450 [fs] is taken when the probe pulse precedes the pump and therefore is representative of the unexcited sample. The spectrum labeled 0 [fs], where pump and probe pulses overlap in time, exhibits a red shift of the absorption edge at wavelengths in the vicinity of the bandgap. The 200 [fs] spectrum shows a prominent shift of the absorption edge toward higher energies completely dominating the redshift. This suggests that carriers have relaxed to the bottom of the band in a few hundred femtoseconds and initiated bandfilling. The blue shift takes its maximum value in 400 [fs]. The spectra taken from 400 [fs] to 1 [ps] are similar with same magnitude of the blue shift. Finally, after a few picoseconds the band-edge shift begins to recover via electron-hole recombination with nearly complete recovery after +40 [ps]. We observe a small residual shift which persists for times in excess of 500 [ps] attributed to carriers confined to traps.

We model the system by solving the time-dependent carrier-concentration rate equation for 200 [fs] duration hyperbolic secant excitation and 30 [ps] recombination lifetime. The rate equation for the carrier density, neglecting two-photon absorption and higher order effects, is given simply by

$$\frac{dN(t)}{dt} = \frac{N_0}{\tau_r} \operatorname{sech}\left(\frac{t}{t_0}\right) - \frac{N(t)}{\tau_r} . \tag{1}$$

* Jointly with Physics Department

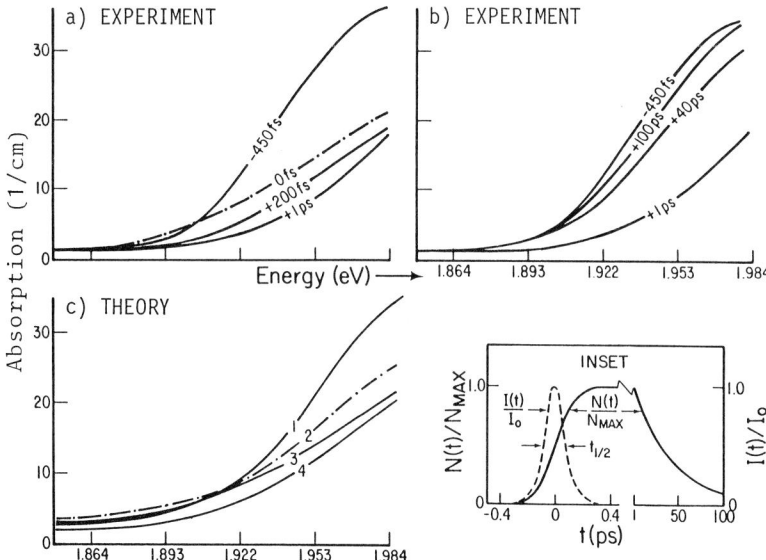

Fig. 1a and b) Room temperature absorption spectra of a 750-μm glass doped with $CdS_{0.24}Se_{0.76}$ microcrystallites for various pump-probe delays. Fig. 1c) Plasma theory calculation, 1) T = 300 [K], N = $10^{16}$ [cm$^{-3}$], 2) T = 550 [K], N = 5 x $10^{17}$ [cm$^{-3}$], 3) T = 450 [K], N = 9.5 $10^{17}$ [cm$^{-3}$], 4) T = 350 [K], N = $10^{18}$ [cm$^{-3}$], Inset; relative temporal evolution of the excitation pulse, I(t), and the carrier concentration, N(t), $N_{MAX}$ = 0.006 $N_o$.

The solution yields a carrier density which initially follows the excitation pulse and then becomes nearly constant for an interval of the first few picoseconds of excitation and then falls off exponentially, as shown in the inset in Fig. 1. The peak carrier density reaches $N_{MAX} = 0.006 N_o$, where $N_o = \alpha I_0 \tau_r / \hbar \omega$ is the steady-state carrier density. Here, $\alpha$, assumed to be constant, is the absorption coefficient at the excitation photon energy $\hbar \omega$ (eventual hole-burning effects are assumed to be weak for the excitation high in the band), $I_o$ is the peak pulse intensity, $\tau_r$ is the recombination lifetime, and $t_0$ is related to the FWHM of the excitation pulse. We model the absorption of the composite by solving for the imaginary part of the average dielectric function of the microcrystallites embedded in the glass matrix using the Maxwell-Garnet equation. According to the plasma theory[3] the dielectric function of the crystallites, $\epsilon(\omega)$, and the absorption, $\alpha(\omega)$, for the above calculated carrier densities are written as

$$\text{Im } \epsilon(\omega) = 8\pi^2 r_{cv}^2 \tanh\left[\frac{\beta}{2}(\hbar\omega - \mu_e - \mu_h)\right] \sum_m \phi_m(0)^2 \delta_\Gamma(\hbar\omega - E_m), \quad (2)$$

$$\alpha(\omega) = \frac{\omega\sqrt{\epsilon_g}}{c} \text{ Im } \left[\frac{\epsilon(\omega)(1 + 2p) + 2\epsilon_g(1 - P)}{\epsilon(\omega)(1 - P) + \epsilon_g(2 + P)}\right], \quad (3)$$

where $\epsilon_g$ = dielectric function of the host glass, p is the volume fraction occupied by microcrystallites, $r_{cv}$ is the conduction-valence band matrix element (assumed here to be independent of energy), $\mu_{e,h}$ are the e-h chemical potentials, and the sum in Equation [2] runs over all e-h pair states (bound and unbound). The tanh term describes filling of the band. The $\phi_m$ (r=0) are solutions of the modified Wannier equation, with the screened Coulomb potential, the $E_m$ are the corresponding energy eigenvalues, and $\delta_\Gamma$ is the phenomenologically introduced broadening. $\phi_m(0)$ and $\delta_\Gamma$

are implicitly functions of the carrier density N and the plasma temperature Tp. In Fig. 1c for comparison we show the computed absorption spectra for various N and Tp. Curve 1 is the computed low excitation $N = 10^{16} [cm^{-3}]$ absorption spectrum (linear spectrum) which is scaled to the $-450 [fs]$ curve of Fig. 1a. Curve 2 is computed for $T_p = 550 [K]$ and $N = 5 \times 10^{17} [cm^{-3}]$. This curve should be compared with the spectrum labeled 0 [fs] in Fig. 1a. However, a calculated spectrum with a nonthermal carrier distribution also gives a spectrum similar to curve 2. This suggests that spectrum labeled 0 [fs] either originates from a distribution of carriers at quasi-equilibrium with a high plasma temperature, or a nonthermal distribution. Curves 3 and 4 are computed for different plasma temperatures $T_p = 450$ and $350 [K]$, respectively, with $N = 9.5 \cdot 10^{17}$ and $10^{18} [cm^{-3}]$, respectively, and should be compared with spectra labeled 200 [fs] and 1 [ps] in Fig. 1a. These results are in good qualitative agreement with the experimental results, indicating that plasma cooling by LO phonon scattering and the corresponding thermalization of the carrier temperature to lattice temperature takes place in 400 [fs] (the 400 [fs] spectrum is similar to the 1 [ps] spectrum). This also implies that band-filling is mainly responsible for the steady-state optical nonlinearity of these materials.

In another series of experiments, the time delay between the pump and probe pulses was kept constant at 1 [ps] and the pump intensity (and thereby the carrier density) is varied. The delay of 1 [ps] was chosen because the plasma temperature almost reaches the lattice temperature before that time (as was described in the last paragraph) and therefore a steady state can be assumed. (The final approach towards the lattice temperature takes place by accoustic phonon scattering and is a comparatively slow process [5]). The results of these measurements are shown in Fig. 2a where we observe a blue shift of the absorption spectrum as the pump intensity is increased. The calculated spectra which are shown in Fig. 2b display a good agreement between the experiment and the theory. These results reconfirm that bandfilling nonlinearity is the dominant mechanism in the steady state, consistent with the results of our nanosecond experiments in a $CdS_{0.9}Se_{0.1}$ doped glass [4].

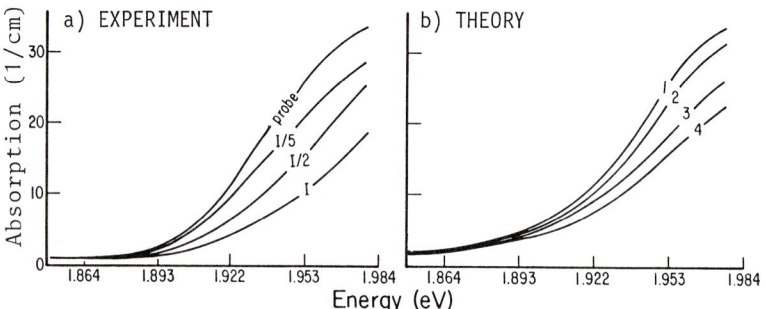

Fig. 2a) Experimental absorption spectra at a fixed pump-probe delay of 1 [ps] for various pump intensities. 2b) Theoretical calculations, for $T_p = 350 [K]$ curve 1) $N = 10^{16} [cm^{-3}]$, 2) $N = 2 \times 10^{17} [cm^{-3}]$, 3) $N = 5 \times 10^{17} [cm^{-3}]$, 4) $N = 10^{18} [cm^{-3}]$.

In conclusion, we have presented an experimental and theoretical study of the role of plasma screening and bandgap renormalization, broadening of the resonances, carrier relaxation, and band-filling, at various times after initial excitation. Evolution of the hot electron gas from nonthermal equilibrium to quasi-equilibrium within a couple of 100 [fs] and complete thermal relaxation in 400 [fs] have been observed.

We would like to acknowledge support from the National Science Foundation, the Air Force Office of Scientific Research, The Army Research Office and the Optical Circuitry of the University of Arizona. G. R. O. thanks the Newport Research Corporation and the Optical Society of America for the NRC fellowship and award.

S. K. thanks the Deutsche Forschungsgemeinschaft for a Heisenberg fellowship and B. Batdorf for numerical assistance.

1. R. L. Fork, B. I. Greene and C. V. Shank, Appl. Phys. Lett. 38, 671 (1981).
2. W. H. Knox, M. C. Downer, R. L. Fork and C. V. Shank, Opt. Lett. 9, 552 (1984).
3. L. Banyai and S. W. Koch, Z. Physik B63 (1986).
4. G.R. Olbright and N. Peyghambarian: Appl. Phys. Lett. 48, 1184 (1986).
5. See e.g., Proceedings of the International Conference on High Excitation and Short Pulse Phenomena, Trieste, Italy, J. Lumin. 30 (1985).

# High-Contrast Ultrafast Phase Conjugation in Semiconductor-Doped Glass

D. *Cotter*

British Telecom Research Laboratories, Martlesham Heath, Ipswich, Suffolk IP5 7RE, UK

Glasses doped with semiconductor microcrystals (such as $CdS_xSe_{1-x}$) are available commercially as inexpensive steep long-pass colour filters having bandgap energies which span the UV to near-IR spectral region. The discovery by Jain and Lind [1] that these materials exhibit large bandgap-resonant nonlinearity has aroused considerable interest in their possible application for all-optical signal processing. This interest was increased by the observation by Yao et al [2] of fast response times for the optical nonlinearity measured by degenerate four-wave mixing (DFWM); a fast nonlinearity, not resolved by the 30 ps laser pulse-width, was observed superimposed on a strong slow background contribution which was interpreted as being thermal in origin. However, in similar studies by Roussignol et al [3] such a fast nonlinearity was not observed.

This paper reports DFWM in semiconductor-doped glass using sub-picosecond excitation; for the first time the phase-conjugate reflectivity has been time-resolved and the contributions clearly identified. The nonlinearity is found to be entirely electronic in origin; there is no detectable thermal contribution. The risetime for the transient phase-conjugate reflectivity is ~6 ps and the falltime, which is energy-density dependent, spans a wide range from ~10 ns to as fast as ~10 ps. Clear evidence is found for plasma trapping in the semiconductor microcrystals. The most significant new observation is that under appropriate conditions an order-of-magnitude enhancement of the fastest (~10 ps) phase-conjugate signal with respect to the slower background contribution can be obtained, and without saturation.

The experiments reported here were made using Schott OG590 colour filter glass; two different samples of this glass type gave identical results. The excitation source was a synchronously pumped hybrid-mode-locked cavity-dumped R6G dye laser which generated 20 nJ sub-picosecond pulses at 580 nm. A conventional DFWM geometry was used [1] with the laser output split into three beams incident on the sample: two collinear opposing beams (forward and backward beam) and an oblique probe beam incident at an angle of 3° to the forward-beam direction. Figures 1 and 2 show the phase-conjugate signal generated by diffraction of the backward beam from the transient grating formed by the interference of the forward- and probe beams. The inset in fig. 1 shows the DFWM signal as a function of the probe delay; the points describe the field autocorrelation function of the laser and the width (540 fs FWHM) represents the experimental resolution. The DFWM signal as a function of the backward-pulse delay provides a measure of the phase-conjugate response times. The risetime was consistently found to be 6-8 ps, independent of the incident energy density up to the maximum investigated (55 $\mu J/cm^2$). Although fully resolved, this risetime is limited by the beam geometry and 2 mm sample thickness, and is substantially slower than the sub-picosecond intra-band relaxation rates recently reported [4]. At the maximum energy density the initial falltime for the phase-conjugate reflectivity was ~10 ps. A lower-intensity background signal persisted with little change for delay times up to at least ~500 ps. At lower incident energy densities the initial falltime was slower (e.g. 30 ps at 20 $\mu J/cm^2$) and the ratio of peak to background was reduced. At 0.5 $\mu J/cm^2$ (fig. 2) the 6 ps risetime was followed by the slow decay with no observable fast response. Notice however that the phase-conjugate reflectivity is essentially zero after a time corresponding to the inter-pulse separation (260 ns). From the available data the time constant for this slower decay can be estimated as lying in the range 5 - 50 ns, which is consistent with the carrier lifetimes measured by photoluminescence in similar glasses [3]. The observed picosecond risetime and nanosecond decay time eliminate the possibility of a significant thermal contribution.

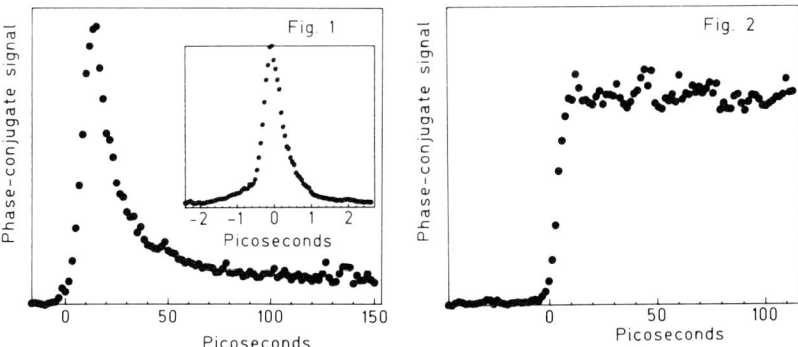

Figures 1,2: Phase-conjugate signal as a function of backward-beam pulse delay at incident energy densities of 55 $\mu J/cm^2$ (fig. 1) and 0.5 $\mu J/cm^2$ (fig. 2). Figure 1 inset: Phase-conjugate signal as a function of probe-beam pulse delay

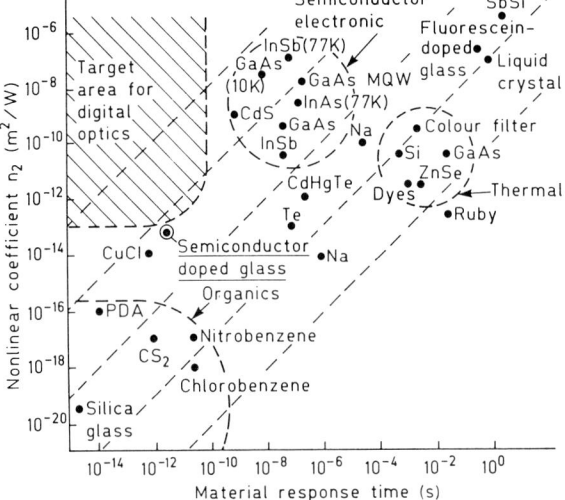

Figure 3: Survey of nonlinear optical materials by Kerr coefficient and dominant material response time

As shown in fig. 1, under appropriate conditions phase-conjugate reflectivity with an overall impulse response time of ~15 ps (FWHM) can be as much as an order-of-magnitude greater than the slower background contribution, and furthermore showed no sign of saturation. The strong saturation effects seen previously with nanosecond excitation [1] were absent from the sub-picosecond-resolved measurements. The observed high contrast greatly increases the attraction of semiconductor-doped glasses for signal-processing devices. The optical energy densities required can be achieved at modest input power by using guided-wave structures.

Figure 3 shows an up-dated survey of materials displaying Kerr nonlinearity; the nonlinear refractive index is plotted against the dominant material response time (generally carrier recombination times in the case of semiconductor materials). Much of the data has been extracted from Gibbs [5]. The general trade-off between nonlinearity and speed of response is clearly apparent from this figure. Notice that in terms of these basic materials parameters, semiconductor-doped glasses closely approach the designated target area for future application in digital optical signal processing devices [6].

References

1. R K Jain and R C Lind: J Opt Soc Am 73, 647 (1983)
2. S S Yao, C Karaguleff, A Gabel, R Fortenberry, C T Seaton and G I Stegeman : Appl Phys Lett 46, 801 (1985)
3. P Roussignol, D Ricard, K C Rustagi and C Flytzanis: Opt Commun 55, 143 (1985)
4. M C Nuss, W Zinth and W Kaiser: this volume
5. H M Gibbs: in Optical Bistability: Controlling Light with Light, (Academic Press 1986)
6. J E Midwinter: Proc IEE 132 J, 371 (1985)

# Femtosecond Vibrational Relaxation of the $F_2^+$ Center in LiF

W.H. Knox, L.F. Mollenauer, and R.L. Fork

AT & T Bell Laboratories, Holmdel, NJ 07733, USA

The change in electronic wavefunction brought about by optical excitation of a color center [1] causes the surrounding ions to seek a new configuration of lower energy. Thus, optical excitation creates a momentary high level of localized vibrational excitation which the center must lose before it can emit in a Stokes shifted band. There are two schools of thought about how the vibrational relaxation takes place. One holds that the "configuration coordinate", which for a color center is the most significant normal mode of vibration, must undergo a number of cycles of damped motion, shedding the excess energy in a sequence of optical phonons. The other holds that the entire relaxation can take place in a time which is less than one optical phonon period , and the energy is released in a shower of coherent phonons [2]. Previous studies of impurities with only moderate coupling between electronic and ionic parts [3] or of a color center with a very large Stokes shift as a result of a configurational relaxation [4] seemed to indicate the former model, since the measured relaxation times were tens or hundreds of phonon periods. We present here the first evidence that a color center (in this case a single electron trapped by a pair of adjacent anion vacancies in an alkali halide crystal) relaxes according to the latter model. That is, we have made measurements indicating a relaxation time of less than one phonon period for that center.

In our experiment, optical pulses of 50 femtoseconds duration at a wavelength of 625 nm were produced in a colliding - pulse modelocked dye laser [5] and amplified at a repetition rate

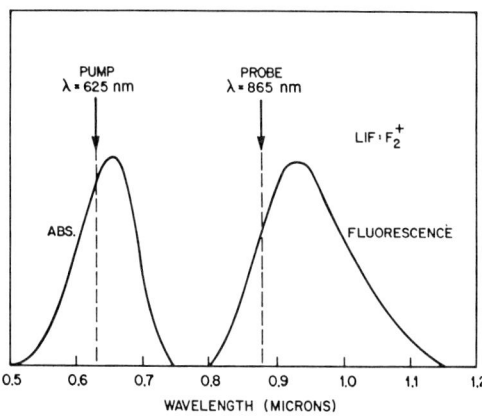

Figure 1. Absorption and emission spectra of color center showing excitation and probing wavelengths.

of 8 kilohertz by a copper vapor laser pumped amplifier to microjoule energies [6]. A fraction of each amplified pulse (~100 nJ) was separated, passed through an adjustable optical delay line, and used to pump the color centers (Figure 1). The sample was 2 mm thick containing centers at an optical density of approximately 1.0 peak and maintained at 77 K. The rest of the energy was focused into a 1 mm jet of flowing ethylene glycol to generate continuum pulses, from which light corresponding to the emission band around 865 nm was selected and combined collinearly with the pump beam to probe the rise of the gain. There is no significant coherent coupling artifact in this experiment. The transmitted probe signal was detected with a photomultiplier and lock-in amplifier and the pump beam was chopped at a frequency of 750 Hz. Gain signals of a few percent were typically obtained.

The measured rise of the gain is shown in Figure 2a. In order to determine the vibrational relaxation time we first measure the cross-correlation function between the 625 nm excitation pulse and the 865 nm continuum pulse by replacing the LiF crystal with a 1 mm KDP crystal and detecting the signal at the sum frequency at 363 nm (Figure 2b). The vibrational relaxation process can be modeled as a three level system with a single exponential response which is convolved with the measured cross-correlation function. The results of such a calculation are shown in Figure 2a assuming response times of 0, 75 and 150 femtoseconds. The comparison of the data with these calculations indicates that the relaxation exhibits a response time of less than 75 femtoseconds. For comparison, the LO and TO phonon periods in LiF are 50 and 118 fs, respectively. Thus, the relaxation does indeed appear to take place in one phonon period or less. We estimate that the excess phonon energy which is dissipated in the vibrational relaxation within the excited electronic state is equivalent to approximately four phonons. Clearly, the energy is dissipated in

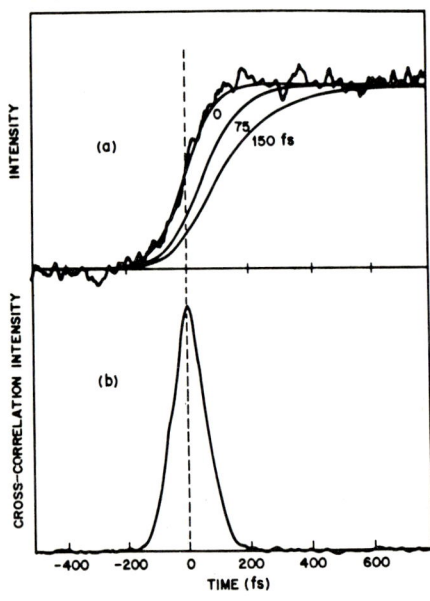

Figure 2. (a) Measured rise of gain and calculated rise assuming relaxation times of 0, 75 and 150 fs, and (b) the measured cross-correlation function of the pump and probe pulses using a 1 mm KDP crystal. The data indicate a vibrational relaxation time of less than 75 fs.

a time which is less than four phonon periods, therefore the relaxation cannot be described in the classical cascade model. At lower temperatures, the relaxation may become slower, as has been observed for another color center [4].

[1] W. B. Fowler, Ed. "Physics of Color Centers" Academic Press, N.Y. 1968

[2] N. Terzi, J. Lumin. 31/32, 194 (1984)

[3] W. H. Knox and K. J. Teegarden, J. Lumin. 31/32, 39 (1984)

[4] J. Wiesenfeld, L. F. Mollenauer and E. P. Ippen, Phys. Rev. Lett. 47, 1668 (1981)

[5] R. L. Fork, B. I. Greene and C. V. Shank, Appl. Phys. Lett. 38, 671 (1981).

[6] W. H. Knox, M. C. Downer, R. L. Fork and C. V. Shank, Opt. Lett. 9, 552 (1984)

# Determination of the Rapid Quenching Rates of Excited State F-Centers by OH⁻ Defects in KCl

*Du-Jeon Jang*[1], *T.C. Corcoran*[1], *M.A. El-Sayed*[1], *L. Gomes*[2], and *F. Luty*[2]

[1]Department of Chemistry and Biochemistry, University of California, Los Angeles, CA 90024, USA
[2]Physics Department, University of Utah, Salt Lake City, UT 84112, USA

## 1. Introduction

Recently, active studies of the interaction and association of substitutional diatomic molecular defects with F-centers in alkali halides have been carried out [1-3]. It was shown [1] that the electronic luminescence and photoionization of the <u>relaxed excited state of F-centers</u> ($F^*$-centers) in KCl reduce drastically with increasing concentration of OH⁻ molecular defects, suggesting a strong coupling between $F^*$-centers and OH⁻ defects. The main purpose of this work is to measure the rates of this quenching using picosecond techniques to determine the type of the electronic coupling (i.e., multipolar vs electron exchange) as well as the phonon assistance in the process. The rates of the quenching are determined by the measurement of the ground state bleach recovery of the F-centers as a function of OH⁻ concentration and temperature using a picosecond streak camera absorption spectrometer [4].

## 2. Experimental

Single wavelength ground state bleach recovery kinetics were obtained on the basis of a single laser shot with a picosecond transient absorption spectrometer utilizing dye emission and a streak camera [4]. The optical setup of the transient absorption spectrometer is shown schematically in Fig. 1. The fluorescence from an organic dye, excited by the third harmonic pulse of a passively/actively mode-locked Quantel 471 Nd:YAG

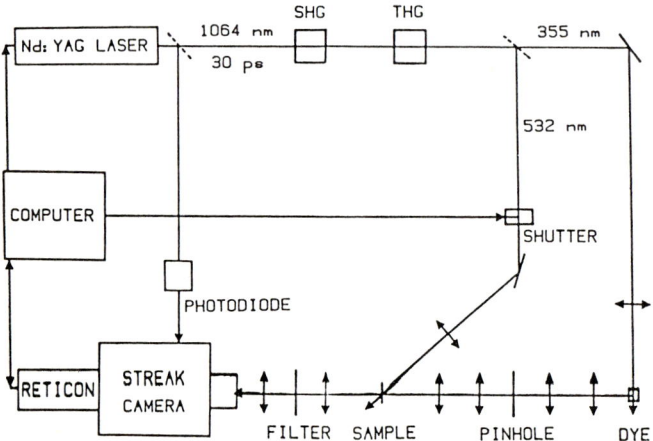

Fig. 1: Schematic diagram of the experimental apparatus. SHG/THG = second/third harmonic generator. Double headed arrow = convex lens.

280

laser, is used for the probe light. The wavelength of the probe light is selected by passing a 5-nm band filter. A Hamamatsu C979 streak camera with a 10-ps time resolution, coupled to a P.A.R. intensified 1420 Reticon with a P.A.R. 1218 multichannel controller, is used as a detector. This is interfaced to a digital LSI 11/23 computer. A ground state bleach in the sample alters the apparent kinetics of the dye emission seen by the streak camera. Comparison of the dye emission kinetics without and with the sample excitation by the second harmonic pulse of the laser yields very accurate bleach recovery kinetics. The single crystals of different $OH^-$ and $OD^-$ dopings were additively colored and quenched. The samples were mounted on the cold finger of a Air Products Liquid Transfer Heli-Tran LT-3-110.

## 3. Results and Discussion

### A. Temporal and Concentration Dependence

Figure 2 shows typical bleach recovery kinetics obtained at different $OH^-$ concentrations (about 2 orders of magnitude higher than the F concentration). The kinetics of the $OH^-$ doped crystals show two distinguishable relaxation processes. The fast process is due to energy transfer from the excited electronic state of the F-centers to the vibrational states of the $OH^-$ stretching mode. The fitting of the observed temporal dependence to the equations derived from multipolar and exchange mechanisms are carried out [5]. The results suggest an exchange type at short times. At long time, all the mechanisms fit the observed results. The slow relaxation process is slightly faster than that observed in the absence of $OH^-$.

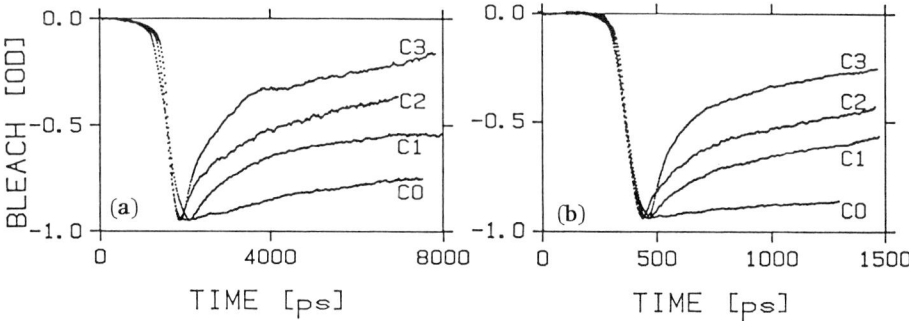

Fig. 2: Typical bleach recovery kinetics of the F-centers in KCl with various $OH^-$ concentrations measured at 120K, using relatively long (a) and short (b) time windows. The samples are excited at 532 nm and probed at 550+5 nm and the $OH^-$ concentrations in mole fractions are 0, $6.1 \times 10^{-4}$, $9.4 \times 10^{-4}$, and $2.3 \times 10^{-3}$ for (C0), (C1), (C2), and (C3), respectively.

### B. Temperature Dependence

The temperature dependence of $OH^-$ quenched fast relaxation process presented in Figs. 3 and 4 shows that the transfer rate increases linearly with temperature below 90K. Above this temperature, it increases exponentially with temperature. Below 90K, the energy transfer is assisted by a one phonon process [6]. Above 90K, activation energies ($\Delta E_a$) of 390 $cm^{-1}$ for $OH^-$ and 270 $cm^{-1}$ for $OD^-$ sample are obtained from the Arrhenius plots shown in Fig. 4. These correspond to the librational energies of $OH^-$

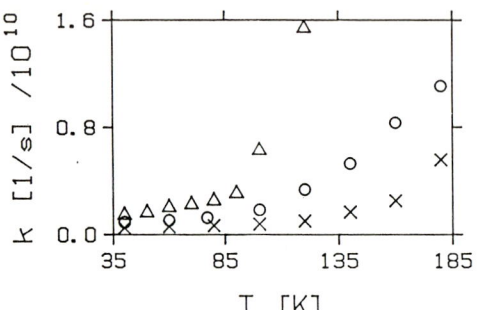

Fig. 3: Temperature dependence of the fast relaxation rate of the F-centers in $OH^-$ doped KCl. The concentrations in mole fraction are $2.3\times10^{-3}$ $OH^-$ for (Δ), $1.8\times10^{-3}$ $OD^-$ for (O), and $9.4\times10^{-4}$ $OH^-$ for (×) respectively.

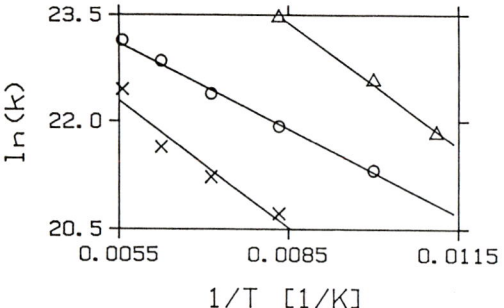

Fig. 4: Arrhenius plots of fast relaxation rates using data taken above 90K. The $OH^-/OD^-$ concentrations are shown in Fig. 3 caption. The activation energies are determined to be 390 $cm^{-1}$ for (Δ) and (×) $OH^-$ samples and 270 $cm^{-1}$ for (O) $OD^-$ sample.

and $OD^-$ respectively in KCl [7,8]. The temperature dependence of the slow relaxation is similar to that of $OH^-$ free samples. The slow relaxation rate is almost independent of temperature below ~100K. Above this temperature, the rate increases exponentially with a corresponding $\Delta E_a$ of 660 $cm^{-1}$. This corresponds to the energy difference between the relaxed excited state of the F-center and the conduction band [9]. This suggests that the mechanism of the slow relaxation process involves the thermal activation of $F^*$ electron to the conduction band followed by rapid trapping leading to bleach recovery as found in pure KCl [9]. One thing to note, however, is that our lifetimes of the slow relaxation below ~100K seen in the time window of ~10 ns are shorter by a factor of ~10 for both pure and $OH^-$ doped samples than the lifetime of $F^*$-centers previously reported for pure KCl [9]. The lifetimes observed in our experiments decrease as the sample excitation intensity increases. The $F^*-F^*$ annihilation interaction could produce in our experiment an additional quenching mechanism due to the large population of $F^*$ generated by the high power used in our picosecond experiment.

Acknowledgement: The work at the University of California was supported by the Office of Naval Research, the work at the University of Utah by NSF grant DMR 81-05532.

## 4. References

1. L. Gomes and F. Luty, Phys. Rev. B $\underline{30}$, 7194 (1984).
2. Y. Yang, W. von der Osten, and F. Luty, Phys. Rev. B $\underline{32}$, 2724 (1985).
3. Y. Yang and F. Luty, Phys. Rev. Lett. $\underline{51}$, 419 (1983).
4. D.-J. Jang and D.F. Kelly, Rev. Sci. Instrum. $\underline{56}$, 2205 (1985).
5. D.-J. Jang, T.C. Corcoran, M.A. El-Sayed, and F. Luty, manuscript in preparation.
6. T. Holstein, S.K. Lyo, and R. Orbach, Phys. B $\underline{16}$, 934 (1977).
7. M.V. Klein, B. Wedding, and M.A. Levine, Phys. Rev. $\underline{180}$, 902 (1969).
8. David Harrison, <u>Ph. D. Thesis</u>, University of Utah, 1970.
9. W.B. Fowler, <u>Physics of Color Centers</u>, Chapter 2, (Academic Press, New York 1968).

# Propagation of Coherent Phonon Polaritons in LiTaO₃ Measured by FIR-Cherenkov-Pulses

*M.C. Nuss and D.H. Auston*

AT & T Bell Laboratories, 600 Mountain Avenue, Room 1D-402, Murray Hill, NJ 07974, USA

We have previously shown that optical rectification of femtosecond laser pulses in LiTaO₃ produces a Cherenkov radiation cone of an extremely short electromagnetic pulse consisting of a single optical cycle and having a period approximately equal to the duration of the input femtosecond pulses [1]. Electromagnetic pulses of only 300 fsec in duration were generated, being the shortest measured electrical waveforms.

On the other hand, the spectral content of these transients is extremely broad, covering a frequency range from 0.1 to 5 THz, corresponding to 3 to 170 cm$^{-1}$, thereby extending from the microwave to the far infrared part of the electromagnetic spectrum. Of equal importance is our ability to detect these electromagnetic transients coherently; i.e. amplitude and phase of the pulse, not just the intensity is measured as is usually the case with most detectors. By retaining the phase information, it is thus possible to retrieve the complete spectral information by Fourier-transforming the waveform. We have demonstrated before, that far infrared spectroscopy can be performed in either reflection or transmission by observing the change in waveform upon reflection off or transmission through the material under investigation [2].

A very interesting application of this technique is to study the propagation of the electromagnetic pulse in the electro-optic material LiTaO₃ itself. In ionic crystals the far infrared pulse is strongly coupled to the transverse optical (TO) lattice modes, thus propagating as a polariton, i.e. a coupled phonon-photon excitation. The dielectric constant $\epsilon(\omega)$ shows strongly resonant behaviour in the range of the TO phonon frequency. Consequently, the polariton character of the electromagnetic pulse is expected to become apparent when the TO lattice reonance is approached [3].

We used a similar experimental setup as previously described [1]. Two femtosecond light pulse trains (pulse duration $t_p \simeq 50$ fsec) were used, one to generate the Cherenkov shock wave by optical rectification of the femtosecond light pulse, the other one to detect the propagating electrical field of the polariton using the electro-optic sampling technique. Both beams were focussed by a common lens to propagate parallel through the crystal. The beam waists were $\sim 1.5$ μm in a LiTaO₃ crystal of 250 μm length. The polarization of the excitation beam is parallel to the optical axis of the crystal, thus producing a Cherenkov wave of the same polarization by optical rectification involving the electro-optic coefficient $r_{33}$.

Several waveforms were recorded as a function of the separation between generating and detecting beams. Fig.1 shows the waveform recorded for a beam separation of 12.5 μm. We would like to emphasize, that due to the superior time resolution and the coherent waveform detection, we resolve the individual oscillations of the excited polariton wave. This has to be compared to standard Raman measurements, where only the envelope of the decay of the polariton excitation can be measured. The envelope of the waveform decays with a time

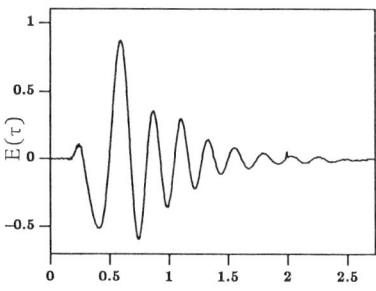

Fig.1
Electric field $E(\tau)$ of the polariton waveform

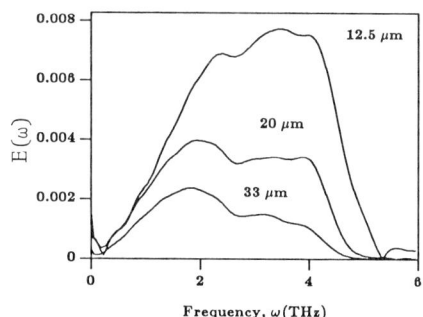

Fig.2
Spectrum $E(\omega)$ of the polariton wave for different beam separations

constant characteristic of the damping of the TO phonon and is measured to be 280 fsec.

The polariton waveforms can be Fourier-transformed to reveal the spectral features. Fig.2 shows the spectra for beam separations of 12.5, 20 and 33 µm, respectively. There is considerable structure in the spectra not expected for a single TO phonon resonance at $\sim$ 6 THz. In particular, when the polariton has propagated 20 µm through the crystal, the spectrum breaks up into an unstructured low frequency component peaked at $\sim$ 2 THz and a pronounced shoulder at frequencies > 2.5 THz. If there were only the known TO lattice resonance at $\sim$ 6 THz, we would expect only an unstructured spectrum shifting to lower frequencies when the wave travels through the crystal [3].

The origin of the waveform distortion can be best understood by evaluating the complex dielectric function $\epsilon(\omega)$ from our data. Both absorption coefficient $\alpha(\omega)$ and refractive index $n(\omega)$ can be derived from amplitude and phase of the Fourier spectrum without need to refer to the Kramers-Kronig relations. The frequency components of the complex electric field amplitude $E(\omega)$ change as a function of traveling distance x according to:

$$E(\omega) = E_0(\omega)/\sqrt{x} \, \exp(-\alpha x) \, \exp(in\omega x/c) \ .$$

Introducing the complex amplitude $q = E(\omega)/E_0(\omega)$, $\alpha$ and n are:

$$-\alpha \cdot \frac{x}{2} = \ln(|q|\sqrt{x}) \ ,$$

$$n\omega x/c = \arg(q) \ .$$

Having q for two different beam separations, $\alpha(\omega)$ and $n(\omega)$ can easily be determined. Fig.3 shows $\alpha(\omega)$ derived from our data up to a frequency of 4.5 THz (150 cm$^{-1}$). A similar structure is seen in the refactive index $n(\omega)$. Both curves obviously cannot be described by a single TO-phonon resonance at $\sim$ 6 THz. Taking the TO-phonon resonance from Raman data to be 6.23 THz and the damping rate from the decay time of the polariton wave to be 1.14 THz (280 fsec decay time), $\alpha(\omega)$ can be calculated for a simple single TO resonance (dotted in Fig.3). Calculating the difference between the experimentally observed absorption coefficient and the one assuming a single phonon resonance, there is strong evidence for the existence of an absorption line at 2.7 THz (90 cm$^{-1}$).

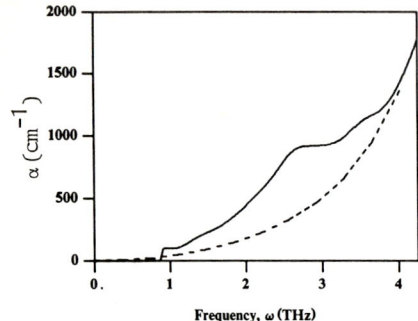

Fig.3

Measured absorption coefficient $\alpha [cm^{-1}]$ (solid) and the theoretical $\alpha$ for a single TO resonance (dashed)

There are essentially no accurate measurements of the dispersion properties of $LiTaO_3$ at these long wavelengths (< 5 THz). But previous data on the spontaneous polarization of $LiTaO_3$ at low temperatures were indicating the presence of a mode at ~ 85 $cm^{-1}$ [4], though it couldn't be resolved in infrared or Raman measurements. A possible explanation for the 90 $cm^{-1}$ mode resolved in the experiment is the ferroelectric nature of the crystal, leading to the softening of the optical phonon resonance and to a certain degree of coupling between $A_1$- and E-symmetry modes. Another candidate is 2-phonon absorption (LA/TA → O) similar to the FIR absorption in GaAs [5].

In summary, individual oscillation cycles of the TO phonon in $LiTaO_3$ could be resolved in time by using femtosecond Cherenkov radiation. Investigating the dispersion of the Cherenkov wavefront as it propagates through the crystal, strong deviations from what is expected from a single transverse optical phonon resonance are observed. We evaluated the complex dielectric function $\epsilon(\omega)$ and find evidence for a phonon mode at 90 $cm^{-1}$ responsible for the anomalous propagation of the Cherenkov wave in $LiTaO_3$. Finally, we would like to emphasize the high resolution and sensitivity of the farinfrared Cherenkov technique compared to other methods in the same frequency range like dispersive Fourier transform spectroscopy.

We appreciate stimulating discussions with K.P.Cheung.

1. D.H.Auston, K.P.Cheung, J.A.Valdmanis and D.A.Kleinman, Phys.Rev.Lett. *53*, 1555 (1984)
2. D.H.Auston and K.P.Cheung, J.Opt.Soc.Am.B *2*, 606 (1985)
3. K.P.Cheung and D.H.Auston, Phys.Rev.Lett. *55*, 2152 (1985)
4. A.M.Glass and M.E.Lines, Phys.Rev.B *13*, 180 (1976)
5. R.H.Stolen, Appl.Phys.Lett. *15*, 74 (1969)
   R.H.Stolen, Phys.Rev.B *11*, 767 (1975)

# Part VI

# Chemical Reaction Dynamics

# Cages, Crossings and Correlations – Theoretical Perspectives on Solution Reaction Dynamics

*J.T. Hynes*

Department of Chemistry and Biochemistry, University of Colorado, Boulder, CO 80309, USA

A theoretical overview is presented on the solvent dynamic influence on radical recombination, activated barrier crossing and time dependent fluorescence.

## 1. Introduction

The experimental and theoretical study of chemical reaction dynamics in solution has undergone a dramatic revolution in the past decade. Here we give a brief overview on some of the central phenomena and issues involved in three areas at the heart of, or closely allied to, this progress. These are the recombination dynamics of photodissociated radicals in solution, the role of the solvent in dynamics of activated barrier crossing, and the character and time scale of the relaxation of polar solvent molecules in the field of ionic or dipolar solutes. Here we give a theoretical perspective; experimental picosecond studies of these questions are described in succeeding papers.

## 2. Radical Recombination Dynamics

The recombination of iodine radicals to form molecular $I_2$ subsequent to photodissociation is perhaps the most studied simple chemical reaction in solution [1]. Classic early studies introduced the concept of the solvent "cage": the reduction of the photodissociation escape quantum yield in solution compared to the gas phase was interpreted via a picture in which the high density solvent molecules "imprison" a large fraction of the incipient radicals, which recoil to form their molecular parent. The behavior of those radicals which initially escape beyond the confines of the "primary" cage and subsequently recombine by some sort of relative diffusive motion was the object of extensive studies in the '50's.

The first picosecond study by EISENTHAL and coworkers [2], for $I_2$ in $CCl_4$ solvent, revealed a time scale of ~ 100 ps for recovery of ground state $I_2$ 532 nm absorption subsequent to photodissociation. This behavior was interpreted in diffusive recombination terms, which actually is only plausible for the small fraction of radicals which have escaped the primary cage. Indeed, early simulation studies [3] placed the time scale for recombination within the cage at about 10 ps.

An alternate and completely different interpretation of this experiment was proposed by NESBITT and HYNES [4], WILSON and coworkers [5,6] and later by ADELMAN and coworkers [7]. Franck-Condon considerations [4] imply that 532 nm absorption by $I_2$ requires extensive relaxation of nascent vibrationally "hot" $I_2$ down the vibrational ladder to levels in the lower part of the well carrying significant oscillator strength. Theoretical calculations of the time scale for this vibrational relaxation range from ~ 100 ps in $CCl_4$ [4,5] to ~ 1 ns in liquid Xe [4,6,7]. In addition to vibration-translation energy transfer, vibration-vibration energy transfer was predicted [4] to be crucial in the 100 ps time scale for $CCl_4$ solvent. This picture proposes vibrational relaxation as a critical slow step in the overall recombination process. Further, slow electronic curve crossing from excited $I_2$ electronic states to the ground state has been suggested to play a role in the dynamics in some solvents [8].

The subsequent experimental study of these questions by the WILSON [9], KELLEY [10] and HARRIS [11] groups has generated considerable controversy. The current evidence favors, in our view, (a) the presence of slow vibrational relaxation, (b) the relative unimportance of diffusive secondary recombination compared to primary cage recombination, and (c) the occurrence of electronic relaxation whose time scale is strongly solvent dependent and whose relevance for the recombination kinetics varies considerably. Important questions that still require considerable elucidation include: (a) What is the detailed character of the vibrational relaxation process and how does it depend on the relaxing molecule and the solvent identity? (b) How much of the recombination occurs in the primary cage, and how much involves longer distance trajectories? (c) What are the mechanisms and time scales for electronic relaxation and how do they depend on the solvent identity?

## 3. Activated Barrier Crossing Dynamics

The recombination reactions discussed above are quite special in that there is no potential energy barrier to the recombination act. More typical chemical reactions require passage over a free energy barrier lying between reactants and products. The traditional description for the rate constant k of such activated reactions is Transition State Theory (TST). This assumes that there is no short time recrossing of the barrier top for a trajectory crossing from reactant to product; rather a stable product is formed directly. Then k has its TST value $k^{TST}$, in which the solvent dynamics play no role [1].

TST is highly successful for gas phase chemical reactions. But in solution, its status is rather less certain. For example, the no recrossing assumption could be violated by solvent collision-induced recrossing of the barrier top, so that an actual k falls below $k^{TST}$ and depends on some measure of the solvent dynamics such as viscosity. In 1940 KRAMERS [12] modeled the reaction as a passage of an effective particle over a parabolic barrier, subject to Brownian friction forces. His classic result is $k^{KR}/k^{TST} = [1 + (\zeta/2\omega_b)^2]^{1/2} - (\zeta/2\omega_b)$ where $\omega_b$ is the barrier frequency (whose square is proportional to the magnitude of the barrier curvature) and $\zeta$ is the friction constant per mass. One extreme solvent effect would obtain at high friction $\zeta/\omega_b \gg 1$. There the solvent buffeting is so intense that there is considerable recrossing of the barrier top. In this regime $k^{KR}/k^{TST} \to \omega_b/\zeta$, whose strong decrease with increasing solvent friction reflects a diffusive barrier passage. There has been explosive recent activity in testing and extending KRAMERS' description [1,13,14]. Some central issues are now briefly discussed.

The KRAMERS treatment is a Markovian, macroscopic Brownian motion treatment in which the entire dissipative effect of the solvent is supposed to act instantly. But there is a fundamental short time scale in the reaction problem, $\omega_b^{-1}$, the characteristic time that the reaction system will remain in the barrier top vicinity in the absence of solvent interactions. For e.g. isomerization reactions, $\omega_b^{-1}$ lies in the range of $10^{-1}$-$10^{-2}$ ps, and the short time solvent forces rather than the long time friction are crucial. This feature was accounted for by GROTE and HYNES [15], who found that $\zeta$ in the Kramers equation should be replaced by the frequency-dependent friction $\hat{\zeta} = \int_0^\infty dt \, e^{-\kappa\omega_b t} \zeta(t)$; $\kappa = k/k^{TST}$, where $\zeta(t)$ is the correlation function of the solvent forces acting on the reaction coordinate. For a sharp barrier reaction, these forces are probed at the high frequency $\omega_b$, rather than at zero frequency. The latter gives the friction constant $\hat{\zeta}(0) = \zeta$, as in $k^{KR}$. Since the finite frequency solvent friction is typically less than $\zeta$, the prediction is that k will be closer to $k^{TST}$ than $k^{KR}$ would say. This feature should be most pronounced at high solvent friction or viscosity. For example, collective hydrodynamic solvent dynamics very important for establishing a large value of solvent viscosity can be completely irrelevant on the short time scale during which the reaction fate is decided.

Experimental evidence on these points is beginning to accumulate. In extensive picosecond photochemical investigations [14,16], FLEMING and coworkers have found that excited and ground state isomerization rates for diphenyl butadiene and DODCI all decrease with increasing solvent viscosity. For low barrier reactions, behavior consistent with the diffusive limit of KRAMERS is observed. But for higher barrier frequency reactions, k decreases less rapidly with increasing solvent viscosity $\eta$ than is predicted from $k^{KR}$ (assuming $\zeta \propto \eta$). Such behavior has been interpreted [16,17] via the GROTE-HYNES theory with frequency dependent friction. Similar behavior has been found by the HOCHSTRASSER [18], BARBARA [19] and other groups [14] in picosecond isomerization studies. One exception is the isomerization of binaphthyl in linear alcohol solvents [20], where $k^{KR}$ is followed.

While the above strongly suggests that frequency dependent friction plays an important role in activated reaction kinetics, in our view there remain outstanding questions: (a) How should the molecular time scale friction on a reaction coordinate be accurately described? While generalized hydrodynamic models are available, we distrust their validity on the short time scales involved in most reactions. (b) What role is played by coupling of the primary reaction coordinate to other nonreactive degrees of freedom? Such coupling can also introduce deviations [21] from KRAMERS behavior. (c) How important are static, as opposed to dynamic, solvent effects [22] on reaction barrier heights? Since k depends exponentially on those heights, this is a source of considerable concern.

There is another regime of interest - that of low friction - for isomerizations. Here energy activation in a reactant well and relaxation in the product well can become rate limiting, so that k will ultimately decrease with decreasing $\zeta$. A few experiments place the locale of this KRAMERS turnover [12] in the liquid phase, but the overwhelming majority of experimental results find that it lies in the high pressure gas phase. [Ref. [14] gives a recent review of the experimental situation.] It has been proposed [23] that the nonreactive internal degrees of freedom are critical here. In a one dimensional view, the solvent must perform the chore of e.g. deactivating the molecule before a recrossing occurs. Internal degrees of freedom coupled to the reaction coordinate can perform this same chore before ultimately giving up energy to the solvent, and this will shift the turnover into the gas phase. For example, a model calculation of stilbene [24] places the turnover in gaseous methane at about 40-70 atm, in reasonable agreement with experiment [14].

Yet the situation is still not very clear, as witnessed by certain discrepancies between high pressure gas and solution rate constants [14] and suggestions of large barrier shifts even in gases of low to modest pressures [25]. Some outstanding questions here are: (a) What is the solvent's role in either assisting or retarding intramolecular energy flow between the reaction coordinate and other internal degrees of freedom? (b) What is the magnitude and origin of low pressure solvent shifts of barrier heights?

## 4. Polar Solvent Relaxation Dynamics

The reaction classes discussed above involve fairly weak coupling to the solvent. But reactions involving ions and/or dipoles in polar solvents should exhibit strong coupling effects. Indeed this is recognized in its static aspects in traditional TST approaches which include in some fashion pronounced solvation effects on e.g. activation free energies $\Delta G^{\neq}$. Here, it is assumed that there is equilibrium solvation, i.e. that the solvent is equilibrated to the charged reaction system at each point along the reaction coordinate [1].

But elementary time scale considerations suggest that there is insufficient time for e.g. solvent dipole reorientation to provide equilibrium solvation for a reaction system in its rapid flight over the barrier. Thus one must be concerned with nonequilibrium solvation effects. Examples [26] include (a) "polarization caging" in which reaction is only possible when the solvent relaxes, leading to a strong inverse dependence of k on e.g. the solvent dielectric relaxation time, and (b) "nonadiabatic solvation" in which the reactant occurs in a nearly "frozen" solvent at the transition state and k is strongly influenced by instantaneous solvent molecule configurations.

However it is unfortunately the case that we know comparatively little about polar solvent relaxation dynamics on the short time scales of relevance of chemical activated barrier crossing. Picosecond studies of time dependent fluorescence (TDF) should provide a much needed window here. When a molecule immersed in a polar solvent absorbs light and changes its charge distribution, the Franck-Condon principle dictates that initially the solvent is out of e.g. orientational equilibrium. The solvent polarization will subsequently relax to solvate the new charge distribution. If the excited state fluoresces during this period, then the TDF shifting frequency will reveal those solvent dynamics, as recently described by BAGCHI et al [27] and VAN DER ZWAN and HYNES [28].

Picosecond studies of TDF are only now beginning to be performed [29]. Important questions to be addressed include: (a) Does the solvent relaxation follow continuum theory predictions? This indicates extremely short times, e.g. ~ 0.25 ps for $H_2O$, and it is unclear just what motions might be involved. (b) Are there inertial or translational effects on the relaxation [28]? This would result in non-Debye-like relaxation behavior. (c) Does the relaxation depend on the excited state charge distribution? Such a nonlinear effect is ignored in current theories.

Support from NSF grant CHE 84-19830 is gratefully acknowledged.

## References

1. J.T. Hynes: Ann. Rev. Phys. Chem. 36, 573(1985)
2. T. J. Chuang, G.W. Hoffman, K.B. Eisenthal: Chem. Phys. Lett. 25, 201(1974)
3. D.L. Bunker, B.S. Jacobson: J. Am. Chem. Soc. 94, 1843(1972); J.T. Hynes, R. Kapral, G.M. Torrie: J. Chem. Phys. 72, 177(1980)
4. D.J. Nesbitt, J.T. Hynes: J. Chem. Phys. 77, 2130(1982)
5. P. Bado, P.H. Berens, K.R. Wilson: Proc. Soc. Photo-Optic. Instrum. Eng. 322, 230(1982)

6. P. Bado, P.H. Berens, J.P. Bergsma, M.H. Coladonato, C.G. Dupuy, J.D. Kahn, K.R. Wilson: In Photochemistry and Photobiology, ed. by A. Zewail, Vol. 1 (Harwood Academic, New York 1983) p. 615
7. C.L. Brooks III, M.W. Balk, S.A. Adelman: J. Chem. Phys. $\underline{79}$, 784(1983).
8. D.P. Ali, W.H. Miller: J. Chem. Phys. $\underline{78}$, 6640(1983)
9. P. Bado, C. Dupuy, D. Magde, K.R. Wilson, M.M. Malley: J. Chem. Phys. $\underline{80}$, 5531(1984)
10. D.F. Kelley, N.A. Abul-Haj, D. Jang: J. Chem. Phys. $\underline{80}$, 4105(1984)
11. A.L. Harris, M. Berg, C.B. Harris: J. Chem. Phys. $\underline{84}$, 788(1986)
12. H.A. Kramers: Physica $\underline{7}$, 284(1940)
13. J.T. Hynes: J. Stat. Phys. $\underline{42}$, 149(1986); P. Hanggi: ibid. p. 105
14. G.R. Fleming, S.H. Courtney, M.W. Balk: J. Stat. Phys. $\underline{42}$, 83(1986)
15. R.F. Grote, J.T. Hynes: J. Chem. Phys. $\underline{73}$, 2715(1980)
16. S.P. Velsko, G.R. Fleming: J. Chem. Phys. $\underline{76}$, 3553(1982); Chem. Phys. $\underline{65}$, 59(1982); S.P. Velsko, D.H. Waldeck, G.R. Fleming: J. Chem. Phys.$\underline{78}$, 249(1983); K.M. Keery. G.R. Fleming: Chem.Phys. Lett. $\underline{93}$, 322(1982)
17. B. Bagchi, D. Oxtoby: J. Chem. Phys.$\underline{78}$, 2735(1983).
18. G. Rothenberger, D.K. Negus, R.M. Hochstrasser: J. Chem. Phys.$\underline{79}$, 5360(1983)
19. S.R. Flom, A.M. Brearley, M.A. Kahlow, V. Nagarajan, P.F. Barbara: J. Chem. Phys. $\underline{83}$, 1993(1985).
20. D. Millar, K.B. Eisenthal: Chem. Phys. Lett. (1985)
21. R.F. Grote, J.T. Hynes: J. Chem. Phys. $\underline{74}$, 4465(1981); G. van der Zwan, J.T. Hynes: ibid. $\underline{77}$, 1295(1982); B. Carmeli, A. Nitzan: Chem. Phys. Lett. $\underline{106}$, 329(1984)
22. B.M. Ladanyi, J.T. Hynes: J. Am. Chem. Soc. $\underline{108}$, 585(1986); D. Chandler: private communication
23. A.G. Zawadzki, J.T. Hynes: Chem. Phys. Lett. $\underline{113}$, 476(1985); M. Borkovec, B.J. Berne: J. Chem. Phys. $\underline{82}$, 794(1985); A. Nitzan: ibid. $\underline{82}$, 1614(1985)
24. A.G. Zawadzki, J.T. Hynes: submitted to J. Chem. Phys.
25. G. Maneke, J. Schroeder, J. Troe, F. Voss: Ber. Bunsenges Phys. Chem. $\underline{89}$, 896(1985)
26. G. van der Zwan, J.T. Hynes: J. Chem. Phys. $\underline{76}$, 2993(1982); $\underline{78}$, 4174(1983); Chem. Phys. $\underline{90}$, 21(1984)
27. B. Bagchi, D.W. Oxtoby, G.R. Fleming: Chem. Phys. $\underline{86}$, 25(1984)
28. G. van der Zwan, J.T. Hynes: J. Chem. Phys. $\underline{89}$, 4181(1985)
29. S.W.Yeh, L.A. Phillips, S.P. Webb, L.F. Buhse, J.H. Clark: private communication; G.R. Fleming: private communication

# Polarity Dependent Barriers and the Photoisomerization Dynamics of Polar Molecules in Solution

*J.M. Hicks, M.T. Vandersall, E.V. Sitzmann, and K.B. Eisenthal*

Department of Chemistry, Columbia University, New York, NY 10027, USA

There has been recent theoretical and experimental interest in potential energy barrier descriptions of molecular structural changes in solution [1-7]. For example, the dependence of photoisomerization kinetics on solvent viscosity has been studied to test barrier crossing models [1-5]. Changes in the photoisomerization rates through a solvent series or as a function of temperature have been attributed to viscosity and thermal effects; other possible effects, such as solvent dependent excited state potential surface variations, have generally not been considered [8]. In particular, strong interactions with polar solvents are expected for molecules which undergo a major charge redistribution on isomerization, whether they be polar molecules such as p-dimethylamino-benzonitrile (DMABN) [7], or nonpolar molecules which pass through very polar intermediate structures such as t-stilbene [3-5]. These polar interactions can be important factors in the dynamics of isomerization, as we will show in this paper.

## 1. DMABN -- Case of a Polar Molecule

DMABN is a classic example of a molecule that exhibits strong solute-solvent interactions, as demonstrated by the fact that its excited state isomerization occurs in polar solvents but not in nonpolar solvents [9]. A general scheme of the dynamics is given in Scheme I [10].

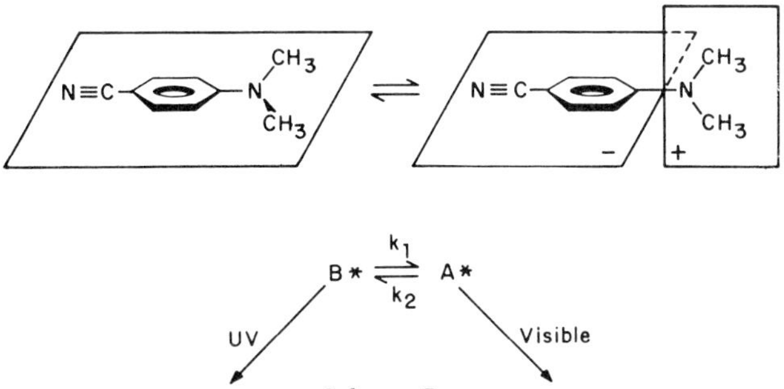

Scheme I

The isomerization takes place on an excited singlet potential surface and involves a 90° rotation of the dimethylamino group about the amino-phenyl bond [10]. This twisting structural change produces an increase in the excited state dipole moment from 6 D to 16 D [11]. In polar solvents, a new visible emission appears due to the solvent stabilized twisted form. Due to a rapid equilibration between the planar ($B^*$) and twisted ($A^*$) structures two emissions are observed. The one in the ultraviolet is due to the planar form and the visible emission is due to the twisted polar form. In alkanes, only the uv emission from $B^*$ is observed. Many time-resolved studies have been conducted on DMABN, yet the role of the solvent is still not clear [7,12-16].

To investigate the effects of the solvent on the photoisomerization kinetics of DMABN, we have carried out studies in a series of linear alcohols, alcohol/alkane mixtures, linear nitriles and nitrile/alkane mixtures (part of the data has been reported [7,12]). A single pulse from the output of a $Nd^{+3}$/glass laser was frequency quadrupled to 265 nm and used to excite DMABN. The fluorescence from the $B^*$ and $A^*$ states was time-resolved with a streak camera detection system which had a resolution of 7 ps [13].

An important aspect of the solute/solvent interactions is the effect of solvent viscosity on the kinetics of a reaction. From studies in neat liquids, we find a viscosity dependence in nitriles of $\eta^{-2/3}$ and in alcohols, $\eta^{-1}$. These can be interpreted as cases of intermediate and strong solute-solvent coupling, respectively. Of key importance in obtaining these viscosity dependences of $k_1$ is the assumption that only the viscosity differences among the various members of the solvent series affect the rate. The possibility that the potential surface may change in different solvents is not considered, nor are the possibilities that boundary conditions, entropy or dynamic polarity effects vary with solvent. For a polar structural transition such as in DMABN the invariance of the potential cannot be assumed. Specifically, the changes in solvent polarity, even within a polar series, can be of sufficient magnitude to produce marked changes in the excited state barrier height. It has long been recognized that there are barrier height changes in going from a nonpolar to a polar solvent. The issue here is that these barrier effects can come into play within a solvent family, e.g. the polar alcohol series, or in one solvent at different temperatures.

To separate the effects of solvent polarity and viscosity, we have used two techniques: (i) $k_1$ was measured at room temperature in isoviscous mixtures of a polar solvent and an alkane, where the polarity of the mixture was controlled by the concentration of the polar solvent, and (ii) the temperature was varied for the neat solvents so that the solutions had the same viscosity but different polarity values. In Fig. 1 we see that the isomerization rate is not a constant at a fixed viscosity and temperature, but indeed varies exponentially with solvent polarity as measured by the widely used [17] empirical solvent polarity parameter $E_T(30)$. The fitting of the

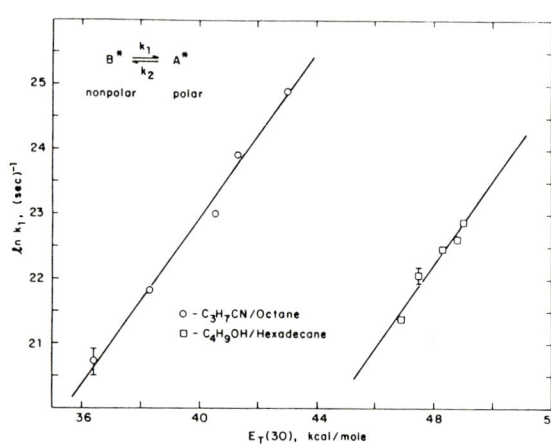

Fig. 1: Plot of $\ln k_1$ for DMABN vs. the solvent polarity parameter $E_T(30)$

rate data with theoretically derived polarity expressions [18] gives the correct trend but the correlations are not as good as when $E_T(30)$ is used.

Further support for the effects of solvent polarity on the isomerization dynamics is obtained from our second technique where the same viscosity is obtained by adjusting the temperatures of the neat liquids. The different rates measured have previously been assumed to be due to the different Boltzmann factors. This can be seen in the following Arrhenius-type expression (1): $k = A\, f(\eta)\, \exp(-E_a/RT)$, where $E_a$ is the barrier height, $f(\eta)$ is the viscosity function (which is a constant for the isoviscosity experiment) and A is the pre-exponential frequency factor. We carried out these measurements in both the neat alcohols and neat nitriles, and found that the rate increases as the temperature is lowered in both solvent families. This result of a "negative" activation energy is contrary to what would be the case if only the Boltzmann factor is changing. To explain these results, we propose that the barrier height is not independent of temperature, but decreases due to a higher solvent polarity at lower temperatures. The polar twisted form of DMABN ($A^*$) is stabilized relative to the initially excited planar form ($B^*$), which thereby leads to a smaller barrier. This decrease in the barrier overcomes the usual Boltzmann effect and is responsible for the increased rate at lower temperature. When we correct for the temperature induced polarity changes, we find that the corrected rate decreases as the temperature is lowered, i. e. a normal positive activation energy is obtained, as seen in Fig. 2.

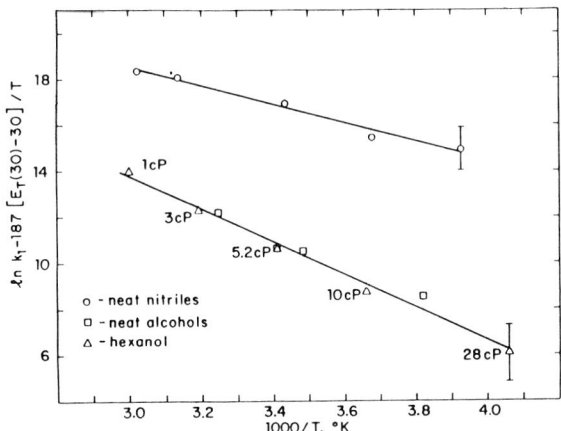

Fig. 2: Arrhenius plot of polarity-corrected rate for DMABN in neat isoviscous nitriles (1.0 cP), neat isoviscous alcohols (5.2 cP), and hexanol at the indicated viscosities

Both the alcohol and nitrile data shown in Figs. 1 and 2 can be explained by introducing a polarity dependent barrier $E_a$: $E_a = E_a^o - A\,[E_T(30)-30]$, where $E_a^o$ is the activation energy in an alkane solvent having an $E_T(30)$ of 30 and A is an experimentally determined factor which determines how strongly the barrier height changes with solvent polarity. The corrected rate, $\kappa$, plotted in Fig. 2, is $k_1 \times \exp -A[E_T(30)-30]/RT = C\exp(-E_a^o/RT)$, where C is the Arrhenius pre-exponential factor. The values of $E_a^o$ are found to be 8.0 kcal/mole in the nitriles and 14.0 kcal/mole in the alcohols. Comparing the nitriles and alcohols at a given solvent polarity, we see that the barrier $E_a^o$ is higher in alcohols by about 6 kcal/mole. We attribute this to the effects of hydrogen bonding between the dimethylamino group of DMABN and the alcohol hydroxy group. The hydrogen bond withdraws electrons

from the electron donating dimethylamino part of DMABN and thereby opposes
the electron transfer to the benzonitrile part of DMABN. For the intramolecular charge transfer in DMABN to occur, the hydrogen bond must be broken.
This increases the barrier for the isomerization in alcohols relative to
nitriles by 6 kcal/mole, roughly the energy of typical hydrogen bonds.

## 2. t-Stilbene -- Case of a Polar Intermediate Structure

The photoisomerization of t-stilbene has been studied by many groups to gain
insight into how the solvent affects this simple chemical change [3-5]. The
observed rate has been discussed in terms of (1), where $f(\eta)$ is either the
Kramers function (which gives poor agreement with the observed rates) or
more often $\eta^{-\alpha}$. This latter form can be obtained from a free volume [19]
or frequency dependent friction model [20-21]. This general equation
predicts that at constant viscosity, a plot of ln k versus $T^{-1}$ should have
the slope $-E_a/R$. We have tested this prediction by studying the rates of
photoisomerization of stilbene in alcohols at various temperatures under
constant viscosity conditions.

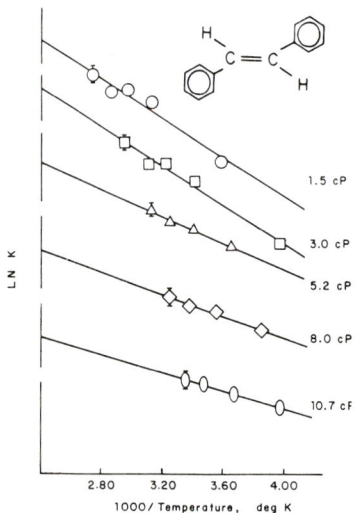

Fig. 3: Plot of ln k vs. $T^{-1}$ for t-stilbene in isoviscous alcohols; the ordinate axis is offset for each viscosity line for viewing purposes

In Fig. 3 it is shown that the slopes of these lines depend on viscosity,
contrary to what the equation predicts. The apparent $E_a$'s vary from 2.6 to
1.2 kcal/mole. A similar deviation from (1) is shown by the dependence of
the rate on viscosity at different fixed temperatures. We find that the
value of $\alpha$ varies by a factor of two over a $50°K$ temperature range. This
result conflicts with the constant $\alpha$ value predicted by the $\eta^{-\alpha}$ dependence
of the rate equation.

Since the isomerization involves an intermediate structure that is highly
polar, we ask whether polarity dependent barrier effects are important in
the isomerization process. Accordingly, we carried out measurements in neat
alcohols at various solvent polarities and temperatures, analogous to our
DMABN study. We found that static polarity effects alone do not explain our
stilbene results in alcohols. In stilbene, dynamic polarity effects could
be important since extensive solvent rearrangements are required in the
isomerization from the initial nonpolar form to the twisted charge separated

form. This is not expected for DMABN because its transition involves an increase in dipole moment along the same axis from a polar (6 Debye) to a more polar (16 Debye) form. The solvent is already arranged about the large dipole in the initially excited DMABN, unlike the case for stilbene, where the initial solvent arrangement is that appropriate to a nonpolar solute molecule. If the solvent motions are not very rapid compared to the isomerization time, then the isomerization dynamics would depend on the dielectric relaxation properties of the solvent [22-29]. The energy separating the trans and intermediate form would therefore depend upon the positions of the surrounding solvent molecules, and thus be dependent on the rate of change of the solvent arrangement. Further work is underway to resolve the role of polarity and other solvent factors in the stilbene isomerization [30].

Conclusions

We find that the isomerization dynamics of molecules that involve a large charge redistribution, such as in DMABN, are strongly dependent on the polarity of the solvent. The solute/solvent interaction can be described in terms of a polarity dependent barrier that separates the two structural forms of the molecule. In addition to the well-recognized effects in going from a nonpolar to a polar solvent, we find that the effects of polarity change within a series of related polar liquids, e.g. linear alcohols or nitriles, and the effects of the change in polarity with temperature are crucial to the observed kinetics. We have also found that hydrogen bonding of DMABN with alcohols impedes the isomerization relative to that of nonhydrogen bonding polar solvents such as nitriles. For t-stilbene in alcohols, we have found from measurements at various viscosities and temperatures that the isomerization cannot be described in terms of frequently applied equations which contain an $\eta^{-\alpha}$ power dependence. Although the t-stilbene isomerization passes through a very polar intermediate structure, we are not able to adequately fit the data using a static polarity correction as we did for DMABN. The possibility of a dynamic polarity effect due to a large solvent rearrangement in going from the nonpolar initially excited t-stilbene to the polar twisted intermediate stilbene structure is discussed.

Acknowledgements: The authors wish to thank the National Science Foundation, the Airforce Office of Scientific Research and the Joint Services Electronics Program 29-85-K-0049 for their support of this work.

References

1. D. P. Millar and K. B. Eisenthal: J. Chem. Phys. 83, 5076 (1985)
2. H. Courtney and G. R. Fleming: Chem. Phys. Lett. 103, 443 (1984); K. M. Keery and G. R. Fleming: Chem. Phys. Lett. 93, 322 (1982); S. P. Velsko and G. R. Fleming: J. Chem. Phys. 76, 3553 (1982); S. P. Velsko, D. H. Waldeck and G. R. Fleming: J. Chem. Phys. 78, 249 (1983)
3. G. Rothenberger, D. K. Negus and R. M. Hochstrasser: J. Chem. Phys. 79, 5360 (1983); M. Lee, G. R. Holtom and R. M. Hochstrasser: Chem. Phys. Lett. 118, 359 (1985)
4. V. Sundstrom and T. Gillbro: Ber. Bunsen. Phys. Chem. 89, 222 (1985); Chem. Phys. Lett. 109, 538 (1984); E. Akesson, H. Bergstrom, V. Sundstrom and T. Gillbro: Chem. Phys. Lett. 126, 385 (1986)
5. G. Maneke, J. Schroeder, J. Troe and F. Voss, Ber. Bunsen. Phys. Chem. 89, 896 (1985); S. H. Courtney and G. R. Fleming: J. Chem. Phys. 83, 215 (1985)
6. J. T. Hynes: Ann. Rev. Phys. Chem. 36, 573 (1985) and references therein

7. J. M. Hicks, M. T. Vandersall, Z. Babarogic and K. B. Eisenthal: Chem. Phys. Lett. 116,18 (1985)
8. Cases where it has been considered include: Ref. 7, J. Troe: Chem. Phys. Lett. 114, 241 (1985); 116, 453 (1985); J. Phys. Chem. 90, 357 (1986), W. L. Hase: J. Phys. Chem. 90, 365 (1986), E. Akesson et al. of Ref. 4, V. Sundstrom and T. Gillbro: J. Chem. Phys. 81, 3463 (1984)
9. E. Lippert, W. Luder and H. Boos: In Advances in Molecular Spectroscopy, ed. by A. Mangini (Pergamon Press, Oxford 1962) p.443
10. K. Rotkiewicz, K. H. Grellman and Z. R. Grabowski: Chem. Phys. Lett. 19, 315 (1973); Z. R. Grabowski, K. Rotkiewicz, W. Rubaszewska and E. Kirkor-Kaminska: Acta Phys. Pol. A54, 767 (1978)
11. W. Baumann: personal communication; Z. Naturforsch. 36A, 868 (1981)
12. Y. Wang and K. B. Eisenthal: J. Chem. Phys. 77, 6076 (1982)
13. Y. Wang, M. McAuliffe, F. Novak and K. B. Eisenthal: J. Phys. Chem. 85, 3736 (1981)
14. D. Huppert, S. D. Rand, P. M. Rentzepis, P. F. Barbara, W. S. Struve and Z. R. Grabowski: J. Chem. Phys. 75, 5714 (1981)
15. F. Heisel and J. A. Miehe: Chem. Phys. Lett. 100 183 (1983); Chem. Phys. 98, 233 (1985); with J. M. G. Martino: 98, 243 (1985)
16. S. R. Meech and D. Phillips: Chem. Phys. Lett. 116, 262 (1985)
17. C. Reichardt: In Molecular Interactions, vol. 3, ed. by H. Ratajcak and W. J. Orville-Thomas (John Wiley, New York, 1982) p.241
18. E. Lippert: Z. Naturforsch. 10a, 541 (1955); N. Mataga, Y. Kaifu and M. Koizumi: Bull. Chem. Soc. Jap. 29, 465 (1956)
19. A. H. Alwattar, M. D. Lumb and J. B. Birks: In Organic Molecular Photophysics, ed. by J. B. Birks, (John Wiley, New York, 1973) ch.8
20. R. F. Grote, G. van der Zwan and J. T. Hynes: J. Phys. Chem. 88, 4676 (1984)
21. B. Bagchi and D. W. Oxtoby: J. Chem. Phys. 78, 2735 (1983)
22. The effects of solvent dielectric relaxation time on processes such as barrierless charge transfer reactions, electron solvation, and rotations of dipolar solute molecules, has been the subject of important experimental and theoretical activity (see refs. 23-29).
23. P. Madden and D. Kivelson: J. Phys. Chem. 86, 4244 (1982)
24. D. Kivelson and K. G. Spears: J. Phys. Chem. 89, 1999 (1985)
25. H. E. Lessing and M. Reichert: Chem. Phys. Lett. 46, 111 (1977)
26. B. Bagchi, D. W. Oxtoby and G. R. Fleming: Chem. Phys. 86, 257 (1984)
27. G. van der Zwan and J. T. Hynes: Chem. Phys. 90, 21 (1984); J. Chem. Phys. 89, 4181 (1985); Chem. Phys. Lett. 101, 367 (1983)
28. G. W. Kenney-Wallace and C. D. Jonah: Chem. Phys. Lett. 47, 362 (1977)
29. L. A. Phillips, S. P. Webb, L. F. Buhse and J. H. Clark: In Ultrafast Phenomena IV, ed. by D. H. Auston and K. B. Eisenthal, (Springer Verlag Berlin, 1984)
30. Collaborative work with Prof. G. Fleming and his group

# Dynamic Solvent Effects on Small Barrier Isomerizations

P.F. Barbara and V. Nagarajan

Department of Chemistry, University of Minnesota,
Minneapolis, MN 55455, USA

1. INTRODUCTION

Ultrafast fluorescence spectroscopy is having a tremendous influence on the understanding of the role of solvent motion in chemical reactions in solution. This paper describes ultrafast measurements on two elementary processes in solution, namely, rotation about a single bond (torsional isomerization) and intramolecular charge separation. We will discuss some exceptionally useful molecular prototypes for each of these processes respectively, derivatives of 2-vinylanthracene 2VA which undergo an s-cis/s-trans torsional isomerization is $S_1$ and bianthryl BA which undergoes a charge-transfer reaction that converts an initially produced, locally excited (LE) form of $S_1$ to a so-called twisted intramolecular charge-transfer form (TICT). Many of the salient elements of the dynamics of solute-solvent interaction appear in an especially simple fashion for these compounds.

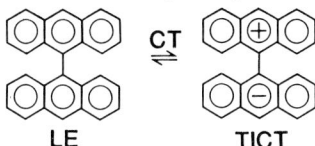

Figure 1. Molecular drawings for 2-vinylanthracenes and bianthryl.

2. TORSIONAL ISOMERIZATION

2-vinylanthracene 2VA ($R_1$=H, $R_2$=H) and derivatives of this compound with alkyl groups for $R_1$ and $R_2$ exist in spectrally distinct s-cis and s-trans conformations in $S_0^1$ and $S_1^2$. Following photon excitation, an adiabatic interconversion of $S_1$, s-cis and s-trans occurs [1]. Spectroscopic and kinetic data show that the reaction coordinate of the isomerization is primarily simple intramolecular bond-torsional motion. The free energy along this coordinate is a double-minimum with very little dependence on the polarity of the solvent. Solvent motion is not a significant component of this reaction coordinate. The time integrated (Fig. 2a) and time gated (Fig. 2b) fluorescence spectra of 2VA and derivatives are essentially a superposition of the spectra of the two conformations,

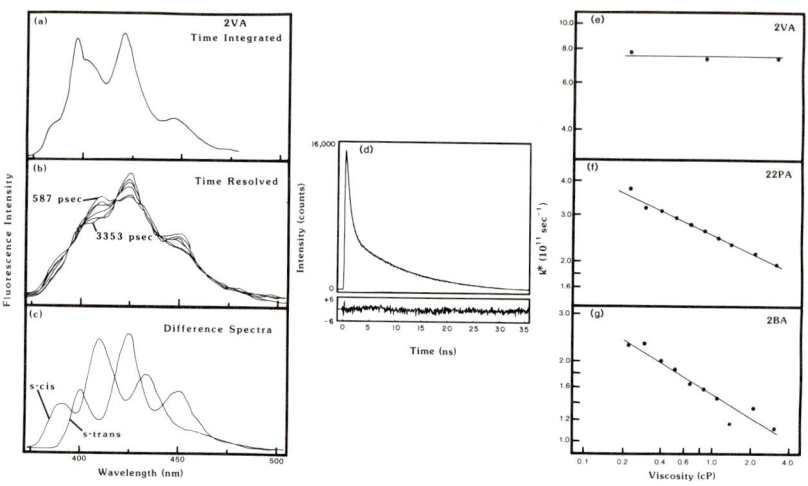

Figure 2. Transient and static fluorescence data on 2VA (Fig. 2a-c) and 22PA (Fig. 2d). Fig. 2e-g are log k* versus log viscosity plots for 2VA, 22PA, and 2BA (see text for further details).

which are individually obtained (Fig. 2c) by analysis of the evolution of the time gated spectra. The individual spectra are not strongly dependent on temperature or solvent polarity. The isomerization is also evident in kinetic traces of the fluorescence in narrow wavelength regions, e.g., Fig. 2d.

The static and dynamic spectroscopy of 2VA is well modeled by a simple phenomenological kinetic scheme in which the s-cis and s-trans forms of $S_0$ and $S_1$ are the only kinetic species. This analysis has allowed us to extract simple first-order, forward and reverse, rate constants for the $S_1$ isomerization, i.e., $k_f$ and $k_r$. The agreement of the model with experiment is demonstrated in Fig. 2d where data and theoretical curves are superimposed. A particularly useful parameter is k*, which is given by the expression $0.5 k_f \exp(E_f/RT)$ where $E_f$ is the barrier energy [1]. The dependence of k* on solvent for 2VA ($R_1$=H), 22PA ($R_1$=methyl, $R_2$=H) and 2BA ($R_1$=methyl, $R_2$=methyl) is shown in Figs. 2e-g in the form of a log k* versus log viscosity plot, where viscosity has been varied by changing the solvent. We have previously shown that the viscosity dependence of k* is due to "solvent friction" induced by solute-solvent collisions. The much greater viscosity effect observed for 2BA and 22PA as compared to 2VA has been taken as evidence for the importance of frequency dependent solvent friction in this reaction.

## 3. INTRAMOLECULAR CHARGE SEPARATION

A comparison of the spectroscopy of BA and 2VA reveals a striking contrast between the photophysical aspects of these compounds. The fluorescence spectral shape of BA as a function of solvent and temperature (Fig. 3a), varies in a complex fashion which *cannot* be accounted for by the simple superposition of two (LE and TICT) distinct solvent-independent spectra, because the fluorescence spectrum of the TICT form is itself strongly solvent dependent [2], due to the large change in dipole moment on emission from TICT to $S_0$. The spectroscopic data and theoretical considerations show

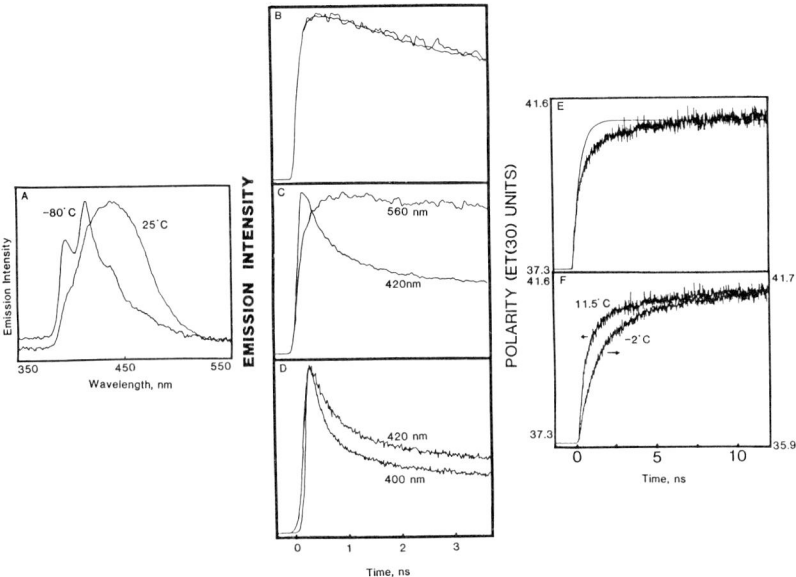

Figure 3. Static and dynamic fluorescence measurements on BA. Fig. 3a: the time-integrated fluorescence spectrum of BA in glycerol triacetate. The remaining panels are described in the text.

that the solution-phase reaction coordinate connecting LE and TICT involves solvent motion due to the large dipole moment of TICT.

Figure 3b-d portrays kinetic traces for BA emission at various wavelengths. The fluorescence kinetics in ethylacetate (Fig. 3b) are apparently singly exponential at all wavelengths because the initially excited form (LE) very rapidly equilibrates with the TICT form. The equilibration in ethylacetate at room temperature is too rapid to resolve with time-correlated single-photon-counting spectroscopy. In contrast, the emission kinetics for BA in the more viscous solvent glycerol triacetate (GTA) is dramatically emission-wavelength dependent (Fig. 3c). The evolution of the spectrum in Fig. 3c-d is *not* due to a simple first-order interconversion (LE ⇌ TICT) as evidenced by the inability to fit the emission kinetics to a simple scheme like that used for 2VA (see above).

The complex time-evolution of the fluorescence spectrum of BA is due to two processes. First, a rapid charge-transfer process converts the non-equilibrium excess of LE molecules into "weakly solvated" TICT species. Second, the instantaneous TICT spectrum changes as the solvent reorients in its approach toward equilibrium polar solvation of TICT. The transient solvation is actually much slower than the initial CT reaction, and the spectral evolution is mostly due to solvation, which is associated with the microscopic dielectric relaxation of the solvent. The transient solvation of the TICT form of BA is analogous to the well known transient Stokes-shift phenomena of molecules like 4-amino-phthalamide 4AP which possess a larger *instantaneous* dipole moment in $S_1$, i.e., only one form of $S_1$ exists. The evolution of the fluorescence spectrum of 4AP is a consequence of transient solvation *only* - not *both* transient solvation *and* (LE ⇌ TICT) interconversion, which is the situation for BA.

We recently developed a theoretical framework for modeling the photodynamics of compounds like 4AP and BA [3]. The basic premise of the model is that the photodynamic parameters (radiative rate constant, spectral shape, and population decay rate constant) are a continuous function of the solvent "polarity" which we associate with the empirical solvent scale $E_T(30)$. We have developed a procedure for experimentally determining, for a specific solute, the dependence of photodynamic parameters on $E_T(30)$. The relevant data are the fluorescence spectrum, quantum yield, and lifetime of the solute in various rapidly relaxing solvents, i.e., solvents for which the observed emission is due predominantly to a relaxed solvent configuration. This procedure allows us to extract for *slowly relaxing solvents* a time dependent polarity parameter from the emission kinetics as shown in Fig. 3e. This time dependent solvent parameter is roughly analogous to the time-dependent Stokes-shift. However, the measurements are made in a different fashion. Both parameters measure the evolution of the solvation energy [3] of the fluorescent probe.

The transient solvation of BA in GTA is highly non-exponential as shown in Fig. 3e where a single exponential change (convoluted with the instrument response function) in $E_T(30)$ is compared to the observed effect. This is consistent with the non-exponential nature of the bulk dielectric relaxation response function [4] of this solvent. We have observed that the time dependence of $E_T(30)$ in GTA at 12°C is experimentally indistinguishable for BA and the charge-transfer compound ADMA. Additionally, the solvation dynamics of 4AP in GTA (as measured by transient Stokes-shift) has been found to be similar to that of BA in GTA. These results suggest that the solvation dynamics in all these cases are only a weak function of solute-solvent interaction and are dominated by solvent forces.

Acknowledgment. Acknowledgment is made to the National Science Foundation (Grant No. CHE-8251158) and to the National Institutes of Health (Shared Instrument Grant No. RR01439).

4. LITERATURE

1. A.M. Brearley, S.R. Flom, V. Nagarajan, P.F. Barbara:
   J. Phys. Chem. **90**, 2092, (1986).

2. W. Rettig, M. Zander: Ber. Bunsenges. Phys. Chem. **87**, 1143, (1983).

3. V. Nagarajan, A.M. Brearley, P.F. Barbara: "Ultrafast Measurements of Transient Solvation: New Methods and Examples", J. Chem. Phys. to be submitted.

4. A.M. Ras, P. Bordewijk: Rec. Trav. Chim. **90**, 1055, (1971).

# Solvation Dynamics in Polar Liquids: Experiment and Simulation

*M. Maroncelli, E.W. Castner, Jr., S.P. Webb, and G.R. Fleming*

Department of Chemistry and James Franck Institute,
The University of Chicago, Chicago, IL 60637, USA

Current theoretical studies by a number of workers have focused attention on the importance of solvation dynamics in determining the rates of reactions in solution [1]. This is especially true of reactions involving substantial charge redistribution in polar solvents, where solvation energies may be quite large. The first step in testing these ideas is to obtain direct, microscopic measures of the kinetics of dipolar solvation. Experimentally, such information is available by monitoring the temporal evolution of the electronic spectrum of a probe solute after instantaneously changing its charge or dipole moment. Spectral shifts as a function of time directly monitor the course of solvation as the system reequilibriates to the new solute charge distribution.

The data available from previous solvation experiments are fragmentary and apparently discordant. In the well-studied case of electron solvation in n-alcohols, the solvation time correlates convincingly with the dielectric relaxation time of the second Debye dispersion region present in these solvents [2]. Since this time scale is close to the single-molecule rotation time the above observation suggests that solvation is dominated by a small number of neighboring molecules. On the other hand, electron solvation in water has been shown to be complete in ~100 fs [3]; orders of magnitude shorter than the water single-particle reorientation time. If the solvent behaved as a continuum dielectric fluid, solvation times would be equal to the solvent "longitudinal" relaxation time $\tau_L$ [4], which for water (490 fs) is much faster than the Debye time (~10 ps) and close to the experimentally determined limit. Solvation times measured for the excited state dipole of 4-aminophthalamide in alcohols [4] and TNSDMA [5] also appear to correlate with longitudinal relaxation times rather than the single-particle rotation times. Thus in the latter two cases the solvation time scale seems to have a macroscopic origin involving long-range interactions and not simply nearest-neighbor effects. Before a unified interpretation of these apparently contrary results can be achieved, measurement of solvation times for a variety of solvent/solute systems is needed.

In this paper we report on two different approaches to studying solvation dynamics, one purely experimental and the other using molecular dynamics simulation methods. The experimental studies involved solvation dynamics of three quite different probe molecules: LDS-750, coumarin 153, and 1-aminonaphthalene. Solvents examined included alcohols, amides, and a number of polar aprotic solvents. Temperature dependence of the solvation dynamics was also examined in some of these solute/solvent systems. Time-correlated single-photon counting and fluorescence upconversion were used to measure the time-dependent Stokes shift of the emission spectrum after excitation. The combined application of these two methods has proven advantageous due to their complementary time resolution and signal-to-noise capabilities. With both techniques, a series of fluorescence decays

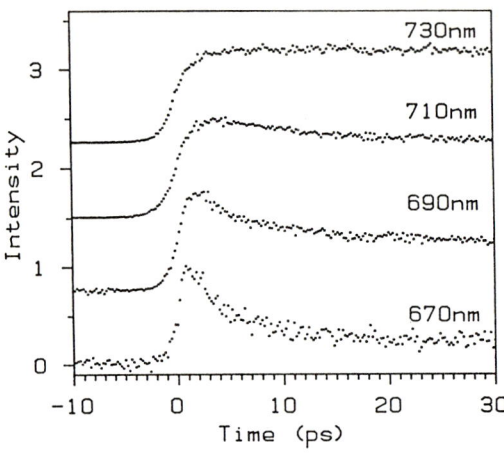

Fig. 1. Typical fluorescence decay series for LDS-750 in dimethyl sulfoxide (20°C) obtained using fluorescence upconversion.

were measured at fixed wavelengths spanning the fluorescence spectrum (Fig. 1). The spectrum at any time can then be reconstructed from a time-slice across this decay series. Finally, peak frequencies of such time-evolving spectra ($\nu(t)$) were used to construct the normalized correlation function $C(t) = (\nu(t)-\nu(\infty))/(\nu(o)-\nu(\infty))$ which provides a probe independent measure of the solvation dynamics.

Experimental results for a wide range of solvent/solute systems are summarized in Fig. 2. Here we have plotted the ratio of the observed solvation times $\tau_{obs}$ to the longitudinal relaxation time $\tau_L$ of the solvent [7] as a function of static dielectric constant $\varepsilon_0$. The observed Stokes shift correlation functions $C(t)$ are not simple exponential decays so that we have chosen an average time constant equal to the first moment of $C(t)$ for this comparison. At the course-grained level of Fig. 2, $\tau_{obs}/\tau_L$ appears to be simply correlated to the macroscopic solvent property $\varepsilon_0$. It should be emphasized that this correlation encompasses three probe solutes in both non-associated and highly associated solvents with $\tau_L$'s ranging between 2 and 1200 ps. The continuum-model prediction that solvation should proceed with time constant equal to $\tau_L$ is upheld in solvents having low to moderate $\varepsilon_0$. In contrast, for solvents with very high $\varepsilon_0$ such as propylene carbonate ($\varepsilon_0$=77, 82, 87 data) and N-methylpropionamide

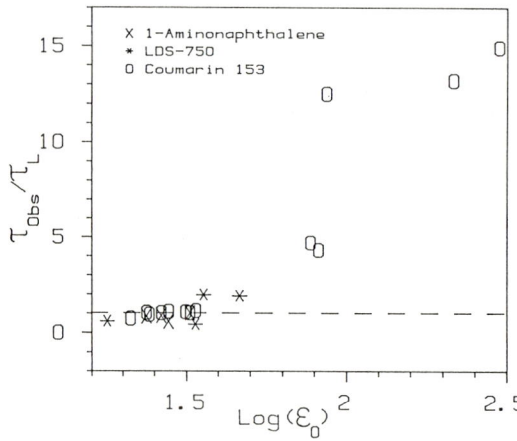

Fig. 2. Summary of observed solvation times $\tau_{obs}$ compared to solvent longitudinal relaxation times $\tau_L$. The ratio $\tau_{obs}/\tau_L$ is plotted versus solvent dielectric constant $\varepsilon_0$ for a range of solute/solvent pairs.

($\epsilon_0$= 215,299 data), the solvation is observed to proceed 5-15 times slower than predicted by $\tau_L$. Thus in this regime continuum models are clearly inadequate. In addition, closer inspection of the results in even the lower $\epsilon_0$ solvents reveals deviations from continuum predictions. As illustrated by the n-propanol data in Fig. 3, although the average time constant is close to $\tau_L$, the observed C(t) functions deviate significantly from the simple exponential decays predicted.

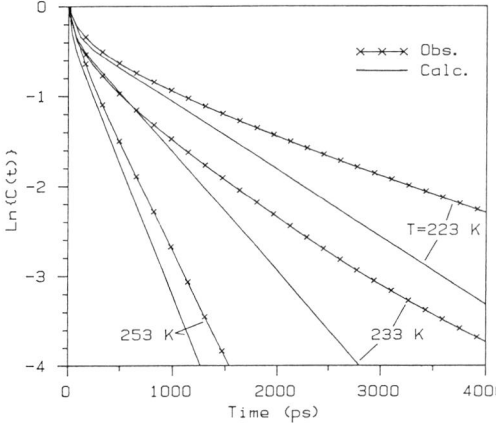

Fig. 3. Observed and calculated fluorescence shift correlation functions C(t) for coumarin 153 in n-propanol. Calculated curves are based on approximating the dielectric response of n-propanol with a Budo form in the manner described in Ref. 4a.

In the molecular dynamics (MD) simulations we have studied spherical clusters of 512 solvent molecules and a Lennard-Jones (LJ) solute in which is embedded some type of charge distribution such as a point charge, point dipole, or dipolar charge pair. Using equilibrium MD calculations, we collect the relevant time correlation functions of electrical properties with which to determine non-equilibrium solvation dynamics via the fluctuation-dissipation theorem. Solvents under study include the ST2 model of water and its isotopic variants, and a model of acetonitrile. Complete simulations have been carried out for ST2 water with an uncharged and a charged LJ solute. Results for the uncharged solute are shown in Fig. 4. The electric-field correlation functions $\langle \underline{E} \cdot \underline{E}(t) \rangle$ plotted here are directly comparable to the Stokes shift correlation function C(t) that would be observed in response to a step change in an embedded point dipole. Fig. 4a shows the contributions to this correlation function due to successive shells centered at the solute, and Fig. 4b shows a semilog plot of the total $\langle \underline{E} \cdot \underline{E}(t) \rangle$ along with the total dipole moment correlation function $\langle \underline{M} \cdot \underline{M}(t) \rangle$ for comparison. Based on Fig. 4 and similar results for a charged LJ solute we observe the following features. The dynamics within the first 25 fs are determined by librational motions of first solvation shell waters. A substantial part (~50%) of the total solvation energy is achieved here. After this initial decay, solvation proceeds on a time scale of ~500 fs. As shown in Fig. 4b, the $\langle \underline{E} \cdot \underline{E}(t) \rangle$ correlation function does not appear to be a single exponential. Continuum models would predict this function to decay exponentially with time constant equal to $\tau_L$ of ST2 water. We have not yet calculated $\tau_L$ for this system but we anticipate a value of 100-300 fs [8], less than the observed solvation time scale. Thus while further analysis is necessary, these simulated results show analogous deviations from continuum model predictions as do our experimental results. Finally, the simulations provide some insight into why solvation times (and $\tau_L$) are so much faster than solvent single-particle rotation times (and $\tau_D$).

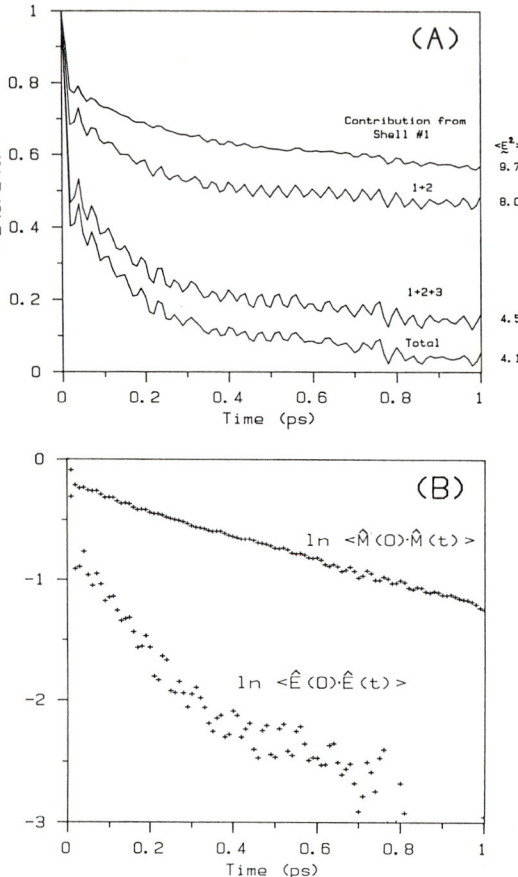

Fig. 4. Electrical field correlation function $<\underline{E}\cdot\underline{E}(t)>$ for a LJ solute in ST2 water calculated from MD stimulation. This correlation function is comparable to the experimentally determined C(t) function. (a) Decomposition of $<\underline{E}\cdot\underline{E}(t)>$ into contributions from different shell regions. These curves have been normalized to unity at time zero; relative amplitudes $<\underline{E}^2>$ are given at the right. (b) $<\underline{E}\cdot\underline{E}(t)>$ plotted on a semilog scale. The correlation function of the total dipole moment $\underline{M}$ of the spherical cluster is shown for comparison. Note the apparent non-exponentialty of $<\underline{E}\cdot\underline{E}(t)>$.

As illustrated by Fig. 4a, the contribution to $<\underline{E}\cdot\underline{E}(t)>$ from any small number of molecules, such as in the first solvation shell, exhibits a long time constant of roughly the single-particle correlation time. Addition of successive shells of solvent makes the response successively more rapid. Thus it is the collective nature of the process which enables the fast solvation response.

In summary, from both experiments and computer simulations we find that $\tau_L$ provides a reasonable guide to solvation times in many solvents. This guide fails however for solvents with very high dielectric constants ($\varepsilon_0 > 70$) where solvation times are much longer than $\tau_L$. Continuum -

based models at best provide a crude qualitative representation of solvation dynamics, and more refined models are needed to help interpret the present results.

References

1. See for example the review by J.T. Hynes,: Ann. Rev. Phys. Chem. 36, 573 (1985) and references therein.
2. G.A. Kenney-Wallace and C.D. Jonah,: J. Phys. Chem. 86, 2572 (1982).
3. Y. Gaudel, A. Migus, J.L. Martin, and A. Antonetti,: Chem. Phys. Lett. 108, 319 (1984).
4. (a) B. Bagchi, D.W. Oxtoby and G.R. Fleming,: Chem. Phys. 86, 257 (1984); (b) G. van der Zwan and J.T. Hynes,: J. Phys. Chem. 89, 4181 (1985).
5. S.W. Yeh, L.A. Philips, S.P. Webb, L.F. Buhse and J.H. Clark,: Ultrafast Phenomena 4, 359 (1984).
6. E.M. Kosower and D. Huppert,: Chem. Phys. Lett. 96, 433 (1983).
7. For solvents with more than one dispersion region $\tau_L$ used refers to the lowest frequency region "1" and is calculated as $\tau_L = (\epsilon_{\infty 1}/\epsilon_{01})\tau_{01}$.
8. This estimate is based on the assumption that ST2 water should reflect the properties of real water except that $\epsilon_\infty$ is less than that of real water due to neglect of vibrational and electronic contributions.

# Femtosecond Study of Electron Localization and Solvation in Pure Water

*Y. Gauduel, J.L. Martin, A. Migus, N. Yamada, and A. Antonetti*

Laboratoire d'Optique Appliquée, INSERM U275, ENSTA, Ecole Polytechnique, F-91128 Palaiseau Cedex, France

## 1. Introduction

The structure and dynamics of formation of the solvated electron is still of considerable theoretical and experimental interest in several areas of physics, chemistry and biology [1-7]. In particular, knowledge of the dynamics of electron localization and solvation is of prime importance for understanding electron transfer processes in condensed matter. In polar media such as alcohols and aqueous solutions, laser photolysis of solute molecules and pulse radiolysis have been extensively used to generate excess electrons and to observe changes in solvated electron spectra with time on nano- to picosecond time scales [4-8]. In such investigations, an electron is used as a microprobe of the local dynamics of molecular reorganization induced by the excess electron. Solvation kinetics have been measured in alcohols by picosecond spectroscopic techniques but in liquid water this kinetics has not been solved. The upper limit most recently established is 0.3 ps [9]. Indeed only ultrafast techniques can help to measure the kinetics of solvation of excess electrons in aqueous solutions. The solvation time in water at ambient temperature, much shorter than the overall rotational motion (10 ps), suggests the existence of pre-existing deep traps [9]. However, it has been argued that in a continuum dielectric theory, the characteristic time for a point charge solvation is given by the longitudinal relaxation time $\tau_L$, of the order of 0.2 ps, which seems compatible with previous experimental results [9,10]. It should then be noticed that if the water solution behaves as a continuum dielectric fluid, a spectral shift of the initial electronic state should reflect the relaxation of the medium around the excess electron [11]. Using femtosecond techniques, we have obtained evidence of the existence of a localization process and we have resolved the spectral evolution of the initially non-hydrated state.

## 2. Experimental

In our experiments, electrons were generated by a two-photon photoionization of water [12,13] and the transient steps implied in the hydrated electron formation were investigated at the femtosecond time scale by absorption spectroscopy techniques . The femtosecond laser pump-probe setup has been previously described [14]. Basically, UV 310 nm pulses of about 100 fs duration and 5µJ energy were focused into a 0.3 mm diameter spot on the sample, which was doubly deionized and bidistilled water. At 21°C, the resistivity is greater than 19 MΩcm and oxygen was removed by bubbling with pure nitrogen gas. The photoionization of water took place in a fixed volume cell moved in a plane perpendicular to the propagation of the light so that each pulse at 10 Hz repetition rate excited a new region of the sample.

A key point in these experiments is the exact knowledge of both the instrumental instantaneous response and the zero-time delay between the pump and probe pulses. This is of prime importance since the time responses are expected to be of the order of the pulse duration, in which case the rise of the signal does not change significantly in shape but is rather time shifted [15]. This information was obtained by inducing a weak instantaneous absorption in a n-heptane solution in the same conditions as in the water experiment.

## 3. Results

Induced absorption kinetics have been recorded over the whole visible spectrum and in the near infrared (IR) up to 1.25 µm. Figure 1 shows the reconstructed transient spectra obtained 2 ps after the femtosecond 310 nm excitation of liquid water at room temperature. This wide structureless band which peaks around 720 nm is identical to the spectrum assigned to the solvated electron in experimental conditions of pulse radiolysis and photolysis of ionic aqueous solutions [1,12].

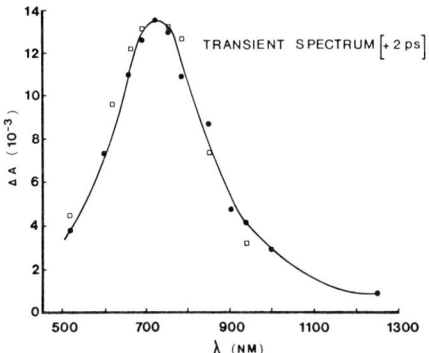

Figure 1 : Transient absorption spectra of pure liquid water recorded 2 ps after excitation with 100 fs pulses at 310 nm. The symbols □ and • represent two separate sets of experiments performed at room temperature.

All the kinetics (see Fig. 2 for two test wavelengths) are in good agreement with a model which takes into acount two absorbing species.
The first one, absorbing in the infrared, appears with a time constant of $T_1$ ($T_1$ = 110 ± 30 fs) and relaxes following a first-order kinetics $A(t) = A_0 \exp(-Kt)$ with the time constant $T_2 = 1/K = 240 \pm 40$ fs towards the solvated species, whose absorption peaks at 720 nm. At this wavelength the induced signal $A(t)$ follows the kinetics equation $A'(t) = A_0 - A_0/(T_2-T_1)(T_2\exp{-t/T_2} - T_1\exp{-t/T_1})$ as expected after a first-order kinetics of appearance of the precursor of the solvated species (with $T_1^1$ = 110 ± 30 fs as observed at 1.25 µm) and a subsequent relaxation also with $T_2^1$ = 240 ± 40 fs. The whole set of kinetics taken at different wavelengths could be fitted equally well assuming both previous absorbing species with, however, different ponderation. For instance we may notice at 1.25 µm a non-complete recovery due to the solvated electron contribution. It is important to point out that this procedure allows us to fully compute the different transient absorption spectra, in particular the IR one which decays in 240 fs.

## 4. Discussion

The photoionization of pure liquid water by high intensity femtosecond UV pulses provides useful information on details of the kinetics of electron solvation in a polar medium which is not perturbed by the presence of solute.

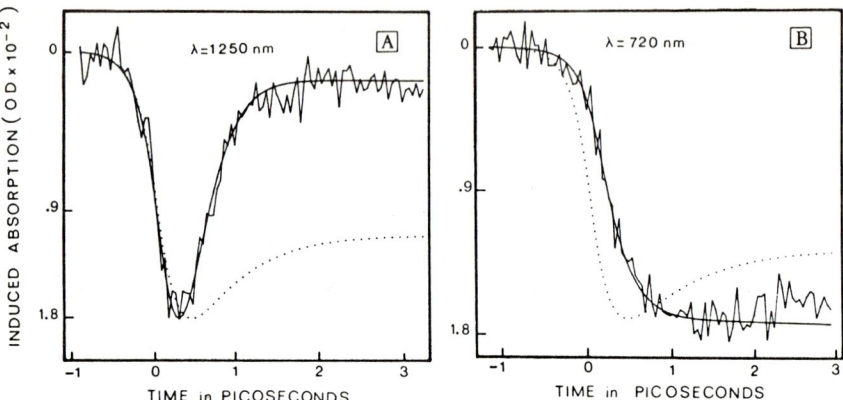

<u>Figure 2</u> : Time-resolved data obtained after femtosecond laser photoionization of pure liquid water.
The instantaneous responses at 1.25 μm and 0.72 μm (.....) are defined in a pure solution of n-heptane. The smooth lines indicate the computed best fit of the experimental curves.
A : Localized electron appearance time constant $T_1$ = 110 ± 30 fs and time relaxation $T_2$ = 240 ± 40 fs.
B : Hydrated electron appearance assuming the same values of $T_1$ and $T_2$.

The data analysis clearly indicates that solvation of electrons in pure liquid water proceeds through an intermediate state which appears with a 110 fs time constant and then relaxes in 240 fs towards the solvated state. Furthermore, the transient IR absorption spectra computed at different delays during and after the UV excitation are all similar in shape, so that we can deduce the existence of two distinct and well-defined transient absorption spectra. These observations do not support the hypothesis of the predominant existence of deep traps into which the electrons fall directly during the solvation. They rather imply at least a two-step process, namely the capture of the quasi-free electron by shallow permanent traps or by transient local fluctuations of the potential surface, and a subsequent 240 fs relaxation towards the solvated state. We propose this precursor of the solvated electron to be a state where the electron is spatially extended. The very first risetime for the appearance of this state (110 fs) implies that efficient mechanisms involved in the localization process do not require large molecular reorientation. The understanding of this mechanism implies the knowledge of the structure and the size of the water clusters. The kinetics reported here suggest that the $e^-(H_2O)_n$ system is stable on the femtosecond time scale but that n, the number of water molecules composing a cluster, is probably small. What is the minimum cluster size for the existence of bound surface states? It is possible that the stability of $e^-(H_2O)_n$ involved in the localization process is dynamic in origin. A water dimer could then stabilize an electron in a diffuse cloud with a large spatial extension. This configuration would be an electronic state distinct from the solvated electron one, which is known to be clearly localized and roughly spherical with a radius of about 3 Å [16,17]. This picture is consistent with our observation of two distinct absorption spectra. We may also speculate that localization of the excess electron in the cluster is promoted by fluctuations in the electronic charge density of the solvent molecules [7,18].
We may wonder whether a long range interaction with the medium, such as a dielectric relaxation, can explain the ultrafast 240 fs decay of the localized

species towards the solvated one. In this respect it is admitted that the solvation characteristic time is given by the longitudinal relaxation time $\tau_L = \tau_D \epsilon_\infty / \epsilon_0$ where $\tau_D$ is the water Debye relaxation time (10 ps) and $\epsilon_0$ ($\epsilon_\infty$) the static (high frequency) dielectric constant [10]. This value of $\tau_L$ in the range 0.2-0.4 ps is in striking agreement with our $T_2$ value. However a dielectric relaxation when treated as a diffusive type of process (Debye relaxation) implies a continuous modification of the medium interacting with the electrons. This should appear as a continuous spectral shift of the electron absorption band from the IR towards the visible [11]. Such an effect is in contradiction with our data from which we rather deduce a well-defined and quasi-invariant absorption IR spectrum for the localized species.

In conclusion, we have experimentally shown that the solvation of electron in liquid water at room temperature proceeds through at least **two transitions** involving in particular a **localized state** absorbing in the IR, with **lifetime 240 fs**. Our observations rule out the assumption of a high density of pre-existing deep traps and do not support a pure dielectric relaxation mechanism for the electron solvation. The exact identification and comprehension of these localized and solvated states need more theoretical work using tools like molecular dynamics simulations and path integral techniques [19] to explore the obvious quantum mechanical character of the structural and dynamical properties of the electron in polar solvents.

References
1. E. Hart, M. Anbar: The hydrated electron (Wiley, New York 1970)
2. J. Jortner, N.R. Kestner (Eds): Electrons in fluids (Springer-Verlag, New York 1973)
3. L. Kevan, B.C. Webster (Eds) Electron-solvent and anion solvent interactions (Elsevier, Amsterdam 1976)
4. P.M. Rentzepis, R.P. Jortner, J. Jortner: J. Phys. Chem. 59 766 (1973)
5. W.J. Chase, J. W. Hunt: J. Phys. Chem. 79, 2835 (1975)
6. G.A. Kenney-Wallace, C.D. Jonah: Chem. Phys. Lett. 47 362 (1977)
7. G.A. Kenney-Wallace, C.D. Jonah: J. Phys. Chem. 86, 2572 (1982)
8. G.E. Hall, G.A. Kenney-Wallace: J. Chem. Phys. 32, 313 (1982)
9. J.M. Wiesenfeld, E.P. Ippen: Chem. Phys. Lett. 73 47 (1980)
10. D.F. Calef, P.G. Wolynes: J. Phys. Chem. 87, 3387 (1983)
11. L.D. Zusman, A.B. Helman: Chem. Phys. Lett. 114 301 (1985)
12. D.N. Nikogosyan, A.A. Oraevsky, V.I. Rupasov: Chem. Phys. Lett. 77 ,131 (1983)
13. H. Miyasaka, H. Masuhara, N. Mataga: Chem. Phys. Lett. 98, 277 (1983)
14. A. Migus, A. Antonetti, J. Etchepare, D. Hulin, A. Orszag: J. Opt. Soc. Am. 2, 584 (1985)
15. Y. Gauduel, A. Migus, J.L. Martin, Y. Lecarpentier, A. Antonetti: Ber. Bunsenges. Phys. Chem. 89, 218 (1985)
16. A.M. Brodsky, A.V. Tsarevsky: Adv. Chem. Phys. 44, 483 (1980)
17. P.J. Rossky: N.Y. Academic Sciences (in press) 1986.
18. A.L. Nichols, D. Chandler: J. Chem. Phys. 84 , 398 (1986)
19. C.P. Jonah, C. Romero, A. Raham: Chem. Phys. Lett. 123, 209 (1986)

# Time-Dependent Fluorescence Shift in Alcoholic Solvents: A Non-Debye Behaviour Related to Hydrogen Bonds

C. Rullière, A. Declémy, and Ph. Kottis

Centre de Physique Moléculaire Optique et Hertzienne,
Université de Bordeaux I, 351 Cours de la Libération,
F-33405 Talence Cedex, France

We have studied strong solute-solvent interactions on a picosecond time scale, via time-resolved emission from electronically excited species in solution. In the case of alcoholic solvents (which are highly aggregated by hydrogen bonds), we have shown a non-exponential behaviour on the Time-Dependent Fluorescence Shift (TDFS) of the excited species. We discuss the different solvent parameters which influence this time evolution. We show that some microscopic parameters such as steric hindrance and proton donor character are more correlated with the observed phenomena than macroscopic parameters such as viscosity, for example.

## 1. Experimental Measurements of Time-Dependent Fluorescence Shift (TDFS)

We used a picosecond spectrometer which has been described already in details elsewhere[1]. Typically, this totally automated set-up allows us to obtain the whole absorption or emission spectrum of an initially excited species in five minutes in the range 4300-8000 Å for time delays t after excitation ranging from 0 to 500 ps (the time resolution is of the order of 10 ps). Using this set-up we recently studied [2] polar compounds in different solvents. These compounds are characterized by a great change of their permanent dipole (about 10 D) upon electronic excitation and by some hydrogen bond acceptor sites (lone pair of heteroatoms). We showed that in non-polar and non-aggregated polar solvents (toluene, dimethylformamide (DMF) (Fig. 1)), the emission spectrum does not change as the time evolves

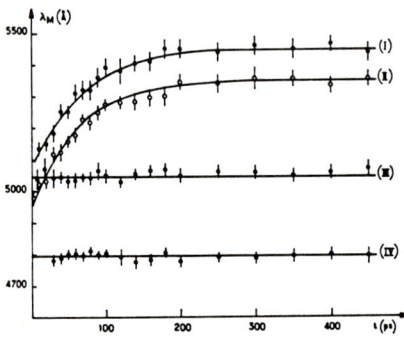

Fig. 1 Time evolution of the emission maximum of solute in I) ethylene glycol ; II) ethanol; III) DMF ; IV) toluene.

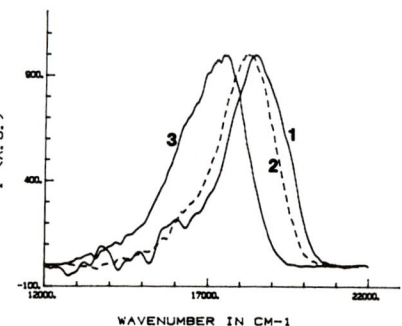

Fig. 2 Time evolution of the solute emission spectrum in pentanol: 1) 10 ps after excitation; 2) 100 ps ; 3) 350 ps.

after excitation, on the time scale available with our set-up (t> 10 ps). So in the case of these solvents, we conclude that the dipolar reorientation of solvent molecules is very fast and unobservable on our time scale at room temperature. On the other hand, in alcohols such as ethanol, ethylene glycol (Fig. 1) and pentanol (Fig. 2) we observe a continuous shift of the emission spectrum toward long wavelengths. This shift is related to the stabilization, in the excited state, of the hydrogen bonds, due to specific solute-solvent and solvent-solvent interactions.

So in the case of hydrogen bonded aggregated solvents such as alcohols, the reorganization of solvent molecules is slower than in non-hydrogen bonded solvents (observable on the picosecond time scale at room temperature). This is due to specific interactions in the medium, related to the hydrogen bond presence.

## 2. Time Evolution of TDFS in Different Alcohols, and Discussion.

In order to evaluate the key parameters which control the relaxation process in aggregated solvents, we have considered different sized solvents such as methanol, propanol, butanol (Fig. 3) and more or less aggregated solvents such as ethanol and trifluoroethanol (Fig. 4). In Fig.3 we observe that in the case of small and weakly aggregated solvent such as methanol the kinetics of TDFS is roughly exponential. But, as the size of the solvent molecules increases (such as for propanol and butanol), the kinetics becomes highly non-exponential and cannot be described by one or a sum of exponential terms as in the Debye model. Moreover, in a weakly aggregated solvent such as fluorinated ethanol ($CF_3-CH_2-OH$), the kinetics is faster than in the parent solvent such as ethanol (Fig. 4) whose macroscopic parameters (dielectric constant, viscosity) are similar, but which is more aggregated by hydrogen bonds. Finally we note from Fig. 1 that similar kinetics are observed in ethanol and ethylene glycol, which have quite different viscosity at room temperature (ethanol: $\eta \sim 1.20$ cp, ethylene glycol $\eta \sim 19.9$ cp).

So we have shown that in alcohols, such parameters as size or degree of aggregation have a strong influence on TDFS kinetics and cannot be taken into account by means of only one macroscopic parameter. Some models [3] have been proposed to treat solvation in polar solvents. But they must be modified in order to integrate hydrogen bonding and specific related effects.

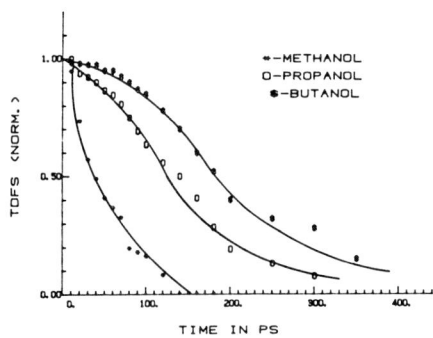

Fig.3 TDFS kinetics of solute in: methanol (*); propanol (O); butanol ($).

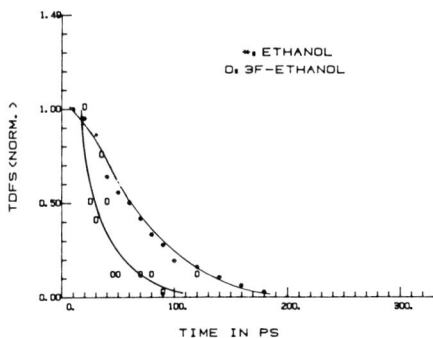

Fig.4 TDFS kinetics of solute in ethanol (*); trifluoroethanol (O).

Now, the important question is : How should one take into account the different interaction terms in the solute-solvent correlation function, which is revealed by the TDFS?

References

1. C. Rullière, A. Declémy, Ph. Pée: Rev. Phys. Appl. 18, 347 (1983).
2. A. Declémy, C. Rullière, Ph. Kottis: Chem. Phys. Lett. 101, 401 (1983).
3. B. Bagchi, D.W. Oxtoby, G.R. Fleming: Chem. Phys. 86, 257 (1984).
   G. Zwan, J.T. Hynes: J. Phys. Chem. 89, 4181 (1985).
   Y.T. Mazurenko: Opt. Spectrosc. 48, 388 (1980).
   Y.T. Mazurenko: Opt. Spectrosc. 55, 277 (1983).

# Picosecond Dynamics of Proton-Anion Ion Pair Geminate Recombination

## D. Huppert and E. Pines

Raymond and Beverly Sackler Faculty of Exact Sciences,
School of Chemistry, Tel Aviv University, 69 978 Tel-Aviv, Israel


Charge separation is induced in solutions of many chemical and biochemical systems by light absorption. The primary step in these reactions is either an electron or a proton transfer from a donor site to a suitable acceptor. Solvated ion pairs, if produced, can either geminately recombine or separate by diffusion. Geminate recombination was recognized to be extremely important in radiation induced electron-cation ion pair generation [1]. As for proton transfer reactions, much less attention has been paid to this phenomenon mainly for two reasons. The first reason is that proton transfer reactions are usually being carried out in aqueous solutions where the coulombic attraction is very efficiently screened. In contrast, electron transfer reactions are usually being carried out in hydrocarbon solutions where the coulombic screening is much less effective. The second and less obvious reason is that geminate electron-excited cation recombination usually quenches the excited state where in many cases proton transfer to an excited anion does not quench the anion [2]. As a result, proton transfer reactions are usually bidirectional both in the ground and the excited state [3]. It means that upon recombination the excited parent molecule can undergo redissociation. Thus, the combination of relatively short range coulombic attraction and consecutive dissociations makes geminate recombination much less apparent in aqueous solutions proton transfer reactions than in electron transfer ones.

The concept of bidirectional dissociation-recombination kinetics within the lifetime of an ion-pair is not a new idea for the chemists. EIGEN [4] in particular was able to explain the essence of diffusion controlled reactions as well as vast experimental data by assuming a two stage reaction kinetics.

$$H^+ + B^- \underset{k_S}{\overset{k_D}{\rightleftarrows}} H^+ \cdots B^- \underset{k_1}{\overset{k_2}{\rightleftarrows}} HB \qquad (1)$$

where $k_D$ and $k_S$ are the diffusion controlled rate constants of formation and separation of the ion pair and $k_2$, $k_1$ are the geminate ion pair recombination and the product dissociation rate constants respectively.

As a model system for the investigation of geminate recombination in proton transfer reactions we chose the excited state dissociation of 8-hydroxy pyrene 1,3,6 trisulfonate (HPTS)* in aqueous solutions. In this case, the formed excited anion is four times negatively charged. The Onsager length [5], $R_c$, which scales the coulomb attraction, is relatively large (30 Å) and approaches the values found for geminate electron-cation recombinations in alcoholic solutions. Direct picosecond measurements have shown that the excited HPTS dissociates within 100 ps. When strong acids were added to the solution the dissociation process becomes reversible and can be approximated by Eq.(1). $k_D$ of the reaction was found to be $(8 \pm 3) \times 10^{10}$ $M^{-1} s^{-1}$ [6]. The apparent effect of the acid on the (HPTS)* decay is the addition of a 6 ns long tail whose amplitude increases with the added acid concentration. In concentrated acid solutions the (HPTS)* biexponential decay coalesces to a single exponential decay with a lifetime found in nonaqueous solvents where no dissociation occurs ($\tau_0$ = 6 ns). In pure aqueous solutions, the decay of the (HPTS)* was found to retain its characteristics in weak acidic media, namely a fast and almost exponential decay followed by a long nonexponential "tail". We attribute this behaviour to a reversible dissociation geminate recombination process.

It is well established that the recombination kinetics of an isolated ion pair is described by the Smoluchowski Equation which considers the motion of a Brownian particle of charge q in the field of an oppositely charged Coulomb center. Although theoretically the equation can be solved [5], a full kinetic solution of a reversible geminate recombination reaction has not as yet appeared in the literature. Instead, we used a time dependent reaction scheme, Eq.(2):

$$HB^* \underset{k_2(t)}{\overset{k_1}{\rightleftarrows}} H^+ \cdots B^{-*} \qquad (2)$$

where HB* is the excited HPTS molecule and $H^+\cdots B^-$ is the excited ion pair. The time dependence of $k_2$ is due to the diffusion which gradually decreases the density of one ion against its counter ion. Assuming free diffusion as a reasonable approximation for long time behaviour the average density of one ion against its partner will be given by $\rho(t) = \Omega(t)/V(t) \propto \Omega(t)/(r(t))^3$ where $\rho(t)$ is the density of the ion, $\Omega(t)$ is the probability that the ion has not reacted (the escape probability), V(t) is the effective volume which encloses the two ions, and r(t) is their separation. For a free diffusion $r(t)=(D_{12}t)^{1/2}$ where $D_{12}$ is the relative diffusion coefficient between the two ions. Hence $V(t) \propto (Dt)^{3/2}$ and

$$\rho(t) \propto \Omega(t) \cdot (Dt)^{-3/2} . \qquad (3)$$

Assuming a diffusion controlled reaction the rate of the reaction will be given by:

$$R(t)=k_D\,\rho(t)=4\pi D_{AB}R_D\,\rho(t)=k_2 t^{-3/2} \quad ; \quad R_c=Z_1Z_2 e^2/\varepsilon_0 \cdot k_B \cdot T \qquad (4)$$

where $Z_1$, $Z_2$ are the charge numbers of the two ions, e the electron charge, $\varepsilon_0$ the static dielectric constant of the medium, $k_B$ is the Boltzmann constant and T the absolute temperature. R(t) will go as $t^{-3/2}$ if $\Omega(t)$ approaches a constant value. At shorter times R(t) will depend on $\Omega(t)$ in accordance with Eq.(2).

Following these considerations we thus adopted for $k_2(t)$ the expression given by Hong and Noolandi for the long time recombination rate of an isolated ion pair after it was generated by a $\delta$ function distribution at a mutual separation distance of $r_0$. For the diffusion controlled limit and when $R_c \gg a$, the contact radius, it takes the form [5]:

$$k_2(t) = k_2^\circ\, t^{-3/2} \quad ; \quad k_2^\circ = R_c\,\exp(R_c/r_0)/2(\pi D)^{1/2} \qquad (5)$$

which has the desired $t^{-3/2}$ dependence.

Formally Eq.5 depends on $r_0$ through the $\exp(R_c/r_0)$ term which is the escape probability when time goes to infinity. $\Omega_{(\infty)}$ thus represents the fraction of the pairs which ultimately remain separate when the back dissociation is not allowed. However in the case of repeated dissociation $\Omega_{(\infty)}$ always approaches unity [7]. Indeed, the dissociation of the HPTS molecule is completed within the lifetime of the excited state. Therefore the calculated $r_0$ values must have limited physical meaning in our approach. Still one may look at it as some average measure for the actual separation between the geminate ions. This is only if the distribution function for the geminate ions separations changes relatively slowly compared to the observed dissociation process. This may be the case if the primary events of proton dissociation and recombination in aqueous solutions are subpicosecond processes. If so, both the diffusion and the molecular parameters control the apparent rate of dissociation. For HPTS this rate is $10^{10}$ s$^{-1}$ and the corresponding diffusion length is $(Dt)^{1/2}$ or roughly $10^{-7}$ cm = 10Å for the proton - HPTS anion system ($D_{12} \sim 10^{-4}$ cm$^2$ s$^{-1}$, $t = 10^{-10}$ s). It may be thus justified to assume that the shape of the distribution function of the protons does not change much until the dissociation recombination cycles are practically completed. The kinetic solution of Eq.2, using expression (5) for k (t) is given by:

$$\frac{[R^*OH]_t}{[R^*OH]_o} = 1 - \frac{k\int_0^t \exp(k_1 t' - 2k_2^o t'^{-1/2})dt'}{\exp(k_1 t - 2k_2^o t^{-1/2})} \qquad (6)$$

At room temperature and in aqueous solutions of various HPTS concentrations ($10^{-3}$ - 4 x $10^{-6}$ M) the convolution of Eq.6 with the laser pulse reproduced the measured decay profile. The best fit was achieved using the values of $k_1 = 1.5 \times 10^{10}$ s$^{-1}$ and $r_o = 35$ Å [8]. The larger apparent value of $k_1$ and the relatively large value of $r_o$ both agree well with the idea that prior to our observation an ultrafast equilibrium is established between dissociation and recombination in which only minute concentration of ion pairs is formed. This equilibrium is then disturbed by diffusion which constantly decreases the concentration profile of one ion against the other. To balance this decrease the molecule continues to dissociate until dissociation is completed. A solid evidence for such a mechanism is still a matter of further investigations. However the deviation from purely exponential decay (the decay "tail") which serves us as a measure for the geminate recombination processes was found to follow in many different reaction environments the proposed reaction mechanism. In particular, when other proton acceptors such as the acetate anion and the pyridine molecule had been introduced to the solution the amplitude of the decay "tail" decreased considerably as a function of the proton acceptor concentration. A similar behaviour was found when salts like LiClO$_4$ had been added to the solution and screened the coulomb attraction. It was also found that the isotopic D to H substitution considerably increases the amplitude of the decay "tail" [8]. Thus, in accordance with Eq.5 the rate of geminate recombination increases as the rate of the free diffusion decreases ($D_H^+/D_D^+ = 1.45$ at $20°C$), where coulombic screening and proton trapping decreases geminate recombination.

The temperature effect, shown in Fig. 1, was found to agree with Eq.5. The temperature dependence of $k_2(t)$ is through the dependence of the Onsager length and the diffusion coefficient. However since the $(\varepsilon \cdot T)^{-1}$ factor in $R_c$ is almost temperature independent between 273-373 K the major contribution to the temperature effect results from the diffusion coefficient. The diffusion coefficient of the proton in water increases by a factor of 4 between -12° and 100°C. A $(D)^{1/2}$ dependence of $k_2(t)$ will thus cause a decrease in $k_2(t)$ by a factor of 2 between -12°C and 100°C. The relative calculated values of $k_2(t)$ in the 3 temperatures shown in Fig.1 are: 1.0, 0.66, 0.48 for T=-12°C, 20°C, 80°C respectively. As it can be seen from Fig.1, geminate recombination does decrease when the temperature is raised. However a quantitative fit was not possible without changing $k_1$ the molecular dissociation constant. Indeed, it is reasonable to assume that $k_1$ is temperature dependent.

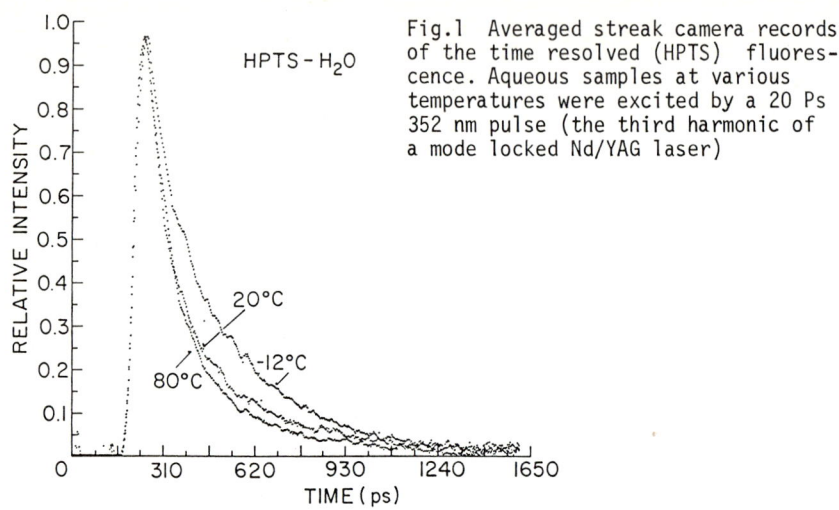

Fig.1 Averaged streak camera records of the time resolved (HPTS) fluorescence. Aqueous samples at various temperatures were excited by a 20 Ps 352 nm pulse (the third harmonic of a mode locked Nd/YAG laser)

To sum up, our treatment cannot be considered a dynamical treatment on the level of the molecular dissociation and recombination processes. However, we do feel that it presents some valuable ideas about the overall mechanism of dissociation and that it can be used for a reasonable averaged description of these processes on the ps-ns time scale. On this time scale a $t^{-3/2}$ dependence of the geminate recombination seems real and the phenomenon itself is proven to play an important role in aqueous solutions proton dissociation reactions.

References

1. C.L. Braun and T.W. Scott, J. Phys. Chem. 87, 4776 (1983).
2. Th. Förster, Pure Appl. Chem. 24, 443 (1970).
3. A. Weller, Progress in Reaction Kinetics 1, 1 (1961).
4. M. Eigen, Angewandte Chemie 3, 1 (1964).
5. K.M. Hong and J. Noolandi, J. Chem. Phys. 68, 5163 (1978).
6. E. Pines and D. Huppert, Chem. Phys. Lett. 126, 88 (1986).
7. N. Agmon, J. Chem. Phys. 81, 2811 (1984).
8. E. Pines and D. Huppert, J. Chem. Phys. 84, 3576 (1986).

# Excited State Proton Transfer in Matrix Isolated Water and Methanol Complexes of 2-Hydroxy-4,5-benzotropone and 3-Hydroxyflavone

D.F. Kelley and G.A. Brucker

Department of Chemistry, Colorado State University,
Fort Collins, CO 80523, USA

The rates and mechanisms of excited state proton transfer reactions have recently been subjects of considerable interest. In particular, the role of the hydrogen-bonding solvent environment in intramolecular proton transfer has been extensively studied. A variety of molecules which undergo excited state proton transfer can be studied in solution with hydrogen-bonding solvents. However, the solution environment produces many different solute-solvent hydrogen-bonding complexes, resulting in unresolved spectra and complicated proton transfer kinetics.

We report here the study of excited state proton transfer reactions in hydrogen-bonded solute-solvent complexes of known stochiometries. Specifically, methanol and water complexes with 3-hydroxyflavone (3HF) and 2-hydroxy-4,5-benzotropone (HBT) can be readily formed in 10K argon matrices.

HBT: normal $\lambda_{em}$ = 420 - 480 → proton transfer → tautomer $\tilde{\lambda}_{em}$ = 480 - 570

3HF: normal $\lambda_{em}$ = 370 - 450 → proton transfer → tautomer $\tilde{\lambda}_{em}$ = 490 - 580

These complexes are formed by codeposition of dilute solvent ($H_2O$ or $CH_3OH$), argon mixtures with a very low concentration of 3HF or HBT. Typical argon:solvent concentrations are 2000:1. Subsequent annealing of the matrix at temperatures where solvent, but not 3HF or HBT diffusion is significant, results in the desired solute·(solvent)$_n$ n=0,1,2 complexes. As annealing procedes, different features appear in the fluorescence excitation and dispersed emission spectra. Analysis of the appearance kinetics establishes the stochiometry of the complex giving rise to each spectral feature. A 3HF/MeOH dispersed emission spectrum resulting from annealing this process is shown in Figure 1. Figure 1 shows that the emission intensities of 3HF·(MeOH)$_1$ and 3HF·(MeOH)$_2$ are about 20% and 4%, respectively, of the uncomplexed 3HF emission. These assignments are consistent with the solution work by Kasha[1]. The 3HF·(MeOH)$_1$ complex presumably corresponds to a cyclically hydrogen-bonded structure, with both the 3HF alcohol and ketone moities hydrogen-bonded to the methanol.

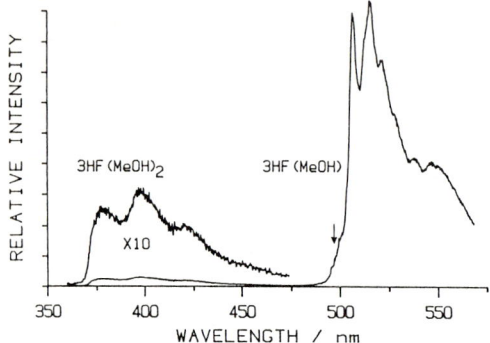

Figure 1. Dispersed emission spectrum of 3HF/MeOH mixture in 10 K argon. The shoulder at 490-505 nm (arrow) is due to 3HF·MeOH complexes and the emission in the 370-450 nm region is due primarily to 3HF·(MeOH)$_2$ complexes.

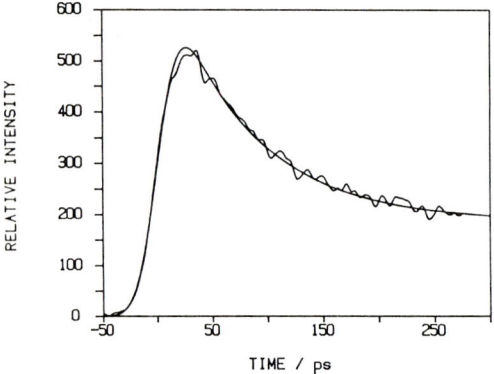

Figure 2. 3HF·MeOH emission kinetics at 498 $\pm$ 2 nm. The solid curve corresponds to the convolution of the temporal instrument response function with an 80 ps and a smaller long lived emission decay.

The dynamics of each of these complexes (n = 0,1,2) has been studied by picosecond emission spectroscopy. The tautomer emission kinetics indicate that uncomplexed 3HF undergoes very rapid proton transfer (<10 ps). A similar result was found for HBT. Linewidth measurements indicate that the HBT proton transfer time is about 30 fs$^2$.

The 3HF·MeOH emission kinetics at 498$\pm$ 2 are presented in Figure 2, and show fast (80ps) and slow (>1ns) components of the emission decay. Static spectroscopic studies permit the assignment of the two temporal components. These studies indicate that the 3HF anion emission spectrum is similar to that of the tautomer, but shifted ~10nm to the blue. We find that the fast emission component is present at wavelengths >480nm, whereas the slow component is negligible at wavelengths <490nm. Based on these results we suggest that the fast and slow components of the emission decay can be assigned to the 3HF anion and tautomer, respectively. From this tentative assignment, we conclude that proton transfer from the 3HF alcohol to the methanol occurs rapidly (<10 ps), forming a 3HF anion. Subsequent methanol to 3HF ketone proton transfer require ~80 ps. Similar results have also been obtained for HBT. The 3HF·$H_2O$ emission spectra and kinetics are similar to those obtained with 3HF/methanol. The kinetics indicate a slightly shorter anion lifetime (~55ps) for the 3HF·$H_2O$ case. However, we note that the amplitude of the rapidly decaying component of the 485-500nm monosolvate emission is much less than in the 3HF/methanol case, resulting in some uncertainty in the lifetime measurements. We tentatively interpret the difference in 3HF anion lifetimes in terms of the relative stabilities of the methanol and water cations.

These studies have also elucidated the proton transfer rates for 3HF and HBT complexes with two solvent molecules. In all cases, the 420 nm emission was found to decay on the nanosecond timescale. No corresponding emission risetime was observed anywhere in the spectrum. These results indicate that n = 2 complexes show little or no proton transfer at 10K.

REFERENCES

1. D. McMorrow and M. Kasha, J. Phys. Chem. **88**, 2235 (1984).

2. D.J.Jang, G. A. Brucker and D. F. Kelley, J. Phys. Chem. - submitted.

# Detection of the Inverted Region in Photo-induced Intramolecular Electron Transfer

R.J. Harrison[1], G.S. Beddard[1,*], J.A. Cowan[2], and J.K.M. Sanders[2]

[1]Chemistry Department, University of Manchester,
Manchester M13 9PL, UK
[2]University Chemical Laboratory, Lensfield Rd.,
Cambridge CB2 1EW, UK

Photo-induced intramolecular electron transfer has been found to occur in molecules where a porphyrin donor and an electron acceptor group are held apart by a double bridge. Both charge separation and recombination are observed and these rates of electron transfer are fastest in solvents of highest dielectric constant. Both the classical and quantum theories of electron transfer describe the general features of our experimental observations. In particular, a decrease in rate is seen at high exoergicity, in the so-called "inverted region".

Introduction

Electron transfer between excited state donor-acceptor (D..A) species is of continuing interest, as a result of their importance as models for parts of the photosynthetic reaction centre. Species in which the donor and acceptor are held apart in relatively fixed orientations within the same molecule are especially useful in elucidating the factors affecting the rate of transfer. For intramolecular electron transfer the rate is not constrained by a diffusional limit imposed by the solvent and competing side reactions such as exciplex formation are avoided. The reaction proceeds according to the following scheme

$$[D..A] \xrightarrow{h\nu} {}^1[D..A]^* \xrightarrow{K_{sep}} [D^+..A^-]^* \xrightarrow{K_{rec}} [D..A]$$

In the molecules studied the donating group is a porphyrin, the excited state spectroscopy of which is well-known [1], the capping accepting groups are benzoquinone ($QP_3$), methylviologen [2] and pyromellitimide ($BPC_3$). We have observed both the rates of charge separation and recombination on the picosecond timescale by following the transient absorption changes of the porphyrin [2]. By varying the polarity of the solvent, the level of stabilization of the charge separated state can be altered by as much as 0.5 eV. In this way, the rate of electron transfer can be measured as a function of free energy change both for the initial separation from $S_1$ and final recombination from the charge separated state back to the ground state. We have also studied the fluorescence yields and decays of the D..A species as a function of temperature in several solvents. This provides corroborative evidence for the absorption data and enables a measure to be made of the activation energy for the initial charge separation process.

Fig. (1) Structure of Capped Porphyrins

The compounds used are shown in fig. (1) and were prepared as previously described [3]. Their redox potentials were determined by cyclic voltammetry with 0.2 M TBAFB as electrolyte. $E^o$ for $QP_3$ was found from measurements of the individual moieties in acetonitrile to be 1.35 eV, whilst $BPC_3$ was measured directly in dichloromethane (1.77 eV) and in N,N-dimethylformamide (1.69 eV). Our system consists of an amplified, synchronously pumped dye laser, tunable between 576-640 nm with typical pulse widths of 0.5ps and power of 0.5mJ/pulse at 10 Hz. Transient absorption measurements were made using the usual "pump and probe" techniques with continuum generation and are described elsewhere [1,2]. Fluorescence decays were determined by single-photon counting and also with a streak camera (Imacon 500) and OMA combination.

Results

In both $QP_3$ and $BPC_3$, the transient spectrum initially formed is characteristic of a porphyrin excited state, growing-in with the laser pulse and showing a bleaching of the intense Soret band at 400 nm and a broad absorbance centred at 460 nm. Both features decay rapidly, as charge separation occurs, with a lifetime of 2 ps in the most polar solvents used and 10 ps in the least polar. Bleachings at 500, 530 and 570 nm (Q bands) recover more slowly with lifetimes ranging from 11 to 250 ps dependent on the polarity of the solvent as before. This longer decay corresponds to the recovery of the ground state, with no longer lived states (e.g. triplets) being observed. The slowest separation and recombination rates were in toluene $\epsilon$ = 2.4 and fastest in MeCN, $\epsilon$ = 37. The fluorescence lifetime of the parent mesoporphyrin is 10 ns in aerated dichloromethane solution thus the measured transient decay rates are very good approximations to the electron transfer rates. The data is shown in fig. (2) together with curves calculated from the Marcus theory [4], inner curve, and from quantum formulae given by Ulstrup and Jortner [5]. Classically electron transfer can be treated as an activated process, the rate of reaction being given by

$$K(r) = 2\pi/\hbar |V(r)|^2 . (4\pi\lambda kT)^{-\frac{1}{2}} . \exp((-\Delta E + \lambda)^2/4\lambda kT)$$

where $V(r)$ is the exchange interaction of D. A at separation r and the activation energy is $(-\Delta E + \lambda)^2/4\lambda$. Clearly the rate is fastest when the free energy change ($\Delta G^0 = \Delta E$ as $\Delta S^0$ is small) equals the total reorganisation energy $\lambda$. A value of $\lambda$ = 0.7 eV is indicated from the experimental results in these and other molecules [2,6], and this value was used in the classical calculation, together with values of the sum of molecular radii (4 Å)and separation (6 Å)and $V(r)$ = 0.0054 ± 0.001 eV. The energy of the charge separated state in each solvent can be calculated using the values for the half-wave potentials measured by cyclic voltammetry in a high polarity solvent (MeCN or DMF) and then applying a correction term, derived by Weller [7], to account for the difference in

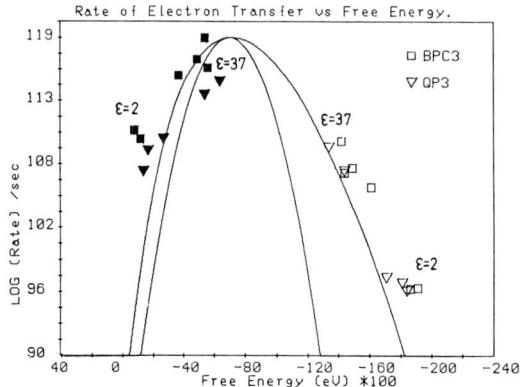

Fig. (2) Filled symbols are for charge separation rates ($K_{sep}$). Dielectric constants increase from left to right for $K_{sep}$ and from right to left for $K_{rec}$. $\epsilon$ values for $BPC_3$ are 2.24, 2.38, 4.8, 9.1, 15.4 and 20.7.

dielectric constant. These corrections seem to over-estimate the polarity effect in low to medium polarity solvents.

The fluorescence quantum yields of $QP_3$ and $BPC_3$ were also found to decrease in solvents of higher dielectric constant, and their small values - ranging from $2.6 \times 10^{-4}$ to $5.0 \times 10^{-4}$ - are indicative of the efficient quenching of $S_1$. At 77 K the fluorescence quantum yield for $BPC_3$ in toluene ($\epsilon = 2.4$) increases by 700 times, whilst in EPA ($\epsilon = 6.6$) it increases by 185 times. The fluorescence lifetime of $BPC_3$ has been measured at room temperature and was found to be bi-exponential. Both short and long components correspond to the charge separation process, but the long component is due to fluorescence from molecules in which the quencher is at unfavourable orientations to and large separations from, the porphyrin. The measured fluorescence lifetimes were 19 ps in toluene, 8 ps in dichloromethane and 10 ps in EPA. The shorter values are slit width limited by the streak camera and appear longer than those measured in absorption. The fluorescence lifetimes as a function of temperature have also been measured in toluene and EPA. All show an increase in lifetime as the temperature falls, the effect being most marked in toluene (the least polar and hence having the highest activation barrier). In each solvent the fall-off in rate begins to decrease at 150 K and the rate becomes effectively constant below 100 K with a lifetime of 2.4 ns in toluene and 1.0 ns in EPA. This deviation from typical Arrhenius behaviour is characteristic of electron tunnelling becoming important at lower temperatures. Activation energies for the charge separation process have been calculated from the higher temperature portions of the data and give values of 0.095 eV in toluene and 0.066 eV in EPA.

References

[1] M.P. Irvine, R.J. Harrison, M.A. Strahand & G.S. Beddard; Ber. Bunsen. Phys. Chem. 89, 226 (1985) and references therein.
[2] M.P. Irvine, R.J. Harrison, G.S. Beddard, P. Leighton & J.K.M. Sanders Chem. Phys. 104, 315, (1986).
[3] J.A. Cowan & J.K.M. Sanders; J. Chem. Soc. Perkin (I), 2435, (1985).
[4] R.A. Marcus & P. Siders; J. Phys. Chem. 86, 622, (1982).
[5] J. Ulstrup & J. Jortner; J. Chem. Phys. 63, 4358, (1975).
[6] M.R. Wasielewski, M. Niemczyk, W.A. Svec & E.B. Pewitt; J. Am. Chem. Soc. 107, 1080, (1985).
[7] A. Weller; Z. Physik. Chem. N.F. 133, 93, (1982).

# Ultrafast Studies Designed to Test the Fundamental Statistical Assumptions Underlying Chemical Reactivity in Liquids

*C.B. Harris, J.K. Brown, M.E. Paige, D.E. Smith, and D.J. Russell*

Department of Chemistry, University of California,
Berkeley, CA 94720, USA

We have been studying the photodissociation and geminate recombination of $I_2$ as a model of a simple chemical reaction in the liquid phase[1,2]. By monitoring the progress of the reaction with picosecond transient absorption/bleach spectroscopic techniques, we have obtained a measure of the ground electronic state vibrational energy distribution as a function of time during recombination in a variety of solvents including alkanes, chlorinated methanes, and most recently liquid xenon. Figure 1 illustrates the results for $I_2$ in $CCl_4$. These vibrational distributions cover a range roughly from v=0 to v=30, which is a range of vibrational energies from 214 $cm^{-1}$ to 173 $cm^{-1}$ assuming a gas phase RKR potential. In this paper we compare these distributions with simple theories for chemical reactions in liquids from two diverse points of view. The first set of theories are based on isolated binary collision ideas, while the second are based on hydrodynamic concepts.

Isolated binary collision (IBC) models assume that "hard" collisions are responsible for vibrational relaxation, and that these hard collisions are both rare and uncorrelated. In this limit the problem may be decomposed into two halves, one is the determination of the collision rate and the second is that of generating a matrix which describes the probability of relaxing from an initial vibrational state i to a final state j. Abul-Hai and Kelley[3] have recently suggested that SSH theory[4] is sufficient to fit the data for this reaction. Although we disagree with this conclusion, we will use this as our starting point for illustrating IBC models. In addition to the assumptions of isolated binary collisions, SSH theory assumes the $I_2$ is a harmonic oscillator, and that the collisions occur through a nearest-neighbor exponentially repulsive potential. A comparison of our data for $I_2$ in "spherical" $CCl_4$, using the parameters reported by Abul-Hai and Kelley is shown in figure 2. As one can see from a comparison to the experimental data, the relaxation rate in this case is much too slow especially in the upper portion of the potential. More realistic IBC calculations based on V-T and V-V have been calculated by Nesbitt and Hynes[5] for spherical $CCl_4$ and an exponential interaction potential. In this case the transition probability matrix was generated from energy loss distributions obtained by integrating the classical equations of motion for "$CCl_4$" colliding with morse oscillator $I_2$. Although the results qualitatively fit in the lower portion of the spectrum, the overall features of the relaxation profile in $CCl_4$ are not in accord with the data. In addition, experimental results in other solvents[2] suggest that V to R may be more important in the lower half of the potential than V to V transfer processes. This idea is currently being investigated by studying the reaction in simple supercritical solvents such as $CO_2$.

Another point of view for condensed phase chemical reaction theories is that based on Langevin dynamics[6] in which the detailed interactions of the solvent are replaced by random and dissipative forces which are connected by the fluctuation-dissipation theorem. Various properties of the solvent are included in this description by appropriate modeling of the correlations in the random force.

The simplest approach is to model the solvent as a viscous medium entirely characterized by its coefficient of friction. In this case the Langevin equation assumes that fluctuations in the force obey Markovian probabilities. A consequence of this type of theory is that excess energy is dissipated to the solvent continuum exponentially for all frequencies of motion. The picosecond transient absorption/bleach measurements which probe vibrational levels as high as half way up

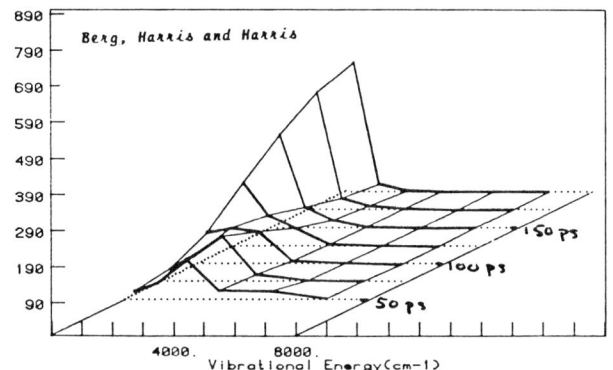

Figure 1: Experimental results in $CCl_4$.

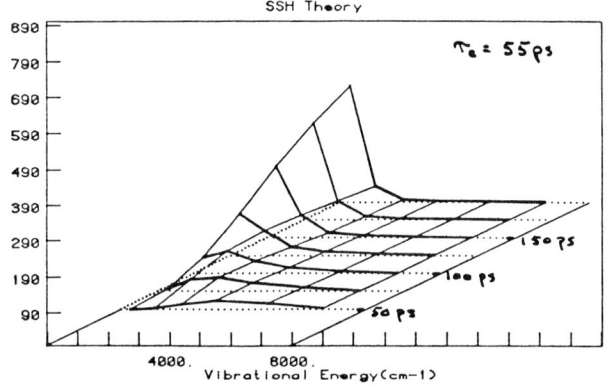

Figure 2: Predictions of SSH theory

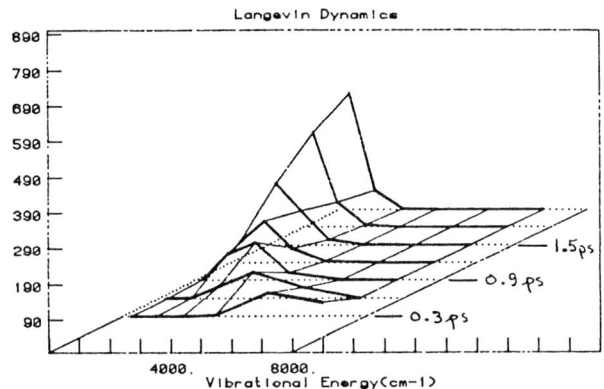

Figure 3: Predictions of Langevin theory

iodine's ground state well, show that the vibrational relaxation of the iodine molecule following geminate recombination does not proceed in an exponential fashion[2]. The Langevin equations for the reaction use a coefficient of viscosity which correlates with the solvent's bulk viscosity. The net result, as shown in figure 3, is that the dissipation of the excess vibration energy is two orders of magnitude faster than is experimentally observed. This discrepancy shows the need to incorporate into the model some information about the availability of the various accepting modes of the solvent for dissipating iodine's vibrational energy.

As an attempt to understand our data in terms of a model explicitly employing a frequency dependent damping we have turned to the generalized Langevin equations[7]. An inherent problem with this method is in determining the appropriate form for the frequency dependent viscosity, $g(t)$. Generally speaking, a rigorous calculation of the memory kernal, $g(t)$, is as intensive as full scale molecular dynamic simulations of the reaction. As an initial attempt to simplify the problem we have chosen $g(t)$ to be of the form $g(t)=a\exp(-bt)$. The constants a and b have been chosen such that the diffusion constant and total force autocorrelation function determined from a Lennard-Jones xenon molecular dynamics simulation is best fit. We have chosen to study the system $I_2$ in liquid xenon since the xenon has no internal vibrational or rotational states to further complicate the friction term and since we have recently experimentally time-resolved the ground state vibrational relaxation process in this system[8]. A comparison of the vibrational energy as a function of time between the Langevin and generalized Langevin models for RKR $I_2$ is shown in figure 4a and 4b respectively. For the simulations it is assumed that $I_2$ is approximately the size as xenon and the $I_2$-Xe potential is similar to the Xe-Xe potential; therefore, iodine would see roughly the same environment as a xenon. Although these assumptions are not strictly valid, they seem reasonable within the crudeness of the model. In figure 4, the exponential nature of the Langevin simulation is immediately noticeable as well as its extremely rapid decay. The rapid decay is attributed to the presence of high frequency "solvent" modes. It should be noted that the $I_2$ vibrational frequencies range from about 100 cm$^{-1}$ near the top of the ground state well to 214 cm$^{-1}$ at the bottom of the well. Putting in the exponential damping term in the generalized Langevin model clearly reduces the intensity of the high frequency modes in the region of the $I_2$ vibrational frequencies, and the relaxation time is seen to drastically lengthen in this case. Also notice the obvious nonexponential character to the decay for this model.

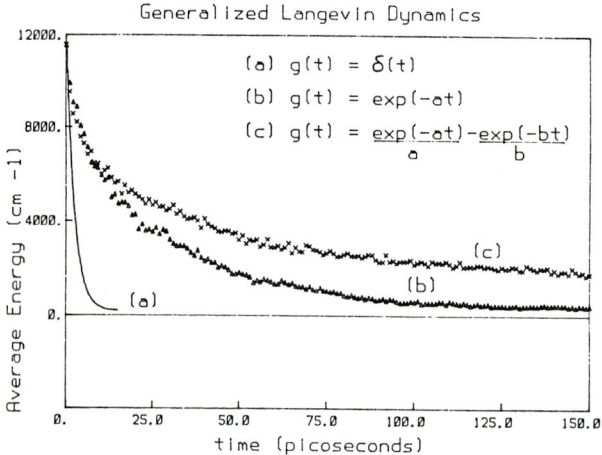

Figure 4: Average $I_2$ vibrational energy vs. time in liquid Xe.
(a) Langevin equations.
(b) Generalized Langevin equations: Single correlation
(c) Generalized Langevin equations: Double correlation

Although the exponential decaying damping case is still far from fitting the experimental results, it represents a significant improvement over the simple Markovian Langevin equations which characterize the solvent with a frequency independent friction. It should be noted however, that the exponential decaying damping function is physically unreasonable due to the discontinuity in the first derivative at zero time. This discontinuity leads to the unphysical presence of large amplitude modes throughout the range of the iodine vibrational frequencies. By using a damping function of the form $(\exp(-at)/a - \exp(-bt)/b)$ the problems of discontinuous derivatives is avoided and consequently the density of high frequency modes is reduced[9]. The results from this function are shown in figure 4c. The parameters are chosen to fit the force autocorrelation function obtained from a full scale molecular dynamics simulation of Xe[10]. From comparison to the experimental data and a full scale molecular dynamics simulation[10] we can conclude that it represents a significant improvement in modeling of the solvent for this reaction. Most importantly it demonstrates that condensed phase vibrational distributions can not be determined from Markovian theories and that at least second order correlations in the memory kernal are necessary to calculate reasonable rates which are in accord with experimental results.

This work was supported by the National Science Foundation.

References:

1.  M. Berg, A. L. Harris and C. B. Harris, Phys. Rev. Lett. 54, 951 (1985).

2.  A. L. Harris, M. Berg and C. B. Harris, J. Chem. Phys. 84, 788 (1986).

3.  N. A. Abul-Haj and D. F. Kelley, J. Chem. phys. 84, 1335 (1986).

4.  R. N. Schwartz, Z. I. Slawsky and K. F. Herzfeld, J. Chem. Phys. 20, 1591 (1952).

5.  D. J. Nesbitt and J. T. Hynes, J. Chem. Phys. 77, 2130 (1982).

6.  S. Chandrasekhar, Rev. of Mod. Phys. 15, 1 (1943).

7.  H. Mori, Progr. Theor. Phys 33, 423 (1965).
    R. Zwanzig, Ann. Rev. Phys. Chem. 16, 67 (1965).
    R. Kubo, Rep. Prog. Theor. Phys. 29, 255 (1966).

8.  M. E. Paige, D. J. Russell and C. B. Harris, J. Chem. Phys. Submitted.

9.  D. E. Smith, J. K. Brown and C. B. Harris, unpublished results.

10. J. K. Brown, C. B. Harris and J. C. Tully, unpublished results.

# Geminate Recombination and Relaxation of Condensed Phase Molecular Halogens

D.F. Kelley and N.A. Abul Haj[+]

Department of Chemistry, Colorado State University,
Fort Collins, CO 80523, USA

## I. Introduction

The photodissociation followed by geminate recombination of condensed phase diatomic molecules is one of the simplest and most extensively studied of all chemical reactions. The low atom reactivities and visible absorption spectrum make the molecular halogens, particularly $I_2$ and $Br_2$, well suited to these studies. Photoexcitation of $I_2$ or $Br_2$ in inert solvents results in rapid (<15 ps) predissociation. Ground vibronic state repopulation occurs following atom recombination, electronic and vibrational relaxation. Atom recombination can occur into the ground as well as several electronically excited states. Previous studies have shown that in both $I_2$ and $Br_2$, recombination results in considerable population in the $^3\Pi_{2u}$ states.[1] Relaxation from this state takes place on a nanosecond timescale, while atom recombination and vibrational relaxation take place much faster. This separation of timescales permits the study of the recombination and vibrational relaxation processes, with a minimum of complications resulting from electronic relaxation. The rates of vibrational relaxation can be determined by time-resolved absorption spectroscopy due to the fact that each vibrational level has its absorption spectrum. Adsorption spectra of each vibrational level are broad and overlapping in the solution phase. However, it is still possible to determine transient vibrational population distributions and hence vibrational relaxation rates, from time-resolved absorption spectra.

In this paper, we present picosecond spectroscopic data and calculational results which indicate the timescale for ground electronic state repopulation and elucidate the rates and mechanisms of vibrational relaxation.

## II. Experimental

The experimental apparatus has been described in detail elsewhere.[2] Briefly, it consists of a streak camera based transient absorption spectrometer, using an active/passive modelocked Nd:YAG laser light source. Samples ($\sim 10^{-3}$M) were excited with 3mJ, 532nm, 25 ps pulses. Sample interrogation was accomplished by passing the pseudo-cw fluorescence of an organic dye through the sample excitation volume. The dye fluorescence was then passed through a 10nm bandpass filter, and imaged onto a Hamumatsu C979 streak camera. The streak camera is interfaced into a PAR 1254E SIT vidicon and a DEC LSI 11/02 minicomputer. Comparison of the apparent dye emission kinetics with and without sample excitation results in very good transient absorption kinetics of the 10nm band.

---

[+]present address: The Aerospace Corporation, Los Angeles, CA 90009.

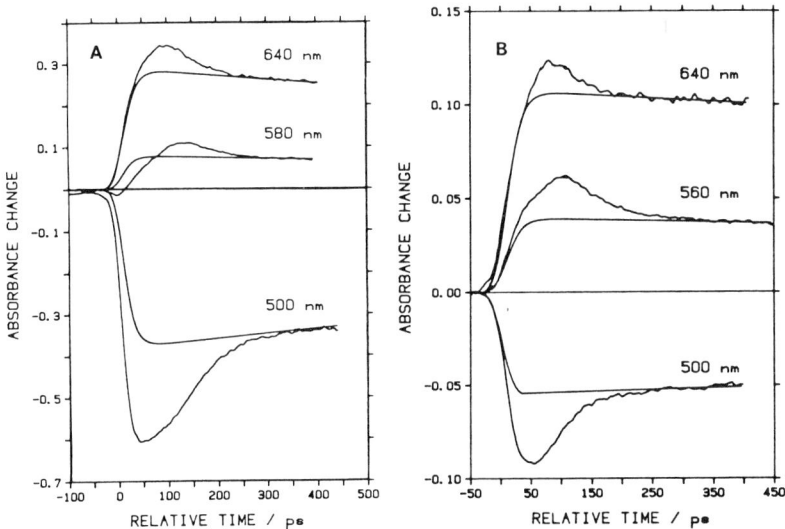

Figure 1. Experimental transient absorption difference (transient minus static absorbance) kinetics of A) $I_2$ in $CCl_4$ and B) $Br_2$ in $CCl_4$ at various wavelengths. Also shown are calculated curves corresponding to the long lived A' state transient.

## III. Results and Discussion

Typical absorption kinetics for $I_2$ in $CCl_4$ and $Br_2$ in $CCl_4$ are shown in figure 1.

In both cases the long lived transients (>250 ps) are due to population in the $A'(^3\Pi_{2u})$ states.[1] Curves fitting these long lived decays were calculated from the convolution of the instrument response function and exponential decays and are also shown in figure 1. There are additional transients at short times (<250ps) which are associated with population in the ground state vibrational manifolds. In the $I_2$ case the 500, 580 and 640 nm absorptions correspond to population in the v=0, v=1-2, and v=4-10 vibrational levels, respectively.

While the data presented in figure 1 reveals the timescale of the recombination/vibrational relaxation process, detailed statements about the relaxation mechanisms require that the data be compared to theoretical results. We have performed dynamic simulations based on an isolated binary collision model for vibration to translation (V-T) energy transfer. These calculations are based Schwartz, Slawsky, Herzfeld (SSH) theory,[3] and ignore vibration to vibration (V-V) energy transfer. The extent to which these assumptions are valid may be assessed by the level of agreement between the calculated and the observed absorption kinetics. The calculational model makes use of two adjustable parameters. One parameter is the time required for repopulation of the X state vibrational manifold, and the other scales the state to state vibrational relaxation rates given by SSH theory. The X state repopulation time, $\tau_x$, may be due to the time required for atom recombination. Alternatively, $\tau_x$ may represent the time required for relaxation from very weakly bound electronics states. The present data does not address this ambiguity. The scaling parameter is related to the average number of collisions required for a v=1 to v=0 transition ($Z_{10}$). The details of this model have been described elsewhere.[4] We note that with $\tau_x$ = 0 and $Z_{10}$ = 550, these calculations accurately reproduce the V-T molecular dynamics calculations by Wilson et al.[5] The simple isolated binary collision model was also found to give quantitative agreement with the data for $I_2$ in weakly interacting solvents such as $CCl_4$ and $CFCl_3$. The model was found to be inadequate for $I_2$ in more strongly interacting solvents such as $CH_2Cl_2$, and for $Br_2$ in $CCl_4$. The kinetic parameters and times at which the maximum absorbances occur for $I_2$ in several solvents and $Br_2$ in $CCl_4$ are presented in Table I.

TABLE I: Kinetic Parameters for $I_2$ and $Br_2$ in Various Solvents

|  | time (ps) to maximum absorption | | $Z_{10}$ | $\tau_x$/ps | static absorption maximum $\lambda$/nm |
|---|---|---|---|---|---|
|  | 580nm | 640nm |  |  |  |
| $I_2/CFCl_3$ | 240 | 160 | 570 | 55 | 519 |
| $I_2/CCl_4$ | 150 | 100 | 180 | 55 | 516 |
| $I_2/CH_2Cl_2$ | 60 | 40 |  |  | 504 |
| $Br_2/CCl_4$ | 100 | 80 |  |  |  |

Several correlations are clear from the data in Table I. The gas phase $I_2$ absorption maximum is at 526 nm. Table I shows that $CFCl_3$ perturbs the gas phase spectrum the least of all solvents studied, and $CH_2Cl_2$ perturbs it the most. Correspondingly, the vibrational relaxation is also the slowest in $CFCl_3$ and the fastest in $CH_2Cl_2$. Furthermore, the isolated binary collision model fails for $CH_2Cl_2$, $CHCl_3$, and other strongly interacting solvents which result in rapid relaxation.

The fact that this simple V-T relaxation model works quite well indicates a minimum of V-V relaxation processes for $I_2/CFCl_3$ and $I_2/CCl_4$, despite a near resonance between the $I_2$ and the lowest $CCl_4$ vibrational frequencies. However, based on a V-T mechanism[3], $Br_2/CCl_4$ would be expected to relax much more slowly than $I_2/CCl_4$. Table I indicates that $Br_2/CCl_4$ relaxes faster than $I_2/CCl_4$. Furthermore, we find that the V-T model cannot adequately explain the temporal evolution or relative intensities of the $Br_2/CCl_4$ spectral transients. Both of these facts strongly suggest that $Br_2/CCl_4$ relaxes by a V-V mechanism. The reason why $Br_2$ but not $I_2$ relaxes by V-V energy transfer can be seen from the following argument: Vibrational energy transfer from an excited diatomic molecule occurs most efficiently in a direct end on collision. The diatomic vibrational motion in such a collision results in a time varying force along the line connecting the diatomic and collider centers of mass. In the case of $CCl_4$ this force results predominately in excitation of C-Cl bond stretch motions. The 214 cm$^{-1}$ $CCl_4$ vibration in near resonance with the $I_2$ stretch, is a pure bending mode, and hence not easily excited in such a collision. However the 314cm$^{-1}$ $CCl_4$ distortion mode does involve C-Cl bond length changes, and can be collisionally excited by the 325 cm$^{-1}$ $Br_2$ vibration. The same arguments also apply to the comparison of $I_2/CFCl_3$ and $Br_2/CFCl_3$ vibrational relaxation rates. As expected on the basis of the above argument, $Br_2/CFCl_3$ relaxation was found to be much faster than was observed for $I_2/CFCl_3$.

Acknowledgement

This work was supported by the National Science Foundation.

References

1. D. F. Kelley, N. A. Abul-Haj and D. J. Jang, J. Chem. Phys. 80, 4105 (1984).

2. D. J. Jang and D. F. Kelley, Rev. Sci. Inst. 56, 2205 (1985).

3. J. D. Lambert, Vibrational and Rotational Relaxation in Gases, (Clarendon, Oxford, 1977).

4. N. A. Abul Haj and D. F. Kelley, J. Chem. Phys. 84, 1335 (1986).

5. P. Bado, P. H. Berens, J. P. Bergsma, S. B. Wilson and K. R. Wilson, in Picosecond Phenomena III., (Springer, Berlin, 1982), p. 260.

# Fast Photochemical Processes of Aromatic Nitro Compounds in Solution

B.B. Craig[1], S.K. Chattopadhyay[2], and J.C. Mialocq[3]

[1] Code 6540, Naval Research Laboratory, Washington, DC 20375, USA
[2] Department of Chemistry, Georgetown University,
  Washington, DC 20057, USA
[3] Département de Physico-Chimie, CEN-Saclay,
  F-91191 Gif-sur-Yvette Cedex, France

## 1. Introduction

Although aromatic nitro compounds exhibit a rich photochemistry, spectroscopic characterization of their reactive excited states has been impeded generally by their low quantum yields of fluorescence and efficient channels of non-radiative deactivation. In this paper, we have employed picosecond absorption spectroscopy to examine processes of photodissociation and photoinduced intramolecular hydrogen abstraction and intramolecular charge transfer for various aromatic nitro compounds in solution. This technique has recently been successfully employed by YIP et al. [1] for identification of the lowest $n\pi^*$ triplet states of nitrobenzene and alkyl nitrobenzenes in solution.

## 2. Experimental

A detailed description of the dual beam picosecond spectrometer is given elsewhere [2]. Briefly, excitation was provided by the fourth harmonic (266 nm, 22 ps, ca. 0.5 mJ) of an active-passive modelocked Nd:YAG laser system with interrogation by a white light continuum pulse (415-900 nm, ca. 60 ps), derived by focusing 30% of the fundamental beam into a cell containing a $D_3PO_4:D_2O$ (1:1 v/v) mixture. The signal and reference beams were focused onto the entrance slit of a spectrometer (Instruments SA, HR 320, 152 gr/mm) and were dispersed onto the face of an ISIT vidicon detection system (PAR 1257, 1216, 1215). Each absorption spectrum was an average of at least 12 laser pulses.

## 3. Results and Discussion

### A. Intramolecular Hydrogen Transfer

The proposed mechanism for intramolecular hydrogen transfer from an ortho substituent to a nitro group is given in (1). The nature of the reactive excited states, the role of a potential biradical intermediate, and the rate of the hydrogen abstraction process are not established. In pursuit of these goals, we have investigated this process for three isomers of dinitrotoluene (DNT) in methanol [2]. The spectra obtained at early times following 266 nm excitation of 2,6-, 2,4-, or 3,4-DNT are very similar (Fig. 1); their risetimes are instrumentally limited. At later times the signals from the 2,6 and 2,4 compounds evolve to yield spectra characteristic of their respective aci-quinoid forms [3]. From the decay of signal in the red, first order rate constants of $2.1 \times 10^9$ s$^{-1}$ and $2.2 \times 10^9$ s$^{-1}$ were determined for removal of the

$$R\underset{}{\overset{R\;H}{\diagup}}NO_2 \xrightarrow{h\nu} R\underset{}{\overset{R\;H}{\diagup}}NO_2^* \longrightarrow R\underset{}{\overset{R}{\diagup}}\overset{OH}{\underset{O^-}{N^+}} \longrightarrow R\underset{}{\overset{R}{\diagup}}\overset{OH}{\underset{O^-}{N^+}} \qquad (1)$$

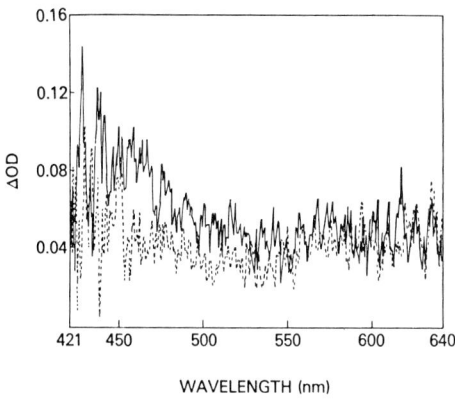
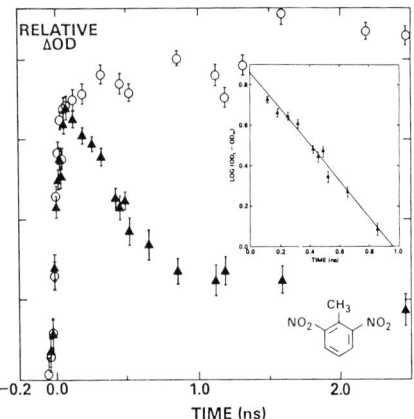

Fig. 1 Transient spectra recorded 33 ps after 266 nm excitation of aerated 5 mM solutions of DNT in methanol. Solid line, 2,6; broken, 3,4

Fig. 2 Time dependence of the signals from 2,6-DNT; (o) 415-460 nm, (▲) 535-625 nm. Inset: semi-log plot of the decay in the red; $k = 2.1 \times 10^9$ s$^{-1}$

initial species for 2,6- and 2,4-DNT, respectively (Fig. 2). In contrast, the 3,4 isomer, which has no nitro group ortho to a methyl hydrogen, exhibits no late term absorption. The initial spectrum simply decays to a zero signal level with a rate constant of $1.7 \times 10^9$ s$^{-1}$. The initial spectra are assigned to the lowest triplet state of each isomer [2]. If we assume that each triplet state has a common radiationless decay rate (the 3,4-DNT value of $1.7 \times 10^9$ s$^{-1}$), then the rate of intramolecular hydrogen abstraction for 2,6- and 2,4-DNT can be estimated to be $0.5 \times 10^9$ s$^{-1}$ and $0.6 \times 10^9$ s$^{-1}$, respectively. A triplet state mechanism as the major pathway (we cannot exclude a minor singlet channel) for production of the aci-species demands the intermediacy of a triplet biradical. From our spectra, we conclude that (i) the biradical collapses to the aci-form very rapidly (< 10 ps) and may not absorb in our probe wavelength region, or (ii) the biradical is spectrally indistinguishable from the aci-form.

B. Photodissociation

Following 266 nm excitation of tetryl (a nitroaromatic nitramine) in methanol, n-hexane, or dichloroethane, we have monitored an intense visible absorption, which is produced within the laser pulse (Fig. 3). The signal shows no decay out to 15 ns. Studies on longer timescales indicate that the decay kinetics are complex and solvent dependent; in n-hexane a second order rate $k/\epsilon = 5 \times 10^5$ cm s$^{-1}$ was determined. Although the spin character of the dissociative excited state is unknown, we tentatively assign the transient absorption to the N-methyltrinitroanilino radical produced following N-NO$_2$ cleavage (2). This would be consistent with the recent observation of rapid NO$_2$ production from 266 nm excitation of gaseous dimethylnitramine [4].

$$RNNO_2 \rightarrow RN\cdot + NO_2 \qquad (2)$$

C. Intramolecular Charge Transfer

In the photodecarboxylation of p-nitrophenylacetate (3), we have identified a long-lived (90 ns) intermediate, which has a $\lambda_{max}$ ca. 290 nm and a weak

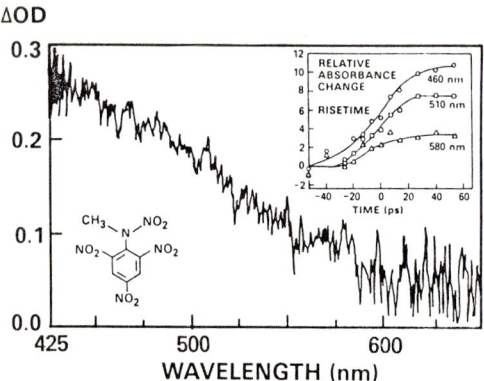

Fig. 3 Transient spectrum recorded 53 ps after 266 nm excitation of an aerated 5 mM solution of tetryl in methanol

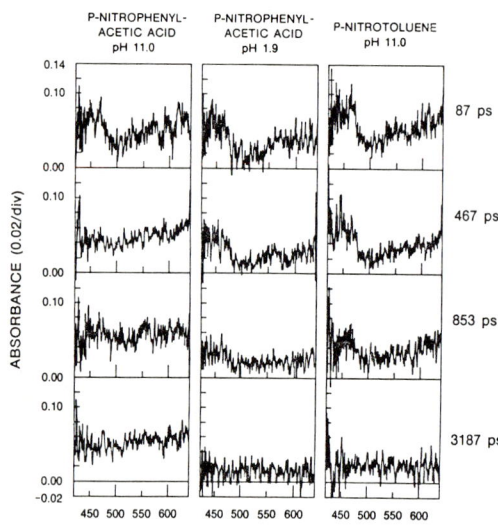

Fig. 4 Transient spectra recorded after 266 nm excitation of aerated 2 mM aqueous solutions of p-nitrophenylacetic acid and p-nitrotoluene

featureless absorption extending throughout the visible [5]. The intermediate appears to be produced through the lowest triplet state of p-nitrophenyl-acetate, which is identified at early times by its characteristic absorption bands at ca. 440 nm and 625-650 nm [1] (Fig. 4). We postulate that the intermediate is formed following intramolecular charge transfer from the carboxylate group to the nitrophenyl ring. It may be viewed as a diradical-like species in which an extra electron is delocalized over the nitrophenyl ring as in the radical anion of p-nitrophenylacetate. The charge transfer step may require a highly specific conformation of the parent molecule; attainment of this conformation could then provide an account of its formation at t > 87 ps. Such a mechanism is consistent with the fact that p-nitrophenylacetic acid, which cannot support the charge transfer step, does not undergo photodecarboxylation. In this case, the triplet state simply decays with a lifetime (ca. 900 ps) characteristic of that for p-nitrotoluene (Fig. 4).

$$p\text{-}NO_2C_6H_4CH_2CO_2^- \rightarrow p\text{-}NO_2C_6H_4CH_2^- + CO_2 \qquad (3)$$

## 4. Acknowledgement

This work was supported by ONR.

## 5. References

1. R.W. Yip, D.K. Sharma, R. Giasson, D. Gravel: J. Phys. Chem. $\underline{88}$, 5770 (1984); $\underline{89}$, 5328 (1985).
2. S.K. Chattopadhyay, B.B. Craig, submitted to J. Phys. Chem.
3. B.B. Craig, S.J. Atherton: In Applications of Laser Chemistry and Diagnostics, ed. A.B. Harvey, Proc. SPIE 482, 1984, p. 96; S.J. Atherton, B.B. Craig, Chem. Phys. Lett. in press.
4. J.C. Mialocq, J.C. Stephenson, Chem. Phys. Lett. $\underline{123}$, 390 (1986).
5. B.B. Craig, R.G. Weiss, S.J. Atherton, submitted to J. Am. Chem. Soc.

# Cage Recombination and Unimolecular β-Scission Reactions of Sulfur Centered Free Radicals

T.W. Scott and S.N. Liu

Exxon Corporate Research Science Laboratories, Route 22 East, Annandale, NJ 08801, USA

It is well known that free radical processes exhibit extremely diverse chemistry through the action of a limited number of elementary steps. Among the most facile elementary reactions of sulfur centered thiyl radicals are cage recombination and reversible addition to carbon-carbon double bonds [1]. Cage recombination is currently an important testing ground for theories of chemical reaction dynamics in liquids. Picosecond studies have focused primarily on the halogen photodissociation reaction [2]. The work described here deals with the recombination reactions of a large organic free radical, the phenyl thiyl radical, in complex hydrocarbon liquids.

$$\text{Ph-S-S-Ph} \underset{\text{Cage Recombination}}{\overset{\text{UV Photolysis}}{\rightleftarrows}} \text{Ph-S}\cdot \| \cdot \text{S-Ph} \quad (1)$$

In a separate series of experiments, the dynamics of unimolecular β-scission reactions are studied through the reversible addition of thiyl radicals to unsaturated hydrocarbons. Previous studies of thiyl radical addition have proven difficult with conventional flash photolysis equipment due to the large rate constant for β-scission [3].

$$\text{Ph-S}\cdot + \text{C=C} \underset{\text{β-Scission}}{\overset{\text{Addition}}{\rightleftarrows}} \text{Ph-S-C-C}\cdot \quad (2)$$

Phenyl thiyl radicals are generated in liquid hydrocarbon solvents by 355 nm irradiation of diphenyl disulfide. Their visible absorption spec-

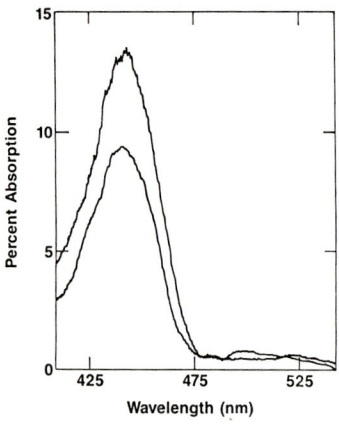

Fig. 1. Transient absorption spectrum of the 355 nm photolysis products of diphenyl disulfide in decalin. The upper curve was recorded during the 35 psec photolysis light pulse. The lower curve was obtained after an 800 psec delay. Both spectra were measured using a white light continuum generated by 1064 nm irradiation of an $H_2/D_2O$ mixture.

trum is characteristic of the lowest electronic transition of benzylic type radicals containing an inductive substituent [4]. Figure 1 shows a series of picosecond absorption spectra following photolysis of diphenyl disulfide in decalin. Spectra recorded during the initial 35 psec photolysis pulse and after an 800 psec delay have similar bandshapes with maximum absorbance at 440 nm. Reaction kinetics are measured at 436 nm by transient optical absorption. Figure 2 (upper) shows the decay of a thiyl radical pair in decalin at two temperatures. High temperature clearly increases the pair survival probability. Within the hydrodynamic model of diffusion controlled recombination, temperature enters principally through the viscosity dependence of the free radical diffusion coefficient. Figure 2 (lower) shows the same recombination reaction when viscosity is altered by changes in solvent composition. The recombination half-life of a thiyl radical pair is typically 125 psec, considerably slower than for the recombination of halogen atoms. Apparently the stringent steric requirements for polyatomic radical recombination and the occurrence of a dipole-dipole reorientation barrier in the case of a polar radical pair have a strong influence on the dynamics of cage recombination.

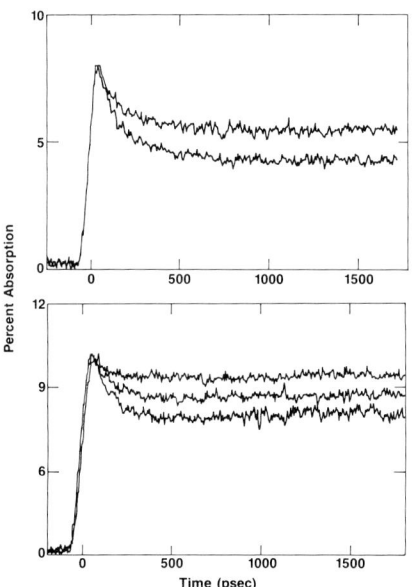

Fig. 2. Time resolved optical absorption decay of phenyl thiyl radical pairs in hydrocarbon liquids. The top figure shows the recombination kinetics in decalin at 22°C (upper) and at -15°C (lower). The bottom figure refers to three different solvents at room temperature: hexane (upper, 0.30 cP), dodecane (middle, 1.38 cP) and decalin (lower, 2.42 cP). All measurements were made at 436 nm by anti-Stokes Raman shifting the Nd:YAG second harmonic in hydrogen gas at 300 psi.

The addition of phenyl thiyl radicals to olefins is a rapid exothermic reaction which is generally reversible. Although the extent of reversibility is thought to depend on the relative stability of the primary sulfur centered radical and the secondary carbon radical, a direct determination of the equilibrium constant for Eqn. 2 has not been reported. Using picosecond transient absorption techniques at high reactant concentrations, both the forward and reverse rate constants can be directly measured. The reaction kinetics of the phenyl thiyl radical are characterized by an initial rapid decay followed by a plateau as the equilibrium is established (see Fig. 3). In liquid hexatriene, we find a 350 psec half-life for addition and a 1.1 nsec half-life for β-scission. A diffusion controlled reaction would occur with a half-life of 100 psec,

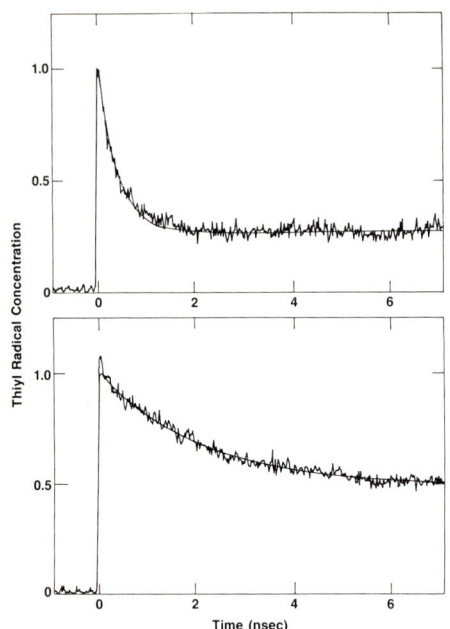

Fig. 3. Decay kinetics of phenyl thiyl radicals in pure liquid 1,3,5-hexatriene (upper) and styrene (lower). The solid lines are two parameter fits to the rate expression derived from Eqn. 2 of the text in a straight forward manner. Transient absorption spectra of the thiyl radical in hexatriene show no change in bandshape between 50 psec and 4 nsec after photolysis. Kinetic measurements were made as described in the caption to Fig. 2.

neglecting steric factors. The experimental addition and β-scission half-lives for reaction with styrene are 2.8 nsec and 3.0 nsec, respectively. The greater addition rate for hexatriene supports the Hammond postulate [5] concerning the relative rates of exothermic reactants. The temperature dependence of thiyl radical chemistry is currently being pursued for more detailed kinetic and thermodynamic information on rapid free radical reactions in solution.

Acknowledgments

The authors are grateful for the advice and encouragement shown by W. N. Olmstead and R. E. Overfield at Exxon Corporate Research.

References

1. P.I. Abell: In Free Radicals, ed. by J.K. Kochi, Vol.II (Wiley-Interscience, 1973) p. 80
2. See for example, T.J. Chuang, G.W. Hoffman and K.B. Eisenthal, Chem. Phys. Lett. 25, 201 (1974); P. Bado and K. Wilson, J. Phys. Chem. 88, 655 (1984); A.L. Harris, M. Berg and C.B. Harris, J. Chem. Phys. 84, 788 (1986)
3. O. Ito and M. Matsuda, J. Phys. Chem. 86. 2076 (1982)
4. E.J. Land: In Progress in Reaction Kinetics, ed. by G. Porter (Pergamon Press, 1965), p.369-402
5. G.S. Hammond, J. Am. Chem. Soc. 77, 334 (1955)

# The Influence of Friction and Deuteration on Stilbene Isomerization

*S.H. Courtney, M.W. Balk, S. Canonica, S.K. Kim, and G.R. Fleming*

Department of Chemistry and James Franck Institute,
The University of Chicago, Chicago, IL 60637, USA

Introduction

   The dependence of chemical reaction dynamics on the surrounding medium is a topic of much current interest. The goal is to understand the effects of the medium upon the reaction kinetics proceeding from the isolated molecule [1] to the low and high pressure gas, and liquid solution [2,3]. We present excited state absorption anisotropy measurements of trans-stilbene-h12 and discuss gas phase and solution results for trans-stilbene-d12.

   Emission lifetimes in both the gas phase and solution were measured by the time correlated single photon counting technique. The absorption and anisotropy studies were performed with a colliding pulse femtosecond ring laser oscillator amplified at 10 Hz using four longitudinally pumped cells. The second harmonic was utilized as the pump pulse and the probe pulse was either the fundamental or was obtained by continuum generation.

Results and Discussion

   The vapor pressure of stilbene at room temperature is less than 1 mTorr. Thus, no collisions occur during the fluorescence lifetime. However, in contrast to the single exponential decays observed in jet cooled stilbene, nonexponential decay is observed in the thermal isolated molecule experiments. This may be attributed to the Boltzmann distribution over rovibrational levels in the ground electronic state. We find that in the collisionless gas phase, the trans-stilbene-d12 also shows nonexponential decay with a time scale similar to the trans-stilbene-h12 while the -d12 has a longer lifetime in solution. The ratio $\tau(d12)/\tau(h12)$ in solution is approximately 1.5 and is insensitive to solvent and temperature (alcohols and alkanes). Figure 1 shows that in the absence of collisions the stilbene-h12 fluorescence decays slightly slower than stilbene-d12, whereas recent jet experiments [1] show the opposite behavior. The jet and thermal isolated molecule rates can be reconciled by considering the difference between the vibrational density of states of the perproto and perdeutero species. The densities of states were calculated for both species and from them the corresponding room temperature Boltzmann distributions. The Boltzmann distribution for d12 is broader and peaks at a higher vibrational energy. Hence, even though the jet experiments show that for h12 and d12 molecules with the same vibrational energy the reaction probability is smaller for d12, the d12 hot levels have a greater population so that the overall reaction rate is faster for d12. However, as methane buffer gas pressure is increased the h12 decay initially increases more rapidly than the d12 decay so that by 1 atmosphere the h12 decay is more rapid than the d12 decay. By 5 atmospheres a ratio of $\tau(d12)/\tau(h12)$ very similar to the solution value ($\simeq 1.5$) is reached. Note, however, that the maximum rate in h12 is not achieved until about 100 atm of methane. The implication is that there are two processes with very different time scales involved in the increase of rate with pressure.

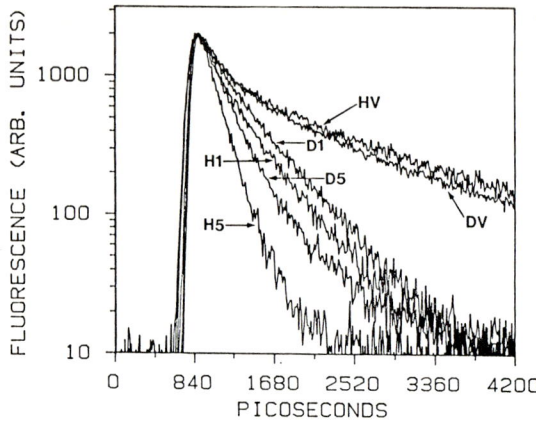

Fig. 1. Comparison of fluorescence decay curves for stilbene h12 and d 12 at 296°K.

HV = stilbene -h12 vapor
DV = stilbene -d12 vapor
H1, H5, D1, D5 at 1 atm and 5 atm pressure of methane.

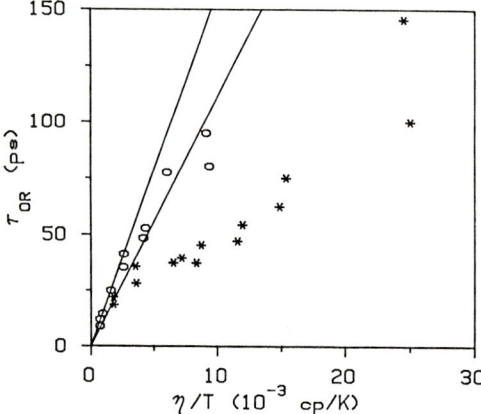

Fig. 2. The viscosity dependence of the trans-stilbene reorientation time in the series of normal alkanes (O) and the series of normal alcohols (*) at approximately 30°C. The solid lines border predicted values using slip boundary conditions.

Figure 2 shows rotational reorientation times for trans-stilbene-h12 obtained from anisotropic absorption measurements. The solid lines give upper and lower limits for the reorientation time calculated from slip boundary conditions assuming trans-stilbene is a prolate rotor with the transition moment along the symmetry axis. It is evident that the reorientation times predicted using slip boundary conditions are in reasonable accord with measurements in alkane solvents. However, even in these solvents the reorientation time exhibits a non-linear increase with viscosity. More striking is the behavior in the series of normal alcohols where the reorientation times (for a given n/T) are shorter than in the corresponding alkane solution. In the alcohols, "isoviscosity plots" imply a continuously varying barrier height. The non-linear viscosity dependence has an implication when attempting to compare isomerization rates to barrier crossing theories as in Figure 3.

Apparently the deviation from linear behavior is sufficient to account for previous qualitative failures of Kramers equations [2]. This method was previously applied in the case of DODCI and it did not improve the quality of the fit [4]. Other mechanisms could also produce the observed deviations, such as a slight decrease in the barrier height as the solvent viscosity increases or the multidimensional character of the isomerization. Work is in progress to further clarify these possibilities.

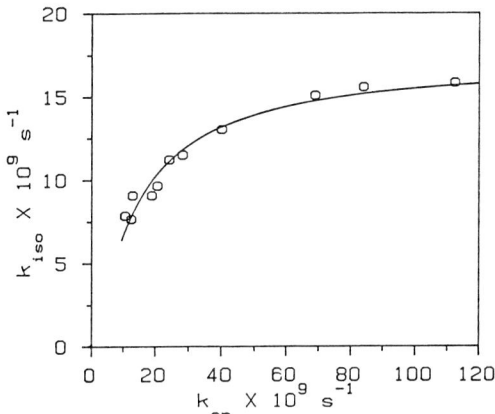

Fig. 3. Plot of the isomerization rates in normal alkanes vs. reorientation rates measured under the same conditions. The solid line represents a fit to Kramers equation assuming the friction felt by the isomerization is proportional to that for the rotation [4].

Acknowledgement

This work was supported by a grant from the National Science Foundation. S. Canonica was supported by the "Stefano Franscini Fonds" of ETH Zurich. We thank A.H. Zewail for supplying the perdeutero-stilbene.

References

1. P.M. Felker and A.H. Zewail, J. Phys. Chem. 89, 5402 (1985).
2. G.R. Fleming, S.H. Courtney and M.W. Balk, J. Stat. Phys. 42, 83 (1986).
3. M. Lee, O.R. Holtom and R.M. Hochstrasser, Chem. Phys. Lett. 118, 359 (1985).
4. S.P. Velsko, D.H. Waldeck and G.R. Fleming, J. Chem. Phys. 78, 249 (1983).

# Kramers-Hubbard Approach to the Solvent Dependence of Isomerization

## M. Lee and R.M. Hochstrasser

Department of Chemistry, University of Pennsylvania,
Philadelphia, PA 19104, USA

The excited states of trans-stilbene and diphenylbutadiene undergo rapid isomerizations (1,2) having rates for different solvents that did not prove to be predictable from a Kramers model (3). The one dimensional Kramers equation in the spatial diffusion regime has the form:

$$k_{iso} = \frac{\omega_a}{2\pi\omega_b} \cdot \frac{\beta}{2} \cdot \left\{ \left[ 1 + \left(\frac{2\omega_b}{\beta}\right)^2 \right] - 1 \right\} \exp(-E_b/kT) \qquad (1)$$

in which the rate of isomerism $k_{iso}$ is given in terms of the angular velocity correlation frequency $\beta$, the barrier height $E_b$ and the potential function along the isomerism coordinate characterized by the reactant and barrier frequencies $\omega_a$ and $\omega_b$. A common procedure is to express $\beta$, via a Stokes relation, in terms of the solvent viscosity. For these two molecules such an approximation does not yield satisfactory agreement between theory and experiment. The agreement was improved by incorporating a frequency dependence (4) into the friction, but unusually shaped potential surfaces seemed to be required (1) even though the interpretation of $\omega_a$ is made complicated by the dimensionality (5).

We have measured the isomerization rates of stiff-diphenylbutadiene (S-DPB) in n-alkane solvents at 19°C by the time correlated single photon counting method. In Figure 1 the rates are plotted as a function of viscosity and compared with the Kramers model. The energy barrier, $E_b$ was measured by an isoviscosity plot, giving 3.3 kcalmol$^{-1}$. Figure 1 shows the high viscosity region where significant deviations are seen to occur in best fits of the DPB and the earlier (1) stilbene data to equation (1).

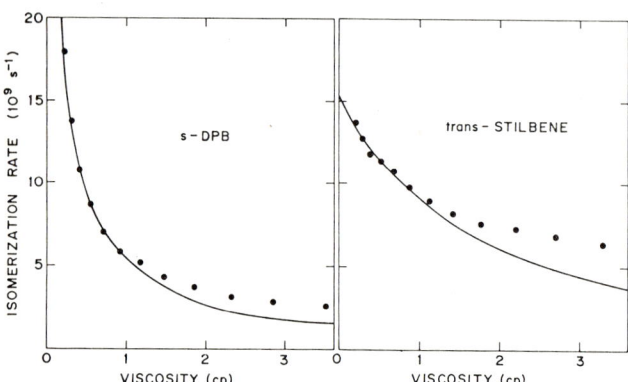

Fig. 1 The Kramers fit of the isomerization rates of diphenylpolyenes vs. viscosity, optimized to the low viscosity region

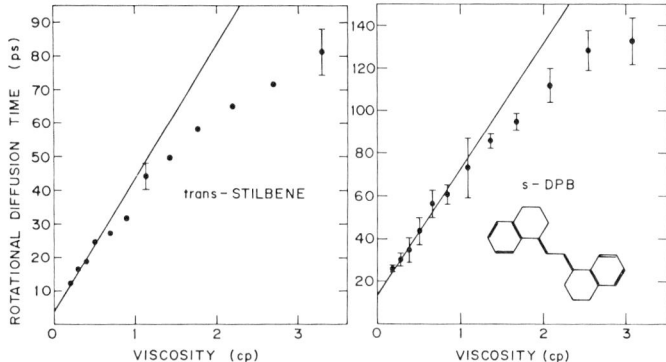

Fig. 2 The rotational diffusion times of stilbene at 22.5°C and S-DPB at 24°C vs. viscosity

In this report we address experimentally the validity of using hydrodynamic variables in the Kramers theory. Such an assumption was not involved in the original development by Kramers. Our analysis was begun by measuring the rotational reorientation times, $\tau_R$, of trans-stilbene and stiff-diphenylbutadiene (S-DPB) in n-alkane solvents, as shown in Figure 2[6]. The solid lines are obtained from the Stokes-Einstein-Debye (SED) equation.

$$\tau_R = C \cdot \eta + \tau_R^0 \tag{2}$$

where C is a constant and $\tau_R^0$ is an empirical zero viscosity intercept [7]. In neither of our examples does $\tau_R$ versus viscosity show SED behavior, demonstrating the breakdown of the hydrodynamics. The deviation from the SED theory was previously explained by the free volume effect [8]. The deviations become most marked when the size of the solvent is comparable with or larger than that of the solute. So it is evident that the friction coefficient in Kramers equation cannot be obtained from the solvent viscosity. As an alternative approach we obtain the (reduced) friction constant or angular velocity correlation frequency directly from the rotational reorientation time by the Hubbard relation [9].

$$\beta = 6pkT \; \tau_R/I_R \tag{3}$$

where we have introduced a constant of proportionality (p) to accommodate the possible differences in the effectiveness of collisions in the isomerization and overall rotation process. This is expected to be an exact proportionality in the regime where the angular momentum correlations decay very rapidly compared with orientational correlations. Fig. 3 shows the variation of the isomerization rates of S-DPB and stilbene with angular velocity correlation frequency obtained from equation (3). By setting $\beta \propto \tau_R$ we are assuming that isomerism and overall rotation are proportional to the collisional frequency. This will be the case when the molecule responds linearly to collisions tending to cause motion along the isomerism coordinate. Values of $I_R = 6.67 \times 10^{-44}$ m² Kg for S-DPB and $3.4 \times 10^{-44}$ m² Kg for stilbene were used.

The optimum fit for stilbene yields $\omega_a/2\pi C_0 = 196$ cm$^{-1}$, $\omega_b/2\pi C_0 = 176$ p cm$^{-1}$ and for S-DPB $\omega_a/2\pi C_0 = 424$ cm$^{-1}$ and $\omega_b/2\pi C_0 = 16$ p cm$^{-1}$. In both cases the Kramers-Hubbard fit to the data is excellent and the $\omega_b$ for stilbene is physically reasonable.

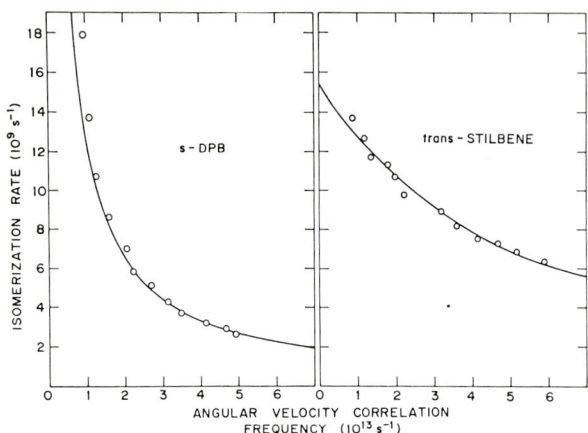

Fig. 3 The Kramers-Hubbard fit of the isomerization rates vs. angular velocity correlation frequency

In conclusion we have avoided using solvent viscosities as approximations for the angular velocity correlation frequency by measuring the overall rotational relaxation times and using them in the Hubbard relation. The assumption that overall motion and isomerization motion have proportional angular velocity correlation times, which is apparently not valid in all cases (10), then allows us to fit the isomerzation data to Kramers equation without using the viscosities. This Kramers-Hubbard fit finesses questions regarding the validity of the one-dimensional Kramers equation and focuses attention on the SED equation.

Acknowledgement: The research was supported by NIH and NSF. We thank A. J. Bain, P. J. McCarthy and C. H. Han for their contribution to the measurements of $\tau_R$ using polarization spectroscopy; and A. B. Smith III and J. N. Haseltine for the synthesis of s-DPB.

References

1. G. Rothenberger, D. K. Negus and R. M. Hochstrasser, J. Chem. Phys. 79, 5360 (1983).
2. S. P. Velsko and G. R. Fleming, J. Chem. Phys. 76, 3553 (1982)
3. H. A. Kramers, Physica 7, 284 (1940)
4. R. F. Grote and J. T. Hynes, J. Chem. Phys. 73, 2715 (1980)
5. L. R. Khundka, R. A. Marcus and A. H. Zewail, J. Phys. Chem. 87, 2473 (1983)
6. M. Lee, A. J. Bain, P. J. McCarthy, C. H. Han, J. N. Haseltine, A. B. Smith III and R. M. Hochstrasser, to be published in J. Chem. Phys.
7. B. J. Berne and R. Pecora, Dynamic Light Scattering (Wiley, New York, 1976)
8. J. L. Dote, D. Kivelson and R. N. Schwartz, J. Phys. Chem. 85, 2169 (1981)
9. P. S. Hubbard, Phys. Rev. 131, 1155 (1963)
10. S. Velsko, D. H. Waldeck and G. R. Fleming, J. Chem. Phys. 78, 249 (1983)

# Photoisomerization Studies of Substituted Stilbenes: 4,4′-Dihydroxystilbene and 4,4′-Dimethoxystilbene

## D.M. Zeglinski and D.H. Waldeck

Department of Chemistry, University of Pittsburgh,
Pittsburgh, PA 15260, USA

Recent studies of photoisomerization in liquid solution have revealed significant modification of isomerization rates by frictional and dielectric solute-solvent interactions. Reported here are initial studies of the isomerization dynamics of 4,4'-dihydroxystilbene and 4,4'-dimethoxystilbene in the normal alcohols. Comparisons are drawn between this work and previous studies of the unsubstituted stilbene [1].

The fluorescence lifetimes of the stilbenes were measured via the time correlated single photon counting method. The excited state decay could be fit to a single exponential over at least five lifetimes (see Fig. 1). Transients were measured with both 313 nm and 325 nm laser excitation and fluorescence decays were monitored across the emission band with no significant variation in the lifetime. Relative quantum yields of fluorescence were measured in the photon counting fluorimeter ($\Phi_F$ = 0.05 for stilbene in methylcyclohexane/isohexane was used as a standard). Isoviscosity studies ($\eta$ = 1.2 cp) of the isomerization rate were performed by variation of the temperature and the solvent at atmospheric pressure.

Arrhenius type plots of the measured rate at constant viscosity were used to obtain a value for the intramolecular barrier to isomerization, approximately 4.9 kcal/mole for 4,4'-dimethoxystilbene and 4.6 kcal/mole for 4,4'-dihydroxystilbene. These barriers are quite high compared to the

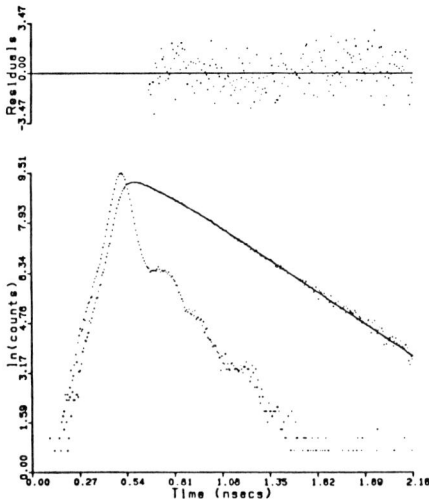

Figure 1. Fluorescence decay of 4,4-'dihydroxystilbene in methanol at 297 K. The single exponential fit gives a lifetime of 264 psec and a chi-squared of 1.33.

internal barrier of the unsubstituted stilbene of 3.5 kcal/mole in alkane solvents and <1 kcal/mole in alcohol solvents [1]. Recent studies [2] have shown that isoviscosity plots for t-stilbene in the normal alcohols change slope at different viscosities. This effect is being examined for the substituted stilbenes as well. These activation energies were used to extract the "reduced" isomerization rates [3] from the measured rates. These reduced rates are plotted versus viscosity in Fig. 2. Each set of data was fit to the functional form

$$k_{iso} = B/\eta^\alpha \qquad (1)$$

with $\alpha$ approximately 0.36 for dihydroxystilbene and 0.27 for dimethoxystilbene. Also shown in Fig. 2 are best fits to the full hydrodynamic Kramers expression,

$$k_{iso} = (A\eta/B)[(1 + (B/\eta)^2)^{1/2} - 1]. \qquad (2)$$

Equation (1) gives a slightly better fit to the data than (2).

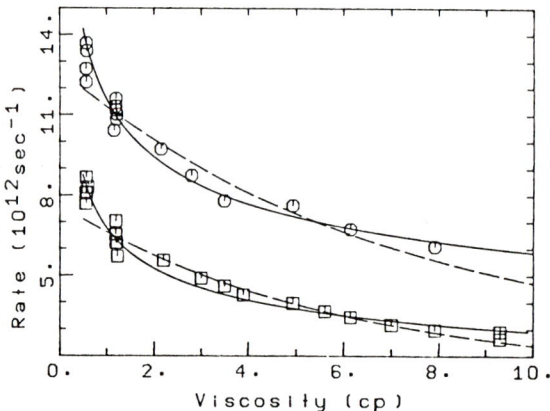

Figure 2. A ○ represents 4,4'-dihydroxystilbene data and a □ represents 4,4'-dimethoxystilbene data. The solid lines are best fits to equation 1 and the dashed lines are best fits to equation 2.

Comparison of our results with the unsubstituted stilbene allows the effect of barrier height on the extent of solute-solvent frictional coupling to be evaluated. These data agree with previous studies of photoisomerization in that a significant internal barrier leads to a weak viscosity dependence of the rate. Previous studies of the unsubstituted stilbene isomerization gave an $\alpha$ of 0.32 in the alkanes and an $\alpha$ of 0.6 in the alcohols [1]. The studies of dimethoxystilbene show a weaker dependence on the solvent viscosity than the unsubstituted stilbene case. This observation is consistent with previous observations because of its high internal barrier and occurs even though twice as much volume is displaced upon isomerization than in the unsubstituted case. However, the studies of 4,4'-dihydroxystilbene give a stronger viscosity dependence than the unsubstituted stilbene in alkane solvents despite the fact that it possesses a larger internal barrier. Recent modeling has assumed that the frictional boundary condition changes only the magnitude of the friction. These initial results suggest that the strength and character of the intermolecular frictional coupling may qualitatively modify the solvent dependence.

Acknowledgements

We thank G. R. Fleming and A. J. Cross for access to their iterative convolution fitting routines. Also, we are grateful to R. L. Garrell for assistance in purifying the substituted stilbenes.

References

1. a) G. Rothenberger, D. K. Negus and R. M. Hochstrasser, J. Chem. Phys., 79, 5360 (1983); b) V. Sundstrom and T. Gillbro, Chem. Phys. Lett., 109, 538 (1984); c) G. R. Gleming, S. H. Courtney and M. W. Balk, J. of Stat. Phys., 42, 83 (1986).

2. a) K. B. Eisenthal, private communications; b) S. H. Courtney, M. W. Balk, S. Canonica, S. K. Kim and G. R. Fleming; Ultrafast Phenomena Conference V, (1986).

3. S. P. Velsko, D. H. Waldeck and G. R. Fleming, J. Chem. Phys., 78, 249 (1983).

# Picosecond Studies of Barrierless Torsional Diffusion

## D. Ben-Amotz and C.B. Harris

Department of Chemistry, University of California at Berkeley, and Materials and Molecular Research Division of Lawrence Berkeley Laboratory, Berkeley, CA 94720, USA

The large amplitude motion of a solute molecule can be viewed as one of the simplest liquid phase chemical reactions. The dynamics of such a process are dictated by the shape of the solute's potential energy surface and its coupling to the stochastic thermal forces of the solvent bath. Numerous systems which undergo large amplitude motion involving barrier crossing have been studied over the past several years. Few systems whose reaction coordinate is barrierless have so far been investigated. The goal of the present experiments is to test the validity of current theories for barrierless liquid phase chemical processes.

The strong dependence of the internal conversion rate of triphenylmethane (TPM) dyes on solvent viscosity, temperature and pressure suggests that large amplitude motion, perhaps torsional twisting of the phenyl rings about the central carbon bond, dominates the internal conversion dynamics. Our observation of a linear viscosity dependence of the excited state lifetime of crystal violet at various fixed temperatures (see Fig. 1) suggests that simple barrierless torsional diffusion is the rate limiting step in the fast internal conversion from the first excited electronic state down to the ground state[1]. The fact that the decay time increases with temperature at constant viscosity is surprising in view of the usual role of temperature in activating chemical reaction rates. In this work we report various consequences of this unusual temperature dependent behavior of TPM dyes and attempt to explain these effects using a simple model.

Experiments were performed using amplified sync-pump-dye-laser pulses (1ps,1mJ,10Hz). Crystal violet, a TPM dye molecule, was excited at 589nm and probed at 460nm in order to measure the transient absorption arising from the electronically excited state. The decay of this excited state absorption signal is presumed to represent the time it takes the phenyl

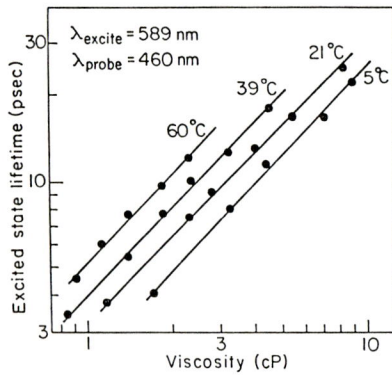

Fig. 1 Viscosity dependence of the excited state decay time of crystal violet in a series of normal alcohol solvents between methanol and octanol.

rings to torsionally diffuse before curve crossing back down to the ground state. Studies were carried out at various temperatures in a series of normal alcohol solvents.

The linear viscosity dependence of the excited state decay time at various fixed temperatures seen in Fig. 1 is qualitatively consistent with the prediction of both a simple Stokes-Einstein-Debye model and the more recent Bagchi-Fleming-Oxtoby[2] model for barrierless diffusion in the presence of stochastic thermal forces. Using a reasonable volume for the diffusing phenyl rings, V=100cc/mole, and using the Stokes-Einstein-Debye equation,

$$\Theta = \sqrt{(RT\tau/3V\eta)}$$

where $\tau$ is the measured relaxation time and $\eta$ is the solvent viscosity, one can calculate the root mean square diffusion angle, $\Theta$, of the phenyl rings at a given temperature. This calculation, when applied to the data in Fig. 1, indicates that the diffusion angle increases from about 7 degrees at 5°C to about 12 degrees at 60°C. This behavior suggests that the increased excited state decay time with temperature may be due to a modification of the equilibrium torsional distribution of the phenyl rings with temperature. If increasing the temperature has the effect of shifting the equilibrium torsional distribution away from the conformation corresponding to a near crossing of the ground and excited state potential surfaces, then one would expect to observe an increase in the internal conversion rate with temperature. This simple suggestion can be used to understand various previously unexplained temperature dependent properties of the internal conversion dynamics of TPM dyes.

Figure 2 displays the temperature dependence of crystal violet's excited state decay time in various normal alcohol solutions (dashed lines)

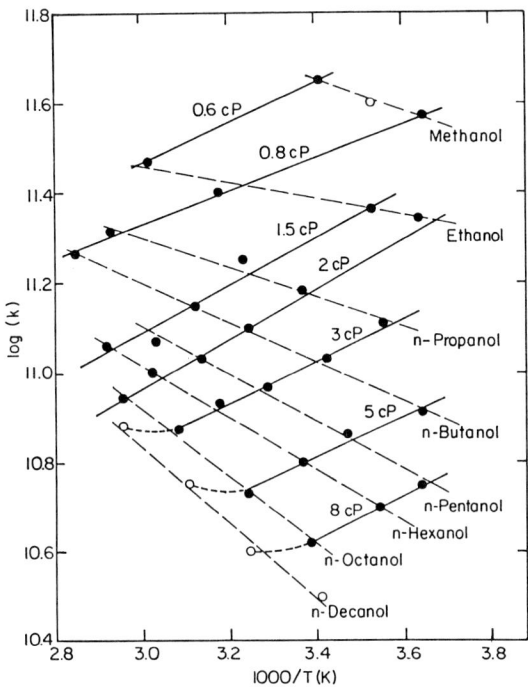

Fig. 2 Arrhenius plots of the excited state decay rate of crystal violet as a function of temperature, in several normal alcohol solvents, and at several constant viscosities. The open circle points were not used in calculating the constant viscosity slopes.

and at various constant viscosities (solid lines). Two features of the observed behavior are unusual: (i) the constant viscosity lines have slopes indicating a negative isoviscosity activation energy in the short chain alcohols, and (ii) the lines for each solvent have slopes which are about half the solvent's viscosity activation energy. For a barrierless process one would expect a nearly zero isoviscosity slope[2] and a slope for each solvent which is equal to its viscosity activation energy. Both of these unusual features can be explained by allowing for a temperature dependent equilibrium torsional distribution as indicated by the Stokes-Einstein-Debye calculation. When the data in Fig. 2 is corrected for the change in diffusion angle with temperature, one obtains isoviscosity slopes which are nearly zero and activation energies for each solvent which are within about 10% of the solvent's viscosity activation energy.

In longer chain alcohol solutions, the viscosity dependence of the excited state decay is found to deviate from the linear dependence seen in Fig. 1. One consequence of this behavior is evident in the isoviscosity plots in Fig. 2, where the decay rates in n-decanol are seen to deviate from the linear trends found in the shorter chain solvents. This nonlinearity has been taken to indicate the formation of a barrier to tortional twisting in the longer chain alcohol sovents[3]. An alternative suggestion is that this deviation from linearity results from a nonlinear relationship between the microscopic friction experienced by the twisting molecule and the macroscopic viscosity of the solvent. Such a saturation effect has previously been proposed to account for similar deviations from linearity of rotational diffusion times measured in long chain linear alkane[4] and alcohol solvents[5].

In conclusion, various apparent anomalies in the internal conversion dynamics of TPM dyes can be explained by allowing for a temperature dependent equilibrium torsional distribution of the phenyl rings. Once these apparent anomalies are understood, TPM dyes emerge as good model systems for testing the detailed predictions of theories for barrierless processes in liquids. Several unique consequences of the barrierless form of the reaction potential, such as a non-exponential excited state decay function and small deviations of the isoviscosity slope from zero in certain limits[2], can now be investigated more quantitatively than has previously been possible. Such studies are currently in progress.

This work was supported by the National Science Foundation.

References
1. D. Ben-Amotz and C. B. Harris: Chem. Phys. Lett. 119, 305 (1985), and references therein.
2. B. Bagchi, G. R. Fleming and D. W. Oxtoby: J. Chem. Phys., 78, 7375 (1983).
3. V. Sundstom and T. Gillbro: J. Chem. Phys., 81, 3463 (1984).
4. W. F. Paynko, J. Yarwood and D. J. Gardiner: Chem. Phys., 78, 319 (1983).
5. T. J. Chuang and K. B. Eisenthal: Chem. Phys. Lett., 11, 368, (1971).

# Time-Resolved Fluorescence Spectra of Ethidium Bromide

*J.H. Sommer*[1,4], *T.M. Nordlund*[1,2], *M. McGuire*[3], *and G. McLendon*[3]

[1] Department of Biophysics and Department of Physics and Astronomy, University of Rochester, NY 14627, USA
[2] Laboratory for Laser Energetics, University of Rochester
[3] Department of Chemistry, University of Rochester
[4] Present address: P.O. Box 6123, Yale Station, New Haven, CT 06540, USA

Isomerizations are thought to play an important role in a variety of biological and chemical systems [1]. The reactions are viewed as activated processes governed by a barrier. The rate differs from that of classical transition-state theory in that it depends upon both the activation energy and friction. The viscosity determines the friction at the barrier. Both very high and very low viscosities hinder isomerizations: high viscosities slow the rate of crossing at the top of the barrier and low viscosities provide insufficient thermal coupling with the solvent for activation to proceed efficiently. The frictional force is not necessarily determined by the static viscosity. Much work has gone into studying dynamic frictional forces [2].

We have used a streak camera based fluorometer [3,4] in order to study relaxation processes in systems ranging from tyrosine in protein molecules to dye molecules in solution. This instrument records approximately 120 nm emission spectra with under 20 ps time resolution. We report here on a study of ethidium bromide dissolved in glycerol, in water and intercalated in DNA. We find a fluoresence red shift with time for ethidium in glycerol which we quantitatively describe using the method of Singular Value Decomposition (SVD) [5]. Time-dependent spectral changes of ethidium in water or intercalated in DNA, on the other hand, are very small.

1. Methods

Fluorescence spectra of ethidium were recorded in 20 ps windows at various times ranging from 0 to 3 ns after excitation by a 20 microjoule, 530 nm pulse from a Q-switched, mode-locked, single pulse selected Nd:YAG laser. The entire spectrum was recorded simultaneously for each shot with a jitter-free streak camera read by a two-dimensional optical multichannel analyzer. The signals from 50 shots were summed at each time delay to increase the signal-to-noise ratio. This is described in detail in Ref. 4. The sample temperature was regulated from -10° to 50° C by a thermostatted brass block surrounding the cuvet. Quantitative analysis using the SVD method determines the minimal, orthonormal basis set which describes all of the spectra collected. It thus determines the minimum number of independent spectral species and the time dependence of their amplitudes. The first basis spectrum is the best least squares fit to the observed set of spectra. The second basis spectrum is orthogonal to the first and, together with the first, gives the best two-component fit to the data. Each successive basis spectrum is orthogonal to the others and provides the best n-component fit. The procedure generates as many basis spectra as input spectra, but the higher components are progressively less necessary for description of the data. Usually the first two or three basis spectra describe the data adequately and the higher components describe noise. Once obtained, the significant basis spectra can be used to generate the spectra as a function of time, as in Fig. 1. The generated spectra are equivalent to the original spectra with components corresponding to random noise removed.

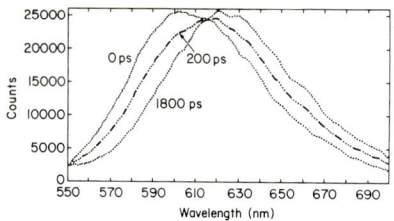

Fig. 1. Reconstructed spectra from the EB/glycerol dataset at 25°C, derived from the SVD analysis of Fig. 2. These spectra were constructed by multiplying the basis spectra by the SVD coefficients corresponding to each time, and then normalizing the resulting specta to each other. The time delays are 0, 200 and 1800 ps.

## 2. Results

The time dependence of the EB emission spectrum in glycerol at 25° C (Fig. 1) shows a red shift with time. The shift is about 20 nm and the 1/e time for this shift is about 250 ps. The SVD analysis of this data set in Fig. 2 shows two significant spectral species. The first basis spectrum corresponds to the average of all the input spectra, less noise. The amplitude is proportional to the total number of excited molecules in solution. Its decay describes the fluorescence lifetime of the dye, which is about 5.9 ns. The red shift of the spectrum is described by the second SVD component. This shift is nonexponential in time, as shown by the superior two-exponential fit in Fig. 2. The temperature dependence of the shift rate yields an apparent activation energy of about 4 kcal/mol. Time dependent spectral changes of EB dissolved in water and intercalated in calf thymus DNA are, in contrast, very small. Replacement of the phenyl group of EB with methyl reduces the spectral shift in glycerol to about 10 nm.

## 3. Discussion

We believe the spectral relaxation of EB in glycerol is in large part due to a forced reorientation of the phenyl ring of EB in the excited state [4], (Fig. 3). Theoretical work indicates that the ground state energy of EB depends upon the phenyl dihedral angle [6]. See Fig. 3. If the equilibrium dihedral angle in the excited state differs from that in the ground state, then the phenyl will be forced to rotate upon excitation. Phenyl rotation also has experimental support [7, and BEN-AMOTZ, HARRIS, this volume]. Since water is a low-viscosity solvent compared to

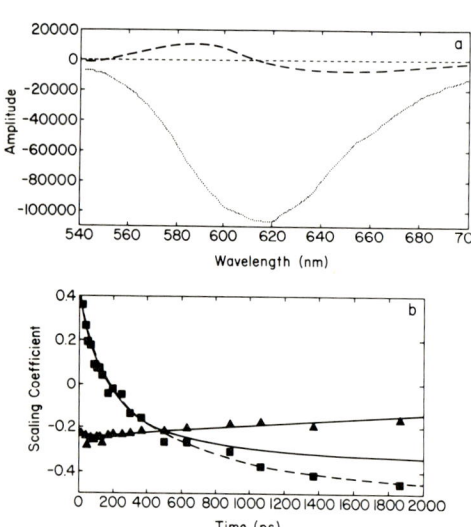

Fig. 2. The results of an SVD analysis of spectra of EB in glycerol. Fig. 2a depicts the two significant SVD-derived basis spectra. The first basis spectrum is drawn with a solid line and is inverted-an artefact of the SVD process. The second basis spectrum is drawn with a dashed line. The coefficients scaling these basis spectra to the observed spectra are drawn in Fig. 2b: ▲, first basis spectrum: ■ , second basis spectrum. The solid lines are single-exponential fits, the dashed are double-exponentials. Note the negative scaling coefficients of the first basis spectrum invert its sense, restoring a positive fluorescence intensity which is slowly decaying to zero.

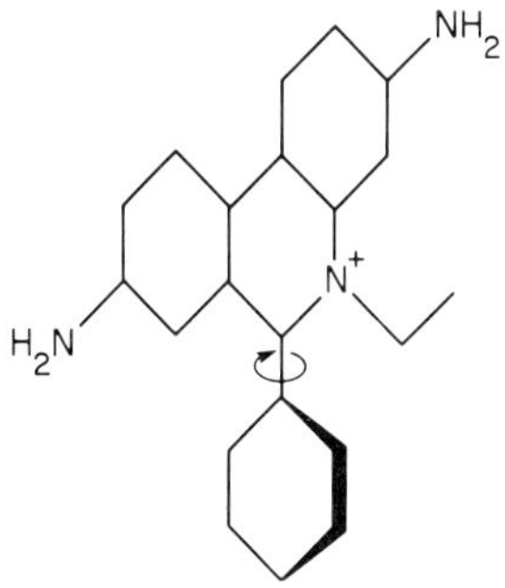

Fig. 3. The ethidium cation.

glycerol, phenyl rotation in water could be faster than the time resolution of our spectrometer. EB in DNA would similarly show a very fast, unresolvable relaxation, since the phenyl group of intercalated EB protrudes into the solvent.

References

1. e.g., see references in this volume: Atkinson et al.; Barbara; Ben-Amotz, Harris; Courtney et al.; Hicks, Eisenthal; Kobayashi, Ohtani; Lee, Hochstrasser; Zeglinski, Waldeck.
2. B. Bagchi, D.W. Oxtoby: J. Chem. Phys., 78, 2735 (1983)
3. G. Mourou, W. Knox: Appl. Phys. Lett., 36, 623 (1980)
4. J.H. Sommer, T.M. Nordlund, M. McGuire, G. McLendon: J. Phys. Chem. (in press, 1986)
5. G.H. Golub, C. Reinisch: Numerical Mathematics 14, 403 (1970)
6. M. LeBret, O. Chalvet: J. Molec. Struct. 37, 299 (1977)
7. G. Laczko, J. Lakowicz: Biophys. Chem. 19, 227 (1984)

This research was supported by NSF grants PCM-80-18458, -80-03004, and -83-02601 and by the sponsors of the Laser Fusion Feasibility Project of the Laboratory for Laser Energetics of the University of Rochester.

# Picosecond and Femtosecond Molecular Beam Chemistry: Coherence and Fragment Recoil Dynamics

*A.H. Zewail*

Arthur Amos Noyes Laboratory of Chemical Physics,
California Institute of Technology, Pasadena, CA 91125, USA

## 1. Introduction

The dynamics of collisionless intramolecular vibrational-energy redistribution (IVR), bond breakage, and bond formation in large isolated molecules is very important and challenging. A fruitful approach to this problem involves the study, via direct measurements in the time domain, of the decay parameters of energetically excited molecules. Using such an approach one can determine unimolecular rate constants as well as study any quantum mechanical coherence phenomena that may be involved in a decay process. Furthermore, in cold supersonic molecular beams these rates and coherence effects can be studied as a function of the *energy* and *character* of individual vibrational modes in a molecule. The results of such studies are important to understanding the nature of energy flow within molecules and to assessing the possibility of vibrational mode-selective laser chemistry [1,2].

Starting in 1980, we have been examining vibrational level-selective rate processes and coherence effects using the technique of picosecond-molecular jet spectroscopy [3,4]. In this technique a picosecond laser selectively excites molecular vibronic levels and allows for time domain studies of rates and coherence effects. The measurements are made on molecules that have been cooled vibrationally and rotationally in a supersonic jet expansion. Such cooling greatly reduces the spectral congestion that arises from thermal effects and allows for the single vibronic level excitation of even very large molecules. With this technique we have studied several problems. These include:

1) Coherence and energy redistribution in isolated large molecules (anthracene, stilbene) [5].

2) Quantum beats and Zeeman effect probing of radiationless processes in pyrazine [6].

3) Intramolecular photoisomerization of t-stilbene [7], 1,4-diphenylbutadiene [8], and styrene [9].

4) Intramolecular proton transfer in methyl salicylate [10].

5) Chromophore-selective excitation and photochemistry of intramolecular charge transfer formation in anthracene-$(CH_2)_3$-N,N-dimethyl aniline [11].

6) Bond breakage and IVR of partially solvated molecules in the supersonic beam [12].

7) Bond rotation dynamics and IVR in bianthryl [13].

Over the past few years we have improved the time resolution to the picosecond and subpicosecond domain, using pump-probe methods, to study coherence and fragmentation. Here, we review the techniques used in this laboratory. The work published in the above-mentioned areas will not be discussed here (see references cited); instead we shall highlight some recent advances.

## 2. The Molecular Beam-Lasers Apparatus

The arrangement we have used to interface the picosecond laser excitation source with the molecular beam involves two laser systems. The first is a synchronously pumped dye laser system, whose coherence width, coherence time, and pulse duration have been characterized [14] by the second harmonic generation (SHG) autocorrelation technique. The pulse width of these lasers is typically 1-2 ps, or, when a cavity dumper is used, 15 ps. The bandwidths of the lasers can be varied using intracavity filters and etalons. Typical bandwidths of the frequency doubled laser output are between 2 and 0.4Å.

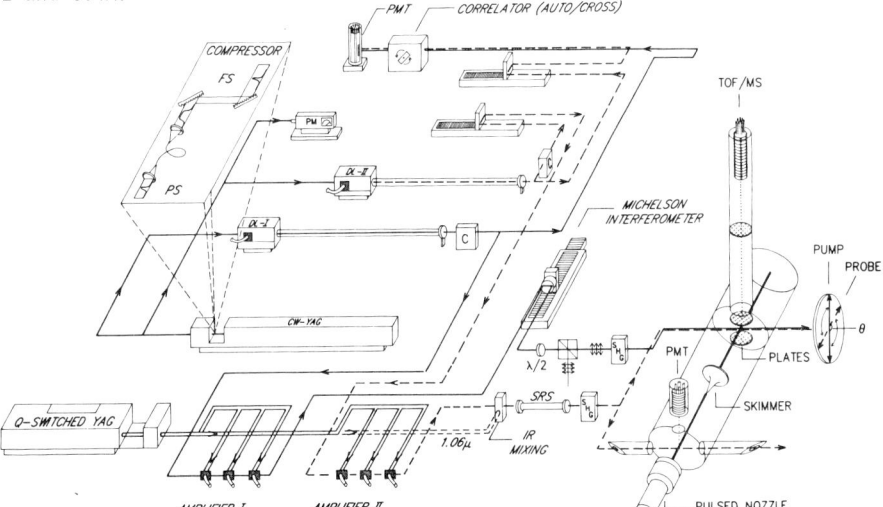

Fig. 1. The experimental setup for obtaining picosecond and femtosecond transients of molecules and reactions in skimmed molecular beams. The pump and probe pulses are different colors ($\lambda$). TOF/MS stands for time-of flight mass spectrometer.

The second laser system used two synchronously pumped dye lasers (oscillators) in connection with two 3-stage amplifiers to increase the pulse energy to the millijoules range (see Fig. 1). For femtosecond studies, we used pulse compression techniques, prior to amplification. The molecular beam apparatus is equipped to allow for the following methods of detection:

1) <u>Time-correlated single photon counting</u> (detecting dispersed or total fluorescence) using a fast photomultiplier (or microchannel plate). This method can achieve a resolution of 40 ps without deconvolution. With deconvolution one can obtain the time constants of single exponential decays having lifetimes of ∼ 10 ps. This is usually done on an unskimmed beam.

2) <u>Pump-probe multiphoton ionization detection</u>. This technique has been developed to measure short-time transient behavior in bulbs [15,4,16] and

beams [17]. Resolution is only limited by the pulse width of the laser and total ion signal is detected.

3) <u>Pump-probe fragment detection: LIF and mass spectrometry.</u> We use this arrangement to resolve the dynamics of molecular fragmentation on the picosecond and femtosecond time scale in the skimmed molecular beams [18].

## 3. Recent Advances

Recently, we have extended the above techniques to allow for (i) direct picosecond (pump-probe) measurements of state-to-state reaction rates; (ii) polarization anisotropy measurements of intramolecular dynamics, which lead to the observation of rotational coherence effects (rephasing!) in large molecules; (iii) detection of ultrafast quantum beats (and their Fourier-transform spectra); and (iv) dynamics of femtosecond fragment recoil in a chemical reaction. Here we briefly discuss some of these studies.

### 3.1 Picosecond time-resolution of IVR and quantum beats

In continuation of our efforts in the area of coherence effects in large isolated molecules (see Section 1), we have recently made two new advances. First, it was shown that nonchaotic *multilevel* vibrational energy flow is present in large polyatomic molecules with excess vibrational energy. The molecules studied are anthracene (at a rotational temperature of < 5K), deuterated anthracenes, and stilbene. The results of this work, completed recently [5], are exemplified in Fig. 2. From these results we obtained matrix elements for vibrational coupling and, more importantly, established the *number* and nature of modes involved in IVR.

Second, at higher excess vibrational energies we observed a "transition" in the behavior of IVR; instead of quantum beats among few levels ($\sim$ 3) at moderate excess energies, a fast decay of $\sim$ 25 ps followed by a modulated

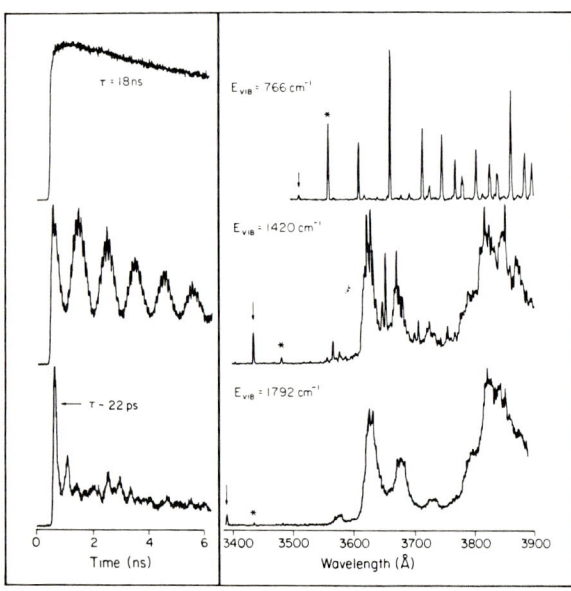

Fig. 2. Vibrational coherence (quantum beats and dephasing) observed in an anthracene beam. Note the excess energy dependence of the transients and the spectral changes of the fluorescence: low excess energy, no IVR; intermediate excess energy, restricted IVR; and high excess energy, dissipative IVR [5].

long time decay was observed at higher excess energy. These observations are interpreted as a manifestation of *dissipative* IVR where the initial deposited vibrational energy at, e.g., 1792 $cm^{-1}$, irreversibly flows to other modes. (We estimate the total number of levels to be $\sim 10$.) The method gives a direct real-time view of this collisionless (beam) dissipative IVR process by the picosecond techniques. The relevance of these studies to mode coupling and selectivity are discussed in a recent review article by BLOEMBERGEN and ZEWAIL [2].

## 3.2. Rotational coherence and rephasing in isolated molecules

In previous work we have used picosecond pump-probe MPI polarization anisotropy techniques [17] to study the rotational/vibrational dynamics in large molecules (stilbene), isolated in bulbs and beams. Subsequently, NEGUS et al. [19] measured fluorescence anistropy of stilbene gas and obtained a decay of $\sim 18$ ps. In both our work and the work of NEGUS et al. the interest is on the effect of excess vibrational energy on the decay of the anisotropy, which has been discussed elsewhere [17,19,20].

Recently, we reported on the observation of purely rotational coherence effect (alignment, dephasing and rephasing) in large molecules (beams) by using a polarized pulse for excitation and an analyzed dispersed fluorescence (Fig. 3). In this case, where the excitation is to the pure electronic origin of the transition, the anisotropy shows recurrences, which are eclipsed or staggered with respect to the original alignment. The period of the recurrence gives the excited state rotational constant directly. This feature makes the technique a very attractive method for high-resolution and Doppler-free studies of large molecules [20].

## 3.3. Fragment recoil

Using picosecond and femtosecond pump-probe techniques we have time resolved the dynamics of fragmentation of reactions on repulsive and bound surfaces: $ICN \rightarrow I + CN$, $NCNO \rightarrow NC + NO$, and $H_2O_2 \rightarrow 2OH$. In these experiments one pulse initiates the bond breakage process, and a second pulse probes the internal energy of the fragment formed. For example, for ICN

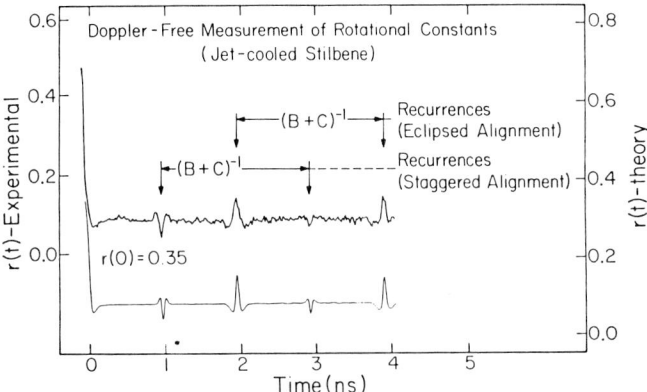

Fig. 3. Rotational coherence (alignment, dephasing and rephasing) observed in a stilbene beam. Note the staggered and eclipsed alignment of the recurrences. The period of the recurrences gives (B+C), the rotational constant of the excited state.

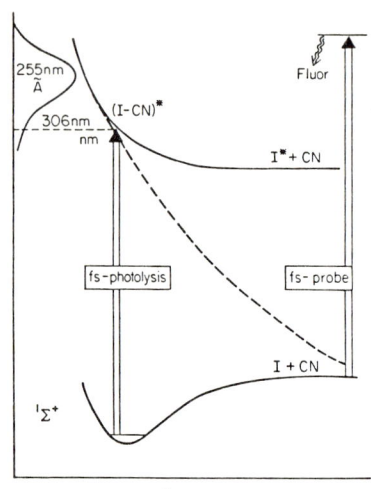

Fig. 4. The femtosecond photofragment experiment of the reaction ICN → I + CN. In this case, the probe pulse monitors the formation of the CN radical using laser-induced fluorescence: $\tau_{rec.} = 600 \pm 100$ fs.

fragmentation we measured $\tau = 600$ femtoseconds for the breaking of the C—I bond. In $H_2O_2$ decomposition, we observe biexponential build-up of the OH radical. Details of these reports are given in Refs. 18 and 21. Much will be learned in the coming years about the dynamics of chemical reactions using these pump-probe techniques in molecular beams--the time scale is just right for the distance scale of recoiling fragments, and for the state-to-state rates.

This work is supported by grants from the National Science Foundation; DMR85-21191 and CHE85-12887. I would like to thank all my students and postdoctoral fellows whose contributions, given in references, were very important to the research. I also would like to thank Marcos Dantus for his kind help with the figures, and in particular his "artistic touch" in Fig.1.

References

1. See, e.g., A. H. Zewail: Physics Today, 33, 27 (1980)
2. N. Bloembergen and A. H. Zewail: J. Phys. Chem. 88, 5459 (1984)
3. For a recent review see P. M. Felker and A. H. Zewail: In Applications of Picosecond Spectroscopy to Chemistry, ed. K. B. Eisenthal (D. Reidel Pub. Co., Holland 1984) p. 273
4. Ahmed H. Zewail: Faraday Discuss. Chem. Soc. 75, 315 (1983)
5. Wm. R. Lambert, P. M. Felker and A.H. Zewail: J. Chem. Phys. 75, 5958 (1981); P. M. Felker and A. H. Zewail: Phys. Rev. Lett. 53, 501 (1984); Peter M. Felker and Ahmed H. Zewail: J. Chem. Phys. 82, 2961 (1985); Peter M. Felker and Ahmed H. Zewail: J. Chem. Phys. 82, 2975 (1985); Peter M. Felker and Ahmed H. Zewail: J. Chem. Phys. 82, 2994 (1985); P. M. Felker, W. R. Lambert and A. H. Zewail: J. Chem. Phys. 82, 3003 (1985); Ahmed H. Zewail: Ber. Bunsenges. Phys. Chem. 89, 264 (1985)
6. P. M. Felker, Wm. R. Lambert and A. H. Zewail: Chem. Phys. Lett. 89, 309 (1982); Peter M. Felker and Ahmed H. Zewail: Chem. Phys. Lett. (in press)
7. J. A. Syage, Wm. R. Lambert, P. M. Felker, A. H. Zewail, and R. M. Hochstrasser: Chem. Phys. Lett. 88, 266 (1982); J. A. Syage, P. M. Felker and A. H. Zewail: J. Chem. Phys. 81, 4685 (1984); J. A. Syage, P. M. Felker and A. H. Zewail: J. Chem. Phys. 81, 4706 (1984); Peter M. Felker and Ahmed H. Zewail: J. Phys. Chem. 89, 5402 (1985)

8. J. F. Shepanski, B. W. Keelan and A. H. Zewail: Chem. Phys. Lett. <u>103</u>, 9 (1983); S. H. Courtney, G. R. Fleming, L. R. Khundkar and A. H. Zewail: J. Chem. Phys. <u>80</u>, 4559 (1984)
9. J. A. Syage, F. Al Adel and A. H. Zewail: Chem. Phys. Lett. <u>103</u>, 9 (1983)
10. P. M. Felker, Wm. R. Lambert and A. H. Zewail: J. Chem. Phys. <u>77</u>, 1603 (1982)
11. P. M. Felker, J. A. Syage, Wm. R. Lambert and A. H. Zewail: Chem. Phys. Lett. <u>92</u>, 1 (1982); J. A. Syage, P. M. Felker and A. H. Zewail: J. Chem. Phys. <u>81</u>, 2233 (1984)
12. Peter M. Felker and Ahmed H. Zewail: Chem. Phys. Lett. <u>94</u>, 448 (1983); Peter M. Felker and Ahmed H. Zewail: Chem. Phys. Lett. <u>94</u>, 454 (1983); Peter M. Felker and Ahmed H. Zewail: J. Chem. Phys. <u>78</u>, 5266 (1983)
13. Lutfur R. Khundkar and Ahmed H. Zewail: J. Chem. Phys. <u>84</u>, 1302 (1986)
14. D. P. Millar and A. H. Zewail: Chem. Phys. <u>72</u>, 381 (1982)
15. B. I. Greene and R. C. Farrow: J. Chem. Phys. <u>78</u>, (1983)
16. J. W. Perry, N. F. Scherer and A. H. Zewail: Chem. Phys. Lett. <u>103</u>, 1 (1983)
17. N. F. Scherer, J. F. Shepanski and A. H. Zewail: J. Chem. Phys. <u>81</u>, 2181 (1984); N. F. Scherer, J. W. Perry, F. E. Doany and A. H. Zewail: J. Phys. Chem. <u>89</u>, 894 (1985)
18. J. L. Knee, L. R. Khundkar and A. H. Zewail: J. Chem. Phys. <u>82</u>, 4715 (1985); J. L. Knee, L. R. Khundkar and A. H. Zewail: J. Chem. Phys. <u>83</u>, 1996 (1985); J. L. Knee, L. R. Khundkar and A. H. Zewail: J. Phys. Chem. <u>89</u>, 4659 (1985); N. F. Scherer, J. L. Knee, D. D. Smith and A. H. Zewail: J. Phys. Chem. <u>89</u>, 5141 (1985)
19. D. K. Negus, D. S. Green and R. M. Hochstrasser: Chem. Phys. Lett. <u>117</u>, 409 (1985)
20. P. M. Felker, J. S. Baskin and A. H. Zewail: J. Phys. Chem. <u>90</u>, 724 (1986); J. S. Baskin, P. M. Felker and A. H. Zewail: J. Chem. Phys. <u>84</u>, 4708 (1986)
21. N. F. Scherer, F. E. Doany, A. H. Zewail and J. W. Perry: J. Chem. Phys. <u>84</u>, 1932 (1986)

# Picosecond Laser Study of the Collisionless UV Photodissociation of Energetic Materials

J.C. Mialocq[1] and J.C. Stephenson[2]

[1] CEN-Saclay, IRDI/DESICP/DPC/SCM UA 331, CNRS,
F-91191 Gif-sur-Yvette Cedex, France
[2] National Bureau of Standards, Molecular Spectroscopy Division,
Gaithersburg, MD 20899, USA

It is of interest to elucidate the fast processes in the initiation of explosive decomposition of energetic materials in order to control their sensitivity and effectiveness. It is accepted that a shock causes bond rupture in a condensed phase energetic molecule leading to the production of free radicals. Explosions may then occur by subsequent radical chain reactions. We have undertaken the picosecond study of the photolysis of nitroalkanes ($RNO_2$) and dimethylnitramine $(CH_3)_2NNO_2$ which are both model compounds and energetic materials and we have examined the breaking of the $C-NO_2$ bond (dissociation energy, $D = (59.4 \pm 1.4)$ kcal/mole) in nitroalkanes and $N-NO_2$ bond ($D = 46.2$ kcal/mole) in dimethylnitramine (DMNA) [1]. The situation is however more complicated because other fragmentation processes have been demonstrated leading to NOH, HNO, HONO or OH elimination via intramolecular rearrangement. Moreover the photodissociation mechanism should be dependent on the excitation wavelength. In this work, the 266 nm picosecond laser excitation affects a $n \rightarrow \pi^*$ transition in nitroalkanes and falls in the envelope of three absorption bands in DMNA [2]. The bond rupture is investigated by observing fluorescence from electronically excited $NO_2^*$ and also by laser-induced fluorescence probing of ground state $NO_2$ fragments formed in the collisionless photodissociation of the gaseous nitrocompounds under study.

The present experimental results are quite relevant to the more general understanding of $NO_2$ emission in shock waves accompanying $CH_3NO_2$ pyrolysis [3].

I - Experimental

The laser was an active-passive mode-locked $Nd^{3+}$ - YAG oscillator-amplifier system which delivered a 30 mJ (1064 nm) single pulse at 10 Hz [4]. The frequency doubled 532 nm probe pulse was 31 ps duration (FWHM) as monitored by a streak camera. After frequency doubling the green pulse, the 266 nm pump pulse was separated from the optically delayed 532 nm probe pulse. The two pulses were then collinearly recombined as in the experiment of Goldberg et al. [5]. The areas of the green and UV laser beams were estimated to be $2.7 \times 10^{-3}$ cm$^2$ and $3 \times 10^{-4}$ cm$^2$ respectively. Maximum energies of 200 µJ (UV) and 1.0 mJ (green) were fairly constant from pulse to pulse. The zero time delay between the pump and the probe pulses was determined by probing with 532 nm pulses the photobleaching of rhodamine 6G caused by the 266 nm pulses. RNO$_2$ or DMNA contained in a Tee shaped 4 cm diameter glass cell equipped with three fused silica windows were excited at 266 nm and the NO$_2$ ground state fragment was probed by laser-induced fluorescence (LIF) using the delayed 532 nm pulse. The fluorescence emitted in the observed 2 mm long interaction region was collected, spatially and spectrally ($\lambda > 580$ nm) filtered to discriminate against scattered laser light and analyzed with a photomultiplier. The signal was averaged with a boxcar averager-integrator, the 50 ns gate being scanned over the NO$_2$* decay curve in 100 - 1000 seconds according to the nitrocompound investigated. NO$_2$ diluted in argon was used as an actinometer. No LIF was observed when this mixture was excited at 266 nm above the NO$_2$ dissociation limit (398 nm). Under 532 nm excitation, the initial NO$_2$* fluorescence was proportional to the gas pressure and to the laser energy. The Stern-Volmer plot of the NO$_2$* decay rate constant versus the pressure is a straight line. The estimation of the NO$_2$ quantum yields is based on the courageous hypothesis that the 532 nm absorption cross section of NO$_2$ is the same for thermal and photoproduct NO$_2$.

## II - Results and discussion

The fluorescence signal from NO$_2$* formed by 266 nm photolysis of CH$_3$NO$_2$ followed by ($t_D$ = 300 ps) 532 nm LIF probing of the ground state NO$_2$ photoproduct is shown in figure.1, for a single shot experiment.

Fig. 1. Single shot LIF signal (100 ns and 50 mV/div.). 1 torr CH$_3$NO$_2$. Ground state NO$_2$ is probed at 532 nm with a 300 ps delay time.

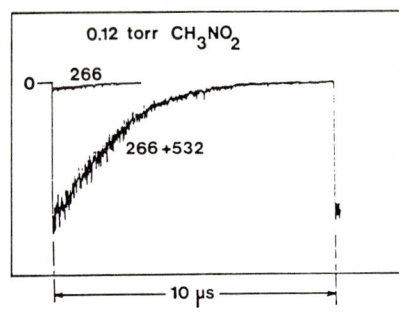

Fig. 2. Plot of the $NO_2^*$ fluorescence intensity as a function of time.

The fluorescence signal delivered by the boxcar averager-integrator after 10 000 laser shots is shown in figure.2. There is a 266-only component of the fluorescence which is about 5 % of the value caused by the 532 nm LIF of $NO_2$.

The picosecond kinetics of the $NO_2$ formation follows closely that of the R-6G photobleaching observed with the same pump and probe pulses (figure.3) showing that the photodissociation occurs within 6 ps. The fluorescence signal from $NO_2^*$ formed by 266 nm photolysis of DMNA and 532 nm probing shows the same behaviour [4]. This is consistent with our RRKM lifetimes for $CH_3NO_2$ (0.21 ps) and DMNA (4 ps) calculated from literature Arrhenius A factors [6,7] and vibrational frequencies [8], by considering that these electronically excited molecules undergo internal conversion to high vibrational levels of the ground electronic state followed by a statistical unimolecular vibrational predissociation.

The quantum yields of ground stage $NO_2$ and excited $NO_2^*$ formation are gathered in the table [4,9].

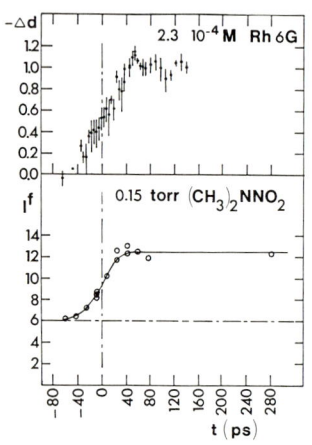

Fig. 3. Photobleaching kinetics of a R-6G solution at 532 nm (upper curve). $NO_2^*$ LIF kinetics (lower curve).

Table : Quantum yields of $NO_2$ and $NO_2^*$ formation in the 266 nm photolysis of $CH_3NO_2$ and DMNA.

|          | $NO_2$              | $NO_2^*$            |
|----------|---------------------|---------------------|
| $CH_3NO_2$ | $0.17 \pm 0.11$   | $< 10^{-3}$         |
| DMNA     | $0.13 < \emptyset < 0.95$ | $0.02 < \emptyset < 0.12$ |

The formation of the ground state $NO_2$ fragment from nitromethane or DMNA is a monophotonic process. Moreover in the $RNO_2$ under study (R = $CH_3$, $C_2H_5$, $n-C_3H_7$ and $i-C_3H_7$), its quantum efficiency does not depend on the nature of the alkyl group.

Formation of $NO_2^*$ from DMNA is also monophotonic but in the case of $CH_3NO_2$ it increases faster than linearly with increasing energy in the UV photolysis pulse. The greater efficiency of $NO_2^*$ formation in DMNA is consistent with the greater available energy after the $N-NO_2$ bond breaking.

Acknowledgement

This work was supported by the Direction des Recherches, Etudes et Techniques under contract N° 84/819. The authors thank Dr. L.S. Goldberg for his advice during the construction of the LIF apparatus and Dr. M.E. Umstead (both from the Naval Research Laboratory, Washington D.C.) for the gift of DMNA.

1. Y.L. Chow : In The chemistry of amino, nitroso and nitrocompounds and their derivatives, Suppl. F, Part 1, ed. S. Patai (Wiley, New-York 1982) pp. 181-290.
2. J. Stals, C.G. Barraclough, A.S. Buchanan:Trans. Far. Soc. 65 (1969) 904.
3. H. Hiraoka, R. Hardwick:J. Chem. Phys. 39, 1963, 2361.
4. J.C. Mialocq, John C. Stephenson:Chem. Phys. Lett. 123, 1986, 390.
5. L.S. Goldberg, M.J. Marrone, P.E. Schoen:"Picosecond Lasers and Applications", SPIE, Vol. 322 (1982) 199.
6. K. Glänzer, J. Troe:Helv. Chim. Acta 55 (1972) 2884.
7. M.E. Umstead, S.A. Lloyd, M.C. Lin: Worskshop on Energetic Material Initiation Fundamentals, Annapolis, MD (USA), (1984).
8. C. Trinquecoste, M. Rey-Lafon, M.T. Forel:Spectrochim. Acta 30A (1974) 813.
9. J.C. Mialocq, J.C. Stephenson: Chem. Phys. (in press).

# Experimental Study of Harmonic Generation with Picosecond 248 nm Radiation

T.S. Luk, A. McPherson, H. Jara, U. Johann, I.A. McIntyre, A.P. Schwarzenbach, K. Boyer, and C.K. Rhodes

Department of Physics, University of Illinois at Chicago, P.O. Box 4348, Chicago, IL 60680, USA

Measurements of the radiation produced by atoms interacting with an ultraviolet field at an intensity of $\sim 10^{16}$ W/cm$^2$ are used to probe the nature of the nonlinear coupling.

## I. Introduction

High power picosecond and subpicosecond excimer laser sources[1-3] have been developed to investigate the electronic behavior of atomic and ionic species exposed to laser field intensities greater than $10^{16}$ W/cm$^2$. With these new sources, one can study ions, electrons, and radiation in the XUV and soft x-ray regions produced by the nonlinear atomic interaction.

## II. Discussion

Investigations of the electron energy and ion charge state spectra using laser intensities of $10^{15}$ to $10^{16}$ W/cm$^2$ and pulse widths of 0.5 to 1.5 psec have been reported.[4-7] These experiments have shown that even for a subpicosecond time scale, electrons interact basically independently and adiabatically with the laser field which results in sequential ionization, at least for the production of the lower charge states of the rare gases, when the laser intensity is of the order of $10^{14}$ to $10^{15}$ W/cm$^2$.

The observation of radiation, produced either as harmonics of the frequency ω of the incident wave or as fluorescence from atomic and ionic species, can be another effective means to study the electronic behavior under intense irradiation. The generation of harmonic radiation using short laser pulses in a homogeneous gaseous medium has the advantage that the electronic response tends to be relatively uncomplicated by density gradient effects[8,9] and large radiation backgrounds caused by inverse bremsstrahlung. This work describes the results of experiments examining the generation of harmonic radiation in He, Ne, and Ar using a 1.5 psec KrF* laser system.

The KrF* laser system, described elsewhere,[2] has been modified for this experiment by omitting the optical fiber and grating compressor yielding a seed pulse of approximately 1.5 psec duration. The seed pulses are amplified by injection into an excimer amplifier chain producing 10-20 mJ laser pulses at 248.4 nm with less than a 10% background of spontaneous emission.

The experimental set-up is illustrated in Fig. 1. A 20 cm (f/10) focal length lens focusses the 248.4 radiation into a target gas supplied by a modified Lasertechnics pulsed valve supersonic gas jet. The gas jet is mounted 10 cm in front of the entrance slit of a 2.2 meter grazing incidence monochromator (McPherson model 247) equipped with a 600 l/mm gold coated spherical grating blazed at 120 nm. With this optical system, the maximum intensity in the focal region is estimated to be $\sim 10^{16}$ W/cm$^2$. Radiation produced in the

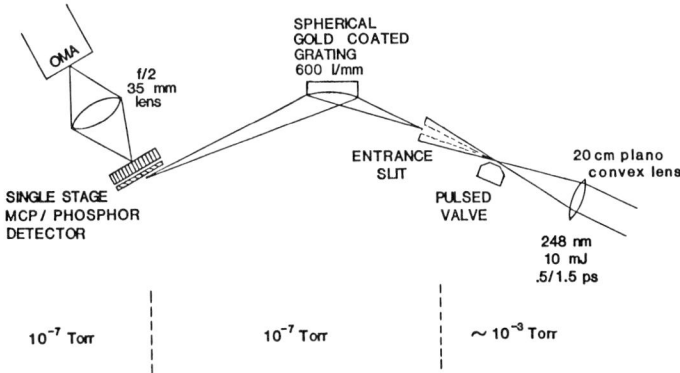

Fig. 1: Schematic of pulsed gas jet and spectrometer assembly

focal volume enters through a 100 µm slit and is dispersed to a position sensitive detector which is mounted tangentially to the Rowland circle of the monochromator. The resolution and accuracy of the monochromator-detector system are typically 0.1 nm. The mounting of the detector allows the observation of radiation between 5 nm and 75 nm. The detector consists of a single stage micro-channel plate with a phosphored fiber optic anode. Detection of photons is achieved by relaying the image of the phosphor screen through a 35 mm Nikon lens to an optical multichannel analyser. Since the detector could not directly see the 248.4 nm radiation of the laser, no filter was used to suppress the background fundamental radiation. Noise subtraction was achieved by determining the background with the gas jet valve closed. The pulsed gas valve, which was modified to allow backing pressure up to 1000 psi, is typically operated at 600 psi with a pulse repetition rate of 2 Hz. Background pressure in various parts of the apparatus are indicated in Fig. 1. The laser is focused to a position a few hundred microns above the nozzle tip which has a diameter of 0.5 mm and a throat depth of 1mm. The estimated gas density in the interaction region is $\sim 3 \times 10^{18}$ cm$^{-3}$.

Helium, neon, and argon have been used as the target gas for the production of fluorescence and harmonic radiation. The highest harmonic observed in He is eleventh (22.6 nm). The seventh harmonic in He appears to be anomalously strong with respect to the intensity exhibited by the fifth, ninth, and eleventh harmonics in that gas. The eleventh harmonic in He represents an energy conversion efficiency of approximately $3 \times 10^{-11}$.

The highest harmonic observed in Ne is thirteenth (19.1 nm), the shortest wavelength produced by this nonlinear means.[10] The relative strengths of the harmonics produced in Ne are shown in Fig. 2. The thirteenth harmonic in Ne represents an efficiency of energy conversion of approximately $5 \times 10^{-11}$.

Of the nonlinear media examined, Ne has proven to be the most effective in producing harmonic radiation. In addition, the results on Ne have shown an unexpected feature. Although the observed intensities of the fifth, seventh, and ninth harmonics decrease in a regular manner, the eleventh and thirteenth harmonics have roughly the same relative intensity, (see Fig. 2). It appears that the observed relative enhancement in both the eleventh and thirteenth harmonics is unlikely to be due to phase matching, since each harmonic was separately optimized by scanning the laser focus through the gas plume to produce the maximum signal.

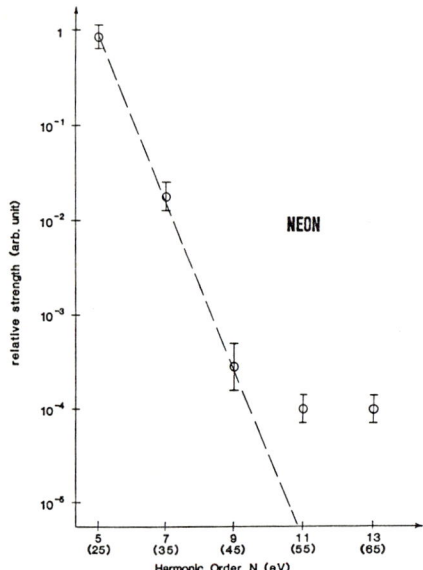

Fig. 2: Relative strength of harmonic generation in Ne using $\sim$ 1.5 psec pulses of 248.4 nm radiation from a KrF* laser at an intensity of $\sim 10^{16}$ W/cm$^2$

Under the experimental conditions in Ne, it is conceivable that self-focusing could enhance the focal intensity which could result in enhancement of the conversion efficiency of higher harmonics. Such a mechanism has been implicated in the generation of harmonics in high density plasmas.[9] However, it is expected that this effect would be accompanied by significant phase modulation and broadening of the harmonic radiation, a feature which is not observed.

Alternatively, it is possible that multiply excited atomic states may enhance[11] the nonlinear susceptibility in certain energy regions in which those states exist. In Ne, doubly excited states begin to occur just above 45 eV. For example, the $2p^4 3d 4p\ ^1P_1$ state[12,13] of Ne is in near resonance with the eleventh harmonic, and likewise, other multiply excited states lying $\sim$ 10 eV higher could be in resonance with the thirteenth harmonic. We note that the electron spectra of Ne exhibit electrons with energies of approximately 6 eV. Although several possibilities exist for the generation of these electrons, the observation of electrons in this energy range is consistent with the excitation of such doubly excited levels.

The highest harmonic observed in Ar is seventh (35.5 eV). However, unlike He and Ne, fluorescence is also observed in the 21.5 - 37.5 nm region. The fluorescence near the seventh harmonic (35.5 nm) consists of indentifiable[14,15] ionic lines coming from $3p^2$ - $3p4s$ transitions in $Ar^{4+}$. In addition, lines which correspond to no known catalogued transitions are observed along with others whose positions coincide with known transitions[16] involving multiply excited levels (e.g. $3s^2 3p^6 \rightarrow 3s^2 3p^4 nln'l'$). The strengths of the fluorescence lines seen generally increased when the laser focus was positioned at a point in the gas plume closer to the spectrometer entrance slit, a position for which the harmonic intensity was decreased.

## III. Conclusions

In conclusion, high order harmonic radiation has been observed using a picosecond 248.4 nm KrF* laser system, and the thirteenth harmonic (65 eV), produced in Ne, represents the shortest wavelength generated by that means. In addition, the fluorescence observed in Ar indicates that a considerable variety of excited electronic states are produced by the interaction with the ultraviolet field at an intensity of $\sim 10^{16}$ W/cm$^2$.

## IV. Acknowledgements

The authors wish to acknowledge the technical assistance of R. Bernico, T. Pack, R. Slagle, and J. Wright. This work was supported by the U.S. ONR, the U.S. AFOSR, the SDIO(ISTO), the DOE, the LLNL, the NSF, the DARPA, and the LANL.

## V. References

1. H. Egger, T.S. Luk, K. Boyer, D.R. Muller, H. Pummer, T. Srinivasan, and C.K. Rhodes: Appl. Phys. Lett. **41**, 1032 (1982).
2. "Subpicosecond KrF* Excimer Laser Source," A.P. Schwarzenbach, T.S. Luk, I.A. McIntyre, U. Johann, A. McPherson, K. Boyer, C.K. Rhodes: Opt. Lett. (in press).
3. J.H. Glownia, G. Arjavalingham, P.P. Sorokin, and J.E. Rothenburg: Opt. Lett. **11**, 79 (1986).
4. T.S. Luk, U. Johann, H. Egger, H. Pummer, and C.K. Rhodes: Phys. Rev. A**32**, 214 (1985).
5. "Rare Gas Electron Energy Spectra Produced by Collision-Free Multiquantum Processes," U. Johann, T.S. Luk, H. Egger, and C.K. Rhodes: Phys. Rev. A (in press).
6. "Multiphoton Ionization in Intense Ultraviolet Laser Fields," U. Johann, T.S. Luk, I.A. McIntyre, A. McPherson, A.P. Schwarzenbach, K. Boyer, and C.K. Rhodes: In *Proceedings of the Topical Meeting on Short Wavelength Coherent Radiation*, ed. by J. Bokor and D. Attwood, (AIP, New York, to be published).
7. "Subpicosecond Studies of Collision-Free Multiple Photon Processes in the Ultraviolet," T.S. Luk, U. Johann, I.A. McIntyre, A. McPherson, A.P. Schwarzenbach, K. Boyer, and C.K. Rhodes: In *Proceedings of the High Intensity Laser Processes Conference*, Quebec, Canada, June 1986 (SPIE, Bellingham, WA, to be published).
8. S. Jachel, B. Perry, and M. Lubin: Phys. Rev. Lett. **37**, 95 (1976).
9. R.L. Carman, C.K. Rhodes, and R.F. Benjamin: Phys. Rev. A**24**, 2649 (1981).
10. J. Bokor, P.H. Bucksbaum, and R.R. Freeman: Opt. Lett. **8**, 217 (1983).
11. C.K. Rhodes: "Ordered Many-Electron Motions in Atoms and X-Ray Lasers," In *Proceedings of Giant Resonances in Atoms, Molecules, and Solids*, Les Houches, France, 1986, ed. by J.P. Connerade et. al. (Plenum Press, New York, to be published).
12. K. Codling, R.P. Madden, and D.L. Ederer: Phys. Rev. **155**, 26 (1967).
13. U. Becker, R. Hölzel, H.G. Kerkhoff, B. Larger, D. Szostak, and R. Wehlitz: Phys. Rev. Lett. **56**, 1120 (1986).
14. R.L. Kelly and L.J. Palumbo: *Atomic and Ionic Emission Lines Below 2000 Angstroms*, NRL Report 7599 (USGPO, Washington, D.C., 1973).
15. S. Bashkin and J.O. Stoner, Jr.: *Atomic Energy Level and Grotrian Diagrams*, Vol. II, (North-Holland, Amsterdam, 1978).
16. R.P. Madden, D.L. Ederer, and K. Codling, Phys. Rev. **177**, 136 (1969).

# Time-Resolved Measurement of Laser-Induced Desorption of a Molecular Monolayer

*G. Arjavalingam, T.F. Heinz, and J.H. Glownia*

IBM Thomas J. Watson Research Center, Yorktown Heights, NY 10598, USA

Recently considerable interest has developed in the use of intense laser radiation to desorb molecular monolayers at surfaces. In addition to the importance of this effect for fundamental investigations of energy flow in absorbed molecules, laser-induced desorption has been applied to the study of surface reactions and surface diffusion. To date, however, no direct time-resolved measurements of the dynamics of the process of laser-induced desorption have been performed on a rapid time scale. In this paper, we present experimental results on the desorption of a monolayer of dye molecules monitored on a picosecond time scale by means of the surface-specific technique of second-harmonic generation (SHG).

The basic scheme of the time-resolved desorption measurement is as follows. A region of the substrate overcoated with a monolayer of dye molecules is exposed to an intense pulse of laser radiation. A second, delayed pulse, focused within the exposed area, serves as a probe beam to generate SH radiation. For an unperturbed dye monolayer strong SHG is expected, since the presence of the substrate creates an asymmetric environment for the adsorbed molecules [1]. After the molecules are desorbed from the surface, their contribution to the SH signal decreases, vanishing as the molecules leave the surface and become randomly oriented.

In our experiment, a monolayer of rhodamine B was prepared on a fused silica substrate by means of a spinning technique [1]. The desorption of the dye monolayer was effected with 6 ps light pulses at a wavelength of 560 nm, resonant with the $S_0 \rightarrow S_1$ transition in the adsorbed molecules. A fluence of ~ 0.5J/cm$^2$ was found to be sufficient to desorb most of the dye monolayer with a single shot of the laser. The desorbing beam was s-polarized and directed on the sample at 45° angle of incidence. In regard to the final state of the system, the nature of the molecular species resulting from laser-induced desorption of rhodamine dye adlayers has been investigated by Letokhov et al. [2] using time-of-flight mass spectrometry. Under their experimental conditions, it was determined that resonant excitation of the $S_0 \rightarrow S_1$ transition by intense picosecond pulses led to a high yield for the desorption of intact ions of the dye molecules.

In order for us to monitor the dynamics of the desorption process with the second-harmonic technique, probe pulses were formed by splitting off a small fraction of the pump beam and passing the light through an optical delay line. The p-polarized probe beam was incident on the sample at a slight angle with respect to the pump, thus permitting the reflected SH radiation generated by the probe to be isolated. Although the probe beam yielded an easily detectable SH signal, it did not lead to any perceptible molecular desorption by itself. Our measurements were performed by recording the SH intensity as the arrival time of the probe pulses was varied with respect to the desorbing pulses. A fresh region of the sample surface was exposed during each laser shot. The point of zero time delay was determined by measuring the SH intensity from the mixing of the pump and probe beams in a thin quartz plate positioned in the plane of the sample.

Figure 1 displays the SH intensity from the dye monolayer as a function of delay time relative to the desorbing pulse. The observed decrease in SHG appears on the time scale of a few picoseconds,

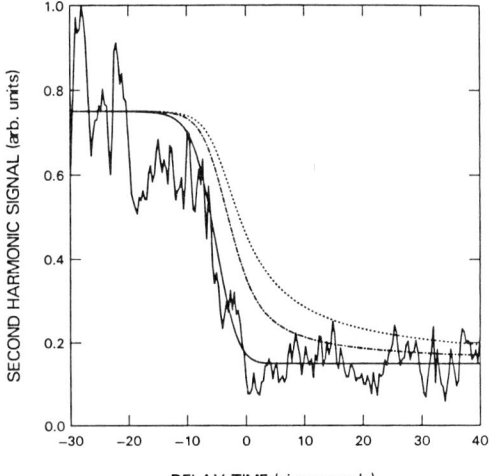

Fig. 1: SH intensity from a rhodamine adlayer as a function of the delay of the probing pulse relative to the intense desorbing pulse. In addition to the experimental data, predictions of the model described in the text are shown for molecules leaving the surface with effective temperatues of 1,000 K (dotted curve), 5,000 K (dash-dot), and infinite temperature (solid).

indicating a very fast response. In order to interpret our results more quantitatively, we will consider the process within the framework of a thermal desorption model. In this model, the rhodamine molecules contribute to the nonlinear response until they become separated from the surface and randomly oriented. For the influence of the substrate to vanish, the molecules must travel a distance D away from the surface. We take D = 10Å, comparable to the size of the rhodamine molecules. A smaller value for D is physically implausible; a larger value will only strengthen the conclusions we draw from the analysis. The decay in the effective nonlinear polarizability of the molecules (the molecular nonlinear polarizability averaged over orientations) is represented by a simple exponential function of distance from the surface. If we denote by $f_T(\vec{v})$ the Maxwellian velocity distribution for molecules with a translational temperature T, then a component of the surface nonlinear susceptibility tensor will vary as

$$\chi_s^{(2)}(t) \sim N_0 + N_s(t) - \int d\vec{v} f_T(\vec{v}) \int_{-\infty}^{t} dt' \frac{dN_s}{dt'} \exp[-(t-t')v_z/D]. \tag{1}$$

Here $N_s(t)$ corresponds to the surface density of molecules at a time t and $N_0$ accounts for the residual contribution to the nonlinearity from the bare surface and from molecules not desorbed by the laser pulse.

It is still necessary to give a relation for the rate at which molecules leave the surface, i.e., for $N_s(t)$. A complete understanding of the steps through which energy from the desorbing laser pulses, initially inducing electronic transitions in the adsorbed dye molecules, is converted into translational energy would be required in order to develop a reliable theory to predict the desorption rate. The time-resolved fluorescence measurements of Kemnitz et al. [3] for rhodamine B adsorbed on fused silica indicate that the relaxation rate from the $S_1$ state decreases into the subnanosecond regime for monolayer adsorbate coverages, deactivating far more rapidly than would an isolated molecule. The characteristic decay times are still appreciably longer than the observed decrease in our SHG signal. This implies that a significant role is played by higher lying states with faster decay rates or direct desorption from excited electronic states. At present, no detailed description of this process can be given. We can, however, observe that within the framework of our model, the rapid fall in the experimental data implies a prompt desorption of the dye molecules from the surface. Consequently, we have chosen to represent the desorption rate as being proportional to the product of the density of molecules remaining on the surface at a given time and the intensity of desorbing laser pulse raised to a positive power. In work on the desorption of organic molecules from thin films, Egorov et al.

[4] have observed a power law relation between the amount of desorbed material and the laser fluence, with an exponent ranging from 2 to 9. In our calculations, the value of the exponent did not strongly influence the rate of decay of the SH signal for time scales longer than or comparable to that of the desorbing pulse and a linear intensity dependence was assumed.

In Fig. 1 curves based on the desorption model are shown for various translational temperatures. As can be seen, the model reproduces the experimental data only for translational temperatures in excess of 5000 K. Such a high translational temperature is perhaps surprising and suggests that full thermal equilibrium between the dye molecules and the substrate has not been attained. Our inferred translational temperature is not, however, at variance with translational temperatures characteristic of products in other laser induced surface processes. For example, recent time-of-flight measurements on the $C_2$ fragments arising from the ablation of polymeric material by ultraviolet laser radiation indicate effective translational temperatures as high as 40,000 K [5]. In the case of desorption of copper atoms from a copper substrate, translational temperatures in excess of 25,000 K have been measured [6]. Our data, pertaining only to the first few picoseconds of the process, will minimize the influence of any laser-induced plasma above the surface, which might alter the final translational energy of the desorbed material.

An alternative explanation of the rapid fall in the SHG, which has not been considered in the model developed above, is the possibility of photofragmentation of the adsorbed molecules on the surface. Such a process could occur on a picosecond time scale and would lead to a dramatic drop in the surface SH signal. Although this mechanism appears to be incompatible with the mass spectra of [2], in those studies, a different substrate material was utilized and the desorption was induced by longer, less intense laser pulses. Further clarification of this issue will require additional measurements of the mass distribution of the desorbed molecules. Finally, it should be noted that a simple bleaching of the adsorbed molecules (i.e., population of the $S_1$ state) is not expected theoretically to give rise to a sharp drop in the SH signal from an otherwise unperturbed molecular adlayer.

In summary, we have performed time-resolved measurements of the laser-induced desorption of a molecular monolayer on a picosecond time scale. The fast response observed in the SH signal can be interpreted either as arising from molecules leaving the surface with a very high translational temperature ( > 5000 K) or from a surface photofragmentation process.

References

1. T.F. Heinz, C.K. Chen, D. Ricard, and Y.R. Shen, Phys. Rev. Lett 48, 478 (1982).
2. V.S. Letokhov, V.G. Movshev, and S.V. Chekalin, Zh. Eksp. Teor. Fiz. 81 480 (1981) [ Sov. Phys. JETP 54, 257 (1981)]; V.G. Movshev and S.V. Chekalin, Kvantovaya Elektron. (Moscow) 10, 1425 (1983) [ Sov. J. Quantum Electron. 13, 925 (1983)].
3. K. Kemnitz, T. Murao, I. Yamazaki, N. Nakashima, and K.Yoshihara Chem. Phys. Lett 101, 337 (1983).
4. S.E. Egorov, V.S. Letokhov, and A.N. Shibanov in Surface Studies with Lasers, edited by F.R. Aussenegg, A. Leitner, and M.E. Lippitsch, (Springer, Berlin, 1983), p. 156.
5. R. Srinivasan, B. Braren, D.E. Seeger, and R.W. Dreyfus, Macromolecules 19, 916 (1986).
6. I. Hussla and R. Viswanathan, Surface Sci. 145, L488 (1984).

# Part VII

# Dynamics of Biological Processes

# Picosecond Electron Transfer and Stimulated Emission in Reaction Centers of *Rhodobacter sphaeroides* and *Chloroflexus aurantiacus*

M. Becker[1], D. Middendorf[1], N.W. Woodbury[1], W.W. Parson[1], and R.E. Blankenship[2]

[1]Department of Biochemistry, University of Washington, Seattle, WA 98195, USA
[2]Department of Chemistry, Arizona State University, Tempe, AZ 85287, USA

Photochemical reaction centers isolated from purple photosynthetic bacteria such as *Rhodobacter sphaeroides* contain 3 protein subunits, 4 molecules of bacteriochlorophyll (BChl), 2 of bacteriopheophytin (BPh), 2 quinones ($Q_A$ and $Q_B$), 1 carotenoid, and 1 non-heme iron atom [1]. The crystal structure of the reaction center of *Rhodopseudomonas viridis* has been solved to 3 Å resolution [2,3]. The pigments lie on either side of a pseudosymmetry axis that passes through the iron atom (Fig. 1). At one end of the pigment cluster is a closely associated pair of BChls (P). The other BChls and the BPhs are located between P and the quinones, which are on either side of the iron.

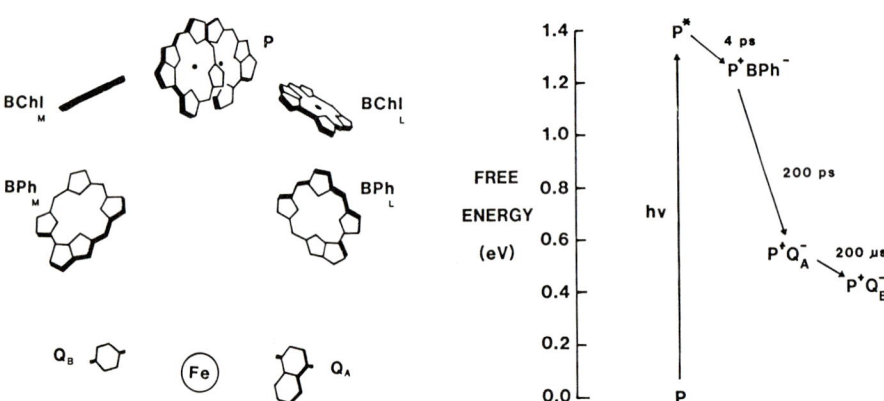

Figure 1 Arrangement of chromophores in *Rp. viridis* reaction centers. Redrawn from [2]. $Q_A$ is a menaquinone in *Rp. viridis* and *Cf. aurantiacus*, and ubiquinone in *Rb. sphaeroides*. $Q_B$ is a menaquinone in *Cf. aurantiacus*, and ubiquinone in *Rp. viridis* and *Rb. sphaeroides*. A BPh probably replaces $BChl_M$ in *Cf. aurantiacus*.

Figure 2 Kinetics and thermodynamics of electron-transfer steps in *Rb. sphaeroides* reaction centers [1,8].

When the reaction center absorbs a photon, the special pair of BChls is raised to an excited singlet state (P*). P* transfers an electron to one of the BPh molecules ($BPh_L$ in Fig. 1) with a time constant of 3 to 4 ps [4-7]. This creates a radical-pair ($P^+$ $BPh^-$), which decays in about 200 ps as $BPh^-$ passes an electron to $Q_A$ (Fig. 2). Electrons move from $Q_A^-$ to $Q_B$ in about 200 μs, and eventually return to $P^+$ by way of other carriers. Photosynthetic bacteria use this cyclic flow of electrons to drive the movement of protons across the cell membrane. The free energy stored in this way ultimately supports the growth of the cells. Although $BChl_L$ is located between P and $BPh_L$ (Fig. 1), recent work indicates that a $P^+BChl^-$ radical-pair is not seen as a kinetically resolvable intermediate prior to the formation of $P^+BPh^-$ in either *Rp. viridis* or *Rb. sphaeroides* [5-7].

The kinetics of the inital electron transfer step between P* and BPh can be measured from the decay of stimulated emission from P* [5-7]. Stimulated emission occurs near 1000 nm in *Rp. viridis* and near 920 nm in *Rb. sphaeroides*. The amplitude of the emission should be proportional to the concentration of P*, provided that the excited state relaxes vibrationally on a time scale that is short, relative to the rate of electron transfer. If nuclear relaxations occur on the same time scale as electron transfer, a shift of the emission spectrum with time could distort measurements of the emission intensity at a fixed wavelength.

Figure 3A shows spectra of the absorption changes that occur near 900 nm, when reaction centers of *Rb. sphaeroides* are excited with flashes lasting approximately 1 ps. The 5 spectra were measured at 1, 2, 3, 5 and 15 ps after the center of the excitation flash. All include an absorbance decrease centered at 865 nm, due to the bleaching of the main ground-state band of P. This bleaching occurs within the response time of the apparatus, and remains when P* is converted to $P^+$. Stimulated emission is seen as a shoulder on the long-wavelength side of the band. The shoulder cannot represent the bleaching of an absorption band, because it is larger than the initial absorbance of the reaction centers in this region [5]. Previous measurements showed that the signal near 920 nm decays with a time constant of 3 to 4 ps, matching the kinetics with which BPh receives an electron, as judged from the bleaching of the BPh's ground-state band at 755 nm [5,6]. Figure 3B shows spectra of the stimulated emission, obtained by subtracting the signals measured at 1, 2, 3 or 5 ps from the signal measured at 15 ps. After normalization, the spectra obtained at the different times are indistinguishable within the experimental noise.

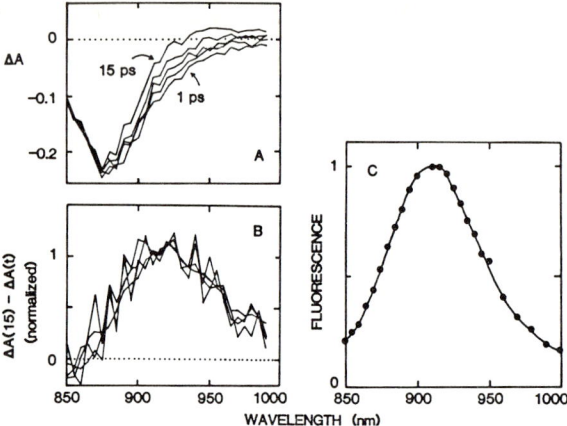

Figure 3  A: Absorbance changes and stimulated emission at 1, 2, 3, 5 and 15 ps after *Rb. sphaeroides* reaction centers (A = 1.9 at 802 nm; 1mm light path; approx. 10° C) were excited at 610 nm. Apparatus and conditions as described [5]. Each point represents 900 laser flashes. B: Normalized difference spectra obtained by subtracting the signals measured at 1, 2, 3 and 5 ps from that at 15 ps. C: Spontaneous fluorescence emission spectrum, measured by single-photon counting [8].

Although the data do not rule out a shift by a few nm between 1 and 5 ps, such a shift would not significantly alter the decay kinetics of the emission measured at 920 nm. Vibrational equilibration apparently occurs more rapidly than the electron transfer reaction. Indications of a relaxation on the 50-fs time scale have been obtained from recent "hole-burning" experiments [9,10].

The spectrum of stimulated emission should be similar to that of spontaneous fluorescence, weighted to slightly longer wavelengths by the ratio of the Einstein B and A coefficients: $B/A = \lambda^3/8\pi h$. The spectra in Fig. 3B are indeed similar to the spectrum of the spontaneous fluorescence from reaction centers (Fig. 3C), but appear to include an additional negative component on the short-wavelength side. The negative component could represent an absorbance change resulting from a difference between the absorption coefficients of P* and $P^+BPh^-$ [6].

Reaction centers of the thermophilic bacterium *Cf. aurantiacus* differ from those of *Rb. sphaeroides* and *Rp. viridis* in containing 3 BChls and 3 BPhs, instead of 4 BChls and 2 BPhs. The chromophore arrangement in *Cf. aurantiacus* probably is similar to that in *Rp. viridis* (Fig. 1) except that $BChl_M$ is replaced by BPh [11,12]. However, electron transfer from P* to $BPh_L$ appears to be slower in *Cf. aurantiacus* than it is in the other species [12,13]. Shuvalov et al. [13] have presented evidence that a $P^+BChl_L^-$ radical-pair is formed as a kinetically resolvable intermediate in *Cf. aurantiacus*. They suggest that P* transfers an electron to $BChl_L$ with a time constant of about 10 ps, and that $BChl_L^-$ reduces $BPh_L$ with a time constant of about 3 ps. Because this work was done with flashes lasting about 33 ps, it required deconvolution of small absorbance changes seen early during the flash. We have therefore reexamined *Cf. aurantiacus* reaction centers using flashes of about 1 ps. Preliminary results are presented in Fig. 4.

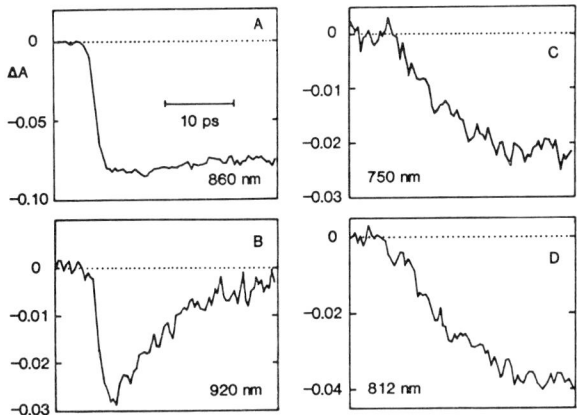

Figure 4    Absorbance changes and stimulated emission in *Cf. aurantiacus* reaction centers.    The sample (A = 0.64 at 812 nm) was prepared as described [14], and examined as in Fig. 3A.   Each data point represents  865 to 1375 flashes.

Figure 4A shows the bleaching of the 865-nm band of P, which reflects the instrumental response time. The development and decay of stimulated emission from P* is shown in Fig. 4B; the bleaching of the 750-nm ground-state band of $BPh_L$, in Fig. 4C. The emission at 920 nm decays with a time constant of approximately 8 ps, and electron transfer to $BPh_L$ occurs with essentially the same kinetics. At 812 nm (Fig. 4D) there is an absorbance decrease that develops with a similar time constant, but there also may be a small bleaching that is coincident with the time of the flash. The major (slower) part of the absorbance decrease at 812 nm probably reflects a blue-shift of the absorption band of $BChl_L$, due to the formation of $P^+$ and $BPh^-$. Although the earlier component could reflect the formation of a small amount of $P^+BChl_L^-$, we believe that it more likely is due to the transient presence of an excited state of $BChl_L$ during the flash [5,7].

377

Acknowledgements: This work was supported by NSF grant PCM-8316161 and USDA grant 84-CRCR-1-1523. We thank C. Kirmaier and D. Holten for helpful discussion.

References

1. W. W. Parson: in Photosynthesis, ed. J. Amesz (Elsevier, Amsterdam, 1986), in press
2. J. Deisenhofer, O. Epp, K. Miki, R. Huber, H. Michel: J. Mol. Biol. 180, 385 (1984)
3. J. Deisenhofer, O. Epp, K. Miki, R. Huber, H. Michel: Nature 318, 618 (1985)
4. D. Holten, C. Hoganson, M. W. Windsor, C. C. Schenck, W. W. Parson, A. Migus, R. L. Fork, C. V. Shank: Biochim. Biophys. Acta 592, 461 (1980)
5. N. W. Woodbury, M. Becker, D. Middendorf, W. W. Parson: Biochem. 24, 7516 (1985)
6. J.-L. Martin, J. Breton, A. J. Hoff, A. Migus, A. Antonetti: Proc. Natl. Acad. Sci. U. S. A. 83, 957 (1986)
7. J. Breton, J.-L. Martin, A. Migus, A. Antonetti, A. Orszag: Proc. Natl. Acad. Sci. U. S. A. in press
8. N. W. T. Woodbury, W. W. Parson: Biochim. Biophys. Acta 767, 345 (1984)
9. S. R. Meech, A. J. Hoff, D. A. Wiersma: Chem. Phys. Lett. 121, 287 (1985)
10. S. G. Boxer, D. J. Lockhart, T. R. Middendorf: Chem. Phys. Lett. 123, 476 (1986)
11. H. Vasmel: Doctoral Thesis, Univ. of Leiden (1986)
12. C. Kirmaier, D. Holten, R. E. Blankenship: Biochim. Biophys. Acta, in press
13. V. A. Shuvalov, H. Vasmel, J. Amesz, L. N. M. Duysens: Biochim. Biophys. Acta, in press
14. C. Kirmaier, D. Holten, R. Feick, R. E. Blankenship: FEBS Lett. 158, 73 (1983)

# Femtosecond Spectroscopy of the Primary Events of Bacterial Photosynthesis

W. Zinth, J. Dobler, and W. Kaiser

Physik Department der Technischen Universität München,
Arcisstraße 21, D-8000 München 2, Fed. Rep. of Germany


Recent investigations of bacterial photosynthesis have shown that the primary steps occur on the time scale of one picosecond /1-3/. These processes can only be studied by means of ultrafast optical techniques. The standard experimental methods yield absorption changes induced by a short excitation pulse measured as a function of time and wavelength. Experiments are now possible with the temporal resolution of better than 100 fs. They present interesting information on the early dynamics of the primary processes. Optical techniques alone provide insufficient understanding of the molecular microscopic processes of the primary steps. It is important to add information from other experiments, e.g. from structure analysis, resonance Raman scattering, molecular dynamics calculations or experiments on model compounds.

In this note we are concerned with time-resolved experiments of the primary steps of photosynthetic units such as bacteriorhodopsin and bacterial reaction centers. We combine the results of ultrafast spectroscopy with other available data in order to gain a better insight into the molecular processes of the primary events.

### Experimental

The experiments are performed using amplified pulses ($t_P$=100 fs, repetition rate 7.5 kHz) from a colliding pulse mode-locked laser operating at 620 nm for the excitation of the samples. Probing pulses are obtained by continuum generation in a jet of ethylene glycol. A narrow fraction of the continuum ($\Delta\lambda \simeq$ 10-15 nm) was selected. The probe pulses monitor the absorbance changes of the sample as a function of time delay between exciting and probing pulses. In order to avoid high exposure the sample was kept in a spinning cuvette. In this way, it was ascertained that each photosynthetic unit absorbs one photon only every second. There was no indication of any photodecomposition of the samples at the excitation densities used in the experiments (less than 10% of the molecules absorb one photon per excitation pulse).

### Bacteriorhodopsin

Bacteriorhodopsin (BR) is a membrane protein contained in the cell membrane of Halobacterium halobium. It acts as a light-driven proton pump building up a proton gradient across the cell membrane upon illumination.

The absorption properties of BR are determined by the only pigment molecule retinal. Retinal is bound in BR via a Schiff base to one lysine of the polypeptide chain. In the light-adapted form of BR, the retinal molecule has the all-trans configuration and the Schiff base is protonated. During the photochemical cycle the absorption properties of BR change substantially. Intermediate states named J,K,L,M,... have been

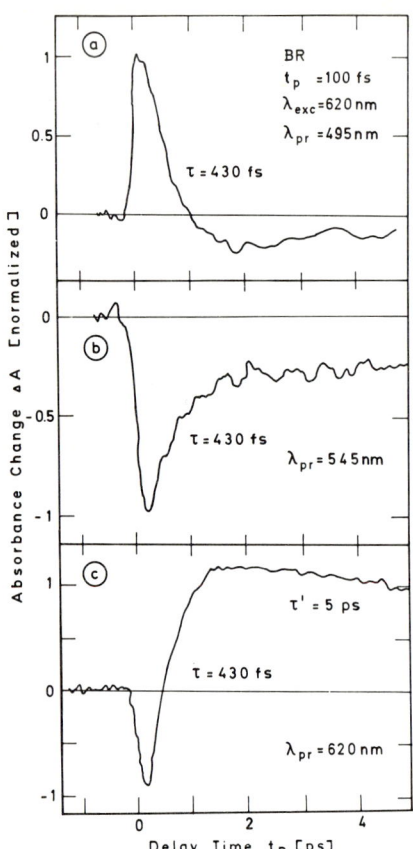

Fig.1  Bacteriorhodopsin: Absorption changes induced by 100 fs excitation pulses at 620 nm

identified. During the course of the cycle the retinal adopts the 13-cis configuration and the Schiff base loses its proton. The photochemical cycle is completed after $\sim$ 10 ms. The first events of the photocycle of BR, where the optical energy is stored in the molecules, have been studied in a number of publications. It was not until recently, that a coherent picture of the primary molecular process was presented /1,4,5/.

Fig.1 shows absorption changes measured as a function of time delay after excitation of the BR molecules at 620 nm. When the molecules are probed at 490 nm, a strong induced absorption is found which builds up with the time resolution of the experiment of 100 fs. In a first relaxation process this absorption change decays with a time constant of 430 fs into a state of reduced absorption from which a slow 5 ps kinetic leads to a partial recovery of the absorption, which stays constant for times longer than 12 ps (not shown in Fig.1). The same time constants (< 100 fs, 430 fs, and 5 ps) are found at the other probing wavelengths of 545 nm and 620 nm in Fig.1b and c, respectively.

The results presented above together with additional information from femtosecond experiments on deuterated BR /1/, from picosecond data on BR containing a sterically fixed retinal molecule /6/, and from resonance Raman scattering /7/ yield the following microscopic picture of the primary molecular processes: Light promotes BR to a Franck-Condon state on the excited-state potential surface. From there a very fast ($\tau < 100$ fs) molecular motion leads to the bottom of the $S_1$ potential surface. Internal conversion with $\tau' = 430$ fs leads to the ground state intermediate J which contains the isomerized 13-cis retinal. Rearrangements of the protein surrounding causes the slower 5 ps kinetics leading to the intermediate K which is stable for the longer time of 300 ps.

Reaction Centers

Photosynthesis in green plants and in most bacterial systems uses chlorophyll or bacteriochlorophyll molecules arranged in so-called reaction centers (RC) where a primary charge separation is initiated. Reaction centers can be isolated from some bacteria. The reaction centers of Rhodopseudomonas viridis or Rhodopseudomonas spheroides contain six pigments absorbing in the visible and near infrared: 4 bacteriochlorophyll (BCl) and 2 bacteriopheophytin (BPh) of type b and a for R-viridis and R-spheroides, respectively. Two of the BCl form the excitonically coupled special pair, from which the charge separation originates. A major progress in the understanding of the molecular processes in the RC was recently achieved when the reaction centers of R-viridis were crystallized /8/ and when a structural analysis of these RC was completed /9/. The arrangement of pigments obtained by structure analysis allowed claculation of the excitonic coupling: it was shown that there is strong excitonic mixing between the various pigments of the RC, substantially influencing the absorption in the near infrared spectral region ($\lambda > 750$ nm) /10/.

It is the purpose of this chapter to compare the dynamic absorption changes in R-viridis - where the pigment arrangement is known - with those of R-spheroides. In addition, the data give information on excitation transfer between the pigments in the RC. Because of the strong excitonic interaction which influences the IR absorption bands we concentrate here on the visible absorption where the $Q_x$ transitions are located. BCl absorbs around 600 nm, BPh around 540 nm and - in the RC of R-viridis - the special pair (P) exhibits a shoulder in the absorption around 620 nm.

Time-resolved absorption measurements are shown in Fig.2 for R-viridis (a,b) and R-spheroides (c,d) after excitation at 620 nm. First we probed around 545 nm at a frequency where the BPh absorbs. The data show at later delay times, i.e. for $t_D > 200$ fs a decay of the absorption with a time constant of 2.8 ps for both types of RC. The absorption decreases at 545 nm, where initially the BPh's absorb strongly suggesting that BPh is reduced to form $BPh^-$. Additional experiments at $\lambda = 675$ nm, a wavelength where $BPh^-$ absorbs, support this interpretation. Figs. 2b and d show the rise of the absorption change on an expanded scale. In both cases the rise follows closely the integrated cross-correlation curve between exciting and probing pulses. We may deduce from these data that the absorption appears within our time resolution of 100 fs for the RC of R-viridis and R-spheroides. The molecular nature of this early state requires some consideration. It may simply be the excited electronic state P* of the special pair or it is a state $P^{+-}$ which contains considerable contribution from a charge transfer state /2/. Both interpretations are possible within the scope of present knowledge.

The experimental findings of the two time constants of < 100 fs and 2.8 ps allow us to draw the following conclusions: After absorption of a photon at 620 nm the excitation is rapidly transferred, $\tau < 100$ fs, to

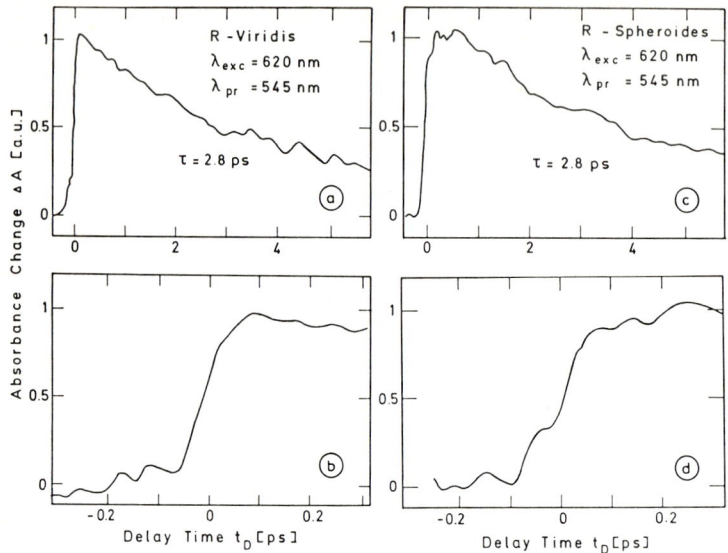

Fig.2 Reactions centers: Transient absorption changes measured at the probing wavelength 545 nm, where initially the BPh absorb. The same time dependence is found for the formation of BPh⁻ in the two reaction centers.

the special pair to form the primary state P* or P⁺⁻. Subsequently, reduction of BPh to BPh⁻ occurs within 2.8 ps. Since both systems, the RC of R-viridis and of R-spheroides, behave in the same manner, we believe that their pigment arrangements and the interactions between the pigments are very similar.

Conclusions

The present investigations of the bacterial photosynthesis of two very different photosynthetic systems - bacteriorhodopsin and bacterial reaction centers -suggest very high reaction rates of the primary events. This common property can be well understood with the help of the reaction scheme shown in Fig.3. After optical excitation, fast molecular rearrangements occur on the excited-state potential surface. From there two pathways are possible: internal conversion to the initial ground

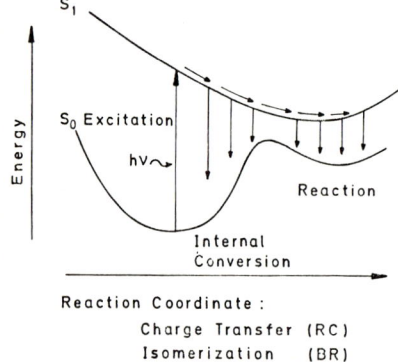

Fig.3 Schematic of the energy surfaces indicating the primary reactions in the investigated photosynthetic systems

state and the desired photochemical reaction. The internal conversion rates for most large pigment systems are very rapid. For this reason, the reactive channel must be faster in order to maintain reasonable quantum efficiencies in the first photosynthetic steps.

Acknowledgement
The authors gratefully acknowledge valuable contributions from H.Michel, D.Oesterhelt, and H. Scheer.


1  M.C. Nuss, W. Zinth, W. Kaiser, E. Kölling, D. Oesterhelt, Chem. Phys. Lett. 117 (1985) 1
2  J.-L. Martin, J. Breton, A. J. Hoff, A. Migus, A. Antonetti, Proc. Nat. Acad. Sci. USA 83 (1986) 957
3  W. Zinth, M.C. Nuss, M.A. Franz, W. Kaiser, H. Michel, in "Antennas and Reaction Centers of Photosynthetic Bacteria", ed. M.E. Michel-Beyerle, Springer, Berlin 1985, p.286
4  H.J. Polland, M.A. Franz, W. Zinth, W. Kaiser, E. Kölling, D. Oesterhelt, Biophys. J. 49 (1986) 651; H.J. Polland, W. Zinth, W. Kaiser, in "Ultrafast Phenomena IV", ed. D.H. Auston, K.B. Eisenthal, Springer Series in Chem. Phys., Springer, Heidelberg 1984, p. 456
5  A.V. Sharkov, A.V. Pakulev, S.V. Chekalin, Y.A. Matveetz, Biochim. Biophys. Acta 808 (1985) 94
6  H.J. Polland, M.A. Franz, W. Zinth, W. Kaiser, E. Kölling, D. Oesterhelt, Biochim. Biophys. Acta 767 (1984) 635
7  M. Braiman, R. Mathies, Proc. Nat. Acad. Sci. USA 79 (1982) 403
8  H. Michel, J. Mol. Biol. 158 (1982) 567
9  J. Deisenhofer, O. Epp, K. Miki, R. Huber, H. Michel, J. Mol. Biol. 180 (1984) 385
10 W. Zinth, E.W. Knapp, S.F. Fischer, W. Kaiser, J. Deisenhofer, H. Michel, Chem. Phys. Lett. 119 (1985) 1

# An Accumulated Photon Echo Study of Sub-picosecond Processes in Photosynthetic Reaction Centers

*S.R. Meech* [1,4], *A.J. Hoff* [2], *and D.A. Wiersma* [1,3]

[1] Picosecond Laser and Spectroscopy Laboratory, Department of Chemistry, University of Groningen, NL-9747 AG Groningen, The Netherlands
[2] Department of Biophysics, Huygens Laboratory of the State University, P.O. Box 9504, NL-2300 RA Leiden, The Netherlands

Abstract

Sub-picosecond photon echo spectroscopy of the near IR transitions of photosynthetic reaction centers reveal an ultrafast (ca. 25 fsec.) decay process. This is discussed in terms of a decay of the excited Frenkel dimer state into a number of molecular eigenstates having intra-molecular and interpigment charge transfer character. Spectroscopic evidence for a coupling between the special pair states and an accessory bacteriochlorophyll is presented.

Introduction

The extremely efficient electron transfer reactions which occur in photosynthetic reaction centers (RC) have long been of interest to ultrafast spectroscopists. These studies, when combined with a knowledge of the RC structure, which was recently determined for *Rhodopseudomonas (Rps.) viridis* [1], can provide a detailed insight into the mechanism of photosynthesis. In [1] it was shown that the pigments are arrayed in two chains with an appropriate twofold rotational symmetry. At the apex are two bacteriochlorophyll (BC) ~ 3Å apart (the special pair, P). 11 Å from these are two so-called accessory BC monomers, and 10 Å from these are two monomeric bacteriopheophytin (BP), which are themselves 17 Å from P.

In a recent accumulated photon echo study of *Rhodobacter (Rb.) sphaeroides* we reported that excitation of the P band with sub-picosecond pulses revealed an unresolvably short (< 200 fsec.) decay time. This was confirmed by stochastic excitation and hole burning measurements, which suggested that the relaxation occurred on a ca. 25 fsec. timescale [2]. We assigned this very rapid process to a charge separation at the dimer. Using 50 fsec. pulses from a CPM laser, at 620 nm, Zinth and co-workers observed a ca. 100 fsec. transient which was assigned to a P→BC electron transfer [3]. Martin et al. [4] used the transient absorption method, with 100 fsec. resolution, to pump *Rb. sphaeroides* at 870 nm and probe between 650 and 1250 nm. They observed that P was oxidised and a single BP was reduced with a rate of $(2.8 \times 10^{-12} s)^{-1}$. No reduction of BC was observed. Further, stimulated emission from the P state was found to decay at the same rate.

In this work we extend our previous observations to other wavelengths and reaction centers and outline a more detailed mechanism for the ultrafast decay.

---

3 Author to whom correspondence should be addressed.
4 Present address: Department of Chemistry, Heriot-Watt University, Riccarton, Edinburgh EH14 4AS, U.K.

## Experimental

The accumulated three pulse stimulated echo is a technique which employs low power pulses for the measurement of optical $T_2$'s on a sub-picosecond timescale. The experiment has been described in great detail elsewhere [5,6]. The essential point is that repetitive excitation by pulse pairs results in a grating being accumulated on the optically pumped transition by means of storage of population in a long-lived bottleneck state, which can be a triplet or a charge separated state. All the measurements reported here were made at < 2K, where pure dephasing processes are frozen out on a picosecond timescale, thus $T_2 = 2T_1$.

To obtain sub-picosecond pulses in the near IR we employed the technique of active/passive mode locking, in which a saturable absorber is mixed with the gain medium of the synchronously pumped dye laser. The mixtures used were R6G/DQOCI, Pyridine 2/HITC, and Styryl 9M/IR140, which gave 0.2, 0.5 and 0.6 psec. pulses respectively [7]. The pulses deviated from transform-limited behaviour by a factor of ca. 1.5. Input powers, tuning elements and output coupling were all critical in obtaining the best pulses. Shorter pumping pulses at high powers and higher output coupling would probably improve the performance. For excitation at 990 nm an IR140 dye laser was tandem pumped by a mode-locked styryl-9 laser to produce ca. 1.5 psec. pulses. For stochastic excitation the dye laser was adjusted to give an output with a broad spectral width of ca. 100 cm$^{-1}$ at 870 nm.

## Results and Discussion

Displayed in figure 1 are the results of some of the accumulated echo experiments. The conditions are given in the legend. In all cases it is clear that the decay is unresolvably short. This result is found for a wide range of experimental conditions [8]. It may be contrasted with measurements on a BC monomer in a 2MTHF glass, which showed an echo decay time of ca. 100 psec. [2]. The rapid decay is observed for both *Rb. sphaeroides* and *Rps. viridis*. For *Rb. sphaeroides* it was established that the decay was independent of excitation wavelength within the P band. The ultrafast decay was confirmed by stochastic echo measurements (Yajima, this meeting) with radiation of 150 fsec correlation time, and by hole burning [2,8]. These measurements suggest a decay time of ca. 25 fsec.; the P transition is essentially homogeneously broadened.

In figure 2 we show the result of an echo excitation study of *Rb. sphaeroides*. Here the echo intensity is plotted as a function of wavelength for constant excitation powers. The sample had a low OD. The rationale for this is as follows. Optical excitation of a transition which carries some of the P oscillator strength will result in accumulation of a grating, and will yield an echo. However, if the laser is resonant with a state which subsequently undergoes radiationless energy transfer to P no grating is accumulated on the pumped transition, so no echo is observed, although the bottleneck is still populated. Thus we can deconvolute from the absorption spectrum components of the special pair. The whereabouts and intensity of the upper part of the P transition ($P^{(+)}$) have been the subject of much debate.

The echo spectrum in figure 2 is particularly interesting because it reveals a double-peaked structure under the 800 nm absorption band of *Rb. sphaeroides*. A similar result is obtained for *Rps. viridis* [8]. This shows
a) that there is significant intensity of P (presumably $P^{(+)}$) at 812 nm (850 nm in *viridis*), and b) that there is a significant and spectroscopically

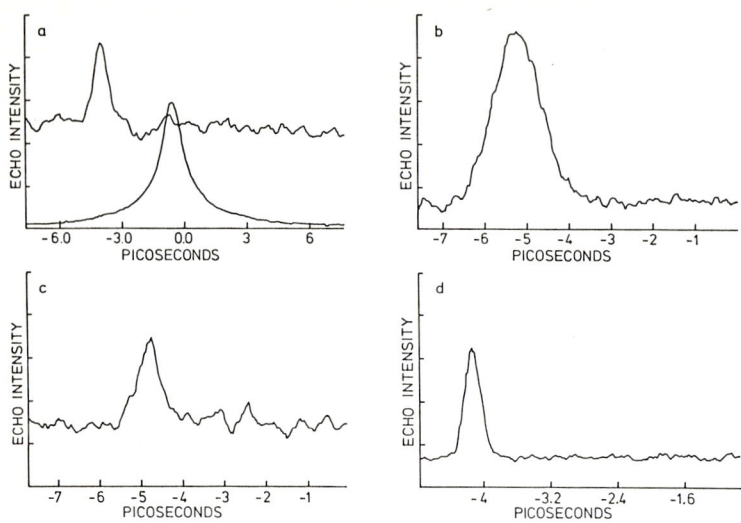

Fig. 1. a) Coherent echo decay (upper) and laser autocorrelation (lower) for *Rb. sphaeroides* at 865 nm. Offset due to nd filter in one beam. b) Coherent echo trace for *Rps. viridis* at 990 nm. c) Coherent echo trace for *Rb. sphaeroides* at 913 nm. d) Stochastic echo trace for *Rb. sphaeroides* at 880 nm. All traces recorded with < 100 µW of c.w. power, spot size 0.01 mm$^2$.

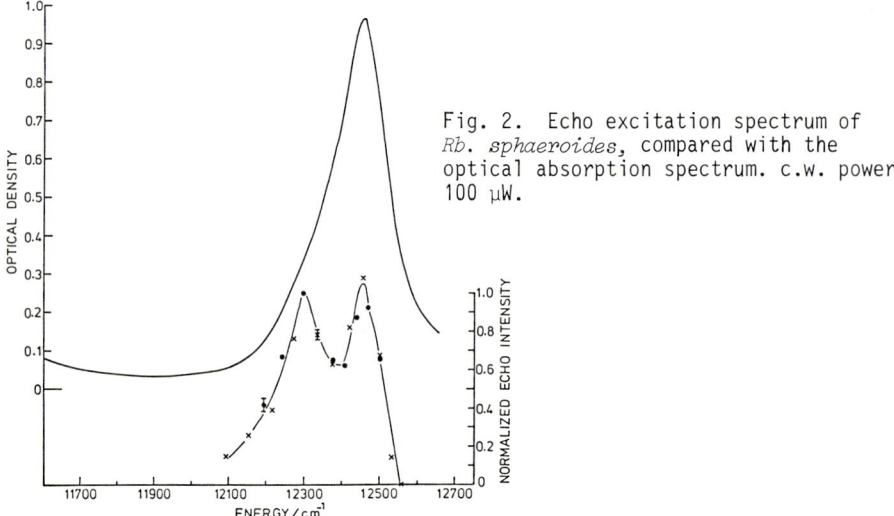

Fig. 2. Echo excitation spectrum of *Rb. sphaeroides*, compared with the optical absorption spectrum. c.w. power 100 µW.

observable interaction between P and at least one of the accessory BC. This represents direct experimental evidence of interpigment coupling within the RC.

Such interpigment coupling can also explain the ultrafast wavelength-independent decay kinetics described above. Specifically we ascribe the ultrashort $T_2$ to a decay of the initially excited Frenkel dimer state into molecular eigenstates arising from mixing of the dimer with an intra-dimer charge transfer state. This implies that P must exist in an asymmetric protein environment. Other CT states (e.g. $P^+$-$BC^-$) are possible and may play a role in the observed coupling between P and the accessory BC. The timescale of the dephasing process is of the order of the inverse linewidth (ca. 25 fsec.). It is not clear that such a process can be associated with an exponential decay constant, which may explain the complex lineshape observed for the P transitions [2, 8, 9].

This mechanism is an appealing one because such an initial charge separation provides a role for the dimer in an RC structure, and from it the one-sided nature of the electron transfer reactions arise quite naturally. Further, as proposed by Warshel [10], the charge separation at the dimer will suppress back electron transfer reactions. The data of figure 2 suggest a role for the accessory BC in the electron transfer reaction. We suggest that the mixed state formed in ca. 25 fsec. contains major contributions from the Frenkel dimer and intradimer charge transfer states, with a minor contribution from a P-BC charge transfer state. It is from this state that the stimulated emission arises [4]. After the formation of the mixed state the electron transfer to the BP can proceed at a more leisurely pace, in the absence of competing back reactions.

Conclusion

Accumulated photon echo and photon echo excitation spectroscopy revealed respectively a very rapid decay process and inter-pigment coupling in photosynthetic RC's. It was proposed that the initially excited dimer state rapidly (25 fsec.) evolves into a state with considerable CT character, primarily from intra-dimer CT, but also from P-BC CT states. This coupled state then decays more slowly into the charge separated $P^+BC$ $BP^-$ state. We note that CT states have high dipole moments, so interactions between them and the protein environment are likely to be important in gaining a detailed understanding of electron transfer dynamics within RC's.

Acknowledgements

We are grateful to L. Nan for the RC preparations. SRM thanks the Royal Society /S.E.R.C. for a fellowship. These investigations were supported by the Netherlands Foundation for Chemical Research (SON) with financial aid from the Z.W.O.

References

1. J. Deisenhofer, O. Epp, K. Miki, R. Huber and H. Michel, J. Mol. Biol., 180 (1984) 385.
2. S.R. Meech, A.J. Hoff and D.A. Wiersma, Chem. Phys. Lett., 121 (1985) 287.
3. W. Zinth, M. Nuss, M. Franz, W. Kaiser in "Springer Series in Chem. Phys." 42 (1985) 286.
4. J.L. Martin, J. Breton, A.J. Hoff, A. Migus and A. Antonetti, Proc. Nat. Acad. Sci., 83 (1986) 957.
5. W.H. Hesselink and D.A. Wiersma, Phys. Rev. Lett., 43 (1979) 1991.
6. W.H. Hesselink and D.A. Wiersma, J. Chem. Phys. 75 (1981) 4192.
7. S.R. Meech, A.J. Hoff and D.A. Wiersma "Proc. UPS 85" (GDR) Eds. E. Klose and B. Wilhelmi, Teubner, Leipzig 1986.
8. S.R. Meech, A.J. Hoff and D.A. Wiersma, Proc. Nat. Acad. Sci., submitted.
9. S.G. Boxer, D. Lockhart and T. Middendorf, Chem. Phys. Lett., 124 (1986) 476.
10. A. Warshel, Proc. Nat. Acad. Sci., 77 (1980) 3105.

# Ultrafast Electron and Energy Transfer in Reaction Center and Antenna Proteins from Photosynthetic Bacteria

*M.R. Wasielewski[1], D.M. Tiede[1], and H.A. Frank[2]*

[1] Chemistry Division, Argonne National Laboratory,
Argonne, IL 60439, USA
[2] Department of Chemistry, University of Connecticut,
Storrs, CT 06268, USA

## 1. Introduction

Reaction Centers from Rps. viridis

The advent of an x-ray structure of reaction center protein crystals from the purple photosynthetic bacterium Rhodopseudomonas viridis has ended a long period of speculation as to the detailed placement of the chromophores within the protein.[1] Spectroscopic studies have shown that excitation of the dimeric bacteriochlorophyll b, BChl b, P960 electron donor results in formation of P960$^+$ and bacteriopheophytin b$^-$, BPheo b$^-$, in < 8 ps followed by electron transfer from BPheo b$^-$ to a menaquinone, $Q_A$, in about 230 ps.[2] The role of the intermediary BChl b molecule, which is positioned between P960 and the BPheo b in the protein, has remained unclear due to an absence of experiments with sub-picosecond time resolution in which only P960 is initially excited.[3] One hypothesis for its role is that the BChl b is reduced by P960 and acts as an intermediate electron carrier prior to the reduction of BPheo b. Here we present transient absorption change data and the corresponding kinetics of electron transfer from P960 to BPheo b in Rps. viridis reaction centers for which the primary donor P960 has been directly excited with 950 nm, sub-picosecond pulses.

Antenna Protein from Rps. acidophila 7750

Carotenoids serve as light-harvesting pigments and as photoprotective agents in photosynthetic organisms.[4] Their role as antenna pigments involves absorption of photons in the blue-green spectral region followed by highly efficient singlet-singlet energy transfer to a neighboring chlorophyll. The dependence of both the rate and mechanism of energy transfer on carotenoid-chlorophyll distance and orientation is unknown. In order to address these questions we have directly measured the rate of singlet energy transfer from the carotenoid to the BChl a in the B800-850 antenna protein from Rps. acidophila 7750 by selective excitation of the carotenoid with 4 ps, 515 nm laser pulses.

## 2. Materials and Methods

Reaction centers from Rps. viridis and the B800-850 antenna protein from Rps. acidophila 7750 were isolated as described previously.[5,6] 1 mM sodium ascorbate was added to the reaction centers to reduce the associated high potential cytochromes $c_{553}$.[7] The protein samples were placed in 1 mm pathlength cells. The absorbances of the reaction centers were 0.8 at 830 nm and that of the antenna protein was 0.8 at 860 nm. A 2 mm diameter spot on the sample cell was illuminated with the pump and probe beams of the transient absorption apparatus. The 514 nm output of a mode-locked Ar$^+$ laser operating at an 82 MHz repetition rate was used to synchronously pump a rhodamine-6G dye laser. Addition of the saturable absorber dye DQOCI [8] to the rhodamine dye solution resulted in 611 nm, 0.4 ps, 0.5 nJ pulses from the dye laser. These pulses were amplified to 1.5 mJ using a 4-stage rhodamine-640 dye amplifier pumped by a frequency-doubled Nd-YAG laser operating at 10 Hz. The resulting 0.45 ps amplified laser pulse was split with a dichroic beam splitter. A 611 nm, 0.45 ps, 0.7 mJ pulse was used to generate a 0.45 ps white light continuum probe pulse. To generate

950 nm pulses a 611 nm, 0.45 ps, 0.8 mJ pulse was focused into a 75 cm long high pressure gas cell containing a 0.5 mm id diameter capillary waveguide and 900 psi of $CH_4$ gas. A long-pass filter was used to isolate the resulting 90 $\mu$J, 160 fs, 950 nm second Stokes Raman line which emerged from the gas cell. To generate 515 nm pulses a 610 nm, 0.45 ps, 0.8 mJ pulse was focused into EtOH to generate a 5 $\mu$J anti-Stokes Raman-shifted pulse at 515 nm. This pulse was amplified to 150 $\mu$J using a 2-stage coumarin-500 dye amplifier pumped by a frequency-tripled Nd-YAG laser. During amplification the 515 nm pulse broadened to 4 ps. The total instrument response function was 0.45 ps for the experiments using 950 nm excitation and 4 ps for those using 515 nm excitation. A 10 $\mu$J 950 nm or a 20 $\mu$J 515 nm pulse was used to excite the reaction center and antenna proteins, respectively. Thus, the samples were excited with at most 1 photon per protein molecule. Pulse lengths were determined by autocorrelation techniques. Typically, 256 laser shots were averaged to obtain the data presented here. Absorbance measurements were made with a double beam spectrometer which employed optical multichannel detection. Time delays between pump and probe pulses were accomplished with an optical delay line. Time constants for kinetic data were determined by iterative reconvolution using the Grinvald-Steinberg method.[9]

## 3. Results

Reaction Centers from Rps. viridis

The ground state optical absorption spectrum of Rps. viridis reaction centers is shown in Fig 1. Excitation of these reaction centers with 950 nm laser pulses results in bleaching of the BChl b dimer band at 960 nm within 0.45 ps. Since depletion of ground state P960 leads to bleaching of this band, the formation time of the bleach is indicative of the 0.45 ps instrument response time.

Figure 2 shows the transient absorption changes that occur in the near infrared and blue-green region of the spectrum 25 ps after excitation. The infrared absorbance changes are dominated by a strong positive absorbance change at 805 nm and a strong bleach at 835 nm. These changes involve both the intermediary BChl b and the BPheo b. On the other hand, the bleach at 543 nm is assigned solely to reduction of the BPheo b by electron transfer from P960.[2]

Figure 3a presents the kinetics for the formation of the bleach at 543 nm following 950 nm excitation. The data is fit smoothly with a single exponential function with $\tau$ = 6.0 ± 0.9 ps. There is no indication of additional kinetic processes occurring within the 0.45 ps time resolution of the measurement. Figures 3b and 3c show respectively

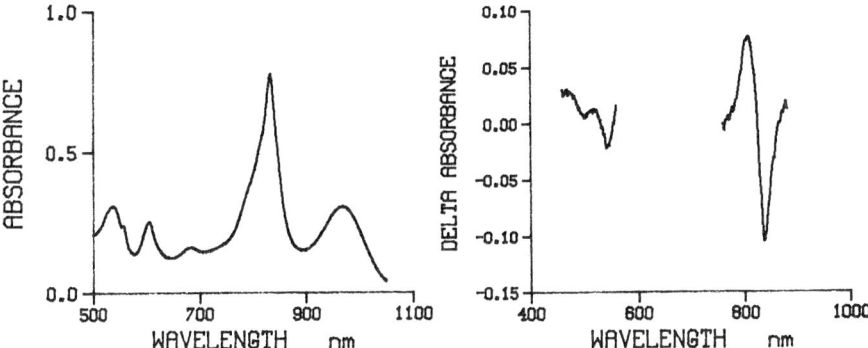

Fig. 1. Ground state optical absorption spectrum of Rps. viridis reaction center

Fig. 2. Transient absorption spectrum of Rps. viridis reaction centers obtained 25 ps after 950 nm, 160 fs laser pulse excitation

the positive absorbance change at 805 nm and the appearance of the bleach at 835 nm following the 950 nm laser flash. These data can also be fit quite well with single exponential functions that yield $\tau = 6.1 \pm 0.9$ ps at 805 nm and $5.8 \pm 0.9$ ps at 835 nm. Thus, the observed absorption changes at 543 nm, 805 nm, and 835 nm exhibit the same kinetics within experimental error.

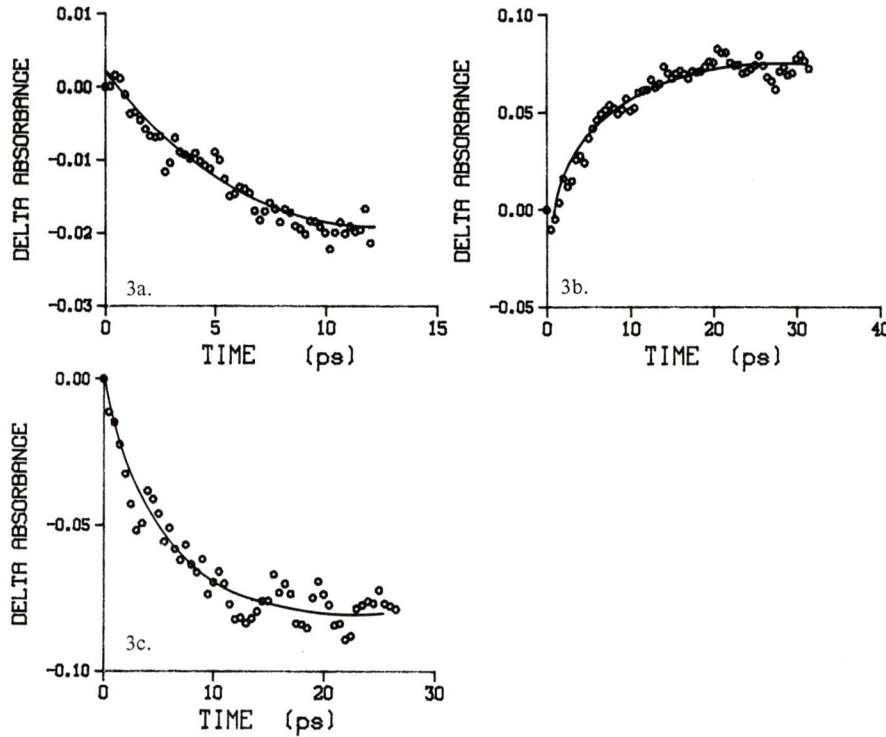

Fig. 3. Transient absorption changes of Rps. viridis reaction centers following 950 nm, 160 fs laser pulse excitation monitored at a) 543 nm, b) 805 nm, and c) 835 nm

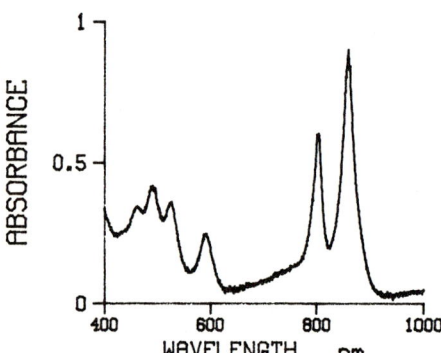

Fig. 4. Ground state optical absorption spectrum of the B800-850 antenna protein from Rps. acidophila 7750

Antenna Protein from Rps. acidophila 7750

Figure 4 shows the ground state absorption spectrum of the B800-850 antenna protein. The carotenoid absorbance was excited at 515 nm and the resultant bleach shown in Figure 5a recovered with a 5.6 ± 0.9 ps time constant. Singlet energy transfer from the carotenoid to BChl a was confirmed by monitoring the bleach of the near-infrared bands. Only the 860 nm band bleached with a 6.1 ± 0.9 ps time constant, Figure 5b. This time constant was the same within experimental error as that for the recovery of the carotenoid bleach. At no time does the 800 nm BChl a band bleach.

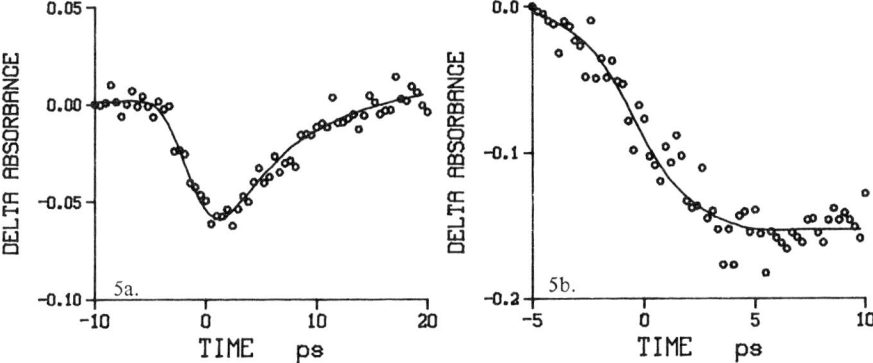

Fig. 5. Transient absorption changes of the B800-850 antenna protein from Rps. acidophila 7750 following 515 nm, 4 ps excitation monitored at a) 480 nm, and b) 860 nm

## 4. Discussion

Reaction Centers of Rps. viridis

The infrared absorption changes observed with 950 nm excitation are all very similar to those observed for Rps. viridis reaction centers previously.[10] The absorption change at 543 nm due to reduction of BPheo b has never been reported on this time scale. However, the position of the bleach agrees well with that reported on a nanosecond time scale.[2]

The synchronous appearance of the bleach at 543 nm and the spectral changes at 805 nm and at 835 nm suggest that the near-infrared spectral changes at 835 nm are primarily due to perturbation of the electronic structure of BChl b by the formation of $P960^+$ BPheo $b^-$. The perturbation of the optical absorption spectrum of the BChl b molecule by the formation of BPheo $b^-$ has also been seen in steady-state spectra obtained for the trapped, reduced acceptor obtained previously.[11] In these spectra a large 835 nm bleach or bandshift is observed adjacent to a much smaller bleach at 800 nm due to formation of BPheo $b^-$.

The role of the intermediary BChl b molecule in Rps. viridis reaction centers may be to lower the overall energetic requirements for electron transfer by strong mixing of its $\pi-\pi^*$ states with those of P960 or by the formation of charge transfer states involving P960. Similar interactions may occur between the BChl b and BPheo b molecules. The exact nature of these interactions remains to be determined. Nevertheless, while our data suggest that the BChl b and BPheo b molecules are strongly coupled electronically, direct excitation of P960 results in reduction of BPheo b without the intermediate formation of BChl $b^-$ as a distinct chemical intermediate with a lifetime longer than 0.45 ps.

Antenna Protein from Rps. acidophila 7750

The lifetimes of the lowest excited singlet states, $S_1$ of carotenoids including $\beta$-carotene are on the order of 10 ps (e.g. 8.4 ps for $\beta$-carotene) [12]. The short lifetimes of these states suggest that the rate of energy transfer from carotenoids to chlorophylls that are required to achieve very efficient energy transfer must be about $10^{11}$ sec$^{-1}$. This requirement further implies that the electronic interaction between the carotenoid donor and the chlorophyll acceptor must be very strong.

Both the rate and efficiency of singlet energy transfer from a carotenoid covalently linked to pyropheophorbide a (PPheo a) in two model compounds have been measured.[13] In one model the $\pi$ systems of the carotenoid and PPheo a possess a maximum edge-to-edge distance of 5 Å, while in the other model this distance is only 2 Å. Energy transfer occurs from the carotenoid to PPheo a at the 2 Å distance with a rate constant of $8.0 \pm 0.2 \times 10^{10}$ sec$^{-1}$ and $53 \pm 5$ % efficiency, while energy transfer at the 5 Å distance occurs at a rate constant of $< 3 \times 10^9$ sec$^{-1}$ and $< 5$ % efficiency.

The efficiency of singlet energy transfer from the carotenoid to the BChl a in the B800-850 antenna protein from Rps. acidophila 7750 is about 50 % as determined from its fluorescence excitation spectrum.[14] Our transient absorption results show that the sum of the rate constants for energy transfer and the intrinsic decay of the carotenoid singlet to ground state is $1.7 \pm 0.5 \times 10^{11}$ sec$^{-1}$. Thus, using a 50% energy transfer efficiency the rate of energy transfer from carotenoid to BChl a in the protein is $8.5 \pm 0.5 \times 10^{10}$ sec$^{-1}$. This rate is comparable to that for the model system with a donor-acceptor distance of only 2 Å. Interestingly, the carotenoid transfers its singlet energy only to the BChl a which absorbs at 860 nm. These results provide evidence that close distances and strong electronic interactions between carotenoids and chlorophylls are necessary to achieve the high energy transfer efficiencies observed in vivo.

5. Acknowledgement

Work at ANL was supported by the Division of Chemical Sciences, Office of Basic Energy Sciences of the Department of Energy. Work at Univ. of Conn. was supported by the NSF under grant no. PCM-8408201.

6. References

1. J. Deisenhofer, O. Epp, K. Miki, R. Huber and H. Michel: J. Mol. Biol. 180, 385-398 (1984)
2. D. Holten, M.W. Windsor, W.W., Parson and J. P. Thornber: Biochim. Biophys. Acta 501, 112-126 (1978)
3. C. Kirmaier, D. Holten and W.W. Parson: FEBS Lett. 185, 76-82 (1985)
4. D. Siefermann-Harms: Biochim. Biophys. Acta 811, 325-255 (1985)
5. H.J. den Blanken, P. Gast and A. J. Hoff: Biochim. Biophys. Acta 68, 365-374 (1982)
6. R. J. Cogdell, I. Durant, J. Valentine, J.G. Lindsay and K. Schmidt: Biochim. Biophys. Acta 722, 427-455 (1983)
7. R.K. Clayton and B.J. Clayton: Biochim. Biophys. Acta 501, 478-487 (1978)
8. G.A. Mourou and T. Sizer: Optics Comm. 41, 47-49 (1982)
9. A. Grinvald: Anal. Biochem. 75, 260-280 (1976)
10. C. Kirmaier, D. Holten and W.W. Parson: Biochim. Biophys. Acta 725, 190-202 (1983)
11. R.C. Prince, D.M. Tiede, J.P. Thornber and P.L. Dutton: Biochim. Biophys. Acta 462, 467-490 (1977)
12. M.R. Wasielewski and L.D. Kispert: Chem. Phys. Lett. (in press) (1986)
13. M.R. Wasielewski, P.A. Liddell, D. Barrett, T.A. Moore and D. Gust: Nature (in press) (1986)
14. A. Angerhofer, R.J. Cogdell and M.F. Hipkins: Biochim. Biophys. Acta 848, 333-341 (1986)

# Femtosecond Spectroscopy of Excitation Energy Transfer and Initial Charge Separation in the Reaction Center of the Photosynthetic Bacterium *Rhodopseudomonas sphaeroides*

*J. Breton[1], J.L. Martin[2], A. Migus[2], A. Antonetti[2], and A. Orszag[2]*

[1] Service de Biophysique, CEN-Saclay, F-91191 Gif-sur-Yvette Cedex, France
[2] Laboratoire d'Optique Appliquée, ENSTA, INSERM,
F-91128 Palaiseau Cedex, France

The initial separation of electric charges, which constitutes the key process of photosynthesis, occurs in a transmembrane chlorophyll-protein complex named the reaction center (RC). RCs from photosynthetic bacteria can be isolated in a functionally intact state and contain three polypeptides, four bacteriochlorophylls, two bacteriopheophytins and at least one quinone ($Q_A$). In the case of the RC from Rps. sphaeroides R-26, the main absorption bands of the pigments are located at 865, 800, 760, 600 and 540 nm (Fig. 1a). The 865-nm band, which bleaches upon (photo)oxidation of the RC, is ascribed to the primary donor (P), a dimer of bacteriochlorophyll. The 800-nm band is assigned to the $Q_y$ transition of the two other "accessory" bacteriochlorophylls (B) while the 600-nm band corresponds to the $Q_x$ transition of all four bacteriochlorophylls. The 760 and 540-nm bands are attributed to the $Q_y$ and $Q_x$ transitions of the two bacteriopheophytins (H), respectively. The X-ray structure of the RC from a related bacterium (Rps. viridis) shows that the two bacteriochlorophylls constituting P, the two B and the two H molecules are organized with $C_2$ symmetry, thus defining two "branches" of pigments extending from P, with only one of them directed towards $Q_A$ (1). We will denote by $B_A$ and $H_A$ the B and H molecules associated with the latter branch.

Picosecond spectroscopy on Rps. sphaeroides RCs has revealed the presence of short-lived intermediates in the electron transfer process (2). Previous studies using 530 or 610-nm excitation have demonstrated the appearance of $P^+$ and $H_A^-$ within 5-10 psec after the excitation pulse. This step is followed by electron transfer to $Q_A$ in about 200 psec. However, the occurrence of earlier electron transfer step(s) and the involvement of $B_A$ in these processes is still strongly debated (3-5). This appears to be due to a combination of

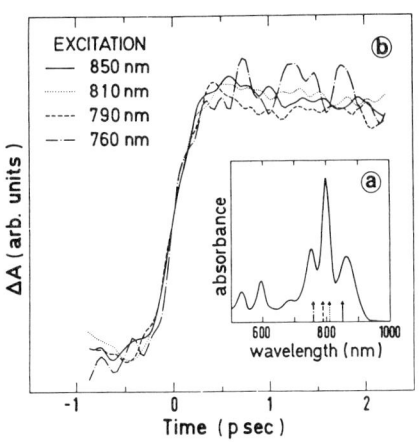

Fig. 1 : a) Absorption spectrum of reaction centers from Rps. sphaeroides. b) Kinetics of the rise of the bleaching observed at 860 nm upon excitation with 150 fsec pulses at the indicated wavelengths demonstrating ultrafast (<100 fsec) energy transfer from H and B to P (see text for details)

unsatisfactory experimental conditions such as ill-suited excitation wavelengths in which the excitation is not directly created on P but rather on H or B, excessive excitation energies which can lead to non-linear processes and pulses of a duration much longer than the phenomena under investigation. In the present study, 80-fsec pulses at 620 nm produced in a passively mode-locked CW dye ring laser were amplified at 10 Hz to generate 150-fsec excitation pulses in the spectral range 760-850 nm using amplification of a continuum in Styryl 9. Under our experimental conditions each laser pulse excited a new region of the sample and photooxidized about 20 % of the RCs.

Upon excitation in H at 760 nm or in B at ∿800 nm, the bleaching of the ground state of P, monitored at 860 nm, occurs in < 100 fsec, i.e. as fast as upon direct excitation of P at 850 nm (Fig. 1b). This ultrafast process, one or two orders of magnitude faster than previously estimated, leads to the formation of an excited state of P (called P*) characterized by (i) an "instantaneous" absorbance increase extending over the whole spectral range investigated (545-1240 nm) and (ii) stimulated emission with a spectrum (860-1000 nm) resembling the fluorescence one (6). This state P* is tentatively assigned to an internal charge transfer state $P^{\pm}$ in the dimer constituting P. Its decay, which can be best monitored by the decay of the stimulated emission at 930 nm (Fig. 2a), is characterized by a 2.8 ± 0.2-psec time constant (6).

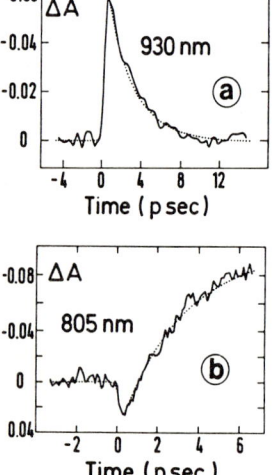

Fig. 2 : Kinetics of absorbance changes of reaction centers upon excitation at 850 nm. a) Decay of the stimulated emission of P* with a 2.8 psec time constant.
b) After an initial absorbance increase at 805 nm the bleaching develops with a 2.8 psec time constant

The appearance of $P^+$, monitored in the radical cation band at 1240 nm, occurs simultaneously with the reduction of $H_A$ monitored at 545 nm (bleaching of $H_A$) and at 675 nm (appearance of $H_A^-$) with the same time constant of 2.8 psec. Upon excitation at 850 nm, observation in the 800-nm region reveals the blue-shift of the 800-nm band, which has been assigned to an electrochromic effect of $P^+$ and $H_A^-$ on the B molecules. After the initial absorbance increase attributed to P*, this band-shift develops with a 2.8-psec time constant (6). The absence of a fast transient bleaching in this spectral range (Fig. 2b) excludes the participation of $B_A$ as a transient electron acceptor operating between the appearance of P* and the reduction of $H_A$. This

394

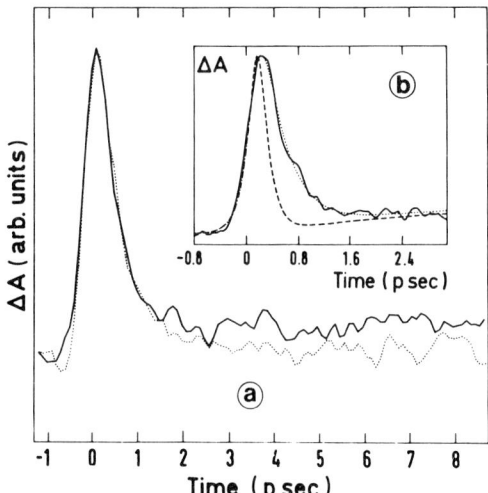

Fig. 3 : Kinetics of absorbance changes at 800 nm following excitation at 807 nm: a)... : reaction centers in the state $P^+H_AQ_A$ prior to the excitation; b) fits assuming a 400 fsec (...) or a 20 fsec relaxation kinetics for the fast transient bleaching

conclusion contrasts the assertion in (3) that $B_A$ is reduced in <1 psec and reoxidized in 7 psec. The presence of $B_A$ in close proximity between P and $H_A$ might serve to facilitate electron tunneling from P* to $H_A$ by lowering the energy barrier between them as suggested in (7).

Upon excitation and observation in the absorption band of B around 800 nm a fast transient bleaching, the recovery of which can be fitted with a 400 ± 100-fsec time constant, is observed (Fig. 3). Such a fast transient bleaching of B can be expected before the excited species B* transfers its excitation energy to P. This interpretation has been previously proposed in (7) to explain a transient bleaching observed in the 800-nm band upon excitation at 610 nm. However, Fig. 1b clearly demonstrates that the energy transfer from B* to P takes place in < 100 fsec. Although the signal to noise ratio in Fig. 1b does not allow us to exclude that about 10 % of P* is generated in approx. 500 fsec, we note that a quantum yield of $P^+$ formation of 0.93 has been reported for Rps. sphaeroides RCs excited at 800 nm, compared to a yield of essentially 1.0 when P is excited directly (8). This decreased yield compared to that observed upon direct excitation of P could correspond to the small loss of B* described in our scheme. The fast transient bleaching would then include the contributions (i) of a small fraction of the B* population relaxing to the B ground state in about 500 fsec and (ii) of most of the B* states transferring to P in ∼50 fsec. Due to the fact that the kinetics are measured with pulses longer than this characteristic time, the maximum amplitude of the ∼50-fsec contribution is attenuated by roughly a factor of 6 while the 500-fsec component is almost not affected. In addition, it seems likely that the transient signal also includes a contribution of stimulated emission from B* (9). An alternative to the competitive channel scheme discussed above would be a situation where a small fraction ( < 10 %) of photooxidized RCs is responsible for the fast transient bleaching. However, this interpretation can be ruled out in view of the observation of this transient relaxing in ∼ 400 fsec even after chemical oxidation of the RCs (state $P^+H_AQ_A$, Fig. 3). This striking observation can be rationalized with our model which primarily involves excited states of B and not of P or H. It thus appears justified to ascribe the transient bleaching around 800 nm previously reported upon excitation of Rps. sphaeroides RCs at 610 nm with 0.7-0.8-psec pulses (2,7) to the same effect as discussed here. The 600-nm region corresponds to the $Q_x$ transitions of the four bacteriochlorophylls. A bleaching of the $Q_y$ transitions of the B molecules should thus be accompanied by a corresponding bleaching of their $Q_x$ transitions. In view of the identical behaviour of Rps. viridis and Rps. sphaeroides RCs upon excitation in their respective $Q_y$ transition of B (9), this effect constitutes in our view the simplest interpretation of the fast transient bleaching at 620 nm reported in (10) for Rps. viridis.

Chemically modified Rps. sphaeroides reaction centers

RCs (in which the LDAO detergent had been exchanged for cholate) were treated with dithionite to reduce $Q_A$. As clearly indicated in Fig. 4 the decay of P* is lengthened from 2.8 ± 0.2 psec to 4.4 ± 0.4 psec. A similar trend has also been reported in (7).

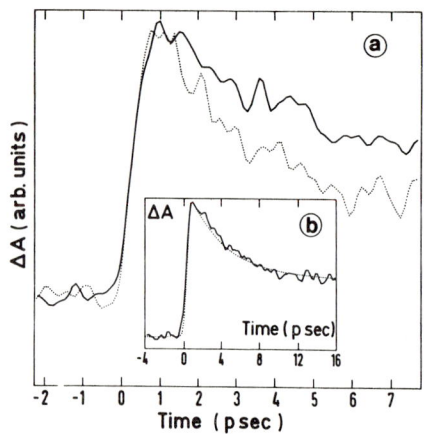

Fig. 4 : Kinetics of absorbance changes at 900 nm following excitation at 807 nm for reaction centers in the state $PH_A Q_A$ (...in a) or $PH_A Q_A^-$ (——— in a and b). In b the fit assumes a 4.4 psec time constant for the relaxation

By treating RCs with borohydride, it is possible to remove the "accessory" B molecule which is not on the $Q_A$ branch. Such modified RCs have recently been reported (11) to exhibit a fast transient bleaching in the 800-nm band upon excitation at 880 nm with pulses of 33 psec duration (but with only partial temporal overlap between the pump and probe pulses). This fast transient bleaching around 800 nm has been taken as a proof of the existence of the $P^+B_A^-$ state (11). We have also investigated the kinetics of such modified RCs (kindly provided by V. Shuvalov) upon excitation at 870 nm (the dye LDS 867 was utilized) with pulses of 150 fsec duration. The absorbances at 870 and 805 nm were 0.32 and 0.56 OD units respectively in a 1 mm cuvette and less than 20 % of the RCs were excited. Under these conditions the kinetics at the three investigated wavelengths of 860, 930 and 805 nm are identical to those observed with unmodified RCs (Fig. 2 and ref. 6). More specifically, no fast transient bleaching could be detected at 805 nm and the kinetics are well fitted with a component of 2.8 psec time constant (to be published). It thus appears that the observations reported in (11) using partial temporal overlap of long duration (33 psec) pulses could not be confirmed upon excitation with ultrafast (150 fsec) pulses at the same wavelength.

**REFERENCES**
1. J. Deisenhofer, O. Epp, K. Miki, R. Huber and H. Michel (1984) J. Mol. Biol. 180, 385-398

2. D. Holten, C. Hoganson, M.W. Windsor, C.C. Schenck, W.W. Parson, A. Migus, R.L. Fork, and C.V. Shank (1980) Biochim. Biophys. Acta 592, 461-477
3. V.A. Shuvalov and V.A. Klevanik (1983) FEBS Lett. 160, 51-55
4. A.Y. Borisov, R.V. Danielus, S.P. Kudzmauskas, A.S. Piskarskas, A.P. Razjivin, V.A. Sirutkaitis, and L.L. Valkunas (1983) Photobiochem. Photobiophys. 6, 33-38
5. C. Kirmaier, D. Holten and W.W Parson (1985) FEBS Lett. 185, 76-82
6. J.-L. Martin, J. Breton, A.J. Hoff, A. Migus, and A. Antonetti, (1986) Proc. Natl. Acad. Sci. USA 83, 957-961
7. N.W. Woodbury, M. Becker, D. Middendorf and W.W. Parson (1985) Biochemistry 24, 7516-7521
8. C.A. Wraight and R.K. Clayton (1973) Biochim. Biophys. Acta 333, 246-260
9. J. Breton, J.-L. Martin, A. Migus A. Antonetti and A. Orszag (1986) Proc. Natl. Acad. Sci. USA, in press
10. W. Zinth, M.C. Nuss, M.A. Franz, W. Kaiser and H. Michel (1985) in Antennas and Reaction Centers of Photosynthetic Bacteria, ed. Michel-Beyerle, M.E. (Springer, Berlin), pp. 286-261
11. V.A. Shuvalov and L.N.M. Duysens (1986) Proc. Natl. Acad. Sci. USA 83, 1690-1694

# Picosecond Transient Absorption Spectroscopy of Green Plant Photosystem I Reaction Centres

B.L. Gore, L.B. Giorgi, and G. Porter

The Royal Institution, 21 Albemarle Street, London W1X 4BS, UK

1) INTRODUCTION

The time-resolved absorption spectra of Photosystem 1 (PS1) reaction centres, isolated from pea chloroplasts, have been measured, at room temperature, following excitation by a 10 ps optical pulse at 600 nm. Energy transfer within PS1 occurs, initially, from excited antenna chlorophyll molecules to P700 (the primary electron donor in the PS1 reaction centre). P700 then undergoes photooxidation and initiates a chain of charge transfer reactions that may be summarised as

$$P700 \longrightarrow A_0 \longrightarrow A_{\overline{1}} \longrightarrow A_{\overline{2}} \longrightarrow P430,$$

where $A_0$, $A_1$ and $A_2$ are believed to a chlorophyll molecule, a quinone species and a specialised iron-sulphur centre, respectively, and P430 is believed to be two iron-sulphur centres. We have studied both the excitation and decay kinetics of the antenna chlorophyll associated with the PS1 reaction centre and the grow in of the P700 signal.

2) EXPERIMENTAL

A conventional Coherent synchronously pumped dye laser system, CR-12 Ar-ion laser pumping a CR-590 folded dye laser, produces 1 nJ, 7 ps pulses at 600 nm and 75 MHz. A Quantel Q-switched Nd:YAG laser is used to amplify these pulses by means of a 4-stage amplifier chain to provide 2 mJ, 10 ps pulses at 10 Hz. Each amplified pulse is split to provide both the excitation pulse, up to 100 µJ, and, by means of continuum generation in a 3 cm cell containing an $H_2O/D_2O$ mixture, the probe pulse. Both pulses are combined on a beam splitter and subsequently pass along the sample and reference arms of a dual beam spectrograph. The pump and probe beams are collinear; they are focused into the sample cell, whilst they pass unfocused through the reference cell. The transmitted beams are then focused into an 0.5 m spectrograph and imaged onto the two halves of a vidicon camera. Spectra are summed and processed in a microcomputer to yield absorption difference spectra. Spectra in this investigation cover the region 625-765 nm, with an overall time resolution of 15 ps. Each spectrum is composed of 250 data points and is the average of 2400 laser shots.

PS1 reaction centres, prepared with a chlorophyll/P700 ratio of 50 [1], were suspended in a medium containing 50mM Tris/HCl, pH 8.0, at a chlorophyll concentration of 25 µM, giving an optical density of 1.4 in a 1 cm cell at the absorption maximum (673 nm). Ferricyanide (1 mM) was added to chemically oxidise, and ascorbate (4 mM) plus 2,6-dichlorophenolindophenol (0.3 mM) was added to chemically reduce, P700 in the dark.

## 3) RESULTS and DISCUSSION

Figure 1 shows the series of seven transient absorption spectra recorded over the spectral region from 625 nm to 765 nm. At each time delay two spectra are shown. The solid line shows the spectrum of the sample with P700 chemically reduced in the dark and the dashed line shows the spectrum of the sample with P700 chemically oxidised in the dark.

Two main spectral features occur, one at 690 nm and one at 700 nm. The one at 690 nm is dominant at early times and decays, in both oxidised and reduced samples, with a life time of approximately 15-20 ps. As can be seen from the series of oxidised spectra this feature undergoes a blue shift as it decays, finally being centred at 675 nm. Previous experiments [2] have shown that the residual bleach at 675 nm decays with a lifetime of between

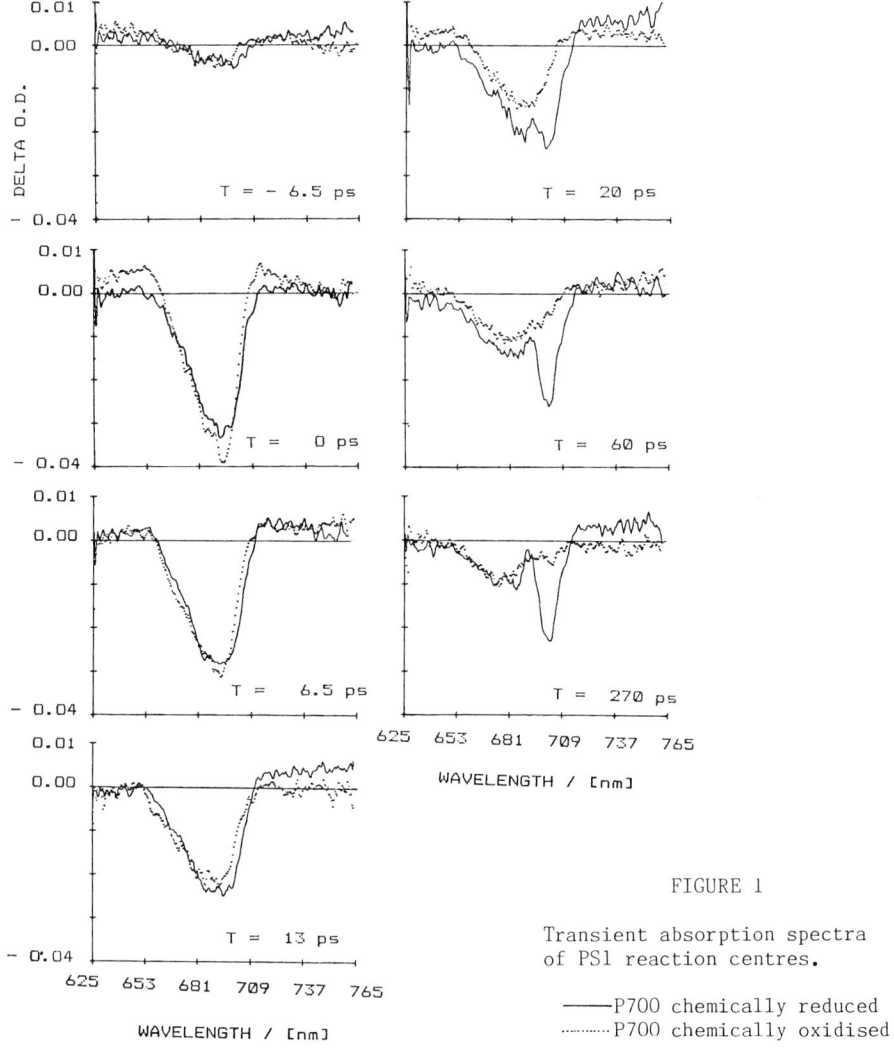

FIGURE 1

Transient absorption spectra of PS1 reaction centres.

———— P700 chemically reduced
·········· P700 chemically oxidised

1 and 2 ns and it is believed that this signal is due to residual triplet chlorophyll molecules which are quenched by carotenoid molecules within the antenna. The decay of the 690 nm signal is, within the resolution of these experiments, independent of the redox state of P700. The second main spectral feature, centred at 700 nm, is very much narrower than the 690 nm signal and occurs only in the spectra of the chemically reduced samples. These two main features may be attributed to the excitation of the antenna chlorophyll to the singlet state and to the photooxidation of P700 molecules, respectively.

By studying the difference between the oxidised and reduced spectra at each time delay it is possible to follow the rise of the P700 signal as the 690 nm signal decays. Despite the shortness of the antenna decay time and the spectral overlap of the 690 nm and P700 signals, there is a clear delay in the appearance of the P700 signal relative to the rise of the 690 nm signal. The rise time of the P700 signal is similar to the decay time of the 690 nm signal, with the P700 signal reaching 80% of its final value within 20 ps.

Throughout these experiments it has been necessary to avoid annihilation processes within individual antenna systems caused by multiple excitation of the antenna chlorophylls. If multiple excitation of a single antenna system does occur this leads to a rapid initial decay (this initial decay is so fast that the bleach follows the time profile of the pulse). In order to obtain some estimate of the degree of initial excitation three approaches have been taken:

1) Reduction of the pump intensity until the initial fast decay component is no longer detected. Experimentally, this fast decay component is not observed if the magnitude of the initial 690 nm signal is less than 0.1 OD.

2) By interpreting the 690 nm signal as a combination of ground state bleach and stimulated emission from the antenna chlorophyll, an estimate of the fraction of chlorophyll molecules excited can be made. However, since the maximum ground state chlorophyll absorption occurs at 673 nm and steady state fluorescence maximum is at 675 nm, we do not feel that the large red shift of the initial 690 nm signal with respect to the ground state absorption can be explained solely as a combination of ground state bleach and stimulated emission from the singlet chlorophyll. Instead, it may be indicative of sub-picosecond energy transfer, within the antenna, to core chlorophyll molecules that absorb and emit significantly redder than the majority of chlorophyll molecules. As a result, we have been unable to use this as a reliable method of estimating the initial degree of antenna excitation.

3) By measuring the magnitude of the P700 signal at late times, e.g. 270 ps, it is possible to determine the fraction of P700 molecules that have been photooxidised. At high pumping intensities, when the 690 nm signal is greater than 0.1 OD, the P700 signal becomes insensitive to the pump power and has a limiting value of -0.04. In the experiment presented in Fig.1 we have ensured that the P700 signal is approximately half of the limiting value and, from a simple analysis of the Poisson distribution of the number of excitations in each antenna system, it can be shown that less than 30% of excited antenna systems receive multiple excitation.

4) CONCLUSION

We have successfully measured the transient absorption spectra of green plant PS1 reaction centres, with a time resolution of 15 ps. By combining a low antenna chlorophyll to reaction centre ratio (50:1) with a low level of pump intensity, we have reduced the occurrence of excitation annihilation within the antennae to a level where it is no longer detectable. Under these experimental conditions we cannot detect any difference in the decay kinetics of samples containing chemically reduced, or chemically oxidised, P700. In the samples containing chemically reduced P700 there is a significant delay between the grow in of the excited antenna chlorophyll signal and the signal due to the excitation of P700, indicative of energy transfer occurring over 15-20 ps. The apparent constancy of the antenna chlorophyll lifetime, with respect to the redox state of the P700, is not easily explained by a simple Förster energy transfer mechanism, but we hope that by improving the time resolution of the current apparatus the causes of this behaviour will be determined by future experiments.

We gratefully acknowledge the financial support of British Petroleum P.L.C and the S.E.R.C.

REFERENCES

1. R.S.Alberte, J.P.Thornber: FEBS Lett. 91, 126 (1978)
2. L.B.Giorgi, T.Doust, B.L.Gore, D.Klug, G.Porter, J.Barber: Biochem. Soc. Trans. 14, 47 (1986)

# Femtosecond-Pulse Spectroscopy of Primary Photoprocesses in Reaction Centers of *Rhodopseudomonas sphaeroides* R-26

S.V. Chekalin, Yu.A. Matveets, and A.P. Yartsev

Institute of Spectroscopy, USSR Academy of Sciences,
SU-142092 Troitsk, Moscow Region, USSR

A bacterial RC consists of four bacteriochlorophyll molecules (two of them being monomers, B, and the other two forming a dimer, P), two bacteriopheophytin molecules, H, one or two molecules of a quinone nature, Q, and also a non-heme Fe atom [1]. In the past decade, the primary processes in RC have been studied with the aid of the picosecond laser spectroscopy techniques, which essentially reduce to the excitation of an RC with a short laser pulse and subsequent probing for absorption changes in them by means of the same pulse but delayed and frequency-converted. In most research works, the observed changes were interpreted as charge transfer from P* to H during 3-8ps and then from $H^-$ to Q during around 200 ps. At the same time, a number of indications were obtained pointing to the fact that there is a fast stage of electron transfer between P* and B that precedes the transfer to H [2,3]. The poor agreement between experimental data obtained by different authors can be due to two causes. Firstly, in most experiments, excitation and probing were effected with pulses from 7 to 40 ps in duration. If this time is long enough for several stages of the process under investigation to occur, the exciting pulse energy will be absorbed not only by the initial state, but also by intermediate ones, and this may change the entire process. Secondly, even where sufficiently short pulses are used, conditions may occur in which the RC's being excited absorb more than one quantum. As demonstrated in the text, this also alters the transfer processes.

The picture of the arrangement of pigments and protein chains in RC's of *Rhodopseudomonas viridis* [1] recently obtained with the aid of X-ray structural analysis shows that the RC pigments are located symmetrically about the P-Fe axis in the form of two "branches" each consisting of B and H molecules. One of these branches terminates with the ubiquinone molecule Q, and it is this branch that the electron transport goes through. The purpose of the other branch is not clear. There is a material difference in the arrangement of the phytol "tails" of the P,B, and H molecules between the operative and inoperative RC branches. The isoprenoid chains of the operative branch are so arranged that they can serve as mediators in the charge transfer between P and H and between H and Q, especially at low temperatures. Also it can be concluded from the X-ray structural picture that the P dimer molecules are fairly closely surrounded on all sides by a protein shield. For this reason, the possibility that the antenna pigment energy can be effectively transferred directly to P seems very dubious. In the protein shield around the P molecule, there are two symmetrical "windows", and the nearest to them are the B molecules and, slightly farther away, the H molecules. It is felt that the B (and possibly H) molecules play the part of intermediate traps to catch the excitation coming to the RC from the antenna before it reaches the dimer P. On the other hand, the same molecules can effect the reverse energy transport from the RC to the antenna when the former is "closed". Based on these assumptions, to establish the role of the various RC molecules, it is necessary to investigate the ways of energy and charge transfer between the RC pigments both in the case of excitation intensities too low to saturate the P dimer ("open" RC's) and in that of saturating intensities ("closed" RC's).

In our experiments, we studied RC specimens of Rhodopseudomonas sphaeroides prepared at the Institute of Soil Sciences and Photosynthesis of the USSR Academy of Sciences (Pushchino). The specimens were excited with a 300-fs laser pulse at 620 nm, both the P dimer and B molecule absorbing at this wavelength. The optical density of the specimens at the excitation wavelength was 0.5. Probing for the induced absorption in the range 700-900 nm was effected with various delays using the same pulse converted into a broadband continuum radiation and delayed relative to the excitation event [4]. Prior to experiment, the optical density $\Delta A$ at 875 nm was plotted against the exciting pulse intensity I at a delay time of 0.5 ps to find the saturation intensity $I_s$. Figures 1 and 2 show absorption variation kinetics and differential spectra plotted for various delay times at $I < 0.3 I_s$. The kinetics near the absorption band maximum of the B molecule were plotted especially thoroughly (Fig.1). The fast bleaching at the beginning of these curves are indicative of the formation of the excited state B* at the moment the exciting pulse arrives. The subsequent decrease in bleaching is due to the exciting energy migrating from B* to P in $(150\pm100)$ fs [5]. The end section of the curves corresponds to the transition $P^* \longrightarrow P^+$ (the 790-nm point is isobestic for this process). It can be seen from the curves of Fig.1 that the energy migration between B* and P ceased by the end of the first picosecond. Therefore, the differential spectrum plotted with a delay time of 1 ps is largely due to P*. As the charge transfer process takes its course, the greatest changes occur in the bands with maxima at 760, 780, and 810 nm (Fig.1). The absorption kinetics measured within these bands are presented in Fig.2. The bleaching kinetics of $\Delta A_{875}$ reflects the process of formation of P*. The appearance of absorption in the initial sections of the bleaching kinetics of $\Delta A_{755}$ and $\Delta A_{785}$ is also due to this process. The sub-

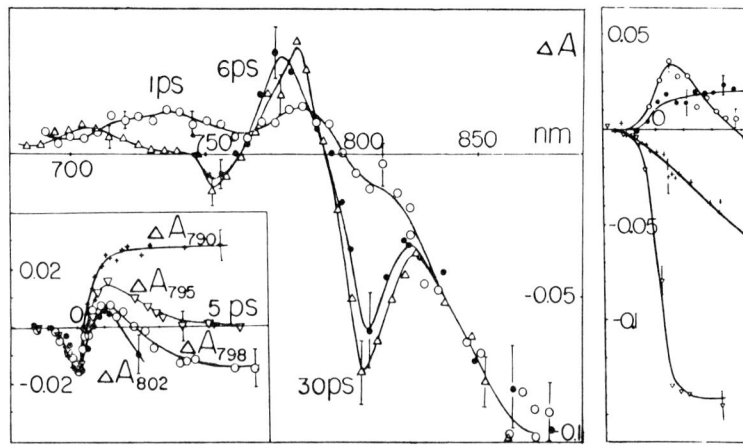
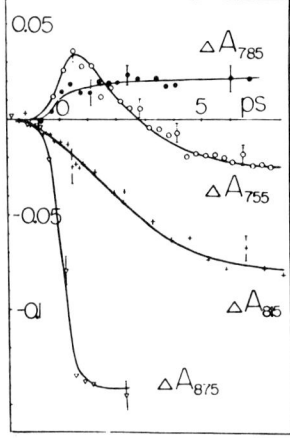
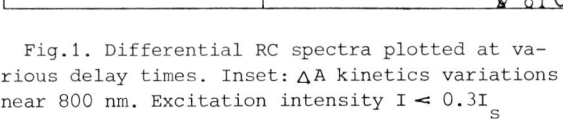

Fig.1. Differential RC spectra plotted at various delay times. Inset: $\Delta A$ kinetics variations near 800 nm. Excitation intensity $I < 0.3 I_s$

Fig.2. $\Delta A$ kinetics variations at various wavelengths at $I < 0.3 I_s$

403

sequent changes of $\Delta A_{755}$ and $\Delta A_{785}$, as well as the kinetics of $\Delta A_{815}$ have one and the same characteristic time of $3.2^{+0.3}_{-0.5}$ ps accurate to within experimental errors. The same time is characteristic of the end sections of the curves in Fig.1 and also of the decay kinetics of P* at 930 nm (not shown in the figure). Hence it can be concluded that variations occurring with a characteristic time of 3.2 ps reflect the process of charge transfer from P*. The bleaching of the 760-nm band is due to the appearance of the H⁻ anion-radical. The bleaching of 810-nm band takes its course at a rate an order of magnitude slower than that of the bleaching of the P band at 875 nm (Fig.2), and we therefore believe that the changes at 810 nm are due to the B monomer and not to the P dimer. The bleaching of the 810-nm band in the course of charge transfer can result from both the blue shift of the B band in the field of P⁺, an absorption peak appearing in this case in the neighborhood of 780 nm, and the formation of the B⁻ anion-radical. The predominance of the bleaching at 810 nm over the absorption at 780 nm (Fig.1) cannot be explained by the shift alone, even if account is taken of the fact that the absorption at 780 nm is partially compensated for by the bleaching of the H bands at 760 nm. For this reason, we explain the additional bleaching at 810 nm by the localization of the electron on B. It thus follows from the experimental data that the charge is transferred from P to both B and H, both these processes being described by a uniexponential function with a characteristic time of $3.2^{+0.5}_{-0.3}$ ps. In the model elaborated in [3], it has been suggested that the electron is transferred from P first to B in a time of ⩽ 1 ps and then to H in 5-7 ps. The kinetics plotted in the vicinity of the maximum of the B band (Fig.1) show no such consecutive electron transfer. This has also been borne out by the recent work reported in [6] wherein the P dimer has been selectively excited at 850 nm. As to the experiments pointing to successive charge transfer, what was observed in them were apparently the processes occurring in overexcited RC's (see below).

The electron transfer process following the excitation of P may be described as the formation in 3.2 ps of the combined state [P⁺H⁻BQ][P⁺HB⁻Q] which then transits into the state [P⁺HBQ⁻] during around 200 ps:

$$[P^*HBQ] \xrightarrow{3.2 \text{ ps}} [P^+H^-BQ][P^+HB^-Q] \xrightarrow{200 \text{ ps}} [P^+HBQ^-]$$

When the charge is transferred to Q, the additional bleaching at 810 nm caused by B⁻ must vanish, and the spectrum in the region 780-810 nm must assume a shape characteristic of a B-band shift. And this is exactly what is observed in the spectra measured at long delay times. The shape of the spectra measured at 76 K [7] can be explained by the B-band shift alone in the case of short delay times (~ 10 ps) as well. This is consistent with the proportion of [P⁺H⁻BQ] in the combined state growing higher with decreasing temperature [3], thus making the charge localization on B practically absent at 76 K.

Figure 3 presents spectra and kinetics obtained with the exciting pulse energy increased 9 times (I ~ 3I$_s$). Their shape differs substantially from that of the curves of Figs.1 and 2. In the $\Delta A_{795}$ kinetics, a rapid bleaching is observed which relaxes in 1.4 ps [5]. The increase of the relaxation time of B* by an order of magnitude compared to that in the case of linear excitation (Fig.1) is apparently due to the appearance of a large number of RC's having their P and B molecules excited concurrently. Also observed is an almost instantaneous bleaching of the H band at 760 nm, which relaxes with the same characteristic time. It is clear from the spectra and kinetics presented that the processes of energy and charge transfer at excitation intensities exceeding the saturating one differ considerably from those in the case of linear excitation. Even with the delay time as long as 30 ps, when all fast processes

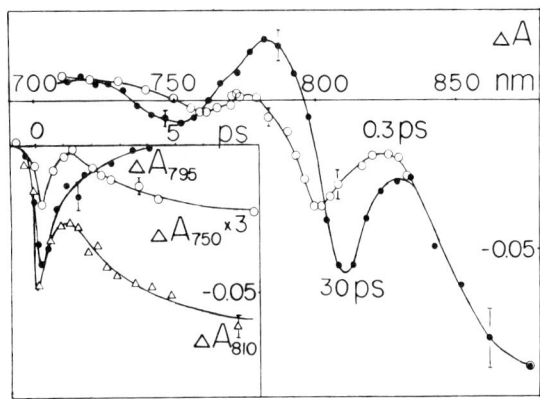

Fig.3. Differential RC spectra and △ A kinetics variations (inset) at $I \sim 3I_s$

have come to an end, the differential spectra show material differences, especially in the vicinity of 780 and 680 nm. Such nonlinear processes are apparently the cause of the considerable differences in the RC spectra and kinetics observed by different authors. Classed with artefacts can be, for example, the observation of the instantaneous bleaching of the B band near 800 nm, followed by relaxation in a time exceeding the pulse duration (Fig.3) which is interpreted as a result of a fast electron transfer from P* to B [2,3].

The authors are grateful to V.A.Shuvalov, A.V.Klevanik, A.Ya.Shkuropatov, and A.V.Sharkov for their letting them have specimens and for many fruitful discussions.

1. J.Deisenhofer, O.Epp, K.Miki, R.Huber, H.Michel, Nature, 318, 618 (1985)
2. V.A.Shuvalov, A.V.Klevanik, A.V.Sharkov, Yu.A.Matveetz, P.G.Kryukov. FEBS Lett. 91, 135 (1978)
3. V.A.Shuvalov, W.W.Parson. PNAS, USA, 78, 957 (1981)
4. Yu.A.Matveetz, S.V.Chekalin, A.V.Sharkov. JOSA B-2, 634 (1985)
5. Yu.A.Matveetz, S.V.Chekalin, A.P.Yartsev. Pis'ma ZhETF 43, 546 (1986)
6. J.L.Martin, J.Breton, A.J.Hoff, A.Migus, A.Antonetti, PNAS, USA, 83, 957 (1986)
7. C.Kirmaier, A.Holten, W.W.Parson. FEBS Lett. 185, 76 (1985).

# Detergent Effects upon the Picosecond Dynamics of Higher Plant Light Harvesting Chlorophyll Complex (LHC)

*J.P. Ide*[1], *D.R. Klug*[1], *W. Kuhlbrandt*[1,2], *G. Porter*[1], *and J. Barber*[1,2]

[1]The Royal Institution, 21 Albemarle Street, London W1X 4BS, UK
[2]Department of Pure and Applied Biology, Imperial College, Prince Consort Road, London SW7 2BB, UK

## 1. Introduction

The main functional role of the LHC *in vivo* is to transfer energy from sunlight to the reaction centres of the thylakoid membrane. The mechanism by which this occurs is thought to be a Forster-like process, each transfer step between chlorophylls being of the order of a picosecond. In this work we investigate the effect of changing the chlorophyll:detergent molar ratio on the resolubilised LHC energy transfer kinetics. Steady-state and time-resolved fluorescence techniques have been used to observe changes in kinetics; the state of the protein has been deduced from circular dichroism spectra which have been interpreted in light of the known geometry of the crystallised LHC (1).

## 2. Experimental

Excited singlet state lifetimes were measured using the technique of time-correlated single photon counting (2). The excitation source consisted of an Argon-ion laser mode-locked at 514.5nm synchronously pumping a tunable Rhodamine 6-G dye laser system (3). This provided pulses of ca.15ps duration (FWHM) in the range 590-630nm. Our time resolution was estimated from the measured instrument response function to be approximately 120ps. Steady-state fluorescence (SSF) spectra were recorded on an MPF4 Perkin-Elmer spectrafluorimeter and circular dichroism (CD) spectra on a Jasco J40CS spectrapolarimeter. The LHC was extracted from pea thylakoid membranes (4) and resolubilised in low ($0.2 \times 10^{-3}:1$), medium ($0.46 \times 10^{-3}:1$) and high (0.025:1) chlorophyll: n-octylglucoside (detergent) molar ratio conditions. The corresponding conditions for Triton X-100 detergent were found to be low ($1.4 \times 10^{-3}:1$), medium ($22.4 \times 10^{-3}:1$) and high (0.24:1). All samples contained 10µg chlorophyll-a/b per ml of detergent.

## 3. Results and Discussion

Figure 1 shows steady-state and time-resolved fluorescence data, together with the circular dichroism spectrum for the medium molar ratio case. The data is qualitatively the same for both detergents under the same chlorophyll:detergent molar ratio conditions.

Under low molar ratio conditions (data not shown) we observe: (1) Two peaks in the SSF at the chlorophyll-b and chlorophyll-a emission maxima; the shape of the SSF is time and excitation wavelength dependent. (2) Complex decay kinetics; main component ca.5.7ns. (3) The soret CD shows initially a chlorophyll-a/b excitonic feature, which disappears after ca.1hr.

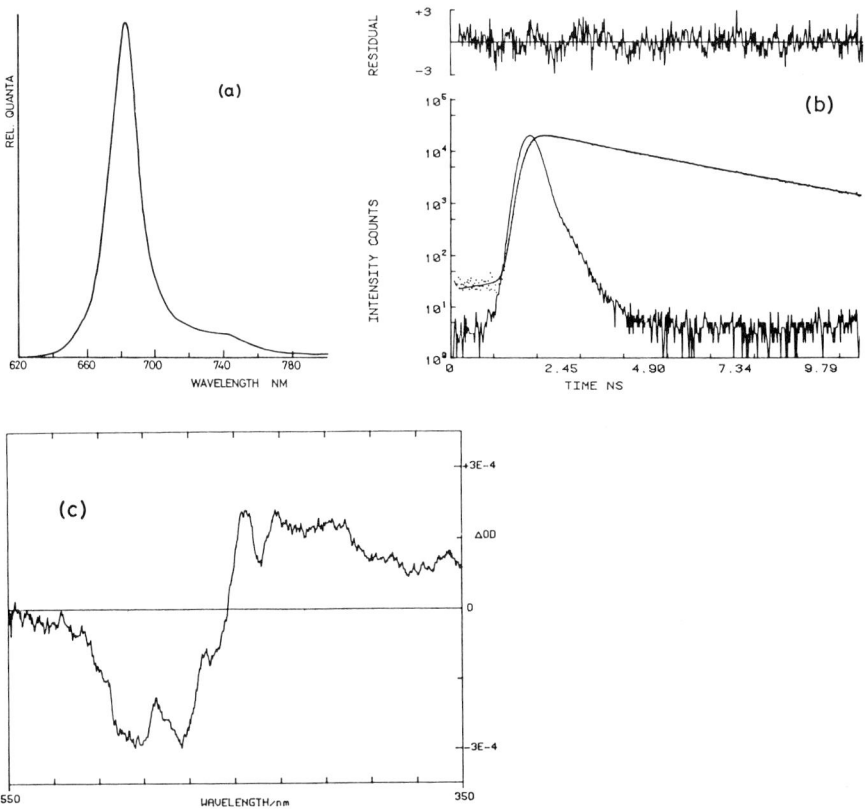

Fig.1.Spectroscopic data for the LHC under medium molar ratio resolubilisation conditions:
(a) steady-state fluorescence emission spectrum; excitation wavelength 470nm.
(b) single photon counting data; reduced chi-square = 1.19, biexponential fit with ($^1$/e) lifetimes of 3.6 and 1.2ns and pre-exponential factors in a ratio of ca. 4:1 respectively; excitation and emission wavelengths 605 and 675nm respectively.
(c) soret circular dichroism spectrum.

Under medium molar ratio conditions (see Fig.1) we observe: (1) Single chlorophyll-a emission peak in the SSF; shape of SSF spectrum is excitation wavelength and time invariant. (2) Biexponential decay kinetics; lifetimes of 3.53ns (±0.04) and 1.12ns (±0.12) and pre-exponential factors in a ratio of ca.4:1 respectively. (3) Soret CD shows a large amplitude chlorophyll-b excitonic feature; a smaller chlorophyll-b feature with reversed polarity with respect to the previous chlorophyll-b excitonic feature and a chlorophyll-a/b exciton; all are conservative. (4) Excitation at the chlorophyll-b Qy absorption maximum leads to greater depolarisation of the chlorophyll-a SSF emission than excitation at the chlorophyll-a Qy absorption maximum.

Under high molar ratio conditions (data not shown) we observe: (1) A single peak in the SSF at chlorophyll-a emission maximum. The integrated area under this peak is considerably less than for the medium molar ratio condition despite identical chlorophyll concentrations. (2) Complex subnanosecond decay data. (3) Identical CD features to the medium molar ratio condition except that the "small" chlorophyll-b feature has acquired considerable amplitude; we refer to this as the quenching exciton. (4) A very close similarity exists between the data obtained under these conditions and that obtained for the suspensions of crystalline LHC.

From the results we deduce the following: (1) Under high molar ratio conditions, extensive aggregation of protein trimers has occurred in a qualitatively similar way to salt-induced crystallisation. The quenching chlorophyll-b CD feature results from a new chlorophyll-b interaction arising through the aggregation process, leading to quenching of both chlorophyll-a fluorescence lifetime(s) and quantum yield. (2) Under medium molar ratio conditions we have predominantly non-aggregated trimers, which are associated with the long decay component (3.53ns). The short decay component (1.12ns) and the small amplitude chlorophyll-b exciton feature are associated with the presence *in vitro* of a small amount of aggregated trimers. Forster energy-transfer appears efficient and the fluorescence depolarisation/CD implies a $C_3$ symmetric chlorophyll-b interaction together with a dimeric chlorophyll-b /a interaction. It seems probable that the $C_3$ symmetry of the protein trimers is a prerequisite for the $C_3$ symmetric chlorophyll-b interaction/ excitonic feature. (3) Under low molar ratio conditions the $C_3$ symmetric chlorophyll-b exciton feature is lost, suggesting that the $C_3$ symmetry of the protein trimers has also been lost. The presence of a chlorophyll-a/b exciton feature suggests that the protein monomers are initially intact, however at longer times the protein appears to denature resulting in the detachment of chromophores from their binding sites on the protein. The decay data supports this assertion on the basis that free chlorophyll-a in detergent micelle has a lifetime of ca.5.7ns (5).

In summary, the high molar ratio condition leads to aggregation of trimer functional units, i.e. crystallisation, the medium molar ratio leads to predominantly trimer functional units with some aggregate present, while the low molar ratio condition leads to loss of trimer geometry and eventual denaturing of the protein monomers.

4. Acknowledgements

J.P. Ide wishes to thank the U.S. Army and D.R. Klug the SERC for financial support during this research.

5. References

1. W. Kuhlbrandt: Nature 307, 478-480 (1984)
2. A.E.W. Knight and B.K. Selinger: Australian J. Chem. 26, 1 (1973)
3. D.V. O'Connor and D. Phillips in 'Time-Correlated Single-Photon Counting', Academic Press, London (1984)
4. W. Kuhlbrandt, T. Thaler and E. Wehrli: J. Cell Biol. 96, 1414-1424 (1983)
5. J.P. Ide et al: Biochem. Soc. Trans. 14, 34 (1986)

# Picosecond Conformational Intermediates in the Bacteriorhodopsin Photocycle

G.H. Atkinson, T.L. Brack, D. Blanchard, G. Rumbles, and L. Siemankowski

Department of Chemistry and Optical Sciences Center,
University of Arizona, Tucson, AZ 85721, USA

The biochemical activity of the bacterium Halobacterium halobrium is initiated through light-driven chemical changes occurring in its purple membrane [1]. The purple membrane is comprised of the protein bacteriorhodopsin (BR) which contains a single retinal chromophore. The structural and conformational changes in this protein-bound retinal caused by light absorption are thought to initiate the proton and ion transport across the membrane that is essential for biochemical activity. The value of BR as a model for the molecular dynamics associated with the retinal chromophore in visual pigments and for proton and ion transport across membranes is well established [1-2]. Although many aspects of both the dynamics in the BR photocycle and the structure of BR intermediates have been reported [2], significant parts of the molecular mechanism underlying its biochemical function remain either unknown or only partially characterized. The initial molecular changes occurring during the first few picoseconds after excitation are of particular interest since they involve the molecular mechanisms by which chemical energy is stored to drive the biochemical activity.

The vibrational degrees of freedom in a complex molecular system such as BR are especially sensitive to changes in structure and conformations. As a consequence, vibrational Raman scattering has been used as the basis for spectroscopic techniques designed to monitor such reactions. Resonantly enhanced Raman spectroscopy provides the additional advantages of increased sensitivity and chromophoric selectivity. By tuning the wavelength of the excitation laser into resonance with the vibronic absorption spectrum of the retinal chromophore, one obtains both several orders of magnitude of increased detection sensitivity and selective vibrational Raman scattering only from the retinal chromophore. A variety of experimental methods have been successfully used to record the resonance Raman (RR) spectra of BR intermediates [3-5]. The results presented here have been obtained with a two laser, pump-probe configuration using an actively mode-locked cw Nd:YAG laser to synchronously pump two cavity dumped dye lasers [6]. This experimental approach provides the opportunity to independently control laser wavelengths, intensities, pulse duration, and focusing parameters. Such experimental versatility is fundamental to optimizing the initiation of the BR photocycle and the recording of picosecond time-resolved resonance Raman ($PTR^3$) spectra of the intermediates. Earlier results using single lasers to record $PTR^3$ spectra confirm this point [7].

The $PTR^3$ spectra of BR intermediates formed within 40 ps of excitation are of principal interest here. The time resolution of the measurements is 5 ps (autocorrelation pulsewidths of $\sim$7 ps). With 570 nm excitation (5ps) and RR probing between 590 nm and 610 nm (5-8 ps), several different molecular transformations in retinal are observed to occur before 40 ps. Two of the largest changes observed are treated in this paper. Specifical-

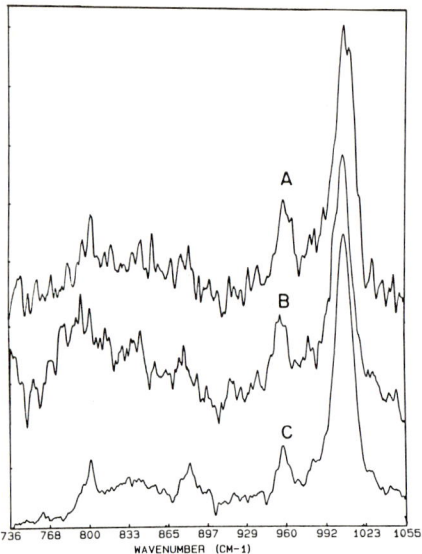
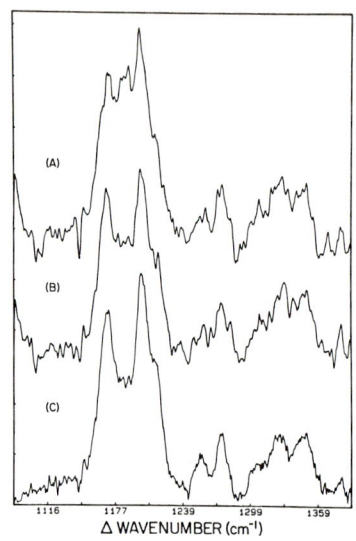

Fig. 1 PTR$^3$ spectra of the BR HOOP region: (A) 40 ps delay, (B) 0 ps delay, (C) probe laser alone. The increased intensity at 959 and 978 cm$^{-1}$ is the same in spectra A and B, but is different than that found in spectrum C

Fig. 2 PTR$^3$ spectra of the BR fingerprint region: (A) 40 ps delay, (B) 0 ps delay, (C) probe laser alone. The growth of the 1194 cm$^{-1}$, 13-cis marker band is apparent

ly, it is observed that the intensities of RR bands in both the hydrogen-out-of-plane (HOOP: 800-1100 cm$^{-1}$) and C-C stretching (fingerprint: 1130-1240 cm$^{-1}$) regions increase during the initial 40 ps interval. Data from the HOOP and fingerprint regions are presented in Figs. 1 and 2, respectively. RR spectra in these spectral regions are shown for the probe laser only (C) and for PTR$^3$ experiments with 0 ps (B) and 40 ps (A) time delays. Since the 0 ps experiments use a time coincidence between the pump and probe laser pulses, the 5 ps duration of each laser pulse determines the time resolution of these data.

Attention is focused here on the time-dependent intensity changes in the 959 cm$^{-1}$ band and the spectral region near 978 cm$^{-1}$ (Fig. 1) and in the 1194 cm$^{-1}$ band (Fig. 2). For the HOOP region, these bands change intensity significantly between the probe-only spectrum and 0 ps delay. No additional intensity change occurs during the initial 40 ps. By contrast, the 1194 cm$^{-1}$ band exhibits a small, (but distinct) intensity change when the probe-only and 0 ps spectra are compared. The intensity of the 1194 cm$^{-1}$ band continues to increase throughout the initial 40 ps period (Fig. 2). These intensity changes can be quantitatively displayed by integrating band profiles as a function of delay time. These data are presented in Fig. 3 for the HOOP and fingerprint regions.

Previous RR studies involving isotopic substitution of $^{13}$C at different sites in retinal [8] provide the basis for assigning bands to specific molecular motions within the chromophore. These assignments derive from a normal coordinate analysis. The changes in Raman band intensities appear

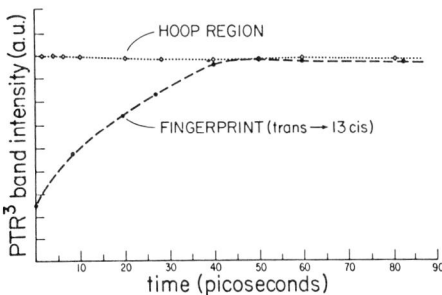

Fig. 3 Resonance Raman band intensity is plotted as a function of the delay between pump and probe lasers for BR HOOP bands at 959 and 978 cm$^{-1}$ and the 1194 cm$^{-1}$, 13-cis marker band.

as continuous functions of reaction time and can be assigned to twisting of the hydrocarbon chain and isomerization at the $C_{13}$-$C_{14}$ position [9-11]. It is clear from Fig. 3 that the isomerization marker band (1194 cm$^{-1}$) shows a time dependence notably different from that of the HOOP bands. The 1194 cm$^{-1}$ band intensity increases smoothly as the delay between the pump and probe lasers is increased from zero to 40 picoseconds while the increased intensity observed in the HOOP region appears immediately at zero delay (i.e., within 5 ps) and remains constant for more than 40 picoseconds.

A molecular model consistent with these PTR$^3$ data involves at least two types of conformational changes in the retinal chromophore. The excited vibronic levels of retinal populated by optical excitation decay to the ground-state potential surface into an all-trans conformation which has twisted along the polyene chain. Such twisting gives rise to increased RR band intensity in the HOOP region in less than 5 ps. The twisted polyene chain appears to remain unchanged for more than 40 ps. The isomerization of retinal around the $C_{13}$-$C_{14}$ bond occurs more slowly and is not completed until 40 ps. Although the precise mechanism remains unknown, it appears that the twisting of the polyene chain facilitates the formation of 13-cis retinal.

Independent of a specific molecular mechanism, these PTR$^3$ data resolve two distinct types of molecular motion within the retinal that occur during the 40 ps following optical excitation. Furthermore, these conformational changes occur at room temperature and under chemical conditions that are consistent with proton and ion pumping in BR.

The extension of RR spectroscopy to the detection of BR intermediates formed on the picosecond time scale makes it feasible to monitor changes in retinal structure and conformation directly. Comparisons with transient absorption and fluorescence data recorded on the same time scale should permit a more detailed and complete view of the dynamics underlying BR chemistry to be formulated.

References

1. D. Oesterhelt and W. Stoeckenius, Proc. Natl. Acad. Sci. USA, 70, 289 (1973)
2. W. Stoeckenius and R. A. Bogomolni, Ann. Rev. Biochem., 51, 587 (1983)
3. G. H. Atkinson, Advances in Infrared and Raman Spectroscopy, ed. by

R.E. Hester and R.J.H. Clark, 9, 1 (North-Holland Publ. London, 1981)
4. G. H. Atkinson, Advances in Laser Spectroscopy, ed. by B.A. Garetz and J.R. Lombardi, 1, 8 (Heyden and Sons, Inc., 1982)
5. G. H. Atkinson, in Chemical Dynamics: NATO Advanced Study Institute, ed. by P. Rentzepis and C. Capellos, (Academic Press, New York, N.Y., in press)
6. G. Rumbles, T. Brack, D. Blanchard, and G. H. Atkinson, Opt. Com., (submitted)
7. I. Grieger and G. H. Atkinson, Biochemistry 24,5660 (1985)
8. M. Braiman and R. Mathies, Proc. Natl. Acad. Sci. USA, 79, 403 (1982)
9. G. H. Atkinson, I. Grieger, and G. Rumbles, Time-Resolved Vibrational Spectroscopy, ed. by M. Stockburger and A. Laubereau, 255 (Springer-Verlag, 1985)
10. G. H. Atkinson in Proceedings of the SPIE Conference: Laser Applications in Chemistry and Biophysics, ed. by M. El-Sayed, 620, 82 (1986).
11. G. H. Atkinson, T. L. Brack, D. Blanchard, I. Grieger, G. Rumbles, and L. Siemankowski, in Time-Resolved Vibrational Spectroscopy, ed. by G. H. Atkinson (Gordon and Breach, New York, N.Y., in press).

# Electron Transfer and Rapid Restricted Motion in Homologous Azurins

*J.W. Petrich[1], J.W. Longworth[2], and G.R. Fleming[3]*

[1] Laboratoire d'Optique Appliquée, ENSTA Ecole Polytechnique,
INSERM U275, F-91128 Palaiseau, France
[2] Department of Physics, Illinois Institute of Technology,
Chicago, IL 60637, USA
[3] Department of Chemistry, The University of Chicago,
Chicago, IL 60637, USA

Electron Transfer from Excited-State Tryptophan to Cu(II)

Using time-correlated single photon counting [1], we have studied the electron transfer rates, $k_{ET}$, and the fluorescence anisotropy decays in a series of homologous blue-copper proteins, azurins, obtained from Pseudomonas aeruginosa (Pae), Alcaligenes faecalis (Afe), and Alcaligenes denitrificans (Ade). The salient difference among these proteins lies in the position and number of their tryptophyl residues. See Figure 1. W48 is buried in the hydrophobic core of the protein; while W118 lies on the protein surface, exposed to the solvent [2].

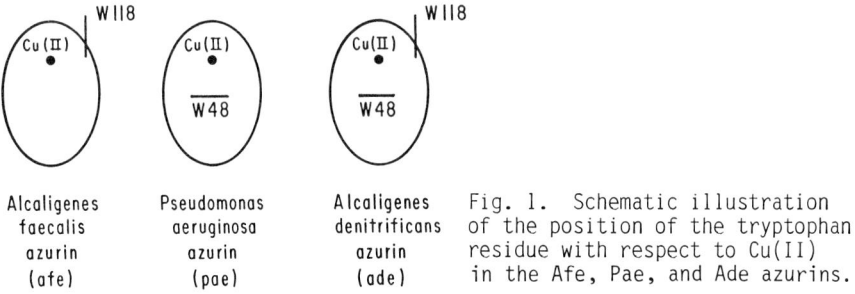

Fig. 1. Schematic illustration of the position of the tryptophan residue with respect to Cu(II) in the Afe, Pae, and Ade azurins.

Figure 2 illustrates the dramatic effect of copper on the fluorescence decay of the azurins. We have attributed the fluorescence quenching by copper to electron transfer from excited-state tryptophan to Cu(II) [3,4].

From our lifetime data, $k_{ET}$ for the reaction $W^* + AzCU(II) \rightarrow W^{*+\bullet} + AzCu(I)$ was determined to be $1 \times 10^{10}$ s$^{-1}$ and $0.5 \times 10^{10}$ s$^{-1}$ for Pae and Afe, respectively; i.e., $k_{ET}(W48)/k_{ET}(W118) = 2$. Using reduction potentials for the donor (D) and the acceptor (A) [4], the appropriate excited-state energies [5], and x-ray crystallographic data [2], the ET reactions were analyzed in terms of Marcus theory [5,6]. If we assume that the D-A distance is the same in Pae and Afe, Marcus theory yields $k_{ET}(W48)/k_{ET}(W118) = 2$, which agrees with experiment. A complete evaluation of $k_{ET}$, however, requires an accurate measurement of the D-A separation, since $k_{ET}$ decreases exponentially with D-A separation [5]. Because the ET most likely occurs not between two point sources but between the delocalized π clouds of tryp-

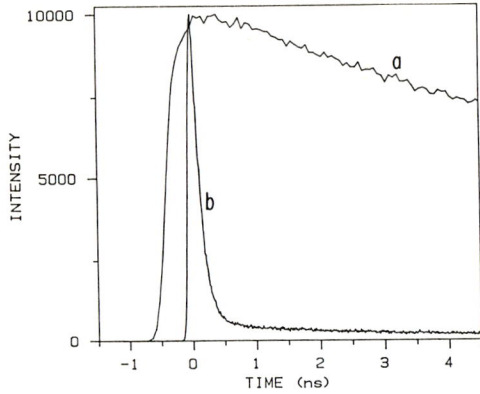

Fig. 2. Fluorescence decays of apo (a) and Holo (b) azurin Pae, pH 5.0 50 mM acetate, 20°C. (a) $\tau_f$ = 5.16 ns  (b) $\tau_f$ = 102 ps. The apparent noisiness of the apo azurin decay reflects the fact that only a few of the channels composing it are displayed. The instrument function for both decays is ~100 ps FWHM. Zero time has been defined arbitrarily.

tophan and the copper ligands, an accurate determination of the D-A separation is difficult. If we measure the D-A separation as the closest distance between aromatic atoms of tryptophan and of a histidyl ligand of copper (9.6 and 7.3 Å [2]), we obtain $k_{ET}(W48)/k_{ET}(W118) = 0.1$. These calculations, however, have not taken into account the influence of the intervening protein matrix and D-A orientation on the reaction rate. For example, the calculations of Ohta et al. [7] have shown that rate enhancements of approximately $10^4$ can be effected by connecting D and A with σ-bonds and that large differences in $k_{ET}$ can be found for different D-A orientations where there is no change in D-A separation.

## Fluorescence Anisotropy Decay of Buried and Exposed Residues

Within our experimental resolution, we find that the W118 has a relatively fast 160 ps component in its fluorescence anisotropy decay in addi-

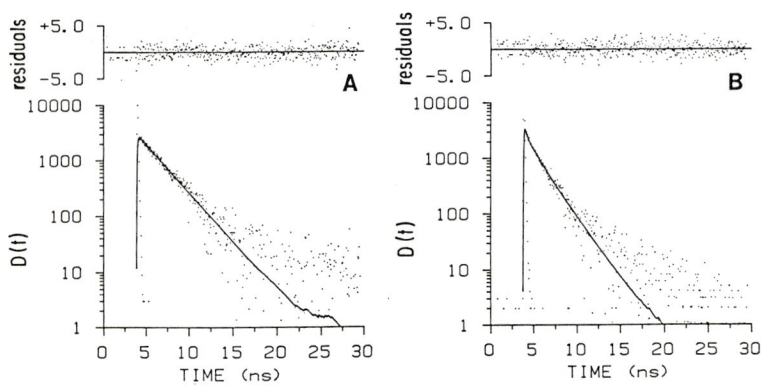

Fig. 3. (a) D(t) for apo Pae azurin. $D(t) = r(t) K(t)$. The fit of D(t) corresponds to an anisotropy decay of 4.81 ns with an r(0) = 0.27. $\chi^2$ = 0.995 and $Z_{runs}$ = -0.064.  (b) D(t) for apo Afe azurin. The fit of D(t) corresponds to an anisotropy decay with $r_1(0) = 0.10, \tau_1 = 0.15$ ns and $r_2(0) = 0.22, \tau_2 = 7.22$ ns. $\chi^2 = 1.075$ and $Z_{runs} = 1.19$. See [1] and [4] and references therein for a discussion of the fitting procedure and definitions of the statistical parameters.

tion to an overall reorientation time of several nanoseconds. The W48 only displays the overall reorientation time. This latter result is in contrast to that of Munro et al. [8] who found a fast component in the anisotropy decay of W48. The extent of the angular restriction of the Wl18 rapid motion can be described by the order parameter, $S^2$ [9]. $S^2$ can be related to a hypothetical cone semi-angle, $\theta_0$, that the transition dipole of the Wl18 can diffuse within. From our data for azurins Afe and Ade, we find that $\theta_0 = 38°$ and $26°$ respectively.

Molecular dynamics simulations using the program CHARMM have been carried out. Stochastic boundary conditions were used which enabled the inclusion of 223 water molecules in the simulations. The results were qualitatively in accord with the experiments and will be described in detail elsewhere [10].

References

1. M.C. Chang, S.J. Courtney, A.J. Cross, R.J. Gulotty, J.W. Petrich, and G.R. Fleming, Anal. Instrum. 14, 433 (1985).
2. G.E. Norris, B.F. Anderson, and E.N. Baker, J. Mol. Biol. 165, 501 (1983).
3. J.W. Petrich, M.C. Chang, D.B. McDonald, and G.R. Fleming, J. Am. Chem. Soc. 105, 3824 (1983).
4. J.W. Petrich, J.W. Longworth, and G.R. Fleming, Biochemistry. Submitted.
5. J.R. Miller, J.A. Peeples, M.J. Schmitt, and G.L. Closs, J. Am. Chem. Soc. 104, 6488 (1982).
6. R.A. Marcus and N. Sutin, Biochim. Biophys. Acta 811, 265 (1985).
7. K. Ohta, G.L. Closs, K. Morokuma, and N.J. Green, J. Am. Chem. Soc. 108, 1319 (1986).
8. I. Munro, I. Pecht, and L. Stryer, Proc. Natl. Acad. Sci. USA 76, 56 (1979).
9. G. Lipari and A. Szabo, J. Am. Chem. Soc. 104, 4559 (1982).
10. L. X-Q. Chen, R.A. Engh, G.R. Fleming, A. Brunger, M. Karplus, J.W. Petrich and J-L. Martin, in preparation.

# Primary Process of Vision: Hypsorhodopsin

*T, Kobayashi*[2], *H. Ohtani* [1;2], *and M. Tsuda*[3]

[2]Department of Physics, Faculty of Science, University of Tokyo, Tokyo 113, Japan

[3]Department of Physics, Sapporo Medical College, Sapporo 060, Japan

Photobleaching process of rhodopsin is initiated by the 11-cis→all-trans photoisomerization of retinal chromophore and is followed by sequential thermal reactions [1]. KOBAYASHI [2] found a bathorhodopsin-like red-shifted species (X) formed within 6 ps in the photolysis of bovine rhodopsin at room temperature. The species was considered by Kobayashi to be the lowest excited singlet state in rhodopsin cies, but the 15-ps lifetime of X is much longer than that expected from the upper limit of the fluorescence quantum yield of rhodopsin. At the present stage, we consider it to be the precursor of bathorhodopsin (B) which was conjectured and called Batho' by HONIG et al. [3]. Four years after the finding by Kobayashi, the red-shifted intermediate was also found in the photolysis of bovine, squid, and octopus rhodopsins [4] and called photorhodopsin. However, Matuoka et al. [5] claimed that hypsorhodopsin (H) was formed solely in multiphoton process. In this work, we utilized a 461-nm picosecond pulse which excited octopus rhodopsin 2.7 times more efficiently than X, the absorption spectrum of which is similar to that of B. Steady-state spectroscopy at low temperatures was also studied. Present results show that H is formed via a thermal process from X and that the formation from the excited state of X also takes place at higher excitation densities.

The block diagram of the apparatus is shown in Fig. 1. The first anti-Stokes Raman scattering (461 nm) was generated by focusing the 532-nm light into acetone. The 461-nm pulse was amplified by coumarine 440 in methanol pumped by the 355-nm pulse.

Sample suspension (Mizudako, <u>Paroctopus defleini</u>) was prepared by a method described previously [6].

Figure 2 shows the difference absorption spectra following the 461-nm excitation of octopus rhodopsin in both the $H_2O$ and $D_2O$ suspensions at 8°C. The large absorbance change around 460 nm was due to the scattering of excitation light. The decrease in absorbance in 415-445 nm region at 0 ps in both the $H_2O$ and $D_2O$ suspensions is caused by the formation of X, the absorption spectrum of which lies in the 490-600 nm region. The spectra 30 ps after excitation clearly indicate the formation of H in both the $H_2O$ and $D_2O$ suspensions. The formation of H is slower in $D_2O$ suspension than in $H_2O$.

The excitation photon density dependence of the absorbance change at 430 nm 30 ps after excitation due to the formation of H is shown in Fig. 3. Open circles and solid curves were obtained by experiments and numerical calculations, respectively. The calculations were based on the following reaction scheme.

---

[1] Present address: Hamamatsu Photonics K.K. Tsukuba Research Laboratory, Ibaraki 300-26, Japan

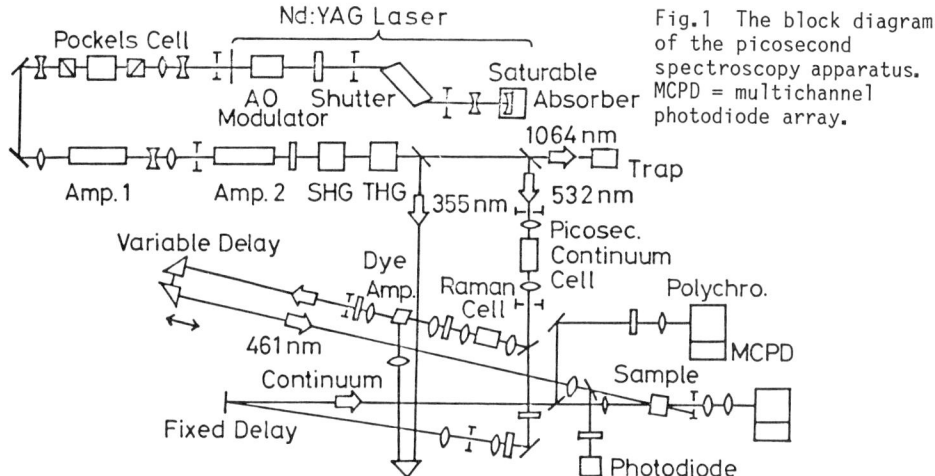

Fig.1 The block diagram of the picosecond spectroscopy apparatus. MCPD = multichannel photodiode array.

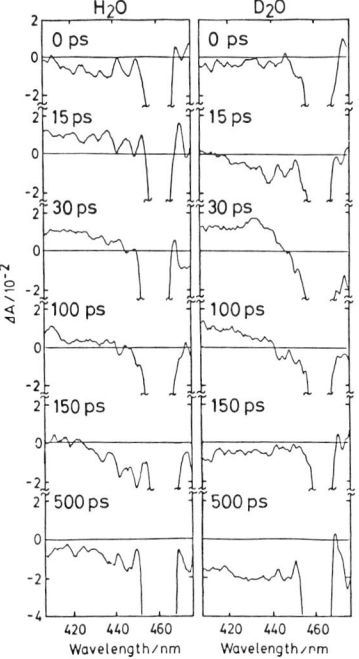

Fig. 2 Picosecond difference absorption spectra of octopus rhodopsin in the $H_2O$ (left) and $D_2O$ (right) suspensions at 8°C following the excitation by 20-ps pulse at 461 nm. The excitation photon density is $9.3(\pm1.3)\times10^{16}$ photons/$cm^2$.

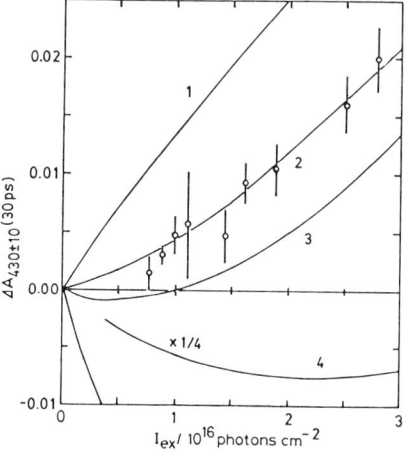

Fig. 3 The excitation photon density dependence of the absorbance change at 430±10 nm 30 ps after excitation at 8°C. The values of efficiency $\phi_H$ [$k_H/(k_H+k_B)$] were set at 1, 0.8, 0.7, and 0 for curves 1, 2, 3 and 4, respectively. The quantum yield of the X formation was 0.5. Lifetimes of X [$1/(k_H+k_B)$] and H ($1/k_B'$) were 15 and 70 ps, respectively.

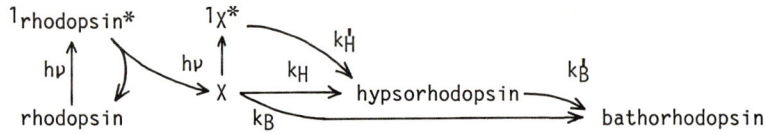

The rate constants for the thermal formations of B and H from X are denoted by $k_H$ and $k_B$, respectively, and $k'_H$ and $k'_B$ are the rate constants of the corresponding processes shown above. Figure 3 shows the calculated absorbance change for the formation efficiency of H in the thermal process $\phi_H$ [$=k_H/(k_H+k_B)$]= 1 (curve 1), 0.8 (curve 2), 0.7 (curve 3), and 0 (curve 4). The calculated value for $\phi_H=0$ correspnds to the case when B is directly formed from X in the thermal process and that H is formed only through the photoreaction of X. An intense bleaching at 430 nm is predicted for excitation photon densities lower than $3\times10^{16}$ photons/cm$^2$. The experimental result cannot be reproduced by the calculation with $\phi_H=0$. Therefore, the formation of H from X must be considered. The observed values were best fitted by the calculated value for $\phi_H=0.8$.

The formation of H was found in the early stage of the irradiation by 480-nm light at 10 K. Photochemical formation of H from X does not take place because of the low excitation power and the short lifetime of X even at 10 K (bovine Batho'; 29±2 ps [7]). The efficiency (about 0.25) of the H formation from X at 10 K was lower than that at 8°C.

PETERS et al. [7] reported that the lifetime of bovine Batho' is about 7 times longer in $D_2O$ suspension than $H_2O$. The result and our data mentioned above show that the decay rate of X ($k_H+k_B$) is affected by $H_2O/D_2O$ exchange. PANDE et al. [8] reported that the formation yield of H is 1.4-1.8 times larger in $D_2O$ suspension than in $H_2O$ at low temperature (12 K). The $H_2O/D_2O$ exchange effect on $\phi_H$ was not found at 8°C. The main decay process of X at 8°C is the X→H conversion (i.e. $k_H>k_B$). Therefore the $H_2O/D_2O$ exchange affects the $\phi_H$ value less efficiently at 8°C than at 12 K.

The primary processes of the photoreaction of octopus rhodopsin at physiological temperature are the thermal formation of both hypsorhodopsin and bathorhodopsin from X. The rate determining step of the X→bathorhodopsin conversion is a proton transfer.

References
1. T. Yoshizawa: In Handbook of Sensory Physiology VII/1, ed. by H. J. A. Dartnall, (Springer, Berlin 1972).
2. T. Kobayashi: FEBS Lett. 106, 313 (1980); Photochem. Photobiol. 32, 207 (1980).
3. B. Honig, T. Ebrey, R. H. Callender, U. Dinur, and M. Ottolenghi: Proc. Natl. Acad. Sci. USA, 76, 2503 (1979).
4. Y. Shichida, S. Matuoka, and T. Yoshizawa: Photobiochem. Photobiophys, 7, 221 (1984).
5. S. Matuoka, Y. Shichida, and T. Yoshizawa: Biochim. Biophys. Acta, 765, 38 (1984).
6. M. Tsuda: Biochim. Biophys. Acta, 545, 537 (1979).
7. K. Peters, M. L. Applebury, and P. M. Rentzepis: Proc. Natl. Acad. Sci. USA, 74, 3119 (1977).
8. A. J. Pande, R. H. Callender, T. G. Ebrey, and M. Tsuda: Biophys. J. 45, 573 (1984).

# Reactivity and Dynamics of Hemeproteins in the Femtosecond and Picosecond Time Domains

*D. Houde, J.W. Petrich, O.L Rojas, C. Poyart[1], A. Antonetti, and J.L. Martin*

Laboratoire d'Optique Appliquée, Ecole Polytechnique, ENSTA,
INSERM U275, F-91128 Palaiseau Cedex, France
[1]INSERM U299, F-92150 Suresnes, France

## 1. Introduction

It has been suggested that the doming of the heme plane in the tetrameric hemoglobin (Hb) molecule induced by ligand dissociation is ultimately responsible for the switch between the R ("relaxed" and oxygenated) and the T ("tight" and deoxygenated) states of normal adult human Hb which enables Hb to bind oxygen cooperatively [1]. In the two types of experiments that we shall discuss here, transient absorption and time-resolved resonance Raman spectroscopy, we consider two questions. First, what transient species are created in the photodissociation process? More specifically, can we discriminate between short-lived excited-states of the heme and structural changes of the protein which affect the heme absorption spectrum? Second, what is the nature of the structural changes that occur subsequent to photodissociation and can they be related in a straightforward manner to the reactivity of Hb?

## 2. Transient Absorption Measurements of Hemeproteins

The experimental apparatus used for transient absorption studies has previously been described [2,3]. Our measurements indicate that photodissociation of the ligand from the heme occurs in less than 50 fs. On the same time scale there appears a species, $Hb^*_I$, which decays in approximately 300 fs to a deoxy-like photoproduct, $Hb^+$, whose spectrum resembles that of the equilibrium deoxy Hb spectrum except in that it is slightly red-shifted (1-3 nm depending on the ligand). Competitive with the formation of $Hb^+$ is the production of another photoproduct, $Hb^*_{II}$, which returns in 2.5 ps to the ground-state ligated species. Although the amount of $Hb^*_{II}$ is sensitive to the nature of the ligand, it has been assigned to an excited state of the heme due to its presence in unligated protoheme and in deoxy Hb and Mb [3,4]. One mechanism by which $Hb^+$ converts to the ground-state ligated species is geminate recombination. The kinetics of geminate recombination on the picosecond time scale are interesting in that they provide a means of studying how the return of the ligands (which have not escaped to the solvent) to their binding sites is influenced by the heme pocket and those portions of the protein that are coupled to the heme pocket. An intriguing example is provided by the geminate recombination in the HbNO system where the kinetics can be fit well (though not necessarily accurately [4]) to two exponentials with 13 and 200 ps time constants. Similar results were obtained earlier by Cornelius et al. [5] and the two components were suggested to arise from differences in the $\alpha$ and the $\beta$ Hb subunits. Figure 1, however, illustrates the differences in the geminate recombination kinetics among HbNO, $\alpha^2(NO)\beta^2(+CN)$, $\alpha^2(+CN)\beta^2(NO)$, and MbNO. The cyanomet subunits cannot be photodissociated, and hence these compounds display geminate recombination only with the $\alpha$ or the $\beta$ subunits. Geminate recombination in these hybrids is, however, modulated by the $\beta$ cyanomet or the $\alpha$ cyanomet subunits of the tetramer. While it can be concluded from Fig. 1 that the $\alpha$ and $\beta$ subunits do behave differently, it must also be noted that the tetrameric structure of Hb significantly influences the kinetics of geminate

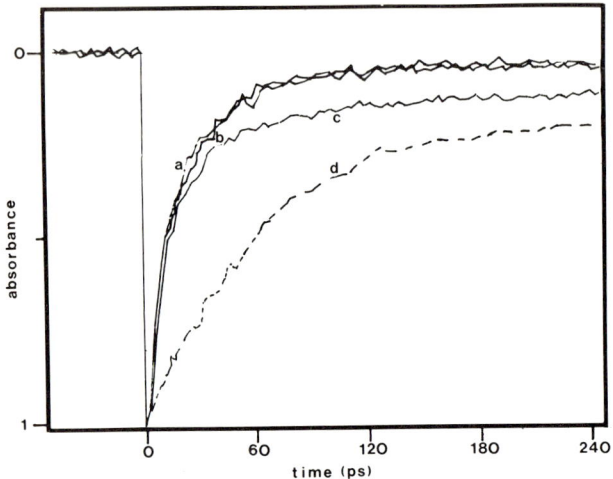

Fig. 1. Geminate recombination of NO with hemeproteins. a) HbNO; b) $\alpha^2(NO)\beta^2(+CN)$; c) $\alpha^2(+CN)\beta^2(NO)$; d) MbNO. The pump wavelength is 580 nm; the probe wavelength, 435 nm. The transient absorption signal is normalized to unity.

recombination. In each instance the absorption recovery at 435 nm can be fit well to a double exponential with 13 ps and 200 ps components where the weight of the 13 ps component decreases in the order: HbNO (85%), $\alpha^2(NO)\beta^2(+CN)$, $\alpha^2(+CN)\beta^2(NO)$, MbNO (15%). The fact that the cyanomet subunits cannot be dissociated and that Mb is a monomer strongly suggests the inadequacy of deriving physical significance from a fit of the geminate recombination kinetics to two exponentials. The transient absorption measurements discussed here provide interesting information concerning the species present upon ligand photodissociation and the roles of the constituents of the Hb tetramer on the ligand rebinding process; they do not, however, directly report on the nature of the specific structural changes experienced by the protein. Time-resolved resonance Raman spectroscopy is a technique which has the potential to yield such structural information [6,7].

3. Experimental

Here we describe a time-resolved resonance Raman spectrometer that we are currently developing (Fig. 2). The production of GW pulses of 100 fs duration has been described elsewhere [2,3]. The 10 Hz train of amplified 620 nm pulses creates two white-light continua. One of the continua is used to produce the 500 µJ photodissociation pulse of 150 fs duration centered at 575 nm with a bandwidth of 50 Å [3]. In order to monitor the vibrational modes directly associated with the response of the heme pocket to ligand photodissociation, Soret enhancement is required; and this is most efficiently done at the maximum of the deoxy Hb spectrum (435 nm). To produce a probe beam at 435 nm, we first select a small bandwidth of 870 nm from the second continuum and amplify this with LDS 867/MeOH pumped with YAG second harmonic. The 870 nm beam is then frequency-doubled with KDP or lithium iodate to 435 nm and amplified to 100 µJ in a coumarin 440/MeOH cell pumped with YAG third harmonic. The resulting probe beam has a bandwidth of 5 Å (to provide spectral resolution) and a concomitantly broadened pulse duration of approximately 700 fs. That there is no jitter between pump and probe beams has been established by measuring the

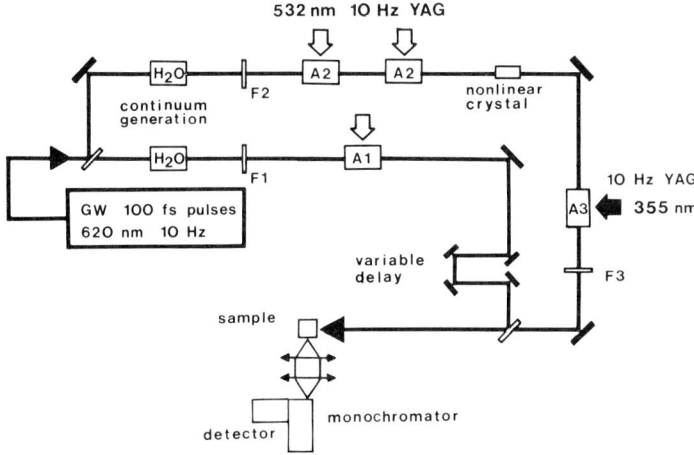

Fig. 2. Subpicosecond resonance Raman spectrometer. F1: 50 Å filter centered at 575 nm; F2: narrow band filter centered at 870 nm; F3: 5 Å filter centered at 435 nm. A1: R6G/(2% ALO/H2O); A2: LDS 867/MeOH; A3: coumarin 440/MeOH.

integral of their cross correlation in a sample that provides an instantaneous response. The pump and probe beams are introduced collinearly into the sample; and the Raman signal is collected with standard optics and coupled into a single stage monochromator (f/8) with a Jobin-Yvon 1200 l/mm grating blazed at 5000 Å. Data is collected with a gated OMA III (EG&G Electronics) whose detector is cooled to maximize uninterrupted integration time. Figure 3 presents preliminary data obtained with our spectrometer. One use for the spectrometer will be to study the time evolution of the Hb Raman spectra in the 200 cm$^{-1}$ region. Findsen et al. [7]

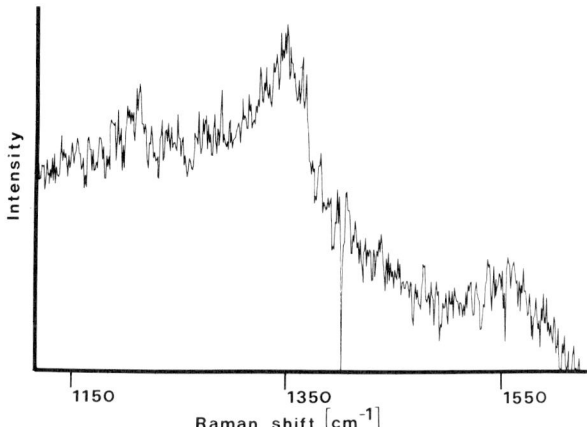

Fig. 3. Resonance Raman spectrum of HbCO 3 ps after photodissociation. The 100 μM HbCO was contained in a cuvette and translated through the pump and probe beams. This spectrum (1 mm slit width) shows no changes from the spectrum taken with no excitation pulse. The bands $\nu_4$ (1355 cm$^{-1}$) and $\nu_{19}$ (1555 cm$^{-1}$) have been determined to be sensitive to the electron density and the size of the porphyrin ring, respectively [6].

have shown for normal adult human Hb that the line at approximately 230 cm$^{-1}$ which has been attributed to the Fe-histidine stretching mode has not evolved in 10 ns from its value at 25 ps.

## 4. Conclusion

We have shown the influence of the β and the α subunits, and the influence of the tetrameric Hb structure itself, on the geminate recombination of NO with hemeproteins. The data indicate that an alternative functional form may be appropriate for fitting the kinetics. Two physically meaningful possibilities are a time-dependent diffusion-controlled process or a power law which can arise from a distribution of barrier heights for the rebinding process [8]. We have also demonstrated the feasibility of a Raman spectrometer with subpicosecond, tunable, independent, and jitter-free pump and probe pulses. Because the probe is independent of the pump and because the pulses are subpicosecond, we preclude the possibility of building up and observing a Raman signal from a steady-state of, for example, short-lived species such as the Hb*$_{II}$ discussed above, and we maintain the capability of obtaining spectra at short times. The spectra obtained, however, must be interpreted carefully. For example, heterogeneity of the Hb subunits may complicate the interpretation of the time evolution of the Raman spectra. Nevertheless, our 3 ps spectrum of photodissociated HbCO indicates that, within the signal-to-noise and spectral resolution, the $\nu_{19}$ and the $\nu_4$ modes have already attained the values observed at 50 ps and 10 ns, respectively [6]. Although these modes may not be the most sensitive markers of changes in the protein near the heme, our Raman data are not inconsistent with the absorption data [3,4] (see above) which suggest that in the vicinity of the heme the protein very rapidly (300 fs) attains the configuration of the relaxed deoxy Hb. Henry et al. [9] have noted that heating of the heme by a photon energy in excess of the Fe-ligand dissociation energy can result, through anharmonic coupling, in downshifts of the Raman lines. Their calculations show that the heme remains hot for tens of picoseconds. Figure 3 indicates that, at least for $\nu_4$ and $\nu_{19}$, the effects of heating are not large.

## 5. Acknowledgements

DH is an NSERC postdoctoral fellow. JWP is the recipient of an NSF postdoctoral Industrialized Countries Fellowship. OLM is a visiting professor from Departmento de Fisica, Centro de Investigacion y Estudios Avanzados del I.P.N., Mexico and is currently supported by La Fondation Nationale de la Recherche Médicale. Parts of this work were funded by INSERM, ENSTA, and Le Ministère de la Recherche et de la Technologie.

1. B.R. Gelin and M. Karplus, Proc. Natl. Acad. Sci. USA. 74, 801 (1977); M.F. Perutz, Ann. Rev. Biochem. 48, 327 (1979).
2. J.L. Martin, C. Poyart, A. Migus, Y. Lecarpentier, R. Astier, and J.P. Chamberet, in Picosecond Phenomena III, Eds. K.B. Eisenthal et al. (Springer, Berlin, 1982) p. 294.
3. J.L. Martin, A. Migus, C. Poyart, Y. Lecarpentier, R. Astier, and A. Antonetti, Proc. Natl. Acad. Sci. USA. 80, 173 (1983); in Ultrafast Phenomena IV, Eds. D.H. Auston and K.B. Eisenthal (Springer, Berlin, 1984) p. 447.
4. J.L. Martin et al. In preparation.
5. P.A. Cornelius, R.M. Hochstrasser, and W.A. Steele, J. Mol. Biol. 163, 119 (1983).
6. J.M. Friedman, D.L. Rousseau, and M.R. Ondrias, Ann. Rev. Phys. Chem. 33, 471 (1982).
7. E.W. Findsen, J.M. Friedman, M.R. Ondrias, and S.R. Simon, Science 229, 661 (1985).
8. R.M. Noyes, Progr. Reaction Kinetics 1, 129 (1961); R.H. Austin; K.W. Beeson, L. Eisenstein, H. Frauenfelder, and I.C. Gunsalus, Biochemistry 14, 5355 (1975).
9. E. Henry, W.A. Eaton, and R.M. Hochstrasser. This volume.

# Picosecond Raman Hole Burning as a Probe of Conformational Heterogeneity: Applications to Oxyhemoglobin

B.F. Campbell and J.M. Friedman

AT & T Bell Laboratories, Murray Hill, NJ 07974, USA

Frauenfelder and coworkers have observed power law rebinding kinetics for photodissociated ligands in hemeproteins at cryogenic temperatures [1]. On the basis of this and other related observations, it was suggested that for a given overall conformation of a protein there is a distribution of very similar subconformations or substates which differ only slightly one from the next. Under liquid solution conditions these substates can interconvert via structural fluctuations; consequently, an instantaneous snapshot of a protein solution would reveal a distribution of conformational substates, whereas a time-averaged picture would reveal an average conformation. If a protein reaction is sensitive to the differences in substate structure, then at low temperatures where each protein is cryogenically trapped in a given non-interconverting substate, each protein molecule will react at the rate characteristic of the respective substate. If at higher temperatures the reaction rate is slow compared to the rate at which a given molecule samples the full distribution of substates, then the reaction will be determined by a single average conformation.

Frauenfelder and coworkers [1] have concluded that in hemoglobin and myoglobin, substates give rise to a distribution of barrier heights at the heme that control the final step in the rebinding of a photodissociated ligand such as CO or $O_2$. Below ~180K the photodissociated ligand is thought to remain trapped within the heme pocket. Rebinding from the heme pocket at these low temperatures is non-exponential because of the distribution of heme associated barriers. The subnanosecond geminate recombination at higher temperatures is thought to be analogous to this cryogenic heme pocket rebinding [2]. The geminate rebinding of $O_2$ at 300 K in Hb is non-exponential on the 200 ps time scale [2]. A possible explanation for this observation is that substate interconversion is slower than 200 ps. Under these circumstances the first $O_2$'s to rebind should preferentially rebind in those protein substates having the lowest heme associated barrier for rebinding. We have initiated a course of study to investigate whether this substate phenomena can be observed on the level of specific structural degrees of freedom.

There is evidence both from cryogenic studies [3] and room temperature [2, 4] transient studies that protein control of the rebinding process occurs primarily through the proximal environment about the heme. In hemoglobin, the frequency of the iron-proximal histidine stretching mode correlates with the yield of geminate recombination [2, 4]. It is therefore likely that changes in protein structure that modify the rebinding process occur at least partially through the heme-proximal histidine linkage. Thus, the heme-histidine linkage is a good first place to look for substate behavior on the structural level.

The heme-proximal histidine linkage can be studied using resonance Raman scattering. Blue excitation wavelengths enhance the Raman mode whose frequency is that of the stretching motion of the iron-proximal histidine linkage. This Raman band is broad (>10cm$^{-1}$ fwhm). A possible source of this broadening is the existence of a distribution of substates having a distribution of iron-histidine stretching

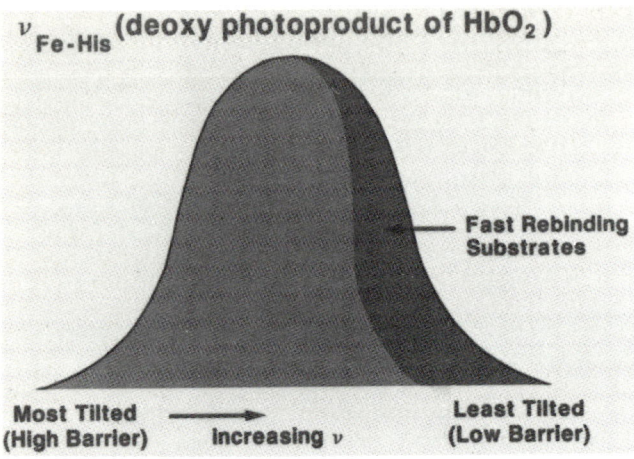

Fig. 1 A schematic of the Raman line shape of the iron-proximal histidine stretching mode showing a possible source of inhomogeneous line broadening due to a distribution of substates

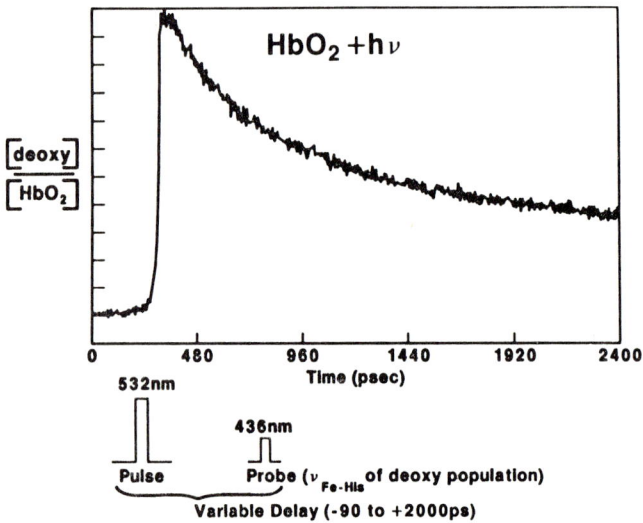

Fig. 2 Experimental set up for monitoring the Raman line shape of the surviving population of photodissociated hemes (deoxy) subsequent to an initial 532 nm pulse. An active-passive mode locked Nd:YAG laser is the source of the 532 nm (30 ps) pulses. The blue pulses are obtained by Raman shifting the 532 nm pulses

frequencies (see Fig. 1). It has been proposed that a decrease in this frequency corresponds to an increase in the tilt angle between the histidine and the heme plane [5]. This tilt would have the effect of making it energetically more costly to move the iron into the plane of the heme upon ligand binding with a net effect of increasing the barrier for ligand rebinding [4b, c, e, f]. Within this model, the lowest and highest frequency components of a transiently inhomogeneous population of photodissociated proteins would correspond to those with the highest and lowest barriers for rebinding, respectively. Thus, if the solution remained inhomogeneous with respect to this degree of freedom over the course of the geminate rebinding, then as the $O_2$'s preferentially rebound to the lowest barrier part of the population, the Fe-His Raman band of the surviving unrebound population should show a change in line shape commensurate with the loss of the low barrier part of the population (see Fig. 1).

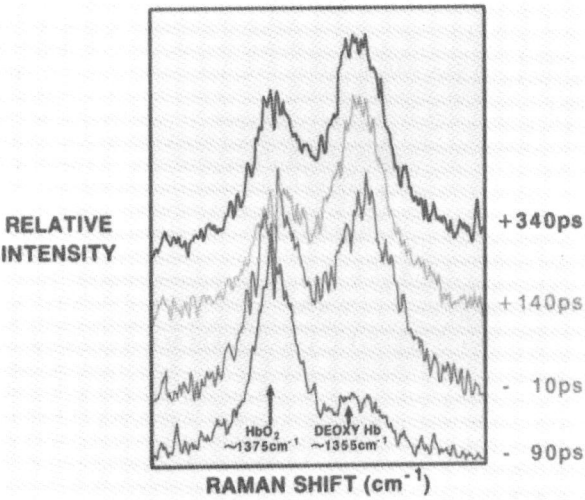

Fig. 3 The relative amounts of oxy and deoxy hemoglobin as a function of pulse-probe delay as reflected in the 1375 (oxy) and 1355 (deoxy) $cm^{-1}$ Raman bands

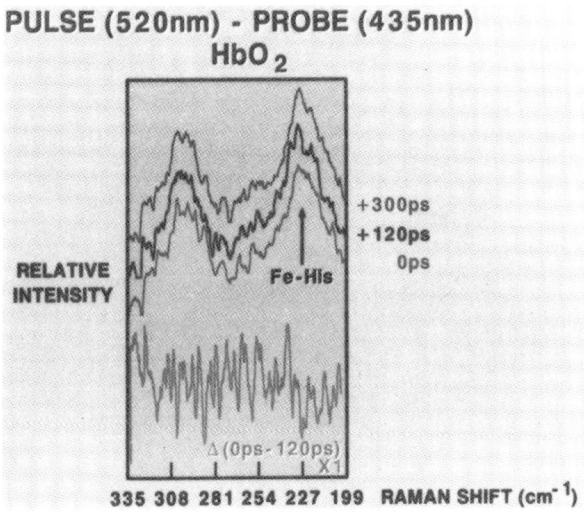

Fig. 4 The intensity normalized evolution of the Fe-His stretching mode subsequent to photodissociation

We have monitored the Raman line shape of the photodissociated population over the course of the ps rebinding using the scheme shown in Fig. 2. In such an experiment it is essential that the probe pulse not rephotolyze the rebound population. Figures 3 and 4 depict the effect of the variably delayed photolysis pulse (532 nm, 30 ps) upon the spectrum generated by a Raman shifted blue (435 nm) probe pulse (25 ps). Fig. 3 shows the relative amounts of photolyzed and unphotolyzed Hb as a function of the pulse probe delay. It can be seen that the probe alone does not cause substantial photolysis. Fig. 4 shows the Fe-His mode under similar circumstances. It can be seen that within our signal to noise there is no indication of Raman hole burning as a function of $O_2$ rebinding. Extending the delay out to 2 ns yielded the same results. An extension of these studies to cryogenic temperatures is currently in progress.

## References

1. R. H. Austin, K. W. Beeson, L. Eisenstein, H. Frauenfelder and I. C. Grunsalus, Biochemistry *14*, 5355 (1975); N. Alberding, S. S. Chan, L. Eisenstein, H. Frauenfelder, D. Good, I. C. Gunsalus, T. M. Nordlund, M. F. Perutz, A. H. Reynolds, and L. B. Sorensen, Biochemistry 17, 43 (1978); D. D. Dlott, H. Frauenfelder, P. Langer, H. Roder, and E. E. Dilorio, Proc. Natl. Acad. Sci. US 80, 823 (1983).

2. E. W. Findsen, J. M. Friedman, M. R. Ondrias, and S. R. Simon, Science **229**, 661 (1985).

3. W. Doster, D. Beece, S. F. Bowne, E. E. DiTorio, L. Eisenstein, H. Frauenfelder, L. Reinesch, E. Shyamsunder, K. H. Winterhalter and K. T. Yue, Biochemistry *21*, 4831 (1982).

4. (a) J. M. Friedman, R. A. Stepnoski and R. W. Noble, FEBS Letters 146, 278 (1982); (b) J. M. Friedman, "Time Resolved Vibrational Spectroscopy," (G. Atkinson, ed.) Academic Press, 307 (1983); (c) J. M. Friedman in *Hemoglobins: Structure and Functions* (A. G. Schneck ed.) Brussels University Press, pp. 269-284 (1984). (d) T. W. Scott, J. M. Friedman, M. Ikeda-Saito and T. Yonetani, FEBS Letters 158, 68; (e) J. M. Friedman, T. W. Scott, R. A. Stepnoski, M. Ikeda-Saito and T. Yonetani, J. Biol. Chem. 258, 10564 (1983); (f) J. M. Friedman, Science *228*, 1273 (1985); (g) J. M. Friedman, S. R. Simon, T. W. Scott, Copeia 3, 679 (1985).

5. J. M. Friedman, D. L. Rousseau, M. R. Ondrias, and R. A. Stepnoski, Science *218* 1244 (1982).

# Ultrafast Studies of Nitrosylmyoglobin

K.A. Jongeward, J.C. Marsters, and D. Magde

Department of Chemistry, University of California at San Diego, LaJolla, CA 92093, USA

1. Introduction

Photodissociation of ligands from hemoproteins may be used to study the dynamics of ligand binding [1-8]. The use of subpicosecond pulses allows processes that occur at the binding site to be singled out for examination.

Although adult human hemoglobin has been characterized in some detail already, important questions about how nature uses the protein environment to control binding will remain unanswered until studies have been completed on a variety of different heme proteins from a range of different species. One important protein is myoglobin, a single chain protein which stores oxygen in muscle tissue. It consists of a heme, which is an iron porphyrin, and approximately 153 amino acids whose sequence varies from species to species. The ligands bind on what is called the distal side of the heme and the amino acids that surround that side of the heme plane are called the distal amino acids. Whale myoglobin has a distal histidine while elephant myoglobin has a distal glutamine [9]. We have chosen these two myoglobins to investigate steric effects on ligand recombination.

2. Experimental

A colliding-pulse mode-locked ring dye laser, pumped by all lines of an argon ion laser, produces output pulses at 628 nm with a duration of 70 fs. Amplification in a four stage Nd:YAG-pumped dye amplifier yields 1.2 mJ subpicosecond pulses at a repetition rate of 10 Hz. Frequency doubling in a 1 mm $KNbO_3$ crystal produces 40 µJ pulses at 314 nm to be used as the photolysis beam. The remaining red light is sent around an optical delay line with a possible 10 ns range and focused into a water cell to create white light for a probe beam. Colored glass filters preselect the wavelength range from 400 to 480 nm for the myoglobin experiments. The probe beam is split into two parts. One beam is used as a reference; the remaining probe beam and the pump beam are recombined and focused to a 1 mm spot at the sample. After the sample, the beams are focused onto the slit of a monochromator and imaged onto the vidicon detector of a PAR OMA-II.

Sperm whale myoglobin was obtained from Sigma; we thank Dr. Vijay Sharma for the elephant myoglobin. The myoglobin samples were made up in a pH 7, bis-tris buffer solution containing 0.1 M NaCl and reduced to the ferrous form with sodium dithionite to give Soret absorptions of approximately 1.0 in a 2 mm quartz cuvette. Care was taken to exclude oxygen and NO was blown over the solutions for an hour. Static absorption spectra were taken before and after experiments to make sure there was no degradation during the runs.

## 3. Results

The transient difference spectra following photolysis of NO from elephant and whale myoglobin at various delay times are shown in Figs. 1 and 2. In both myoglobins, the bleaching at 420 nm corresponds to the disappearance of the ground state six coordinate species and the absorption at 440 nm corresponds to the appearance of the deoxy species. Successive traces in the ElMb-NO figure are taken at 5, 10, 16, 20, and 40 ps while the traces for the WhMb-NO are taken at 4, 10, 20, 80, and 150 ps.

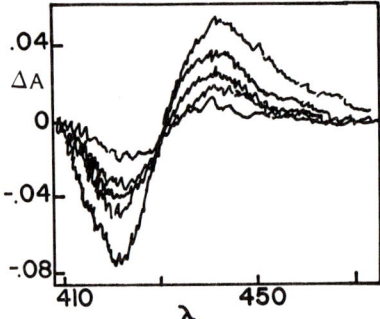

Figure 1. Elephant Mb-NO          Figure 2. Whale Mb-NO

In Figs. 3 and 4, the maximum of the transient absorption is plotted versus time in picoseconds. The crosses are experimental data points and the solid lines are computer fitted curves. For the elephant myoglobin, as the NO rebinds to the iron, the absorption of the deoxy species decays nearly to zero in less than 60 picoseconds and is best fit by a single exponential. In the case of whale myoglobin, the absorption takes longer to decay and appears to have both a fast and a slow component. The data are better fit with a biexponential curve. Our whale myoglobin results are similar to the results for NO rebinding to a different protein, human hemoglobin, reported by CORNELIUS et al. [6]. They report a two component rebinding process with the lifetime of the fast component being 17 ps and the slow around 100 ps.

Rate constants and corresponding decay times for room temperature rebinding of NO to myoglobin are summarized in Table 1. Within experimental error,

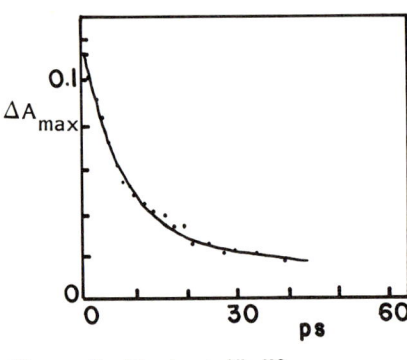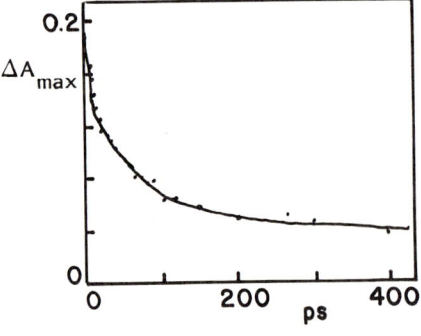

Figure 3. Elephant Mb-NO          Figure 4. Whale Mb-NO

Table 1. Rate constants and lifetimes at 23 C

| | | |
|---|---|---|
| ElMb-NO: | $k = 1.1 \pm 0.1 \times 10^{11}$ $s^{-1}$ | $\tau = 9 \pm 1$ ps |
| WhMb-NO: | $k(fast) = 1.6 \pm 0.9 \times 10^{11}$ | $\tau = 6 \pm 3$ |
| | $k(slow) = 1.4 \pm 0.7 \times 10^{10}$ | $\tau = 71 \pm 35$ |

the fast rate for whale myoglobin is the same as the rate for elephant myoglobin.

## 4. Discussion

The rate of NO geminate recombination to myoglobin depends on the distal environment of the protein. Replacement of the distal histidine residue with glutamine shortens the lifetime of the ligand-heme geminate pair and alters the biphasic nature of geminate rebinding. These results represent the first observation of the effect of changes in the structure of the distal environment on ultrafast ligand recombination from the geminate pair. They show that the kinetics of geminate ligand recombination are sensitive to small changes in protein structure, and illustrate the value of picosecond photolysis experiments in the investigation of protein function.

Both myoglobins show rates of NO rebinding that are much faster than the rates of CO and $O_2$ rebinding to whale myoglobin. CORNELIUS et al. [5] found no evidence of CO or $O_2$ recombination on a subnanosecond timescale. The picosecond geminate recombination of NO to these myoglobins may account for the very low observed quantum yield (Q=.001) for NO photodissociation [10].

This work was supported in part by the NSF and the NIH.

## References

1. C.V. Shank, E.P. Ippen, R. Berson, Science, 193, 50 (1976).
2. J.L. Martin, A. Migus, C. Poyart, Y. Lecarpentier, R. Astier, A. Antonetti, Proc. Natl. Acad. Sci. USA, 80, 173 (1983).
3. B.I. Greene, R.M. Hochstrasser, R.B. Weisman, W.A. Eaton, Proc. Natl. Acad. Sci. USA, 75, 5255 (1978).
4. D.A. Chernoff, R.M. Hochstrasser, A.W. Steele, Proc. Natl. Acad. Sci. USA, 77 5606 (1980).
5. P.A. Cornelius, A.W. Steele, D.A. Chernoff, R.M. Hochstrasser, Proc. Natl. Acad. Sci. USA, 78, 7526 (1981).
6. P.A. Cornelius, R.M. Hochstrasser, A.W. Steele, J. Mol. Biol., 163, 119 (1983).
7. A.H. Reynolds, S.D. Rand, P.M. Rentzepis, Proc. Natl. Acad. Sci. USA, 78, 2292 (1981).
8. J.A. Hutchinson, T.G. Traylor, L.J. Noe, J. Amer. Chem. Soc., 104, 3221 (1982).
9. D.E. Bartnicki, H. Mizukami, A.E. Romero-Herrera, J. Biol. Chem., 258 1599 (1983).
10. E. Antonini, M. Brunori, Frontiers of Biology, 21 (1971).

# Molecular Dynamics Study of Vibrational Cooling in Optically Excited Hemeproteins

*E.R. Henry*[1], *W.A. Eaton*[1], *and R.M. Hochstrasser*[2]

[1]Laboratory for Chemical Physics, NIDDKD, NIH,
Bethesda, MD 20205, USA
[2]Department of Chemistry, University of Pennsylvania,
Philadelphia, PA 19104, USA

1. Introduction: Proteins are disordered materials in certain respects but they can have their equilibrium structures precisely known. Thus, for the case of hemeproteins we know that the heme (metal porphyrin) is nestled within the protein and not significantly in contact with the solvent. Furthermore, the function of the hemeprotein is dependent on the large number of van der Waals contacts between the heme and the protein, all of which are known precisely for the equilibrium structure. Molecular dynamics simulations therefore take on a special flavor for such systems since the "solute" (the heme) is held in a "solvent" (the protein) which has quite limited atomic motion compared with a liquid. We have therefore used the molecular dynamics method (1,2) to explore the timescale and mechanisms of vibrational energy transport out of the heme into the surrounding protein. This is an important problem since it is generally assumed in transient spectroscopic experiments using hemeproteins that only the populations in the various electronic surfaces are forced out of equilibrium by the absorption of light. Clearly it will be important to properly determine the vibrational population distributions and to evaluate their effect on spectra and kinetics of photolyzed hemeproteins. This study represents a first attempt to predict the vibrational dynamics in such systems.

2. Methods: Details of the approach were given elsewhere (3) so we will just summarize here the input to the calculations. The x-ray structures of met myoglobin (4), Mb, and cytochrome-c (5), Cyt-c, were employed along with a potential function for the hemeprotein similar to that used in earlier calculations (6). In addition, electrostatic interactions between partial charges assigned to all the atoms were included. Potential functions for the heme group were parametrized using structural criteria as previously described (6). The potential function is not optimized for the property being examined here so that the results while expected to have predictive value in general terms should not be regarded as other than approximate. For example, we expect the many van der Waals contacts to form the most effective cooling pathway for the heme vibrational energy yet the hydrogens of the heme are not introduced explicitly. The calculation incorporates 1437 atoms for Mb and 969 atoms for Cyt-c.

The absorption of a photon by the heme was simulated by increasing the velocity of each of the 24 atoms in the $\pi$-electron system of the porphyrin instantly and uniformly such that the increase in kinetic energy matched the energy of the photon, which was chosen as either 54 kcalmol$^{-1}$ ($\lambda$ = 530 nm) or 81 kcalmol$^{-1}$ ($\lambda$ = 353 nm).

3. Results and Discussion: Figure 1 shows the averages of eight 60 ps trajectories for Mb and Cyt-c. The temperature was obtained at each 20 fs instant from the kinetic energy of each of the 24 $\pi$-electron associated

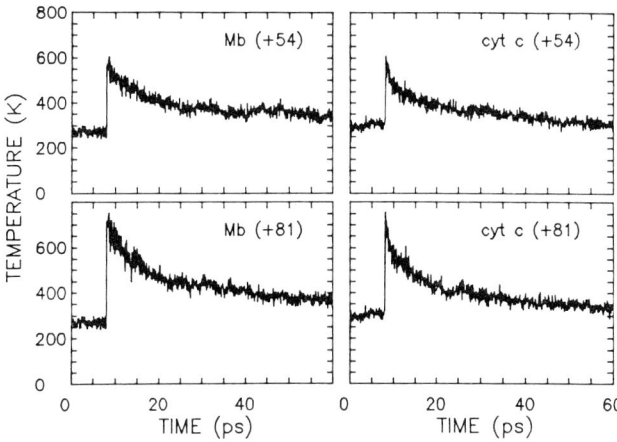

Fig. 1 Average kinetic temperatures of the 24 porphin skeleton atoms as a function of time. The plotted quantity is calculated from the total kinetic energy of the porphin ring atoms using the classical formula $3/2\ Nk_BT = 1/2\ \Sigma\ m_i v_i^2$ (N=24), and the averages are calculated from eight distinct simulations. The actual kinetic temperature exhibits a very abrupt spike at the time of the perturbation (to about 1000 K for the +54 kcal simulations and 1400 K for the +81 kcal simulations) due to the initial deposition of the excess energy entirely as kinetic energy. This spike has been omitted for clarity.

atoms. In both the 530 nm and 353 nm excitation data it is evident that equilibrium is achieved only after many 10's of ps. The data appear to predict a nonexponential cooling. The Figure represents the response to δ-function temperature jumps but the cooling timescale is sufficiently long that about 30% of the peak effect could be sensed by experiments using 30 ps time resolution.

The cooling proceeds in steps that we have just begun to evaluate more quantitatively. The vibrational excitation of the sidechains of the porphyrin occurs almost instantaneously (< 200 fs) in our simulations, resulting in the initial T of Figure 1 being ca. 25% lower than if there were no sidechains: This is basically a single molecule IVR process. We carried out a few experiments which examine the increase of kinetic energy in the protein atoms. The results appeared different for Cyt-c which has covalent connections between heme and protein, and for Mb which has not. In the case of Cyt-c the protein temperature rose sharply during the 5-10 ps following excitation, then it continued to rise slowly. For Mb the protein temperature showed no sharp rise but gradually grew to its equilibrium value over a period of ca. 50 ps. We hope to use these trajectories to answer detailed questions regarding the heat flow in proteins. The dominant mechanism for coupling energy out of the heme appears to involve van der Waals or near neighbor (< 5 A°) contacts (ca. 100 for Mb, ca. 160 for Cyt) between heme and protein but the covalent bonds of Cyt-c might contribute to funneling out a fraction of the energy in Cyt-c very rapidly.

The idea that aromatic molecules may take 10's of ps to cool even in conventional solutions is not new. Studies of transient absorption spectra of excited stilbene molecules were shown to remain heated for ca. 35 ps after laser pulse excitation and the heating effect was shown to be excitation wavelength dependent as expected (7). Recent studies of changes in ground state spectra of aromatics resulting from IR or visible (8,9) excitation have quantitatively evaluated such heating effects and demonstrated cooling times of 10's of ps. The present situation is different because of the specific nature of the binding of the heme to the protein, but the theoretical results are nevertheless in the same time range as found for these prototype systems.

Acknowledgement: We are grateful to M. Levitt for providing us with the original version of the program used in the calculations. A grant to RMH (NIH GM 12592) partly supported this work.

References

1. M. Karplus and J. A. McCammon, C.R.C. Crit. Rev. Biochem. 9, 293 (1981)
2. M. Karplus and J. A. McCammon, Ann. Rev. Biochem. 53, 263 (1983).
3. E. R. Henry, W. A. Eaton and R. M. Hochstrasser, Proc. Natl. Acad. Sci. USA (in press)
4. Brookhaven Data Bank File: 1MBD
5. T. Takano and R. E. Dickerson, J. Mol. Biol. 153, 79 (1981); Brookhaven Data Bank File: 4CYT
6. E. R. Henry, M. Levitt and W. A. Eaton, Proc. Natl. Acad. Sci. USA 82, 2034 (1985)
7. F. E. Doany, B. I. Greene and R. M. Hochstrasser, Chem. Phys. Lett. 75, 206 (1980)
8. A. Sellmeier, P. O. J. Scherer and W. Kaiser, Chem. Phys. Lett. 105, 140 (1984)
9. N. H. Gottfried, A. Sellmeier and W. Kaiser, Chem. Phys. Lett. 111, 326 (1984)

# Chemical Reaction in a Glassy Matrix: Dynamics of Ligand Binding to Protoheme in Glycerol:Water

*J.R. Hill[1], M.J. Cote[1], D.D. Dlott[1], J.F. Kauffman[1], J.D. McDonald[1], P.J. Steinbach[2], J.R. Berendzen[2], and H. Frauenfelder[2]*

[1] School of Chemical Sciences, University of Illinois at Urbana-Champaign, Urbana, IL 61801, USA
[2] Department of Physics, University of Illinois at Urbana-Champaign, Urbana, IL 61801, USA

A fundamental understanding of condensed phase chemical reaction dynamics can be obtained from the study of biomolecules, particularly heme and hemeproteins. The complexity of these systems gives rise to a rich variety of phenomena, allowing many aspects of condensed matter reactions to be examined. The rate theories and puzzles of hemeprotein kinetics have recently been discussed by Frauenfelder and Wolynes [1]. In this work we present new experiments on ligand binding to protoheme (Fe:protoporphyrin-IX) in a glassy matrix. We have studied the rebinding of carbon monoxide after photodissociation over a wide range of time [5ps-10ms] and temperature [300K-70K]. The significance of our results is that (1) the influence of friction and nonadibaticity on a condensed phase reaction can be directly investigated, (2) the influence of an inhomogeneous glassy matrix (glycerol-water 75:25) on the reaction can be studied, and (3) meaningful comparison between protoheme and hemeprotein kinetics isolates the role of the protein relaxation in the reaction.

Temperature-dependent ligand rebinding to protoheme was studied with µs resolution by Alberding et al. [2]. We extended the results to the ps timescale using conventional pump-probe methods with 565 nm dye laser pulses (10µJ, 3ps, FWHM, 80Hz). Figure 1 shows a log-log plot of the time-resolved absorbance change (proportional to $N(t)$, the fraction of protoheme molecules that have not rebound CO at time t) at 300K and 260K. The slow process (Process S) involves binding of CO that have escaped their initial solvent cage by diffusion. This solvent process is exponential in time with a pseudo-first-order rate coefficient that depends on [CO]. As T is decreased, the fraction of CO molecules that escape the cage decreases until it is below 0.01 at 240K [2], well above the glass transition $T_c$ = 173K. Subtraction of the 260K Process S data (solid circles from [2]) yields a nonexponential decay (Process I), the solid rectangles in Fig. 1. The relative fraction of ligands that rebind via Process I increases as T decreases from 300K to 240K. Process I is nonexponential at all temperatures up to 300K.

As T is decreased below 240K, process S vanishes, the rate of process I decreases in a manner consistent with [2], and a new process, I*, is observed. The amplitude of I* increases with further decrease in T until it dominates I. Figure 2 shows the rebinding data at 140K (open rectangles), the temperature at which processes I and I* have equal amplitudes. Subtraction of process I, which is known to have the form of a power-law decay, reveals that process I* (open circles) decays exponentially with a 15 ps lifetime. As T is decreased from 220 to 70K, the rate of process I* decreases linearly with T.

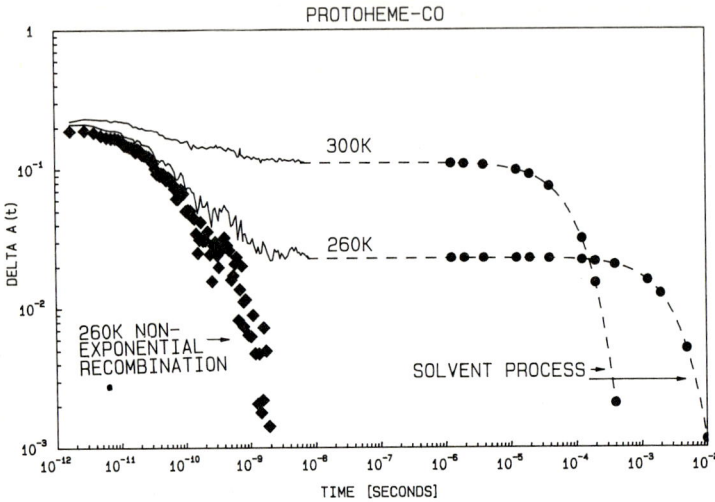

Fig. 1. Ligand rebinding data in the high temperature regime

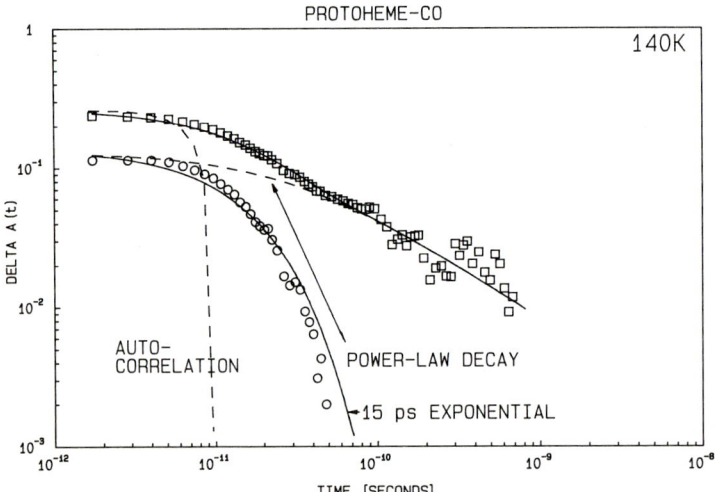

Fig. 2. Ligand rebinding data at 140K where the fraction of ligands that rebind via process I is equal to the fraction that rebind via process I*.

Figure 3 summarizes the temperature dependence of ligand rebinding via these three processes. At the lowest temperatures studied, process I* dominates. As the temperature increases, the fraction of ligands rebinding via process I increases until 240K. At that point the solvent process increases in importance.

In [2], process I was interpreted to arise from recombination of ligands trapped within the solvent cage. The barrier for this process is a property of the heme, and probably arises as a consequence of the transition from the unbound S=2 domed heme structure to the bound S=0 planar structure. The

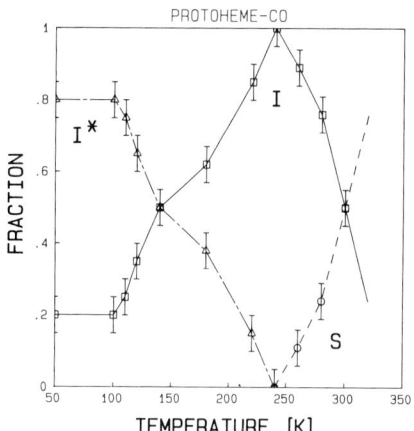

Fig. 3. Relative fraction of ligand rebinding to protoheme via the three processes

rebinding is nonexponential because this barrier possesses distributed activation enthalpy. In the β-chain of hemoglobin, process I becomes nearly exponential above ca. 250K, because the protein structure relaxes significantly on the timescale (1 ns) of the rebinding [3]. In protoheme, this conformational relaxation does not occur on the faster timescale (50 ps) of the rebinding.

The observed (linear) temperature dependence and the large rate of process I* at low temperatures suggest that it does not occur by classical over-the-barrier motion. The exponential time dependence shows that the height of the barrier is not distributed. Process I* may involve recombination to an S=1 heme which has not domed. At 140K, the rebinding rate of 7E10/s is a factor of ~40 slower than kT/h, the frequency of collision between the ligand and the heme. The reduction of the recombination rate can be caused by entropy, friction, and nonadiabatic barrier crossing [1].

This research was supported in part by the National Science Foundation, Solid State Chemistry, grant DMR 84-15070 (DDD), the U. S. Department of Health and Human Services under Grant GM 18051, and the National Science Foundation under Grant DMB 82-09616 (HF), and the National Science Foundation, grant CHE 83-14105 (JDM).

References

1. Frauenfelder, H., Wolynes, P. G., Science, 229, 337 (1985).
2. Alberding, N., Austin, R. H., Chan, S. S., Eisenstein, L., Frauenfelder, H., Gunsalus, I. C., Nordlund, T. M., J. Chem. Phys. 65, 4701 (1976).
3. Ansari, A., DiIorio, E. E., Dlott, D. D., Frauenfelder, H., Iben, I.E.T., Langer, P., Roder, H., Sauke, T. B., Shyamsunder, E., Biochemistry (in press).

# Part VIII

# Energy Transfer and Relaxation

# Energy and Electron Transfer of Adsorbed Dyes on Molecular Single Crystals and Other Substrates

*K. Kemnitz, N. Nakashima, and K. Yoshihara*
Institute for Molecular Science, Myodaiji, Okazaki 444, Japan

## 1. Introduction

Energy and electron transfer frequently occur in the same system. Therefore it is of considerable interest to investigate a system where both phenomena can be observed independently from one another. One such case is dye molecules adsorbed on organic crystals which are ideally suited to serve either as inert or chemically interacting (electron transfer) substrates, depending on the choice of substrates and dyes (Fig. 1) [1]. In order to study electron transfer through the crystal surface, it is necessary to know the competitive contribution of energy transfer on the surface to the overall fluorescence decay. On organic crystals, the competitive processes are electron transfer from the crystal to the excited dye and energy transfer from excited dye monomer to other monomers [2] or to dye aggregates like dimers [3,4,5].

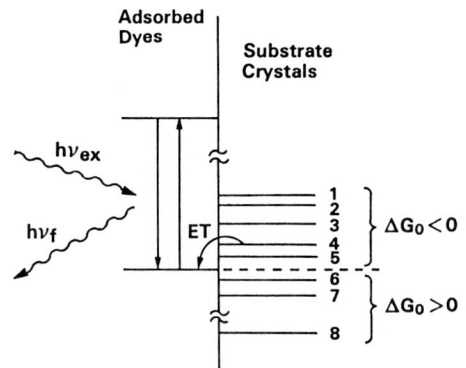

Fig. 1. Energy diagram of the adsorbed dye/substrate system. The substrate materials used are 1. perylene, 2. phenazine, 3. benzoperylene, 4. pyrene, 5. anthracene, 6. phenanthrene, 7. naphthalene, and 8. quartz plate. The dyes used are rhodamine B(RhB), rhodamine 101(Rh 101), and pyronine B (PyB). $\Delta G_0$ is the free energy difference for electron transfer between adsorbed dye and substrate.

We chose to study dry solvent-free systems which show no solvent-solute interaction, allowing the measurement of the temperature dependence and of the vibronic interactions of electron transfer interactions due to the small reorganization energy.

## 2. Experimental

The weak fluorescence of 1/100 of dye monolayer was measured with a time-correlated single photon counting system of subnanosecond time resolution, which has been described elsewhere [6]. The organic single crystals were sublimation grown. The crystals were all freshly grown and stored in the dark before use, in order to minimize a possible surface deactivation due to photo-oxidation.

The dye was adsorbed from aqueous solution, tissue paper was used to soak off the residual dye solution after finishing adsorption in about 10 min. In this way solvent-free system could easily be obtained.

## 3. Results and Discussion

1) Energy Transfer: Energy transfer among the adsorbed dyes can be conveniently studied on chemically non-interacting surfaces of naphthalene, phenanthrene, and glass. In contrast to the three-dimensional case in solution [3], the two-dimensional one on a surface could not simply be fitted with a pure Foerster-type formula of energy transfer (in two-dimensions), but contained substantial exponential contributions, especially at higher coverages. This exponential contribution to the fluorescence decay was explained with fluorescent dimers of RhB [4,5]. An identical decay law was recently obtained for cresyl violet on glass [7]. At very low coverages, the presence of two (or more) monomer species manifested itself in a non-exponential fluorescence decay. A systematic study of the fluorescence decays obtained from dye monomers adsorbed on phenanthrene single crystals of different quality revealed that distorted sites are responsible for the deviation from uni-exponentiality [8,9]. Figure 2 shows the fluorescence decay obtained from two phenanthrene crystal specimens of different quality.

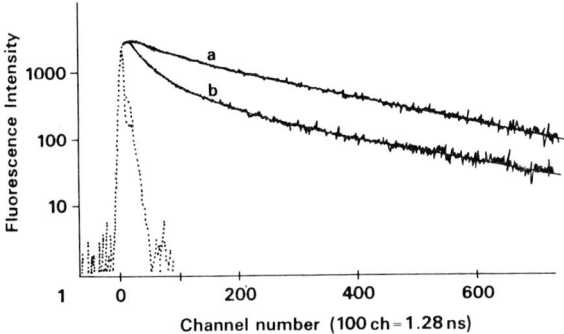

Fig. 2. Effect of crystal quality on decay kinetics: Rh101 adsorbed on phenanthrene. (a) flawless single crystal, (b) specimen covered with many dislocations.

Since all three xanthene dyes in Fig. 3 showed identical fluorescence dynamic behavior on surfaces as described above, in spite of their different molecular rigidity, we concluded that a property of the xanthene skeleton, which is common to all three dyes, might be responsible for the reduced lifetimes. Since RhB and its derivatives can be considered to be an oxygen-bridged triphenylmethane (TPM) dye, we concluded that a twist motion in the skeleton, suffered by excited molecules at distorted sites, might lead to enhanced internal conversion in parallel to the behavior of TPM dyes in solution. In contrast to molecules adsorbed at distorted sites with reduced lifetimes, molecules at ideal sites show an increased lifetime of about 3.5 ns (1.5 ns in water and 4.25 ns for radiative lifetime). Above studies of dyes in the adsorbed state are prerequisite for the determination of the rate of electron transfer.

The phase transition of the substrate single crystals affects the fluorescence decay rates and pre-exponential factors, which is revealed with

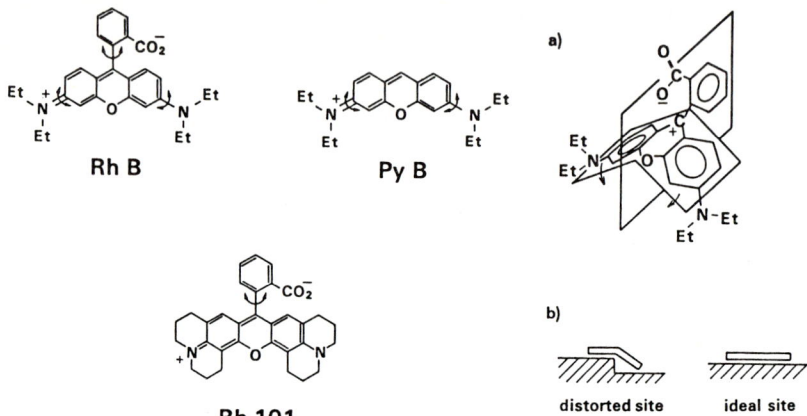

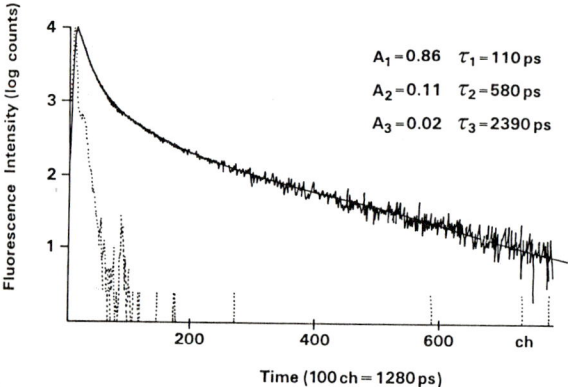

Fig. 3. Molecular structures of dyes and RhB considered as an oxygen-bridged TMP.

$A_1 = 0.86$  $\tau_1 = 110$ ps
$A_2 = 0.11$  $\tau_2 = 580$ ps
$A_3 = 0.02$  $\tau_3 = 2390$ ps

Fig. 4. Fluorescence decay of RhB on anthracene single crystal.

a system of RhB/phenanthrene [8]. Thus dye monomers could be used as sensitive probes in investigation of surface site distribution.

2) Electron transfer: From the coverage dependent fluorescence decays of RhB adsorbed on phenanthrene crystals, we learnt that a coverage as low as 1/25 – 1/100 of a monolayer is necessary in order to prevent energy transfer and dimer fluorescence which would distort our rate constant of electron transfer. We observed fast fluorescence decays with crystals having negative $\Delta G_0$ values. Figure 4 shows a typical fluorescence decay of RhB on anthracene single crystal. We assign the fastest part of the decay due to the electron transfer quenching.

Figure 5 shows the rate constants of electron transfer obtained for five organic single crystals of varying ionization potential. The solid line is one of the fitting calculations using Sarai's version [10] of quantum mechanical theory of electron transfer, involving one intermolecular soft mode ($\omega_s$) and two intramolecular modes ($\omega_i$). Asymmetric decease of the rate constants in the Marcus' inverted region is observed for the first

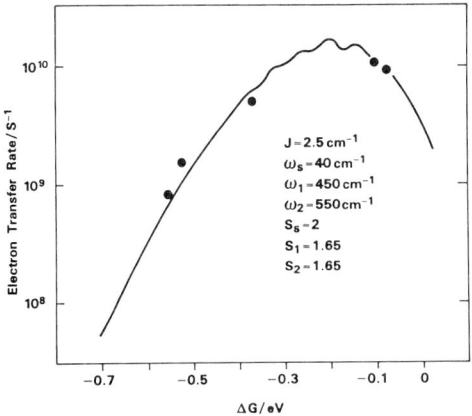

Fig. 5. Rates of electron transfer of RhB adsorbed on organic single crystals of varying ionization potential. Solid line indicates quantum mechanical calculation with parameters described in the figure.

time in solid electron-donor-acceptor systems and indicates the importance of coupling of electron transfer to vibrational modes.

The choice of the parameters is not unique to fit the experimental data, but a fluctuative behavior of the rate constant is begun to be seen as predicted by theory for systems with small reorganization energy. It is about 0.2 eV in our solvent-free system with fixed electron-donor-acceptor geometry. This is considerably smaller than usually encountered in solution. Therefore it is not surprising that addition of water to our dry system slows down the electron transfer rates. This was also observed for RhB adsorbed on inorganic semiconductor crystals like $\alpha$- and $\beta$-SiC and GaP [11].

The fluorescence decay of RhB adsorbed on organic crystals is non-exponential (3-exponential) as was found for semiconductor single crystals. It is explained with adsorption sites with different electron transfer capabilities. In the case of $\alpha$- and $\beta$-SiC and GaP, a thermal equilibrium was discovered among the populations at these sites. With decreasing temperature, the contribution of the fastest component decreases, indicating that the sites with the fastest electron transfer have the smallest binding energy.

1. H. Gerischer, F. Willig: In Topics in Current Chemistry, ed. by F.L. Boschke Vol.61 (Springer, Berlin, 1976)
2. F. Willig, A. Blumen, and G. Zumofen: Chem. Phys. Lett. 108, 222 (1984)
3. N. Nakashima, K. Yoshihara, and F. Willig: J. Chem. Phys. 73, 3553 (1980)
4. K. Kemnitz, T. Murao, I. Yamazaki, N. Nakashima, and K. Yoshihara: Chem. Phys. Lett. 101, 337 (1983)
5. K. Kemnitz, N. Tamai, I. Yamazaki, N. Nakashima, and K. Yoshihara: J. Phys. Chem. (1986) in press.
6. T. Murao, I. Yamazaki, and K. Yoshihara: Appl. Opt. 21, 2297 (1982)
7. P. Anfinrud, P.L. Crackel, and W.S. Struve: J. Phys. Chem. 88, 5873 (1984)
8. K. Kemnitz, N. Tamai, I. Yamazaki, N. Nakashima, and K. Yoshihara: J. Phys. Chem., to be submitted.
9. N. Nakashima and D. Phillips: Chem. Phys. Lett. 97, 337 (1983)
10. A. Sarai: Chem. Phys. Lett. 63, 360 (1979)
11. K. Kemnitz, N. Nakashima, N. Tamai, I. Yamazaki, K. Yoshihara, and H. Matsunami, to be published.

# Optical Pump-Probe Spectroscopy of Dyes on Surfaces: Ground-State Recovery of Rhodamine 640 on ZnO and Fused Silica

*P.A. Anfinrud, T.P. Causgrove, and W.S. Struve*

Department of Chemistry and Ames Laboratory-USDOE,
Iowa State University, Ames, IA 50011, USA

## 1. Introduction

In recent years, time-resolved fluorescence spectroscopy has been investigated for dyes adsorbed onto fused silica [1] and semiconductors [2,3]. On silica, the monomer fluorescence dynamics are nonexponential and tend to be dominated by excitation trapping by dye aggregates; the phenomenological lifetime measured in the limit of low coverage is often comparable to the fluorescence lifetime observed in solution [1]. On ultraviolet-bandgap semiconductors like $TiO_2$, much more rapid fluorescence decay is typically found [2,3], even at low coverage. The channel responsible for this accelerated decay on semiconductors is widely believed to be electron injection into the semiconductor space-charge region. However, the photocurrent efficiencies of liquid-junction solar cells with dye-coated single-crystal semiconductor photoelectrodes are generally small, and the possibility exists that dye electronic excitation may instead decay rapidly and nonradiatively into semiconductor modes [3]. To differentiate between these decay mechanisms, we have done optical pump-probe measurements of ground-state recovery dynamics of rhodamine 640 adsorbed on fused silica and on ZnO at submonolayer coverages.

## 2. Experimental

The synchronously pumped rhodamine 6G laser system has been described previously [1]. The 589 nm pump and probe beams (96 MHz repetition rate) were modulated at 6.5 and 0.5 MHz respectively using Isomet Model 1206C acoustooptic modulators, and variable probe delay was provided using a Micro Controle UT10050PP translation stage. Both beams were focussed with a common 7 cm f.l. lens at the dye-coated surface of a silica or ZnO substrate. An EG&G FOD-100 photodiode detected the probe beam, and phase-locked single-sideband detection was achieved at 7 MHz in a Drake R-7A radio receiver which was modified and augmented with auxiliary frequency-mixing circuitry to provide flexible modulation frequencies in both laser beams [4].

Surfaces were coated with rhodamine 640 by a technique described earlier [1], using water as coating solvent. Dye photooxidation by the pump and probe beams contributed large spurious decays to photobleaching signals exhibited by stationary samples. On silica, signal decay by photooxidation was more rapid at higher dye coverages. Photooxidation decay was independent of coverage on ZnO. Effects of photooxidation were minimized by simultaneously rotating the sample at 12 Hz about its surface normal and translationally cycling it over 2 mm at 1 Hz to provide a raster filling a 0.4 cm$^2$ annular region on the surface.

Uncoated ZnO surfaces exhibited large negative-going transients which were symmetric about t=0 and exhibited ~ 12 ps fwhm, similar to that of zero-background SHG autocorrelation traces. Coated ZnO surfaces exhibited the same transient, superimposed on the dye photobleaching transient to form the total response function R(t). The adsorbed dye response function for t>0 may be obtained in principle by antisymmetrizing R(t). Since the ZnO response function is so large, however, more accurate dye response functions were obtained by first dye-coating half of the substrate surface, and by using a lock-in amplifier to demodulate the radio receiver output at the 12 Hz sample spinning frequency. The use of orthogonal pump and probe polarizations minimized noise arising from surface-scattered beams overlapping near t=0.

## 3. Results

The photobleaching transient decay of rhodamine 640 on silica was strongly accelerated by increased dye coverage, indicating that monomer ground-state recovery on silica occurs via excitation trapping by dye aggregates. At the lowest coverage studied, the transient decay on silica was well simulated by the phenomenological biexponential function f(t) = 0.73 exp(-t/584 ps) + 0.27 exp(-t/19.7 ps). On ZnO, the transient decay was nearly independent of coverage over a range similar to that studied on silica. Typical biexponential fits to antisymmetrized response functions (Figure 1) yielded lifetimes in tens of ps, similar to those observed in fluorescence profiles of dyes on semiconductors. These results support the occurrence of rapid nonradiative dye → surface excitation decay on single-crystal semiconductors.

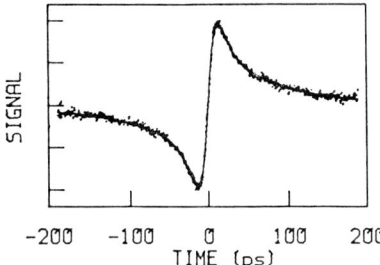

Figure 1. Antisymmetrized response signal for rhodamine 640 on ZnO. The curve is an optimized convolution of a biexponential decay law with the laser autocorrelation function. The decay times are 12.5 and 87.4 ps with preexponential factors of 0.71 and 0.29, respectively.

1. P. Anfinrud, R. L. Crackel, and W. S. Struve: J. Phys. Chem. 88, 5873 (1984).
2. Y. Liang, A. M. Ponte Goncalves, and D. K. Negus: J. Phys. Chem. 87, 1 (1983).
3. R. L. Crackel and W. S. Struve: Chem. Phys. Letters 120, 473 (1985).
4. P. A. Anfinrud and W. S. Struve: Rev. Sci. Instrum. 57, 380 (1986).

# Picosecond Fluorescence Spectroscopy on Molecular Association in Langmuir-Blodgett Films

*I. Yamazaki, N. Tamai, and T. Yamazaki*
Institute for Molecular Science, Myodaiji, Okazaki 444, Japan

The Langmuir-Blodgett (LB) film is a highly ordered molecular assembly of multilayered architecture which is prepared by transferring a compact monolayer spread on a water surface onto a quartz substrate. In this paper, we report new aspects of molecular association of pyrene in LB multilayer films studied by means of a synchronously pumped, cavity-dumped dye laser and a single-photon counting apparatus [1]. The LB films in this study were prepared from stearic acid and small amounts of 16-(1-pyrenyl)hexadecanoic acid (hereafter referred to as PHA). The PHA concentration was varied from 0.01 to 29 mol%, and a concentration effect on molecular association was investigated.

Figure 1 shows typical examples of time-resolved fluorescence spectra of the LB film. Four kinds of fluorescence bands appear depending on the time region: (1) $F_1$ band at 0-200 ps with three peaks at 380, 400 and 421 nm. This band is shifted by 3 nm to the red compared with $F_2$ band. (2) $E_1$ band at 100 ps-7.0 ns. A broad band with a peak at 422 nm. (3) $E_2$ band after 1 ns with a peak at 470 nm, corresponding to the well-known excimer fluorescence of pyrene in solution. (4) $F_2$ band after 3 ns with peaks at 377, 397 and 421 nm, assigned to the monomer fluorescence of pyrene. The stationary-excitation fluorescence spectrum appears to be composed of two bands corresponding to $F_2$ and $E_2$, irrespective of the PHA concentration concerned in this study. The fluorescence excitation spectrum is changed depending on the monitoring wavelength: the excitation spectrum monitored at the monomer region exhibits peaks at 345 and 329 nm for $^1L_a$ absorption band of pyrene, while those of the excimer $E_2$ are at 345 and 329 nm. This means that the pyrene chromophores in the LB film form an associated complex in the ground state, and that the excimer is formed through this associated complex. We assigned the broad band $E_1$ to another excimer band. From the spectral difference between $E_1$ and $E_2$, the conformation of the $E_1$ excimer should be significantly different from that of the $E_2$ excimer which is known to be of a sandwich type. Its conformation might be of a type of partially overlapping pyrene rings.

The fluorescence decay curves of the LB film with various concentrations of PHA are reasonably interpreted as follows: (1) rapid excimer formation in the dimer sites and (2) energy transfer and trapping in the isolated-monomer site. The dimer fluorescence $F_1$ rapidly decays, and its decay time (150 ps) is associated with a rise of the excimer fluorescence $E_2$, indicating that the excimer ($E_2$) forms through the dimer ($F_1$). The excimer fluorescence $E_1$ also shows a rise with much shorter rise time than that of $E_2$. The decay curve after 10 ns monitored at 377-380 nm is recognized predominantly as due to a time behavior of the monomer $F_2$. Its decay curve at low concentration of PHA is almost single-exponential. On increasing the concentration, however, it deviates from the exponential form. These decay curves can be fitted to a decay function of two-dimentional excitation energy transfer and trapping as follows:

$$\rho(t) = A_1 \exp[-t/\tau_D - 2\gamma_A(t/\tau_D)^{1/3}] + A_2 \exp(-t/\tau_D) , \qquad (1)$$

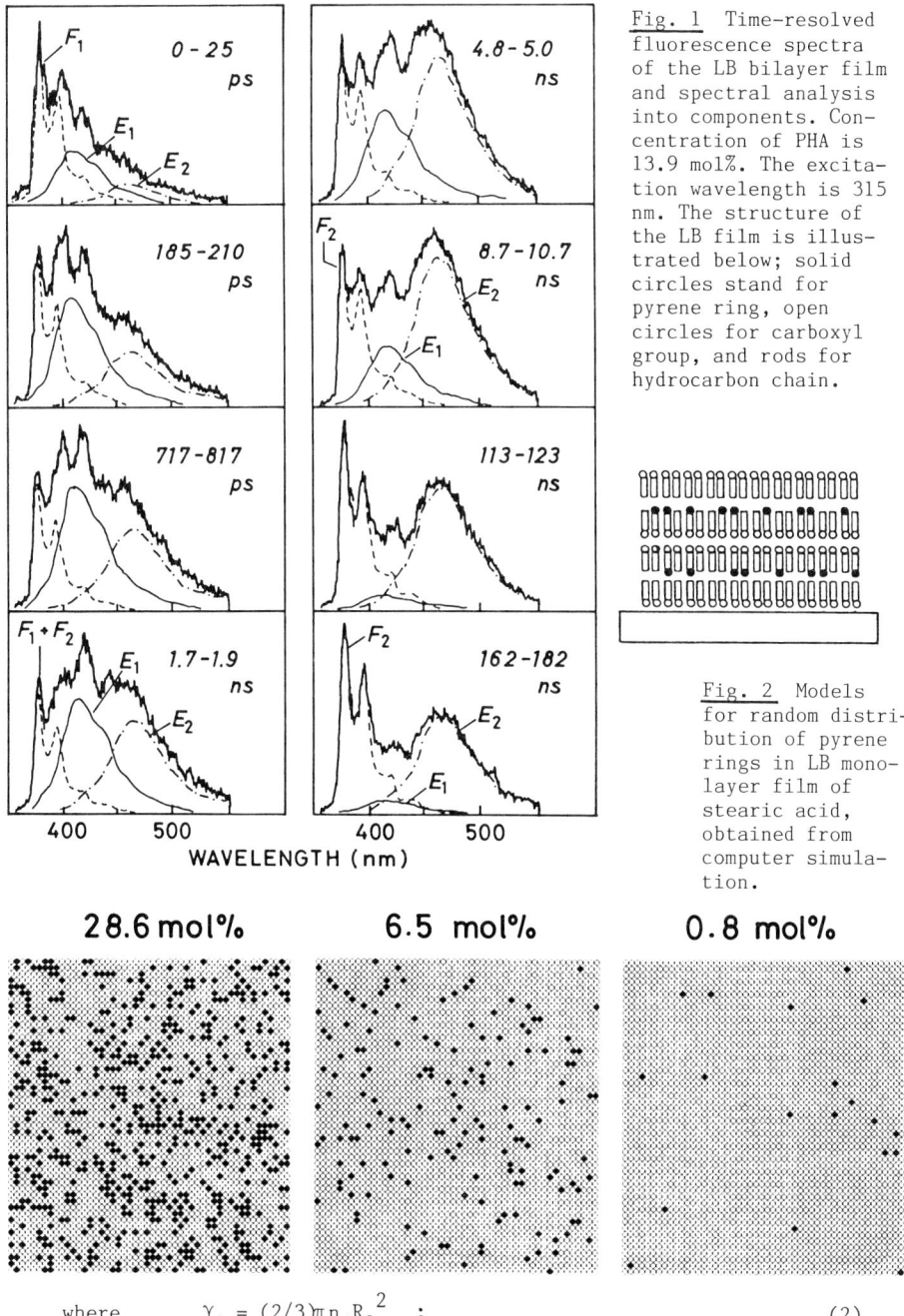

Fig. 1 Time-resolved fluorescence spectra of the LB bilayer film and spectral analysis into components. Concentration of PHA is 13.9 mol%. The excitation wavelength is 315 nm. The structure of the LB film is illustrated below; solid circles stand for pyrene ring, open circles for carboxyl group, and rods for hydrocarbon chain.

Fig. 2 Models for random distribution of pyrene rings in LB monolayer film of stearic acid, obtained from computer simulation.

where
$$\gamma_A = (2/3)\pi n_A R_0^2 \ ; \tag{2}$$

$\tau_D$ is the lifetime of the donor in the absence of acceptor; $n_A$ is the density of acceptor; and $R_0$ is the critical transfer distance. At lower concentrations of PHA, the $A_2$ value is large (40, 20 and 10 % for 0.8, 2.2 and 6.7

mol%, respectively); at higher concentrations above 14 mol%, it becomes negligibly small. The $\gamma_A$ values vary from 0.95 (0.8 mol%) to 1.5 (28.6 mol%).

Suppose that the fluorescence quenching occurs through pyrene and/or its aggregated species. From the decay curve analysis based on (1), the $\gamma_A$ values and therefore $n_A$ values can be estimated for different concentrations of PHA. On the other hand, the trap density $n_A$ can be estimated from a statistical calculation. Fig. 2 shows simulation patterns of random distribution of PHA molecules in a LB monolayer, which were obtained from computer calculation. From these patterns of PHA distribution, the densities of isolated monomer, dimer and higher aggregated species were estimated as a function of PHA concentration. The calculated and experimental values of $n_A$ are compared in Fig. 3. At a high concentration of PHA (28.61 mol%), the experimental and calculated values of $n_A$ are close to each other. On decreasing the PHA concentration, the experimental value of $n_A$ changes slowly, although the calculated values show significant decrease. This may suggest a non-uniform distribution of PHA in the LB films with lower concentrations of PHA. It is probable that PHA molecules form aggregates in the preparation process of a monolayer spread on water surface, leading to an increase in an effective concentration of pyrene chromophores. Such an effect will be pronounced at lower concentrations of PHA.

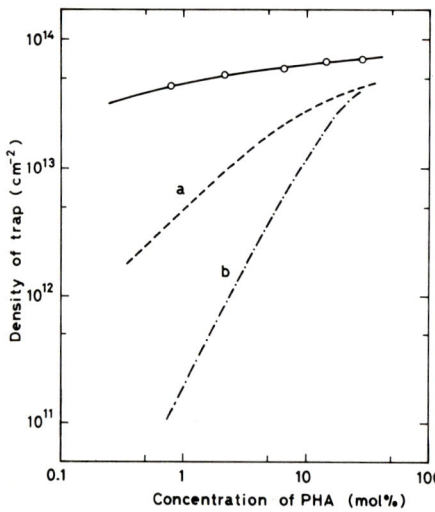

Fig. 3 Results of the decay curve analysis on the basis of two-dimensional energy transfer and trapping. Density of pyrene chromophores ($n_A$) obtained from decay curve analysis is plotted vs PHA concentration. The experimental result is compared with the calculation. In the calculation, (a) pyrene monomer and aggregates and (b) aggregates are taken into consideration as fluorescence quenchers.

The present results have shown that (1) pyrene chromophores in the LB films form two kinds of sites, i.e., isolated monomers and dimers (and/or higher aggregates), with their densities depending on the PHA concentration; (2) in the dimer site, the excimer is formed within 200 ps after excitation; and (3) in the monomer site, when the PHA concentration is higher than 1 mol%, the direct energy transfer and quenching is the dominant pathway of energy relaxation.

Reference

1. I. Yamazaki, N. Tamai, H. Kume, H. Tsuchiya and K. Oba: Rev. Sci. Instrum. **56**, 1187 (1985)

# Fluorescence Concentration Depolarization of DODCI in Glycerol: A Photon-Counting Test of Three-Dimensional Excitation Transport Theory

*D.E. Hart, P.A. Anfinrud, and W.S. Struve*

Department of Chemistry and Ames Laboratory-USDOE,
Iowa State University, Ames, IA 50011, USA


## 1. Introduction

Prior to the development of self-consistent approximations to the solution of the master equation for electronic excitation transport, no analytic theory existed for dealing with the transport dynamics in both the short (non-diffusive) and long (diffusive) time regimes. The probability $G^s(t)$ that the electronic excitation resides on the initially excited molecule at time t is related to the intensities of the polarized fluorescence components via

$$I_\parallel(t) = A\exp(-t/\tau)[1 + 0.8G^s(t)],$$
$$I_\perp(t) = A\exp(-t/\tau)[1 - 0.4G^s(t)], \qquad (1)$$

where $\tau$ is the isotropic fluorescence lifetime. The current highest-order approximation (the 3-body approximation) to $G^s(t)$ was tested by GOCHANOUR and FAYER [2] using polarized fluorescence profiles from rhodamine 6G gated by sum frequency-mixing with 1.06 μm laser fundamental pulses; it yielded good visual agreement with the experimental profiles. In analyzing the self-consistent theory, FEDORENKO and BURSHTEIN [3] have recently noted that it exhibits incorrect long-time behavior and questioned the theory's validity.

In this work, we have obtained polarized fluorescence profiles from DODCI in glycerol, using superior data statistics afforded by time-correlated photon counting ($10^5$ counts/peak channel). A flexible nonlinear least-squares convolute-and-compare analysis is used which avoids parameter correlation in $G^s(t)$ between $\tau$ and the reduced dye concentration $C = 4\pi R_0^3 \rho/3$, where $\rho$ is the dye monomer number density and $R_0$ is the Förster parameter for excitation transfer between two monomers.


## 2. Experimental

The photon-counting apparatus employed for acquisition of fluorescence profiles from DODCI in glycerol is similar to one described earlier [4]. The principal modification was incorporation of a Hamamatsu 1564U microchannel plate phototube with photocurrent pulses amplified by a B&H Electronics AC3011 preamplifier to yield ~ 80 ps instrument function width. The short DODCI lifetime $\tau$~ 1.7 ns in glycerol (vs ~ 3.3 ns for rhodamine 6G) yielded a dynamic range of $0 < t < 5.5\tau$ within the TAC window. Samples were housed in 2 μm cells to minimize self-absorption; isotropic lifetimes $\tau$ from magic-angle profiles were 1.68 ± 0.06 ns at all concentrations except the highest, where excitation trapping by DODCI aggregrates reduced $\tau$ to ~ 1.4 ns at 2.9 mM.

## 3. Results

Single-exponential fits to magic-angle profiles yielded $1.14 < \chi^2 < 2.90$ for most profiles over the concentration range .009 mM < M < 1.82 mM, meaning that these decays are well described by single-exponential functions. $\chi^2 = 9.23$ was obtained at 2.94 mM. Similar $\chi^2$ were obtained when the corresponding polarized fluorescence profiles were fitted using the 3-body function $G^s(t)$ in the model functions of Eqs. (1): $1.42 \leq \chi^2 < 2.80$ for most profiles obtained at the lower concentrations, and $\chi^2 = 7.57$ for the polarized profiles obtained at 2.94 mM. In Fig. 1, we plot optimized reduced dye concentrations C vs actual dye concentrations M. Straight lines show theoretical loci for Förster parameters $R_0$ = 50, 60, and 70 Å. Clustering of data points between $R_0$ = 60 Å and 70 Å for M between 0.31 and 1.82 mM (the latter corresponding to C = 0.93) indicates that the 3-body theory provides an accurate description of transport dynamics in this concentration range, because $R_0$ computed from the DODCI absorption and fluorescence spectra is ~ 68 Å. The divergence at the lowest concentration studied is wholly a consequence of the reduced sensitivity of $G^s(t)$ to C in least-squares fitting when C is small.

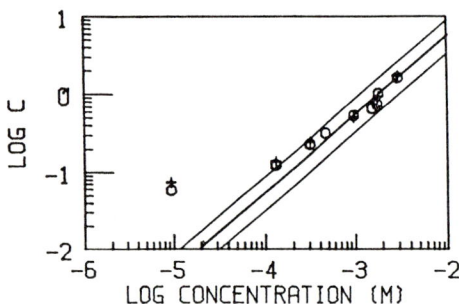

Figure 1. Optimized reduced concentrations C versus DODCI concentrations M in glycerol. Circles and crosses denote data points from different experimental runs; theoretical loci for $R_0$ = 50, 60, and 70 Å are straight lines ordered from right to left.

The optimized reduced concentrations C for 0.13 mM < M < 2.94 mM in Fig. 1 are skewed toward smaller $R_0$ as M increases. Whether this occurs because the 3-body approximation exaggerates the decay in $G^s(t)$ to an extent which increases with concentration, or arises from artifacts in the polarized profiles obtained from photon counting, is currently being investigated.

1. C. R. Gochanour, H. C. Andersen, and M. D. Fayer: J. Chem. Phys. <u>70</u>, 4254 (1979).
2. C. R. Gochanour and M. D. Fayer: J. Phys. Chem. <u>85</u>, 1989 (1981).
3. S. G. Fedorenko and A. I. Burshtein: Chem. Phys. <u>98</u>, 341 (1985).
4. P. Anfinrud, R. L. Crackel, and W. S. Struve: J. Phys. Chem. <u>88</u>, 5873 (1984).

# Fractal Behaviors in Two-Dimensional Excitation Energy Transfer on Vesicle Surfaces

N. Tamai[1], T. Yamazaki[1], I. Yamazaki[1], and N. Mataga[2]

[1]Institute for Molecular Science, Myodaiji, Okazaki 444, Japan
[2]Department of Chemistry, Faculty of Engineering Science, Osaka University, Toyonaka, Osaka 560, Japan

## 1. Introduction

Excitation energy transport and trapping in molecular assemblies have been the subject in recent theoretical and experimental photophysics. For one- and two-dimensional systems, time-dependent equations as expressions of the donor fluorescence decay were given by HAUSER et al. [1]. The kinetic equations for dipole-dipole (Förster type) energy transfer have been formulated under the assumption that the acceptor molecules are randomly distributed. In previous studies dealing with the two-dimensional energy transfer between molecules adsorbed on some solid substrates [2,3], it was demonstrated that the Förster-type kinetics must be modified to obtain good curve fitting for the donor fluorescence decay. The present paper reports on the two-dimensional excitation energy transfer between dye molecules adsorbed on vesicle surfaces by means of a picosecond time-resolved fluorescence spectrophotometer. The fluorescence decay curves of donors are found to fit an equation on the basis of a theoretical framework of "fractals" [4], following the manner of KLAFTER and BLUMEN [5].

## 2. Experimental

Vesicles are static colloidal particles which consist of molecules of two hydrocarbon chains connected to the polar head groups of the surfactants; dihexadecylphosphate (hereafter referred to as DHP) [6] was used in the present study. DHP vesicles were prepared by sonicating the aqueous solution of DHP ($2 \times 10^{-3}$ M) with a sonifier for 30 min at 60 °C. Malachite green was used as acceptor, and two types of dyes, rhodamine 6G and rhodamine B were used as donors. The fluorescence decay curves were measured with a synchronously pumped, cavity-dumped dye laser and a picosecond time-correlated, single-photon counting apparatus [7]. A microchannel-plate photomultiplier (Hamamatu R1564U) was used as a fluorescence detector, so that an instrumental response function of the scattered laser light was measured with 40-ps pulse width (FWHM).

## 3. Theory

"Fractal" denotes a self-similar structure with dilatational symmetry which has potential to describe a multitude of irregular structures [4]. For a two-dimensional system including donor and acceptor dyes, the fluorescence decay function of the donor is given by the following equation [5]:

$$\rho(t) = \exp[-t/\tau_D - \gamma_A (t/\tau_D)^\beta], \quad (1)$$

where $\quad \beta = \bar{d}/s, \quad \gamma_A = x_A (d/\bar{d}) V_d R_0^{\bar{d}} (1 - \beta). \quad (2)$

$\tau_D$ is the lifetime of the donor, and s is the order of the multipolar interaction; $x_A$ is the fraction of fractal sites occupied by the acceptor;

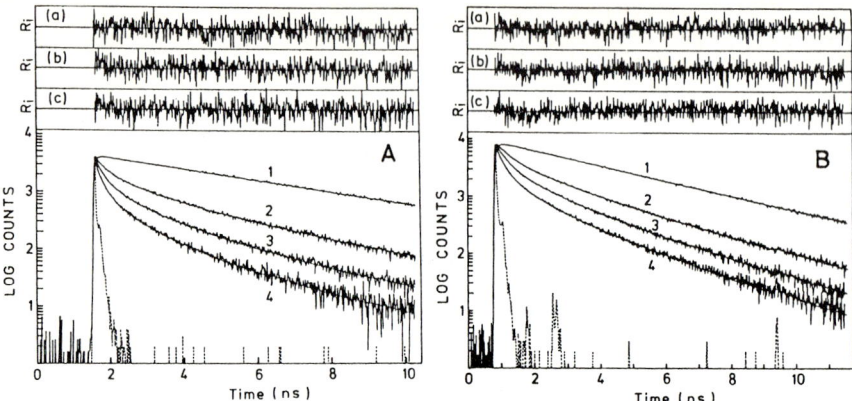

Fig. 1 Fluorescence decay curves of donor, rhodamine 6G (A) and rhodamine B (B), in the presence of acceptor malachite green with concentrations of (1) 0, (2) 2.17, (3) 3.32 and (4) 4.40 x $10^{-6}$ M in (A) and (1) 0, (2) 1.52, (3) 2.32 and (4) 2.98 x $10^{-6}$ M in (B). The theoretical best-fit curves (equation 1) are also shown; (a), (b) and (c) show their weighted residuals for the curves (2)-(4).

d and $\bar{d}$ are the Euclidean and fractal dimensions, respectively; $R_0$ is the critical transfer distance; and $V_d$ is the volume of a unit sphere in d-dimensions.

4. Results and Discussion

Figure 1A shows the fluorescence decay curves of rhodamine 6G (donor) in the presence of malachite green (acceptor) in various concentrations on the DHP vesicle surface. In the absence of acceptor, the fluorescence decay is single exponential with a lifetime of 4.38 ns. By adding the acceptor in as low a concentration as $10^{-6}$ M, the decay profile deviates from an exponential form, indicating that highly effective fluorescence quenching occurs on vesicle surfaces. With another pair of rhodamine B (donor) and malachite green (acceptor), similar behavior can be seen in Figure 1B. In the absence of acceptor, the lifetime of single exponential decay is 3.32 ns.

With the nonexponential decay curves, a simulation calculation by using (1) was carried out by varying values of the parameters $\gamma_A$ and $\beta$. The best-fit curve fitting was obtained by the use of a nonlinear, least-squares iterative convolution method based on the Marquardt algorithm. The results are shown in Fig. 1. It can be seen that the calculated curves on the basis of (1) are well fitted to the experimental curves. Table 1 summarizes the numerical results obtained from the curve fitting. On increasing acceptor concentration, $\gamma_A$ increases linearly with concentration, whereas $\beta$ is approximately constant. The fractal dimension $\bar{d}$ can be derived directly from $\beta$ values through (2); The average values are $\bar{d}$ = 1.31 ± 0.078 for rhodamine 6G and $\bar{d}$ = 1.32 ± 0.049 for rhodamine B. Using calculated values of the critical transfer distance, i.e., $R_0$ = 60 Å for rhodamine 6G - malachite green and $R_0$ = 90 Å for rhodamine B - malachite green, one can estimate the number density of acceptors in the fractal dimension ($x_A$) as shown below.

Table 1  Fractal analysis of the two-dimensional energy transfer by means of (1). Analysis of fluorescence decay curves of donors adsorbed on vesicle surfaces

| Rhodamine 6G | | | | Rhodamine B | | | |
|---|---|---|---|---|---|---|---|
| $n_A$ [a] $[10^{-2} nm^{-2}]$ | $\gamma_A$ | $\beta$ | $\chi^2$ [b] | $n_A$ [a] $[10^{-2} nm^{-2}]$ | $\gamma_A$ | $\beta$ | $\chi^2$ [b] |
| 0.53 | 1.40 | 0.204 | 1.23 | 0.33 | 1.16 | 0.228 | 1.12 |
| 1.07 | 2.75 | 0.210 | 1.09 | 0.65 | 2.20 | 0.220 | 1.10 |
| 1.63 | 4.06 | 0.232 | 1.04 | 0.99 | 3.26 | 0.220 | 1.16 |
| 2.16 | 5.27 | 0.230 | 1.15 | 1.27 | 4.22 | 0.209 | 1.28 |
| 2.59 | 6.25 | 0.217 | 1.27 | 1.57 | 5.10 | 0.221 | 1.35 |

a) Number density of acceptors calculated from the mean surface area of vesicles and the quantity of malachite green.
b) Chi-square values in the curve fitting.

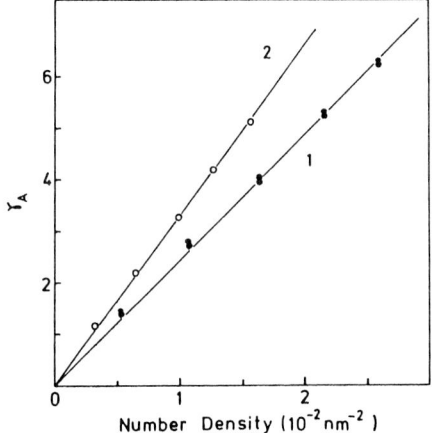

Fig. 2  Plots of $\gamma_A$ against the surface number density of acceptors in the rhodamine 6G – malachite green system (curve 1) and the rhodamine B – malachite green system (curve 2). The surface number densities were calculated from the mean surface area of vesicles and the quantity of malachite green.

Validity of the present analysis was further tested from experiments on (1) the concentration dependence of $\gamma_A$, and (2) the fluorescence quenching measurement in the stationary-excitation fluorescence. Figure 2 shows plots of $\gamma_A$ as a function of number density of acceptors. According to (2), $\gamma_A$ should change linearly with $x_A$ in a fractal dimension. In a system of fractal structure, $x_A$ changes in parallel with the surface number density in a two-dimensional system [5]. It can be seen in Fig. 2 that, in both systems of donor-acceptor pairs, plots of $\gamma_A$ give straight lines against the acceptor density. The difference in the slopes between the two cases is reasonably interpreted as due to the difference of $R_0$. Figure 3 shows plots of the relative fluorescence yield ($I/I_0$) of rhodamine 6G obtained from the stationary-excitation fluorescence intensity measurement. I and $I_0$ are respectively fluorescence intensities with and without acceptor. On increasing the number density of acceptors, the fluorescence intensity rapidly decreases. On the other hand, by integrating (1), we obtain a theoretical expression for the $I/I_0$ ratio as follows:

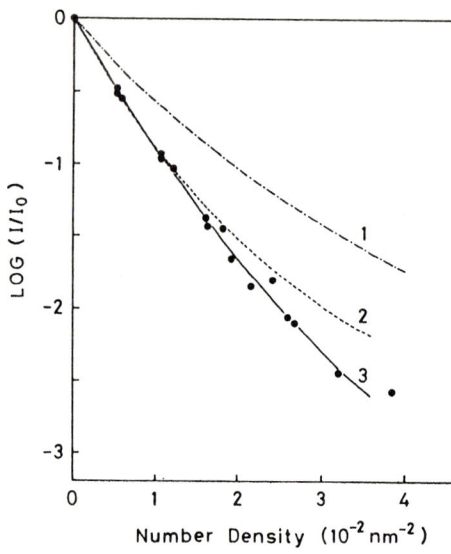

Fig. 3 Plots of the fluorescence quenching ratio ($I/I_0$) of rhodamine 6G versus the surface number density of malachite green. (1) and (2), calculated curves by using the two-dimensional Förster-type equation with $R_0$ = 60 Å and 80 Å, respectively; and (3) calculated curve by using the fractal theory with $R_0$ = 60 Å.

$$I/I_0 = 1 - \gamma_A \int_0^\infty \exp[-x^{1/\beta} - \gamma_A x]dx. \qquad (3)$$

By numerical integration of (3), a calculated curve of $I/I_0$ was obtained as shown in Fig. 3 (curve 3). It can be seen that the experimental curve is well fitted by the theoretical curve.

Apart from the fractal analysis, theoretical $I/I_0$ curves which were obtained from the two-dimensional Förster-type kinetics are also shown in Fig. 3 for rhodamine 6G - malachite green. Since $R_0$ is known to be 60 Å from the spectral overlap, it is seen that the theoretical curve deviates largely from the experimental one. If we fit the theoretical curve to the experimental one, it leads to an unreasonable value of $R_0$ = 80 Å. Thus the usual Förster kinetics on the basis of an assumption of random distribution of acceptors no longer give a correct expression.

The present results have shown that the fractal analysis is applicable to the energy transfer kinetics for dyes adsorbed on vesicle surfaces, and that the fractal dimension of the present systems is $\bar{d}$ = 1.3 irrespective of type of dye molecule. We have not yet attained a simple picture illustrating the dye distribution associated with this fractal dimension. It is worth noting that the fractal dimension 1.3 is identical to the value obtained in the case of the self-avoiding random walk [8]. This might tell us that dye molecules are adsorbed on vesicle surfaces not with a random distribution but with a long-range correlation topologically connecting dye molecules with rods. Recently EVEN et al. [9] studied energy transfer from rhodamine B to malachite green doped into a porous Vycor glass, and found $\bar{d}$ = 1.74. On the other hand, YANG et al. [10] suggested another possibility that this phenomenon results from an excluded volume and is not due to a real fractal structure. The present case of vesicle surfaces provides a two-dimensional system much simpler than the porous glass surface. It is highly probable that the dye distribution exhibits an irregular structure expressed by the fractal.

References

1. M. Hauser, U. K. A. Klein, U. Gösele: Z. Phys. Chem. (Frankfurt am Main) NF. 101, 255 (1976)
2. K. Kemnitz, T. Murao, I. Yamazaki, N. Nakashima, K. Yoshihara: Chem. Phys. Lett. 101, 337 (1983)
3. P. Anfinrud, R. L. Crackel, W. S. Struve: J. Phys. Chem. 88, 5873 (1984)
4. B. B. Mandelbrot: The Fractal Geometry of Nature (Freeman, San Francisco, 1982)
5. J. Klafter, A. Blumen: J. Chem. Phys. 80, 875 (1984)
6. T. Kunitake, Y. Okahata: Bull. Chem. Soc. Jpn. 51, 1877 (1978)
7. I. Yamazaki, N. Tamai, H. Kume, H. Tsuchiya, K. Oba: Rev. Sci. Instrum. 56, 1187 (1985)
8. D. S. McKenzie: Phys. Rep. 27C, 37 (1976)
9. U. Even, K. Randemann, J. Jortner, N. Manor, R. Reisfeld: Phys. Rev. Lett. 52, 2164 (1984)
10. C. L. Yang, P. Evesque, M. A. El-Sayed: J. Phys. Chem. 89, 3442 (1985)

# Transient Vibrational Heating of Molecules After Internal Conversion

*A. Seilmeier, U. Sukowski, W. Kaiser, and S.F. Fischer*

Physik Department der Technischen Universität München, Arcisstraße 21, D-8000 München, Fed. Rep. of Germany

After electronic excitation most organic molecules return to the electronic ground state via radiationless processes. Internal conversion is one of the most important relaxation mechanisms.

A great number of data on internal conversion rates exist, which give information only on the time required by an ensemble of molecules to return from the first excited electronic state $S_1$ to the ground state $S_0$. After internal conversion the energy still resides in the individual molecule.

Recently, we have introduced an ultrafast molecular thermometer which allows us to measure transient internal temperatures of molecules on a picosecond time scale /1/. The excess population of vibrational states in $S_0$ or the internal temperature is monitored via absorption changes at the long-wavelength edge of the $S_1$ absorption. This technique is used to investigate the vibrational state of the molecules after internal conversion in liquid solution. Substantial progress in the understanding of the intermolecular transfer process is made by studying vibrationally hot molecules in different liquid surroundings.

Here, results for azulene in various solvents at room temperature are presented. Azulene was chosen because of its very short lifetime in the $S_1$ state. A large number of vibrationally hot moleculs are rapidly available for investigation.

In our experiment the azulene molecules are excited to the $S_1$ state by the frequency at 18970 cm$^{-1}$. After a very rapid intramolecular redistribution within the $S_1$ state ($\tau \sim 0.5$ ps) the excess energy is transferred to the electronic ground state $S_0$ by internal conversion ($\tau_{ic} \sim 2$ ps /3/). The excess population of vibrational states in $S_0$ is monitored by a second light pulse at a frequency $\nu_2$ which is smaller than the frequency of the pure electronic transition $\nu_{00}$ to the $S_1$ state. Information is obtained on the distribution of the vibrational energy by measuring the absorption change as a function of the probe frequency. The dissipation of the vibrational energy to the surrounding is observed by varying the delay time between the excitation and the probe pulse.

Curve 1 in Fig. 1 shows the long-wavelength tail of the electronic absorption of azulene in CCl$_4$ /2/. The extinction coefficient at frequencies smaller than $\nu_{00}$ (Boltzmann edge)

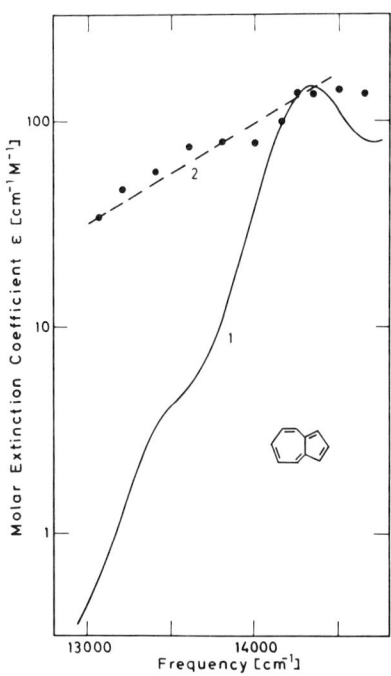

Fig.1 Low-energy absorption tail of azulene at 300K (1) and transient absorption measured 10ps after excitation by photons of energy 19000 cm$^{-1}$ (2).

depends on the population of low-lying vibrational modes. The hot band around 13600 cm$^{-1}$ belongs to a group of vibrations with frequencies of approximately 700 cm$^{-1}$.

Excitation of the sample with picosecond pulses at $\tilde{\nu}_1 = 18970$ cm$^{-1}$ leads to a large change of absorption. The points in Fig.1 show the transient molar extinction coefficient determined by probe pulses 10 ps after excitation of the azulene molecules. We find a broad and strongly increased absorption at long wavelengths. The drastic change of the absorption points to a very rapid intramolecular redistribution and a large temperature rise of the azulene molecules. The broken line through the experimental data in Fig.1 corresponds to a Boltzmann slope of 1200 K. The same internal temperature is calculated when the supplied energy of 19000 cm$^{-1}$ is distributed over all vibrational degrees of freedom of azulene.

How long does the high internal temperature remain in the excited azulene molecules? In Fig.2 the change of absorption is plotted versus time for the solvents CCl$_4$ and CH$_3$OH. The data are taken with probing frequencies of 13600 and 13700 cm$^{-1}$, respectively. The excess absorption rises rapidly with the excitation pulse. Considering the time resolution of the experimental system one estimates an upper limit of 2 ps for the internal conversion and for the intramolecular redistribution of vibrational energy. The data suggest that the redistribution is faster than 1 ps. The intermolecular energy transfer to the solvent, i.e. the cooling of the molecule,

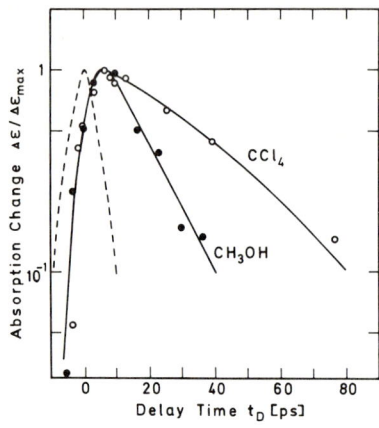

Fig.2 Temporal evolution of the absorption change of azulene in $CCl_4$ and $CH_3OH$, respectively. The curves are calculated using the statistical collision model.

proceeds considerably slower and depends on the solvent. The data in Fig.2 indicate time constants of 40 ps and 13 ps, respectively, for a decrease of the population to a value of 1/e in $CCl_4$ and $CH_3OH$.

Next, we want to discuss a statistical collision model for the intermolecular energy transfer in solutions. The relaxation process is treated in two steps: (i) Isolated binary collisions transfer energy from the hot molecule to a neighboring solvent molecule. (ii) The excess energy of the colliding solvent molecule is transported in the surrounding liquid by macroscopic heat conduction. The second step of the model considers the properties of the liquid.

In the first step the energy transfer is treated in a similar way as in models for the energy distribution in bimolecular reactions. The two collision partners form a short-lived quasi-bound intermediate complex for the duration of the collision. Within the complex, energy equipartition takes place with an exponential time dependence and an equipartition time constant $\tau_e$. The lifetime of the complex, i.e. the duration of the collision $\tau_c$, is much smaller than $\tau_e$. The partitioning of the total energy is incomplete; only a small amount of energy is transferred to the collision partner, a solvent molecule. This first part of the model describes the situation in the gas phase.

In the liquid state there exists a short-range order. Collisions of the hot molecules occur with a few neighboring solvent molecules only. Now we have to discuss the transfer of excess energy to the surrounding liquid, which is approximated by macroscopic heat conduction within the solvent. In solvents with small thermal diffusivity the heat transport from the first shell of solvent molecules to the rest of the liquid and the intermolecular energy transfer is slowed down.

Table 1 shows results for four different solvents, $CCl_4$, $CH_3OH$, $C_6H_6$, and $C_6D_6$. The experimentally determined time constants for the intermolecular energy transfer are smaller when the

Table 1  Experimental and theoretical time constants for the
intermolecular energy transfer in different solvents. The
diffusivity of heat and the energy transferred per collision
in a gas-phase experiment $\Delta E_{gas}$ is used in the calculation.

| Solvent | Transfer time $\tau_{exp}$ [ps] | Tranfer time $\tau_{calc}$ [ps] | Diffusivity of heat [$10^{-3}$ cm$^2$/s] | $\Delta E_{gas}$ [cm$^{-1}$/collision] |
|---|---|---|---|---|
| CCl$_4$ | 40 | 37 | 0.73 | 450 |
| CH$_3$OH | 13 | 15 | 1.0 | 685 |
| C$_6$H$_6$ | 15 | 15 | 0.95 | 920 |
| C$_6$D$_6$ | 10 | 12 | 0.95 | |

thermal diffusivity is large but there is no obvious relation
between the two parameters. The calculated time constants
are determined in the following way: The time constants $\tau_c$ and
$\tau_e$ in our model are determined from gas-phase data /4/, i.e.
they are deduced from the amount of energy $\Delta E_{gas}$ (see Table 1)
transferred in a single collision between an azulene molecule
and the corresponding buffer gas molecule. The diffusivity of
heat given in Table 1 is used to calculate the cooling of the
first shell of solvent molecules. With these parameters curves
are calculated as shown in Fig. 2 for CCl$_4$ and CH$_3$OH. The
calculated time constants correspond very well with the
measured times.

We point to the potential of the model for intermolecular
energy transfer processes in solution. For example, shorter
relaxation times are calculated for smaller (infrared)
excitation energies similar to the experimental observations
/5/.

References
1  A. Seilmeier, P.O.J. Scherer, W. Kaiser, Chem. Phys. Lett.
   105 (1984) 140 ; P.O.J. Scherer, A. Seilmeier, W. Kaiser, J.
   Chem. Phys. 83 (1985) 3948
2  W. Wild, A. Seilmeier, N.H. Gottfried, W. Kaiser, Chem.
   Phys. Lett. 119 (1985) 259
3  C.V. Shank, E.P. Ippen, O. Teschke, R.L. Fork, Chem. Phys.
   Lett. 57 (1978) 433
4  H. Hippler, L. Lindemann, J. Troe, J. Chem. Phys.
   83 (1985) 3906
5  N.H. Gottfried, A. Seilmeier, W. Kaiser, Chem. Phys. Lett.
   111 (1984) 326

# Nonlinear Absorption Spectroscopy of Liquids with Ultrashort IR Pulses

H. Graener, R. Dohlus, and A. Laubereau

Physikalisches Institut, Universität Bayreuth,
D-8580 Bayreuth, Fed. Rep. of Germany

In the past, energy relaxation of excited vibrational states in simple liquids was investigated by Raman probe scattering after stimulated Raman or infrared excitation /1/. These techniques suffer from the smallness of Raman scattering cross section (in the nonresonant case). Valuable data have been accumulated on population life times of higher vibrational states (> 1000 cm$^{-1}$). Direct spectroscopic information on low frequency modes is still lacking, e.g. the recovery time of the vibrational ground state.

Recent progress in the generation of intense, tunable picosecond pulses /2/ has opened up the field of time-resolved nonlinear transmission studies of vibrational absorption bands in the electronic ground state of liquids /3/. In spite of the short molecular time constants involved and the necessary high intensity level of $\sim 10^{11}$ W/cm$^2$, large population changes of several ten per cent can be achieved. Measuring the transmission of delayed weak probing pulses of same frequency as the excitation pulse, the time evolution of the population differences between the upper vibrational level and the vibrational ground state is monitored. For sufficiently short pulses the instantaneous sample transmission

$$T(t) = \exp\{\sigma \ell [N_1(t) - N_0(t)]\} \qquad (1)$$

is observed, where $\sigma$ and $\ell$ respectively are the (effective) absorption cross section and sample length. The number density of molecules in the ground state and upper vibrational level are denoted by $N_0$ and $N_1$. For probe pulses of finite duration the measured energy transmission represents a convolution of $T(t)$ with the pulse shape.

We have studied the transmission changes of small molecules after strong population of CH-stretching modes in the electronic ground state. Experimentally we work with resonantly tuned pulses (2700-4000 cm$^{-1}$) of 15 ps duration and 10 cm$^{-1}$ frequency width which are generated by a mode-locked Nd:YAG laser system and a multiple step parametric generator-amplifier device. An example of our results is presented in Fig. 1. Bromoform in solution of CCl$_4$ is investigated at room temperature. The relative energy transmission $T/T_0$ of the probe pulse is plotted on a logarithmic scale versus delay time (initial transmission $T_0$). The sharp rise of the transmission curve around $t_D=0$ (open circles, solid line) represents the build-up of the population difference $N_1-N_0$ by the excitation pulse. A fast decay follows

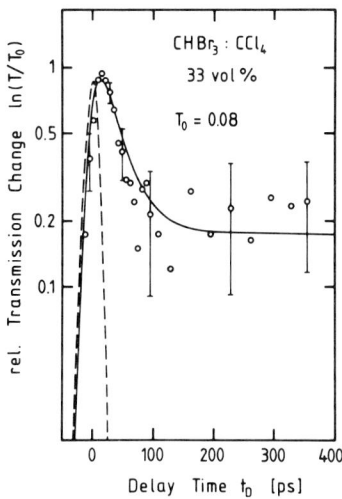

Fig. 1

Relative transmission change ln $(T/T_0)$ of probe energy on a log scale vs delay time for the CH-stretching mode (3020 cm$^{-1}$) of CHBr$_3$

with time constant 35 ± 10 ps, notably slower than the instrumental response curve (broken line). The observed signal decay agrees well with the population life time $T_1$ = 40 ± 10 ps of the v=1 state of the CH-mode known from time-resolved Raman data /4/.

Of particular interest is the signal transient for larger time values. For $t_D \gtrsim$ 150 ps the probe transmission remains almost constant over the observed time interval of ∼.4 ns, well above the steady state value $T_0$. A simple two level approach does not account for the measured time behaviour. Similar results have been obtained for CH$_2$I$_2$ and C$_2$H$_5$OH.

For an explanation of the long time behaviour three possible mechanisms are considered:

(i) thermal redistribution among low-lying vibrational levels ($\leqslant$ 300 cm$^{-1}$)
(ii) transient population of long-lived intermediate vibrational states ($\gtrsim$ 500 cm$^{-1}$)
(iii) photochemical dissociation of vibrationally excited molecules

We have ascertained experimentally that thermal lens effects do not account for the observed long-lived transmission change. Mechanism (i) can be clearly excluded from estimates of the maximum temperature increase ($\leqslant$ 50 K) of the sample and probe transmission measurements (excitation pulse blocked) versus sample temperature in the range 290-360 K. Spectroscopic data of CHBr$_3$ also suggest that the anharmonic shift of the transitions $\nu_3 \to \nu_1+\nu_3$, $\nu_6 \to \nu_1+\nu_6$ is < 2 cm$^{-1}$; i.e. vibrational redistribution among low-lying levels is not detected by the probe transition (see level scheme of CHBr$_3$ in Fig. 2). Mechanism (ii) is well supported by the spectroscopic observation of an anharmonic shift > 5 cm$^{-1}$ of probe transitions $\nu_2 \to \nu_1+\nu_2$, $\nu_4 \to \nu_1+\nu_4$. Transient population of modes $\nu_2$ and/or $\nu_4$ can account for the measured transmission change.

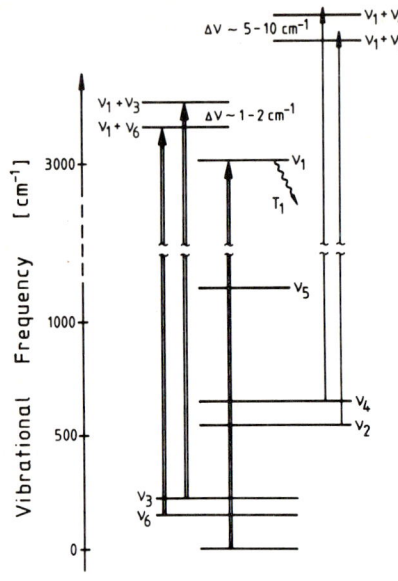

Fig. 2

Normal mode frequencies and several combination tones of $CHBr_3$; double line arrows: excitation process; single line arrows: probing process for possible, long-lived intermediate levels

The role of photochemistry is not settled at the present time. Dissociation of 20 % of the $CHBr_3$ molecules within 100 ps would be required to account totally for the transmission change. Several arguments oppose such a large photochemical effect. The v=1 CH-level is notably populated (several $10^{-7}$) in thermal equilibrium where the solution is stable; i.e. dissociation starting from this level should be too slow by orders of magnitude. A sufficiently large population of higher vibrational levels is unlikely because of the large anharmonic shifts of the v>1 CH-levels and the observation of the v=1 population decay time.

It is suggested that long-lived intermediate levels ($\nu_2$, $\nu_4$) account predominantly for the observed signal transient. In fact a vibrational relaxation time of 2 ns has been observed in ultrasonic dispersion studies for these modes in full accordance with our experimental findings /5/.

1. A. Laubereau and W. Kaiser, Rev. Mod. Phys. 50, 607 (1978)
2. H. Graener and A. Laubereau, Appl. Phys. B 29, 213 (1982)
3. E.J. Heilweil, M.P. Cassasa, R.R. Cavanagh and J.C. Stephenson, Chem. Phys. Lett. 117, 185 (1985)
4. H. Graener and A. Laubereau, Chem. Phys. Lett. 102, 100 (1983)
5. K. Takaji and H. Ozawa, Jap. J. Appl. Phys. 21, 83 (1982)

# Femtosecond Relaxation Dynamics of Large Organic Molecules

*M.J. Rosker, F.W. Wise, C.L. Tang, and A.J. Taylor**

Cornell University, Ithaca, NY 14853, USA
*AT & T Bell Laboratories, Holmdel, NJ 07733, USA

The dynamics of energy relaxation and redistribution in photoexcited molecules is currently a subject of considerable theoretical and experimental interest. Recent developments in apparatus have significantly improved both the temporal and amplitude resolution of these experiments. We have employed the equal-pulse correlation technique to study the ultrafast relaxation of a series of triphenyl methane dyes, which is generally dominated by a strong radiationless process. Representatives of this family of molecules include Malachite Green, Methyl Violet, Ethyl Violet, and Victoria Blue. Data from transmission correlation experiments performed on these dyes consistently exhibit a rapid oscillatory decay following photoexcitation as well as a process with a time constant on the order of a picosecond. We believe that the damped sinusoidal decay may be evidence of quantum beats between molecular eigenstates. Although quantum beats have been observed in the fluorescence decay of large molecules, the oscillation period that we observe is three orders of magnitude shorter than anything which has been reported previously [1]. In addition, measurements of the relaxation of dyes with somewhat different molecular structures reveal rather complex behavior which may or may not include a sinusoidal component. Although these results are not entirely understood, a possible mechanism for the observed features will be presented.

The temporal resolution of our experiment can be attributed to the use of a laser source producing 40 fs pulses with very clean temporal profiles [2]. The emission is peaked at 630 nm (1.97 eV) and the pulse repetition frequency is $10^8$ Hz. Of equal importance, modifications in our data collection technique now allow us to obtain exceptionally high signal-to-noise ratios. Therefore, we are able to deduce time constants from portions of the data which are beyond the coherence length of the pulse, and are thus unaffected by the coherent artifact contribution. We employ a least-squares linear prediction algorithm [3] to extract the amplitudes and time constants of any exponentials present in the data. No manipulation of the data is required before they are submitted to the linear prediction routine.

A typical experimental scan for Malachite Green in ethylene glycol is shown in Fig. 1. For this data the sample was a 30 micron jet of dye dissolved in ethylene glycol to a concentration of $2*10^{-3}$ M. The average power incident on the sample was 1.5 mW per arm of the interferometer and the beams were orthogonally polarized. The initial decay is markedly oscillatory, with a period of 150 fs. The linear prediction method finds three components to this relaxation: exponential decays with time constants of 60 fs and 4.8 ps, and the sinusoid which is exponentially damped with a time constant of 190 fs. The relative amplitudes of these terms are 1, 1, and 0.05, respectively; the fit to the data is shown as a solid line in Fig. 2. The data is an average of $10^4$ scans and we have ruled out the possibility that some experimental artifact is the source of the observed oscillations. Nearly identical results are obtained when the jet thickness is increased to 100 microns, when the power incident on the sample is decreased by up to a factor of three, and

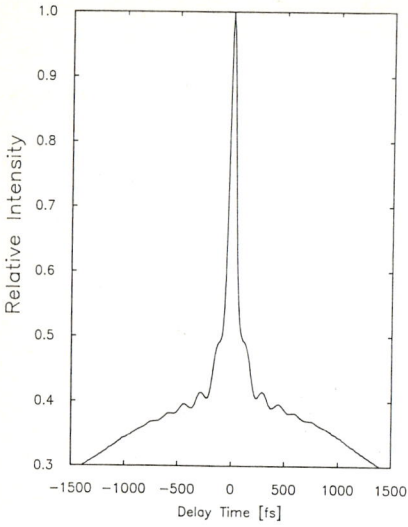
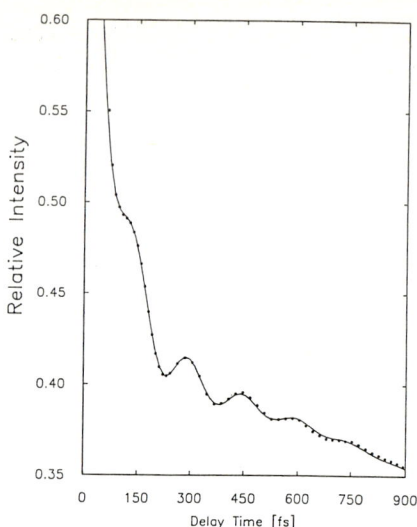

Figure 1. Data for Malachite Green

Figure 2. Fit to the data of figure 1.

for parallel and perpendicular polarizations. The period of the decay is also independent of pulse duration, to the extent that this can be varied while still resolving the oscillatory features.

Qualitatively similar results are found in experiments performed with Methyl Violet, Ethyl Violet, and Victoria Blue dyes. Although the relative amplitude of the sinusoidal component varies somewhat, in all of the triphenyl methane dyes which have been investigated the period of the oscillatory decay is between 155 and 165 fs. In all of these materials the decay also has a large exponential component with a time constant between 30 and 100 fs, as well as a picosecond time scale exponential background.

A similar experimental trace for Nile Blue, taken with the two beams polarized parallel to each other, is shown in Fig. 3. This trace displays very obvious and repeatable non-exponential behavior. Preliminary data, which had a substantially lower signal-to-noise ratio, was fit well by two exponentials with time constants of 80 fs and 520 fs. The 80 fs decay is still quite apparent in the latest pump-probe study of Nile Blue [4]. However, this fit entirely overlooks the structure evident in Fig. 3. This unexpected detail suggests extremely complex decay mechanisms after photoexcitation for Nile Blue.

As a final example, data for the saturable absorber dye DODCI is shown in Fig. 4. Despite the apparant lack of similarity with the triphenyl methane dyes, DODCI also produces an oscillatory signal. Although linear prediction analysis of this data also returns a second sinusoidal component, the principal contribution to the decay has a period of 157 fs, which is remarkably close to the values obtained with the triphenyl methane dyes. To our knowledge this is the first reported study of the energy relaxation of photoexcited DODCI with femtosecond resolution. These results may contribute to our understanding of the operation of passively mode-locked dye lasers.

While the experimental results are unambiguous, the interpretation of the observed relaxation process is not completely clear. The shortest relaxation

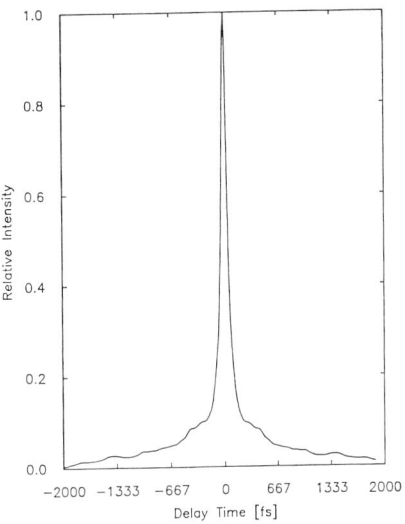

Figure 3. Nile Blue transmission correlation

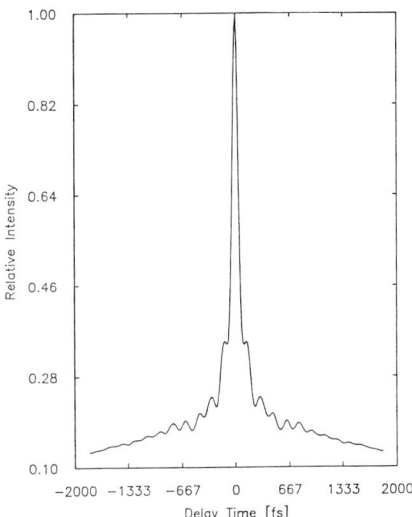

Figure 4. DODCI transmission correlation

times (less than 100 fs in all the dyes that we have studied) may well correspond to a dephasing time for the light-induced coherence between the ground and the excited electronic states of the molecules. It should be mentioned that a previous equal-pulse correlation study of Nile Blue produced data which are consistent with the presence of a fast (<100 fs) relaxation component of small amplitude [5]. Due to the use of much broader pulses and the relatively poor signal-to-noise ratio of those results, however, the authors did not attach significance to portions of the data near the signal baseline.

The damped oscillatory relaxation clearly seen in the decay of the triphenyl methane dyes is puzzling. It is tempting to explain the observed relaxation in terms of a model for quantum beats between two molecular eigenstates of an isolated large molecule. Although the large dye molecules that we study are in liquid solvents, they may be considered isolated on a femtosecond time scale since the collision time with solvent molecules should be on the order of picoseconds. Also, absorption spectra confirm that at the concentrations used in these experiments aggregation of molecules is insignificant. According to this model, only a selected subset of excited states can participate in the initial excitation process. Within this subset there are, say, two states |a> and |b> which are coupled. This leads to two molecular eigenstates,

$|1>\ =\ |a> + |b>$

$|2>\ =\ |a> - |b>$

with an energy separation of $h\Delta_{12}/2\pi$. As an extreme example, if state |a> is optically allowed from the ground electronic state but state |b> is not allowed, the femtosecond laser pulse excites only state |a>, provided that the coherence width of the pulse is greater than $\Delta_{12}$ and spans the split states |1> and |2>, which would be the case in our experiment. Following excitation to |a> at t=0, energy will transfer alternately from |a> to |b> and back, with a damping constant $\Gamma$. This type of oscillation will occur even if |b>

is optically allowed as long as the corresponding optical coupling to the ground state is different from that of |a>. In our experiment, the problem is of course much more complicated than this simple two-mode picture. As discussed by Bloembergen and Zewail, in any particular many-mode case the observation of this type of coherent oscillation will depend very much on 1) whether selectivity in mode coupling is clear enough, and 2) the degree of irreversible relaxation between the subset states and the large number of "bath" modes to which this subset is weakly coupled in such a large system [6]. It is conceivable that the conditions for observing such coherent oscillations happen to be favorable in the case of femtosecond optical excitation of triphenyl methane dyes. Previous observations of quantum beats were made for fluorescence after picosecond excitation in the vapor phase of the large molecule anthracene [1]. This coherence was also found to be related to coupling of molecular states within the vibrational manifold. Our experimental results are distinguished from these not only by the vastly higher beat frequency (6.7 THz, as compared to 11 GHz), but also because the ultrafast time scale allows examination of these molecular levels in the condensed phase.

In conclusion, we have observed the initial relaxation of photoexcited organic dyes with unprecedented resolution. These measurements clearly indicate the presence of complex processes which are extremely fast and substantiate the results of previous work for the slower decays. We believe that we have observed quantum beats in large molecules on a femtosecond time scale for the first time.

This work was supported by the National Science Foundation through the Materials Science Center of Cornell University.

REFERENCES

1. W. R. Lambert, P. M. Felker and A. H. Zewail, J. Chem. Phys. 75, 501 (1981); P. M. Felker and A. H. Zewail, Phys. Rev. Lett. 75, 5958 (1981).
2. J. A. Valdmanis, R. L. Fork and J. P. Gordon, Opt. Lett. 10, 131 (1985).
3. H. Barkhuijsen, R. De Beer, W. M. M. J. Bovee, and D. Van Ormondt, J. Mag. Res. 61, 465 (1985).
4. A. M. Weiner and E. P. Ippen, Chem. Phys. Lett. 114, 456 (1985).
5. A. J. Taylor, D. J. Erskine and C. L. Tang, Chem. Phys. Lett. 103, 430 (1984); 106, 578 (1984).
6. N. Bloembergen and A. H. Zewail, J. Phys. Chem. 88, 549 (1984).

# Population Lifetimes of OH(v=1) and OD(v=1) Vibrations in Alcohols, Silanols and Crystalline Micas

*E.J. Heilweil, M.P. Casassa, R.R. Cavanagh, and J.C. Stephenson*

National Bureau of Standards, Center for Chemical Physics, Room B-268, Building 221, Gaithersburg, MD 20899, USA

1. Introduction

Vibrational energy dynamics of excited fundamental modes of ground electronic state condensed-phase molecules is suspected of playing a role in many bond-breaking physical and chemical processes. As a first step in elucidating this role, it is necessary to compare the vibrational v=1 lifetimes ($T_1$) of a specific functional group in a variety of molecules and chemical environments. It should then be possible to determine whether structural or other vibrational properties of a molecular family influence that particular group's chemical reactivity.

At this time, relatively little $T_1$ lifetime data exists for condensed-phase polyatomic molecular vibrations. Most picosecond measurements have been performed on the CH-stretching modes of a diverse range of hydrocarbon molecules [1]. This paper presents a systematic study of the high frequency OH(v=1) ($\approx$3650 cm$^{-1}$) and OD(v=1) ($\approx$2700 cm$^{-1}$) stretching mode $T_1$ lifetimes of twelve alcohol and eight silanol molecules in dilute room temperature CCl$_4$ solutions. The objective is to determine whether $T_1$ for structurally related compounds is affected by substitution of the hydroxyl linking atom for carbon or silicon. Comparison of these lifetimes to earlier OH on silica results [2-4] and to site specific lifetimes of hydroxyl groups within the crystalline lattice sheets of ten natural micas is also made [5].

2. Method and Results

The technique employed in this work is a picosecond single wavelength infrared saturation/recovery method used previously to measure $T_1$ lifetimes of OH(v=1) on the surface of colloidal silica [2]. Measurements for OH on silica in several solvents [2], inside fused silica [3], and as a function of temperature and isotope [3,4] gave 100$\leq T_1 \leq$250 ps. The stretching frequencies and $T_1$ values for the alcohol and silanol molecules studied here are presented in Table I. One finds that <u>all</u> silanol molecules containing either the SiOH or SiOD group have $T_1$ lifetimes comparable to the above silica results. Apparently $T_1$ is only slightly affected by the substituent groups on the Si atom (i.e., methyl, ethyl, phenyl or amorphous silica). Alcohols of similar structure to the silanols, however, yield $T_1$ lifetimes of 20 ps or less. Micas, composed of OH$^-$ ions octahedrally coordinated to mostly Mg$^{+2}$ (Biotites) and Al$^{+3}$ (Muscovites) cations sandwiched between silica-like sheets, also have $T_1$ lifetimes in a range comparable to the silica/silanol results (see Table II) [5].

3. Discussion

Since SiOH(D) $T_1$ lifetimes appear to be relatively insensitive to the groups attached to the Si atom, it is argued that only those fundamental vibrational

Table I: Vibrational lifetimes for the OH(v=1) and OD(v=1) stretches in alcohols and silanols[a] in dilute $CCl_4$ solution (≤0.007 mole fraction) at 298 K

| Silanol | $\nu_{OH,OD}[cm^{-1}]$ | $T_1$ [ps] | Alcohol | $\nu_{OH,OD}[cm^{-1}]$ | $T_1$ [ps] |
|---|---|---|---|---|---|
| $Me_3SiOH$ | 3690 | 205 ± 21 | $Me_3COH$ | 3614 | < 6 [b] |
| $Me_3SiOD$ | 2722 | 245 ± 25 | | | |
| $Et_3SiOH$ | 3689 | 185 ± 19 | $Et_3COH$ | 3622 | < 6 [b] |
| $Et_3SiOD$ | 2722 | 224 ± 22 | $Et_3COD$ | 2673 | <20 [b] |
| $\varphi_3SiOH$ | 3675 | 206 ± 21 | $\varphi_3COH$ | 3609 | <15 [b] |
| $\varphi_3SiOD$ | 2712 | 292 ± 29 | | | |
| $\varphi_2Si(OH)_2$ | 3610, 3679 | 80 ± 15 | $CH_3OH$ | 3641 | 15 - 30 [b] |
| $\varphi_2Si(OD)_2$ | 2665, 2710 | 134 ± 14 | $CH_3OD$ | 2685 | 52 ± 17 |
| | | | $CD_3OH$ | 3642 | 73 ± 7 |
| | | | $CD_3OD$ | 2690 | 79 ± 17 |
| | | | EtOH | 3625 | 70 ± 10 |
| | | | $\varphi OH$ | 3610 | 5 - 20 [b] |
| | | | $\varphi OD$ | 2665 | 15 - 25 [b] |
| | | | $\varphi F_5OD$ | 2640 | < 15 [b] |

[a] $Me=-CH_3$, $Et=-CH_2CH_3$, $\varphi=-C_6H_5$  [b] Limits deduced from computer simulations.

Table II: $T_1$ vibrational lifetimes of the OH(v=1) stretching mode of hydroxyl ions in ten crystalline micas at 298 K. Muscovite measurements were obtained on either side of the single absorption feature centered at ca. 3630 $cm^{-1}$. Frequencies in wavenumbers $[cm^{-1}]$; uncertainties in $T_1$ are ±1σ

| Muscovite | $\nu_{OPA}$ | $T_1$[ps] | Biotite | $\nu_{OPA}$ | $T_1$[ps] |
|---|---|---|---|---|---|
| 104935 | 3591 | 88 ± 16 | 82063 | 3691 | 198 ± 15 |
| 105013 | 3648 | 90 ± 15 | C3647 | 3691 | 244 ± 15 |
| 105051 | 3591 | 83 ± 18 | C3675-1 | 3691 | 220 ± 18 |
| | 3675 | 85 ± 8 | | 3575 | 125 ± 11 |
| B16862 | 3675 | 105 ± 36 | Ruggles | 3591 | 72 ± 20 |
| Ruggles | 3591 | 79 ± 8 | Unknown | 3540 | 64 ± 5 |
| | 3675 | 114 ± 13 | | | |

modes structurally adjacent to the OH(D) group are involved in accepting the OH(v=1) or OD(v=1) excitation. However, it is conceivable that collisions with the solvent sufficiently perturb the solute to cause the OH(D) (v=1) excitation to leave the OH(D) bond. Because structurally related silanol and carbinol molecules were studied in the same solvent, we assert that the molecule-solvent interactions are similar and that it is the specific intramolecular vibrational energy levels that are responsible for the different $T_1$ lifetimes of these families.

In the absence of any clear theoretical modelling for the vibrational deactivation process, an analysis of all energetically favorable decay channels composed of nearest neighbor fundamental modes has therefore been performed [6]. Assignments from the literature are used for the frequencies of the twelve "methanol-like" vibrational modes to compare, as examples, $Me_3COH$ and $Me_3SiOH$. A subset of these modes includes the MOH bend ($\delta_{MOH}$; M=C,Si), M-OH stretch ($\nu_{MOH}$), symmetric and asymmetric CMC stretches ($\nu^s_{CMC}$, $\nu^a_{CMC}$) and CMO bend ($\delta_{CMO}$). Using the frequencies for these modes, decay processes to nearly resonant ( $|\Delta E| \leq kT = 200$ cm$^{-1}$) harmonic overtone and combination states are obtained. For a particular decay channel involving n vibrationally exchanged quanta (n=order of process), the propensity rule that lower order processes occur more rapidly can then be applied [6].

For the relaxation of $Me_3COH$, one finds 17 nearly resonant n=4 processes, such as:

$$\nu_{OH} \rightarrow \delta_{COH} + 2\ \nu_{COH} + 2\ cm^{-1}$$

and 
$$\nu_{OH} \rightarrow \delta_{COH} + \delta_{CCC} + \nu_{COH} - 76\ cm^{-1}\ .$$

There are also 83 possible n=5 and 495 n=6 pathways. A similar analysis for $Me_3SiOH$ shows there are no allowed n=4 processes, 15 n=5 and 196 n=6 pathways. From this result, the silanol is expected to exhibit a longer vibrational lifetime (as is observed) because it can only relax via a higher order process. This is because the silanol vibrational modes (except $\nu_{OH}$) as a family are significantly lower in frequency than the corresponding alcohol vibrations.

Performing a similar comparison for the isotopic analogs $Me_3SiOH$ and $Me_3SiOD$ is more difficult. Briefly, simple channel counting predicts that the deuterated species would relax more rapidly than the protonated one. This is because lowest order n=4 processes are only found for $Me_3SiOD$. All of these n=4 channels also involve the $\nu_{SiOD}$ mode. Both $Me_3SiOD$ and $Me_3SiOH$ have n=5 channels which include the $\delta_{SiOH(D)}$ bending vibration. The similarity of the observed $T_1$ lifetimes for these two isotopic molecules (with $T_1(OD) \geq T_1(OH)$) favors the interpretation that $\nu_{OH(D)}$ relaxes predominantly to the $\delta_{SiOH(D)}$ mode via a n=5 process and not to the Si-OH(D) stretching motion.

The results for the $T_1$ lifetimes of OH$^-$(v=1) relaxation in Muscovite and Biotite micas yield several interesting observations. As seen in Table II, the hydroxyl ion in the octahedral crystalline environment of $Al^{+3}$ or $Mg^{+2}$ cations produces average lifetimes (by hydroxyl type; see below) in the $90 \leq T_1 \leq 220$ ps range. Again, this range is comparable to the results for all SiOH systems studied so far. The mechanistic arguments (considering neighboring mode frequencies and near resonances) used for the silanol and alcohol systems above also appear to apply to the mica systems. The hydroxyl groups of micas are still coordinated to relatively massive cation centers. The associated vibrational mode frequencies are again found to be lower than those of COH in alcohols and the mica $T_1$ lifetimes are still relatively long ($T_1 \geq$ 90 ps) [5].

The single IR absorption band for Muscovites ($\nu_{OH} \approx 3630$ cm$^{-1}$) and the slightly lower frequency band of Biotites ($\nu_{OH} \approx 3575$ cm$^{-1}$) have both been attributed to OH groups in lattice positions with only two neighboring octahedral cations (dioctahedral sites). These hydroxyls are also oriented in the mica sheet plane [7]. The higher frequency Biotite band ($\nu_{OH} \approx 3691$ cm$^{-1}$) arises from OH groups that are oriented nearly perpendicularly to the mica sheet and are in trioctahedral sites. Because of their specific lattice orientations and relative absorption frequencies, the degree of hydrogen bonding to nearby lattice oxygen atoms is believed to affect the $T_1$ lifetime of these hydroxyls. The more strongly hydrogen bound, in-plane hydroxyls exhibit $T_1 \approx 90$ ps while more weakly bound, out-of-plane Biotite hydroxyl groups have $T_1 \approx 220$ ps. Hydrogen bonding interactions between the hydroxyl group and the local environment may therefore account for the ranges of $T_1$ lifetimes found in the various hydroxyl-containing systems investigated to date [2-6].

While it appears that systems containing the same structural unit (e.g., SiOH or COH) exhibit similar ranges of $T_1$ lifetimes, it remains to be seen whether these results correlate with chemical reactivity. Comparison of alcohol and silanol dehydration or halogenation rates under identical reaction conditions could give insight into this possibility.

This work was supported in part by the Air Force Office of Scientific Research. Special thanks to Drs. P. J. Dunn and J. S. White of the Smithsonian Department of Mineralogy for providing the mica samples used for this work.

References:

1. W. Zinth, C. Kolmeder, B. Benna, A. Irgens-Defregger, S. F. Fischer and W. Kaiser, J. Chem. Phys. <u>78</u>, 3916 (1983) and references cited therein
2. E. J. Heilweil, M. P. Casassa, R. R. Cavanagh and J. C. Stephenson, J. Chem. Phys. <u>82</u>, 5216 (1985)
3. E. J. Heilweil, M. P. Casassa, R. R. Cavanagh and J. C. Stephenson, Chem. Phys. Lett. <u>117</u>, 185 (1985)
4. M. P. Casassa, E. J. Heilweil, J. C. Stephenson and R. R. Cavanagh, J. Chem. Phys. <u>84</u>, 2361 (1986)
5. E. J. Heilweil, Chem. Phys. Letters, in press
6. E. J. Heilweil, M. P. Casassa, R. R. Cavanagh and J. C. Stephenson, J. Chem. Phys., in press
7. S. W. Bailey: In *Reviews in Mineralogy*, ed. by S. W. Bailey, Vol. 13 (Mineralogy Society of America, 1985) Ch. 1 and 2

# $S_0$-$S_n$ Two-Photon Absorption Dynamics of Rhodamine Dyes

P. Sperber, M. Weidner, and A. Penzkofer

Naturwissenschaftliche Fakultät II - Physik, Universität Regensburg, D-8400 Regensburg, Fed. Rep. of Germany

1. Introduction

The intensity-dependent transmission of picosecond ruby laser pulses through methanolic and ethanolic solutions of rhodamine B and rhodamine 6G is studied. The dye molecules are excited by two-photon absorption. The relaxation of the excited molecules involves radiative and radiationless transitions. Besides two-photon absorption the light transmission is affected by stimulated emission at the pump laser frequency, by amplified spontaneous emission, and by excited state absorption.

2. Dye model

The dynamics of the absorption, emission and relaxation processes is described by a realistic multilevel system (see Fig.1). The stimulated emission cross-sections $\sigma_{em}^L$ and $\sigma_{em}^{ASE}$ are deduced from conventional fluorescence spectra. The $S_n$ to $S_1$ relaxation is extremely short compared to the pump pulse duration ($\tau_{ex} \simeq 10^{-13}$s, $\Delta t_L \simeq 30$ ps). The $S_1$-state lifetime $\tau_F$ is obtained by streak camera measurements. The two-photon absorption cross-section $\sigma^{(2)}$ and the excited-state absorption cross-sections $\sigma_{ex}^L$ and $\sigma_{ex}^{ASE}$ are determined in the experiments.

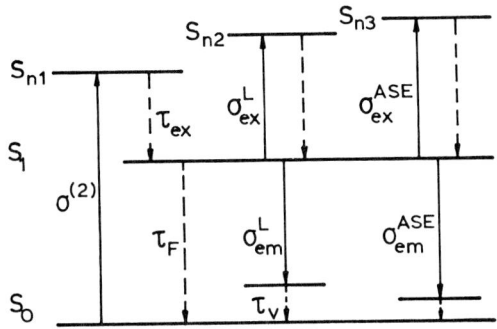

Fig.1 Level system

3. Experiments

The experimental arrangement is sketched in Fig.2. The two-photon absorption cross-section $\sigma^{(2)}$ is determined by measuring the intensity-dependent energy transmission through the sample (photodetectors PD1 and PD3; intensity detection by nonlinear

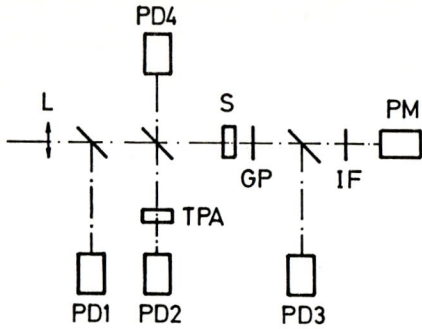

Fig.2 Experimental setup. L, lens; S, sample; GP, glass plate; TPA, CdS crystal

transmission measurement through CdS with detector PD1 and PD2 [1]). The excited state absorption cross-section $\sigma_{ex}^L$ is detected by transmission measurement of a weak back-reflected probe signal (detectors PD3 and PD4, transmission depends on $\sigma_{em}^L - \sigma_{ex}^L$) [2]). The excited state absorption cross-section $\sigma_{ex}^{ASE}$ is deduced from observation of amplified spontaneous emission (photomultiplier PM, signal depends on $\sigma_{em}^{ASE} - \sigma_{ex}^{ASE}$).

## 4. Results

The energy transmission of picosecond ruby laser pulses (duration 30 ps) through rhodamine 6G in methanol (concentration 0.04 mol/l, sample length 2 cm) is shown in Fig.3. The solid curve gives the best theoretical fit to the experimental points. The dashed curves neglect effects of stimulated emission, amplified spontaneous emission and excited state absorption. In Table 1 the obtained absorption cross-section data for rhodamine 6G and rhodamine B are presented. The same data are obtained for the solvents methanol and ethanol.

Table 1: Data for rhodamine dyes

| Parameter | Rhodamine B | Rhodamine 6G |
|---|---|---|
| $\sigma^{(2)}$ | $(1.2 \pm 0.2) \times 10^{-48}$ cm$^4$s | $(1.8 \pm 0.2) \times 10^{-48}$ cm$^4$s |
| $\sigma_{em}^L$ | $(1.9 \pm 0.2) \times 10^{-17}$ cm$^2$ | $(9 \pm 1) \times 10^{-18}$ cm$^2$ |
| $\sigma_{ex}^L$ | $(1 \pm 0.3) \times 10^{-17}$ cm$^2$ | $(5 \pm 1) \times 10^{-18}$ cm$^2$ |
| $\sigma_{em}^{ASE}$ | $(7 \pm 1.5) \times 10^{-17}$ cm$^2$ | $(7 \pm 1.5) \times 10^{-17}$ cm$^2$ |
| $\sigma_{ex}^{ASE}$ | $(1 \pm 1) \times 10^{-17}$ cm$^2$ | $(1 \pm 1) \times 10^{-17}$ cm$^2$ |
| $\lambda_{em}^{ASE}$ | $646 \pm 5$ nm | $626 \pm 5$ nm |

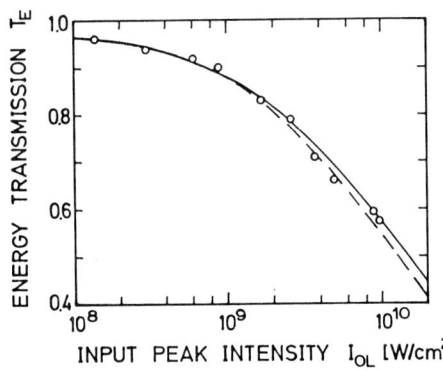

Fig.3 Two-photon transmission

The influence of various dye parameters on the intensity dependent transmission through the dye samples is analysed by numerical simulations. Effective excited state absorption at the pump laser frequency ($\sigma_{ex}^L - \sigma_{em}^L > 0$) reduces transmission while effective stimulated emission at the pump laser frequency ($\sigma_{em}^L - \sigma_{ex}^L > 0$) increases transmission. Amplified spontaneous emission ($\sigma_{em}^{ASE} - \sigma_{ex}^{ASE} > 0$) decreases the $S_1$-state population and therefore reduces the effects of excited state absorption and stimulated emission at the laser frequency. Bottle neck effects occur for the stimulated emission at $\nu_L$ and the amplified spontaneous emission at $\nu_{ASE}$ if the terminal states in the $S_0$ band are not thermalized (time constant $\tau_V$) fast enough.

References

1. W. Blau and A. Penzkofer, Opt. Commun. **36**, 419 (1981)
2. J. Wiedmann and A. Penzkofer, Il Nuovo Cimento **63B**, 459 (1981)

# Nonlinear Optical Response of One-Dimensional Excitons in Polydiacetylene

B.I. Greene[1], J. Orenstein[1], R.R. Millard[1], and L.R. Williams [2],*

[1] AT & T Bell Laboratories, Murray Hill, NJ 07974, USA
[2] Massachusetts Institute of Technology, Cambridge, MA 02139, USA

Interest in the primary excitations of systems exhibiting reduced dimensionality has blossomed over the past few years. Multiple quantum well (MQW) semiconductor structures have proved fertile ground for many new phenomena and theories relevant to excitations confined in 2-dimensions (2D). In the present work, we extend these studies to an archetypical 1-dimensional (1D) system, polydiacetylene-PTS, hereafter referred to as PTS.[1]

The optical absorption of single crystal PTS is dominated by a narrow ($\approx 100$ meV at 300K) peak at $\sim 2.0$ eV polarized along the chain axis, which is excitonic in origin.[2] The enhanced oscillator strength and binding energy of the 1D confined exciton in PTS follows the trend observed in passing from 3D to 2D in the GaAs-GaAlAs MQW system.

Both techniques of transient grating and transient absorption spectroscopy, performed with a 10 Hz amplified CPM dye laser, were utilized. Excitation pulses consisted of 70 fs FWHM pulses at either 1.97 or 3.94 eV. Transient grating measurements were performed in the reflective mode.

When probed in the 1-2 eV region, 3.94 eV excitation resulted in the prompt appearance of a positive absorbance signal followed by a ca. 0.6 ps decay (determined at 1.4 eV) into a metastable state. Transient difference spectra shown in Fig. 1 reveal that this metastable signal corresponds well to the spectrum

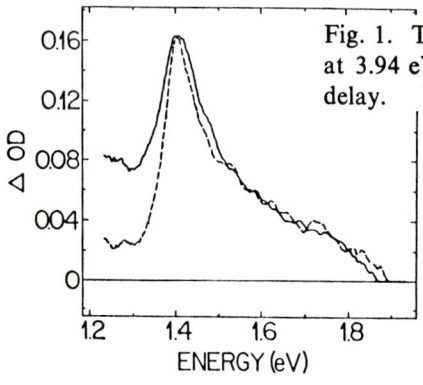

Fig. 1. Transient absorption spectra of PTS excited at 3.94 eV. Solid line 3 ps delay; dashed line 60 ps delay.

---

* AT&T Bell Laboratories, Ph.D. Scholar

of the triplet exciton characterized previously on the microsecond timescale.[3] Excitation at 1.97 eV was also observed to yield triplet excitons, however the excitation in this case was determined to be 2-photon or biexcitonic in nature. We believe the 1-photon triplet quantum yield at 1.97 eV to be negligibly small.

Transient grating measurements performed on resonance at ca. 2.0 eV provide a direct measure of the ground state recovery time subsequent to creation of the singlet exciton. Figure 2 displays the results of such a measurement, yielding a $2.0 \pm 0.2$ ps best fit exponential lifetime. This places an upper limit on the lifetime of the singlet exciton.

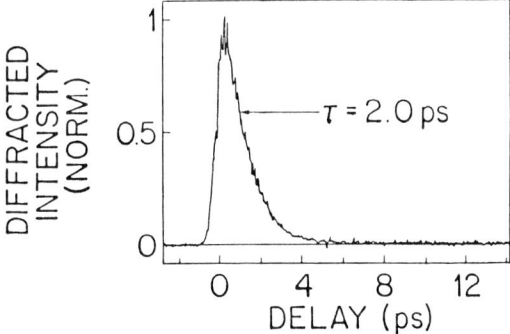

Fig. 2. Normalized diffraction efficiency of PTS excited at 1.97 eV.

The magnitude of the intensity dependent reflectivity change was determined and compared to that calculated using the simple "phase-space filling" model previously utilized in predicting the strengths of optical nonlinearities in GaAs-MQW structures.[4] This model proposes that in the presence of a non-zero exciton population, some of the single-particle band states needed to form the exciton wavefunction are already occupied. This results in a reduction in strength of the exciton transition given by

$$\frac{\delta \chi}{\chi} = -\frac{N}{N_s},$$

where N is the excition density (number per unit length) and $N_s$ is the saturation density. By use of appropriate 1-D exciton wavefunctions, we relate

$$n_s = \frac{N_s}{\sigma} = \frac{2}{3\sigma\xi_o},$$

where $\sigma$ is the cross-sectional area per chain and $\xi_o$ is the exciton length. Our measurement of induced changes in reflectivity together with relations between the complex index of refraction, dielectric constant, reflectivity, and linear susceptibility yield a value of $n_s = 2.0 \times 10^{20}$ cm$^{-3}$.[6] Assuming $\sigma = 100$Å$^2$ gives an exciton length of 33Å, in excellent agreement with recent calculations.[5] We conclude the resonant nonlinear response of PTS is primarily due to this phase-space filling effect.

Previous models explaining the large nonresonant $\chi^3$ in PTS utilize one-electron band theory or three level systems and an anharmonic electron restoring force.[7] Neither model has been completely successful in predicting the sign or magnitude of $\chi^3$. We propose an extension of the phase-space filling model to virtual excitons, in the context of the exciton-polariton picture. We expect that polaritons, through their exciton component, will interact due to the exclusion principle. A simple expression results:

$$\chi^3_{NR} = \frac{\epsilon_1^2 S(\omega)}{32\pi^2 \hbar \omega n_s},$$

where $S(\omega)$ is the exciton structure factor.[8] From our measured value of $n_s$ and a simple approximation for $S(\omega)$, this expression yields results in excellent agreement with experiment,

$$\chi^3_{NR} = \frac{\epsilon_1^2}{32\pi^2}\left[\frac{1}{n_s E_{ex}}\right],$$

where $E_{ex}$ is the exciton energy. The quantitative success of our formalism suggests that it contains the correct physics underlying the optical nonlinearity in PTS.

## REFERENCES

1) J. Orenstein, in "Handbook of Conducting Polymers," T. A. Skotheim, ed. (Marcel Dekker, New York, 1986) p. 1297.

2) M. Pope and C. E. Swenberg, "Electronic Processes in Organic Crystals" (Oxford University Press, New York, 1982).

3) L. Robins, J. Orenstein and R. Superfine, Phys. Rev. Letts., 56, 1850 (1986).

4) S. Schmitt-Rink, D. S. Chemla and D. A. B. Miller, Phys. Rev. B, 32, 6601 (1985).

5) S. Suhai, Phys. Rev. B., 29, 4570 (1984).

6) B. I. Greene, J. Orenstein, R. R. Millard and L. R. Williams, submitted for publication.

7) G. P. Agrawal, C. Cojane and C. Flytzanis, Phys. Rev. B, 17, 776 (1978).

8) V. M. Agranovich and M. D. Galanine, "Electronic Excitation Energy Transfer in Condensed Matter" (North-Holland, Amsterdam, 1982) p. 150.

# Picosecond Photoconductivity and Nonlinear Optical Phenomena in *trans*-Polyacetylene

D. Moses, M. Sinclair, and A.J. Heeger

Department of Physics, Institute for Polymers and Organic Solids,
University of California, Santa Barbara, CA 93106, USA

Transient photoinduced absorption $\delta\alpha(t)$ measurements in trans-$(CH)_x$ have been useful for studying the generation and characteristics of nonlinear excitations in polyacetylene [1]. Here we address the transport properties of these excitations as they are revealed by studies of the transient photoconductivity, $\sigma_p(t)$, and the related nonlinear optical processes, $\chi^{(3)}$. $\sigma_p(t)$ was measured using the Auston switch techniques [2]. The switch consisted of a 50 $\Omega$ microstrip transmission line with a narrow gap (0.2 mm) on which a thin polymer sample was grown. The excitations were produced by uniformly illuminating the sample with 20 ps dye laser pulses (1 µJ/pulse) at 2.1 eV. The overall rise time in our measurements of $\sigma_p(t)$ is ~ 50 ps. Figure 1 shows $\sigma_p(t)$ at a bias field of $1.5 \times 10^4$ V/cm. The fast rise is followed by (approximately exponential) decay with $\tau \sim 300$ ps which is the same magnitude as the rate of decay of $\delta\alpha(t)$. This indicates that mobile charged soliton excitations are produced within picoseconds, consistent with the Su-Schrieffer mechanism [3], and that the decay of photocurrent is primarily due to decay of the number of charge carriers. The magnitude of $\sigma_{p(t)}$ at 50 ps is about 0.3 S/cm for a flux of $10^{15}$ cm$^{-2}$ per pulse, an increase of about six orders of magnitude over the dark $\sigma$ at room temperature! The spectral dependence of $\sigma_p$ at 50 ps is the same as that of the CW $\sigma_p$. The sharp rise in $\sigma_p(t)$ at the photon energy associated with the interband transition rules out the possibility of the direct production of neutral excitons.

The shape of $\sigma_p(t)$ at $t < 10^{-9}$ S as well as the peak in $\sigma_p(t)$ at ~ 50 ps are temperature independent, as is shown in the inset of Fig. 1. This is interpreted as due to the production of "hot" charged soliton excitations that evolved from the initially generated electron hole pairs. The soliton excitations gain an excess kinetic $\varepsilon$ s$\simeq 2\Delta (1 - 2/\pi)$, where $2\Delta$ is the energy gap and $2/\pi$ ($2\Delta$) is the creation energy of soliton pair. On the other hand, the "tail" that follows the peak in $\sigma_p(t)$ at ~ 50 ps is strongly temperature dependent (this tail is not visible on the scale of Fig. 1).

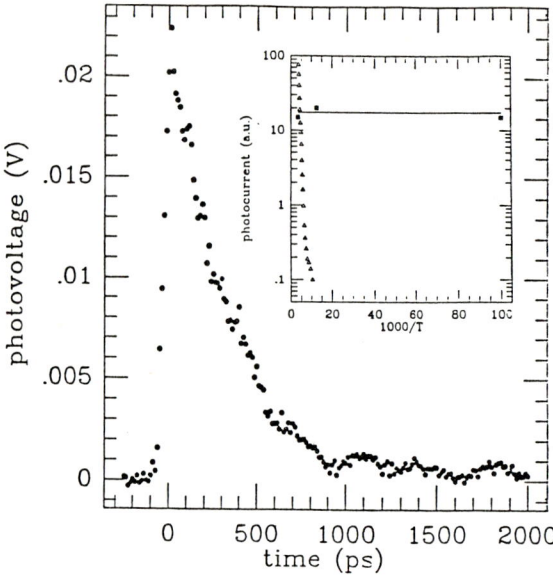

Fig. 1. Transient photovoltage (across 50 Ω Auston) of trans-(CH)$_x$. Inset: Temperature dependence of the magnitude of $\sigma_{ph}$; steady state (△; excited at 500 nm) and picosecond (■; excited at 590 nm). The solid line emphasizes the T-independence of the fast photoconductivity.

The existence of two different photoconduction mechanisms at short and long times is found also in other quasi-1D-systems in which polarons and bipolarons constitute the charge carrier excitations. In initial studies of both polythiophene and polydiacetylene, we find an almost T-independence of the peak in $\sigma_p(t)$, in contrast to the T-dependent tail which follows this peak. The precise T-dependence of the peak is governed by the fraction of the remaining unthermalized carriers at 50 ps after their creation, since this time is the experimental integrating time of our measuring system. For a slow thermalization rate, we expect a T-independent peak in $\sigma_p(t)$. Generally, as the hot excitations are thermalized, their mobility is modified and the existing traps become increasingly more important. Eventually, after a long time, the trapped carriers dominate as indicated by the thermally activated mobility of the steady state photoconductivity.

The transient photoinduced absorption and bleaching measurements demonstrate a major shift of oscillator strength upon photoexcitation occurring at times of $\simeq 10^{-13}$ s. This indicates a large imaginary component of the resonant third-order nonlinear susceptibility which originates in the self-localized photoexcitations (solitons, polarons and bipolarons) which characterize this large class of conducting polymers. The third real, nonresonant order nonlinear optical susceptibility of trans-$(CH)_x$ has been recently measured [4] by third harmonic generation in a thin film. The third harmonic was generated by a pulse train of a mode locked Nd:YAG laser radiation ($\lambda$ = 1.06 µ). Peak pump powers in excess of 10 GW/cm$^2$ were used without damage to the sample. The measured susceptibility is $\chi^{(3)}$ ($3\omega=\omega+\omega+\omega$) = 4 X $10^{-10}$ esu which is comparable to the magnitude of the large nonlinear susceptibilities measured in the polydiacetylene.

## REFERENCES

1. Z. Vardeny, J. Strait, D. Moses. T.-C. Chung and A. J. Heeger, Phys. Rev. Lett. 49, 1657 (1982); C. V. Chank, R. Yen, R. L. Fork, J. Orenstein and G. L. Baker, Phys. Rev. Lett. 49, 1660 (1982).
2. M. Sinclair, D. Moses and A. J. Heeger, Solid State Commun. (to be published).
3. W.-P. Su and J. R. Schrieffer, Proc. Natl. Acad. Sci. USA, 77, 815 (1981).
4. M. Sinclair, D. Moses, A. J. Heeger, K. Vilhelmsson, B. Valk and M. Salour, J. Appl. Phys. Lett. (submitted).

# Singlet Exciton Fusion in Molecular Solids

*R.R. Millard and B.I. Greene*

AT & T Bell Laboratories, Murray Hill, NJ 07974, USA

The dynamics of exciton motion and decay along with energy transfer in molecular solids have been persistently studied over the last twenty years. The phenomenon of singlet-singlet annihilation (fusion) has been demonstrated unequivocally through numerous nanosecond and picosecond time-resolved fluorescence measurements [1-3]. To date, however, no measurements have been presented that show evidence of the microscopically detailed nature of these dynamical processes. In this paper we present and discuss the first such data.

We used the technique of subpicosecond time-resolved absorption spectroscopy to measure the characteristic annihilation rate for singlet excitons on adjacent lattice sites. Measurements were performed with amplified pulses derived from a colliding pulse modelocked dye laser [4]. These pulses were 0.15 ps FWHM, centered at 625 nm (16,000 cm$^{-1}$), and occurred at a repetition rate of 10 Hz. Pump and probe measurements were performed utilizing pulses at the laser fundamental and white light continuum pulses respectively. Details of this experimental technique have previously been presented [5].

Samples consisted of polycrystalline thin films ($\sim 1000$Å) of $\beta$-hydrogen phthalocyanine (H$_2$Pc). This material has several properties important to the success of our experiments. It is highly absorbing at our laser fundamental with $\alpha \approx 2 \times 10^5$ cm$^{-1}$ at 16,000 cm$^{-1}$. Therefore, high excitation densities (ca. $10^{21}$ cm$^{-3}$) were readily attainable with available excitation intensities. H$_2$Pc is readily sublimed to form good optical quality amorphous films, which can be annealed to yield polycrystalline samples. These films have been well characterized, with convincing evidence to support the notion that individual microcrystallites (which have a flake-like morphology and in-plane dimensions of 10-100$\mu$) are isomorphous with the $\beta$ form of the macroscopic single crystal. The films showed remarkable resistance to optical damage. Finally, H$_2$Pc exhibits an intense excited-state absorption spectrum with strong transitions in the visible.

Figure 1 displays the ground state absorbance spectrum of a typical H$_2$Pc sample together with transient absorbance difference spectra taken at three different delay times. The sample was held at 4K. The difference spectra reveal regions of bleaching (due to ground state depopulation) in addition to regions of excited state absorption.

As the delay time was increased, a dramatic evolution of the entire difference spectrum was observed. During the first 10 ps subsequent to excitation, there was

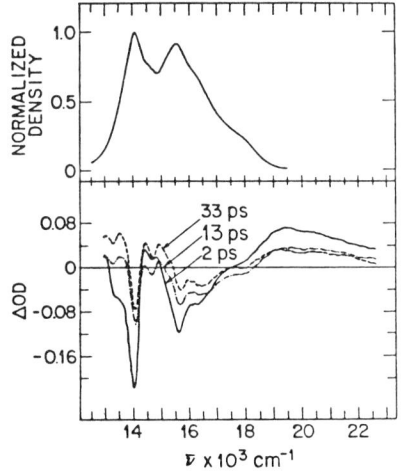
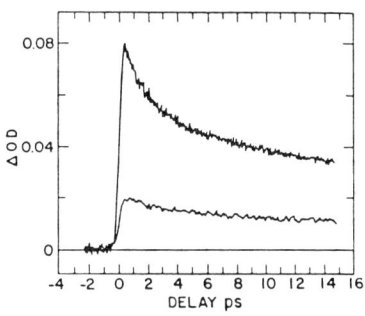

Fig. 1. Upper: Ground state absorption spectrum of thin film sample of H$_2$Pc (4K).

Lower: Transient absorption spectra of H$_2$Pc taken at delay times of 2, 13 and 33 ps (4K).

Fig. 2. Induced change in optical density of H$_2$Pc vs. delay time for an excitation intensity of $5 \times 10^{10}$ W/cm$^2$ (upper) and $5 \times 10^9$ W/cm$^2$ (lower).

a rapid decay of the initially induced transient spectrum. This is due to a nonradiative process which efficiently depopulates the initially prepared exciton population. We chose to probe kinetics at a spectral position where the observed signal could be unambiguously related to the density of excitons. Such a region occurred in the area of excited state absorption between 19,000 cm$^{-1}$ and 22,000 cm$^{-1}$, where minimal ground state absorption insured little contribution from ground state recovery processes manifest in the signal. We observed the spectral shape of the excited state absorption in this region to remain constant throughout the decay. In contrast, the spectral evolution occurring at energies below 16,000 cm$^{-1}$ is quite complex. In particular, a dramatic increase in sample absorption occurred at roughly 13,500 cm$^{-1}$, where an initial net bleaching rapidly evolved into a persistent net increase in sample absorbance lasting longer than 200 ps.

Figure 2 shows single wavelength pump and probe data obtained with pump pulses at 16,000 cm$^{-1}$ and probe pulses at 19,500 cm$^{-1}$. The excitation (pump) intensity was $5 \times 10^{10}$ W/cm$^2$ for the upper trace and $5 \times 10^9$ W/cm$^2$ for the lower trace. These signals decayed in a highly nonexponential fashion. The kinetics were observed to be independent of sample temperature over the range 4 to 300K.

We suggest that the initial rapid decay of the induced absorbance signal is due to singlet exciton-exciton annihilation [6]. The high density of excitons required to make such an observation was achieved by our experimental conditions, where we estimate exciton densities of approximately $1.0 \pm 0.5 \times 10^{21}$ cm$^{-3}$ with our highest intensity excitation pulses. Single crystal specific gravity measurements indicate a value of $1.68 \times 10^{21}$ cm$^{-3}$ for total excitonic saturation [7].

Previous studies have measured time-independent annihilation rates which are comprised of two separable processes: exciton motion and exciton-exciton annihilation. In principle, time-dependent energy transfer rates are expected in measurements of exciton trapping and annihilation. This arises from the fact that subsequent to the initial creation of a spacially homogeneous exciton population, proximate pairs of excitons, or excitons closest to traps, will interact first. Progressively greater mean interparticular distances result in decreasing characteristic interaction rates. Both motion-limited diffusion theory and the Förster long range dipole-dipole interaction picture (in the absence of diffusion) lead to a $t^{-1/2}$ dependence for $\gamma$ (the bimolecular rate constant) [2,8].

For annihilation dominated kinetics, the phenomenological rate equation can be written:

$$dN/dt = -\gamma t^{-1/2} N^2 . \qquad (1)$$

Integration of (1) yields an expression for the exciton number density as a function of time:

$$N(t) = (2\gamma t^{1/2} + 1/N_0)^{-1} . \qquad (2)$$

Our data is observed to obey such a time dependence. By plotting our data in the form $(1/N - 1/N_0)$ vs. $t^{1/2}$, we obtain a straight line whose slope is $\gamma$ (from (2)). We extrapolate a value for $\gamma$, correcting for nonuniform excitation profiles, to be $1.0 \pm 0.5 \times 10^{-16}$ cm$^3$ sec$^{-1/2}$.

Excitonic saturation of the $H_2Pc$ lattice facilitates a clear and simple interpretation of the initial rapid signal decay. Since the majority of excitons will have excitons as nearest neighbors, we propose that a single exponential fit to the data at the earliest decay times corresponds to the lifetime of proximate excitons. Over the first two picoseconds such a fit is quite good, yielding a temperature-independent lifetime of 5.7 ps. This observed lifetime was independent of pump intensity for intensities above ca. $3 \times 10^{10}$ W/cm$^2$. We expect that as the sample becomes saturated, the early time decay rate will be dominated by the single exponential rate associated with proximate exciton pairs. As such, this study presents the only measurement of an exciton annihilation rate free from the ambiguity or possible influence of exciton motion.

Finally we discuss the spectral dynamics in the low energy region of figure 1. We attribute the increase in spectral density at ca. 13,500 cm$^{-1}$ with increasing delay time to the reappearance of vibrationally hot ground state molecules. With the aid of spectral models, we conclude that this cannot be explained by an increase in lattice temperature. Rather, we suggest a vibrationally selective conversion of electronic energy into vibrational bottlenecks as a plausible explanation for these results.

# REFERENCES

[1] A. Bergman, M. Levine and J. Jortner, Phys. Rev. Lett. 18, 593 (1967).

[2] R. C. Powell and Z. G. Soos, J. Luminescence 11, 1 (1975).

[3] D. Schmid, in "Organic Molecular Aggregates,"Springer Series in Solid-State Sciences 49, ed. by P. Reineker, H. Haken and H. C. Wolf, (Springer-Verlag, Berlin, 1983), p. 184.

[4] R. L. Fork, B. I. Greene and C. V. Shank, Appl. Phys. Lett. 38, 671 (1981).

[5] E. P. Ippen and C. V. Shank, in "Picosecond Phenomena," ed. by C. V. Shank, E. P. Ippen and S. L. Shapiro, (Springer-Verlag, Berlin, 1978), p. 103.

[6] B. I. Greene and R. R. Millard, Phys. Rev. Lett. 55, 1331 (1985).

[7] F. H. Moser and A. L. Thomas, "Phthalocyanine Compounds" (Reinhold, New York, 1963), p. 15.

[8] A. Suna, Phys. Rev. B 1, 1716 (1970).

# Matrix Effect on Vibrational Relaxation in Molecular Crystals

*J.R. Hill, E.L. Chronister, J.C. Postlewaite, and D.D. Dlott*

School of Chemical Sciences, University of Illinois at Urbana-Champaign, 505 South Mathews Avenue, Urbana, IL 61801, USA

We have used picosecond coherent Raman scattering (ps CARS) and picosecond photon echo (ps PE) spectroscopies to study vibrational relaxation in low temperature crystals of naphthalene, durene, and naphthalene substituted into durene.

NAPHTHALENE          DURENE

In a crystal of rigid molecules, there are two kinds of mechanical excitations, phonons and vibrons. Phonons involve external (translation and librational) motion, while vibrons are internal to the molecule. In a real crystal, translation and libration are mixed with low frequency vibrations; the mixing is the greatest with floppy molecules. Vibrons also acquire some external character, permitting interaction between neighbors [1]. A third type of excitation, the mixed state, will also occur. A mixed state is a low frequency, large amplitude vibration which behaves like a phonon, or external mode. The lowest frequency vibration in naphthalene, the "butterfly" mode, and the methyl group torsions of durene are classic examples of mixed states [2].

When a vibration is excited in a low temperature crystal, it typically decays to two lower energy modes via cubic anharmonic interactions. In naphthalene, each vibron may decay to any of several lower frequency vibrons (the acceptor states) by emitting a phonon. The rate of emission to each acceptor is mainly determined by the density of states (DOS) [3]. In this case the lifetime is given by

$$(T_1)^{-1} = \frac{36\pi}{\hbar^2} \sum_A \langle B^{(3)} \rangle^2 \rho^P(w_V - w_A) = \frac{36\pi}{\hbar^2} \langle B^{(3)} \rangle^2 D(w_V) . \quad (1)$$

In (1), $T_1$ is the lifetime of the vibron at $w_V$, the B-coefficient is a cubic anharmonic matrix element, $\rho P(w)$ is the phonon DOS at $w$, and A denotes the acceptor states. Equation (1) defines $D(w_V)$, the two-phonon DOS at $w_V$ [3]. Figure 1 illustrates calculation of $D(w_V)$ for the decay of the naphthalene $\nu_9$ mode (w ~510 cm$^{-1}$) in a pure crystal (upper figure) and as a guest in a durene crystal (lower figure). The (laser excited) $\nu_9$ is the heavy bar with arrows. The other heavy bars represent acceptors. The light line is the outline of the experimentally determined phonon density of states of naphthalene [4] and durene [5] at the <u>difference frequency</u> between $\nu_9$ and the acceptors, i.e. it is the DOS of the <u>emitted phonons</u>.

482

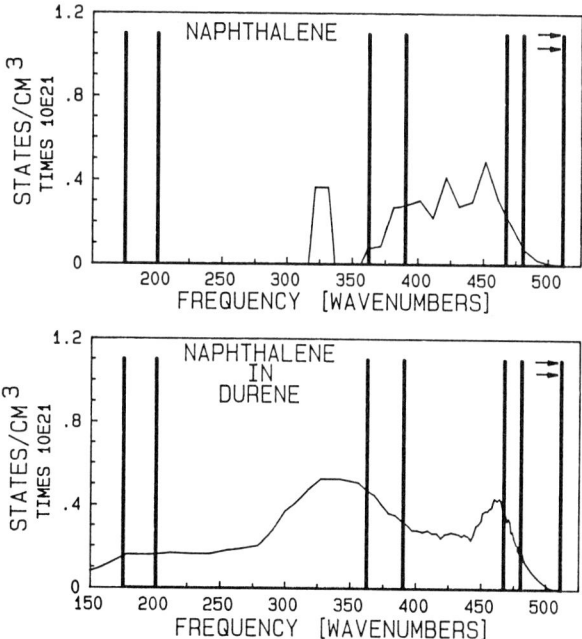

Figure 1. Two phonon density of states for decay of naphthalene $\nu_9$ (heavy line marked by arrows) to a vibrational acceptor (heavy line) and a phonon (light line) in the pure crystal (above) and in durene. Acceptors at a peak of the phonon density of states contribute most to the decay.

In naphthalene, $\nu_9$ may decay to four acceptors via single phonon emission, the largest rate occurring when the emitted phonon is at a peak in the DOS. The 480 and 362 cm$^{-1}$ modes each contribute about 10% and the 467 and 390 cm$^{-1}$ modes each contribute about 40% to the decay [3]. When naphthalene is substituted into the durene host, relaxation involves decay to a naphthalene vibration and a durene phonon. The durene DOS has much larger area than the naphthalene DOS because although both crystals have 12 phonons, the methyl torsions in durene also behave as phonons in the relaxation process.

Table 1 gives measured vibrational lifetimes at low temperature for totally symmetric modes in pure naphthalene (ps CARS [3]), and for naphthalene in durene (ps PE). The two-phonon DOS $D(w)$ for each mode is determined from neutron scattering data and the vibrational spectrum. The anharmonic B-coefficient is calculated from (1). The anharmonic coupling is about a factor of two smaller for naphthalene in the durene matrix. The decrease in anharmonic coupling is due in part to mass mismatch between host and guest, and to the more harmonic nature of the H--H nonbonded interaction relative to C--C interactions.

We have also studied the relaxation of various deuterated naphthalene isomers in the pure crystal [6], and in durene matrix, as well as the vibrons of pure durene [unpublished results].

This research was supported by the National Science Foundation, Solid-State Chemistry Division through grant DMR 84-15070, and an Alfred P. Sloan Fellowship to DDD.

Table 1. $T_1$ is the vibrational lifetime at low temperature
$<B(3)>$ is the cubic anharmonic coupling constant in cm$^{-1}$
$D(w)$ is the density of states in states/cm$^{-1}$

| | PURE NAPHTHALENE[a] | | | NAPHTHALENE IN DURENE[b] | | |
|---|---|---|---|---|---|---|
| | $T_1$ | $D(w)$ | $<B(3)>$ | $T_1$ | $D(w)$ | $<B(3)>$ |
| $\nu_9$ | 140 ps | 0.49 | 0.05 cm$^{-1}$ | 66.4 ps | 1.69 | 0.04 cm$^{-1}$ |
| $\nu_8$ | 62 ps | 0.25 | 0.1 cm$^{-1}$ | * | * | * |
| $\nu_7$ | 19 ps | 1.25 | 0.08 cm$^{-1}$ | * | * | * |
| $\nu_6$ | <10 ps | 0.63 | ---- | * | * | * |
| $\nu_5$ | 92 ps | 0.30 | 0.07 cm$^{-1}$ | 54 ps | 2.17 | 0.04 cm$^{-1}$ |
| $\nu_4$ | * | * | * | 38 ps | 2.96 | 0.04 cm$^{-1}$ |
| $\nu_3$ | 14 ps | 0.78 | 0.12 cm$^{-1}$ | * | * | * |

APPROXIMATE FREQUENCIES (cm$^{-1}$): $\nu_9$ = 511, $\nu_8$ = 766, $\nu_7$ = 1021, $\nu_6$ = 1146, $\nu_5$ = 1385, $\nu_4$ = 1428, $\nu_3$ = 1578.

[a] Picosecond CARS [3]. [b] Picosecond photon echo. *Was not observed.

### References

1. D. D. Dlott, Ann. Rev. Phys. Chem. 37, 157.
2. D. C. Ahlgren and R. Kopelmen, Chem. Phys. 48, 47 (1980); P. N. Prasad and R. Kopelman, J. Chem. Phys. 58, 5031 (1972).
3. C. L. Schosser and D. D. Dlott, J. Chem. Phys. 80, 1394 (1984).
4. E. L. Bokhenkov, I. Natkaniec, and E. F. Sheka, Sov. Phys. JETP 43, 536 (1976).
5. J. J. Rush, J. Chem. Phys. 47, 3936 (1967).
6. E. L. Chronister, J. R. Hill, and D. D. Dlott, J. Chim. Phys. 83, 159 (1985).

# Optical Damage in Molecular Crystals: A Solid State Explosion

*D.D. Dlott, T.J. Kosic\*, and J.R. Hill*

School of Chemical Sciences, University of Illinois at Urbana-Champaign, 505 South Mathews Avenue, Urbana, IL 61801, USA

Optical damage is not always bad, although it might seem so to users of high power lasers. There are many constructive uses for optical damage such as semiconductor annealing, and write-only optical memories. We have recently found a new use for optical damage--simulation of the behavior of energetic solid materials such as rocket propellants or explosives. In our experiments [1], a low temperature crystal of an unreactive material (acetanilide = N-phenyl amide, a molecular crystal consisting of linear chains of benzene groups linked by hydrogen bonds) is intensely irradiated by a tightly focussed ~100 ps pulse ($I \sim 10^9$ W/cm$^2$) which, via multiphoton processes, ionizes a small fraction of the molecules. The ionization process creates a variety of high energy defect centers which are stabilized by the low temperature crystal matrix. The sample is allowed to cool and the pulse is repeated. Eventually defects accumulate in large concentration (a few percent). A considerable density of energy (2000 J/cm$^3$) is stored in a very small volume ($10^{-8}$ cm$^3$). The accumulation process typically takes $10^4$ pulses. The total irradiation time is a few μs.

At this stage, the crystal is ready to begin destruction. Having survived perhaps 10,000 pulses, a violent reaction occurs which irreversibly damages the irradiated volume during the next few pulses. The damage takes the form of an empty cavity or inclusion located in the bulk of the crystal. The destruction period occurs while the sample is irradiated for < 1ns. The forces associated with this damage process are comparable to the explosion of an equivalent amount (10 ng) of TNT [1].

We studied optical damage using a YAG-pumped dye laser [2] in the configuration of Figure 1. The transmission of the 532 nm green pulses was monitored and normalized by twin photodiodes, sample-and-hold circuits and a computer. When the crystal was damaged, the transmission would decrease. The computer determined how many pulses were required to damage the crystal, and kept a record of the transmitted intensity during the destruction period. The damage was initiated with the green pulses alone, or in conjunction with a variable time delay dye laser pulse. Tuning and delaying the dye laser pulse created a sequence which was designed to pump a specific electronic or vibrational state. The defect accumulation was monitored with time resolved CARS [3] (psCARS) from the two dye lasers $\omega_1$ and $\omega_2$. The optical phonons in acetanilide ($\omega \sim 40$cm$^{-1}$) are quite long lived at 10K ($T_1 \sim 1$ns) [3]. As defects accumulate, deterioration of the crystal lattice causes the phonon lifetimes to decrease. Signal averaging was accomplished by destroying 50-100 well-separated regions of each sample.

1. Damage Kinetics

Two representative damage events are shown in Figure 2. The crystals have already been irradiated by thousands of pulses (a few μs of irradiation),

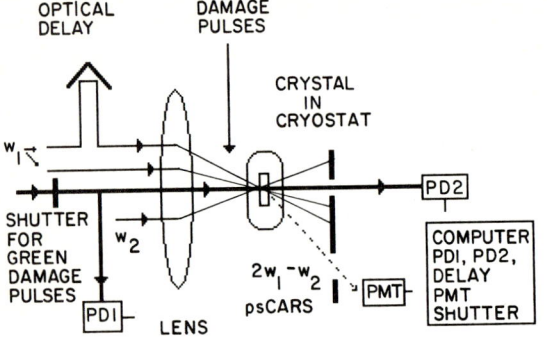

Figure 1. Apparatus for damage studies. A YAG pumped dual dye laser [2] is the pulse source. PD = photodiode. PMT = photomultiplier

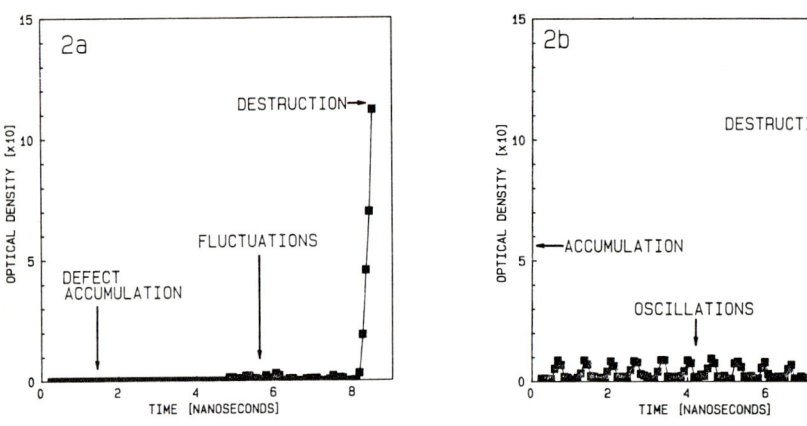

Figure 2. Two representative damage events. In each case the defects were accumulated for tens of thousands of pulses.

during which time there has been no observable change in the optical density. The figures show the final 100 pulses (8.5 ns of irradiation) before catastrophic damage is observed. In 2a, tiny fluctuations in transmitted intensity begin to occur a few ns before the .5 ns of catastrophic damage. In 2b, perfectly periodic fluctuations in the transmitted intensity occur for a few hundred ns before damage. Each damage event is unique. There is wide variation in the duration of the accumulation and destruction period. The magnitude of the fluctuations or oscillations which precede the destruction also varies. However the period of the oscillations does not change much under the various conditions we used [1]. The kinetics of these events seem to be analogous to a chemical relaxation oscillator driven by a noisy source (fluctuations in the laser amplified by multiphoton ionization).

2. Damage-Detected Spectroscopy

In a "damage-detected" picosecond experiment, the rate (averaged over many events) of optical damage is determined for various pulse sequences. In one series of experiments, we have shown that defect production occurs via two successive 2-photon absorption steps which ionize acetanilide. The energy level diagram for acetanilide is shown in Figure 3, where several important 2-photon transitions are indicated [1]. Damage was only observed with pulse sequences which 2-photon excite $S_1$. Dye laser wavelengths between 612 nm and 602 nm cannot excite $S_1$ by any 2-photon process, but they can 2-photon ionize $S_1$. Such pulses increase the damage rate whether simultaneous with $S_1$ excitation, or delayed a small fraction of the $S_1$ lifetime.

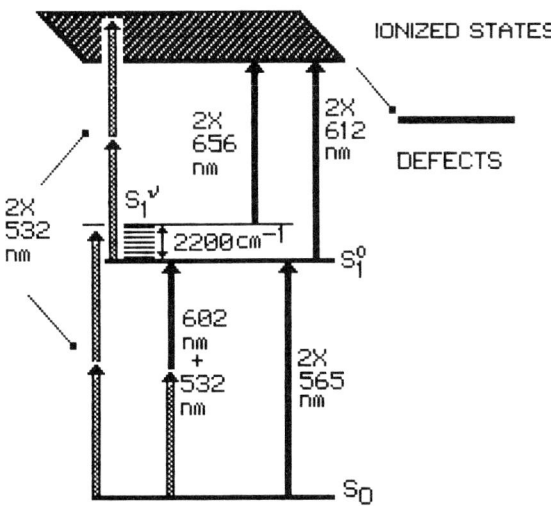

Figure 3. Energy level diagram for acetanilide showing several two-photon transitions.

It was also shown that excess population in a long-lived vibron ($T_1$ = 25 ps, $\omega$ = 1000 cm$^{-1}$) could strongly affect the damage rate. Low temperature crystals were excited by simultaneous 532 nm and 561.9 nm pulses which produced a vibrational coherence in the symmetric ring stretching mode. The equivalent vibrational temperature was T* = 160K. In this case, the damage rate decreased by a factor of 2. The direction of this effect suggests that the excited vibrons may participate in annealing of the defects [1].

## 3. Phonons and Defects

Hydrogen bonded crystals typically have several long-lived optical phonons [3]. In a virgin acetanilide crystal, phonons at ca. 40 cm$^{-1}$ have lifetimes of several hundred ps, corresponding to Raman linewidths of ~10$^{-2}$ cm$^{-1}$. Such sharp phonon transitions may be used to probe crystal lattice destruction via the broadening of the Raman line, or equivalently the increase in psCARS decay rate. Figure 4 shows the results of a psCARS experiment where the crystal was repeatedly irradiated for 58,000 pulses (5μs of irradiation), and the phonon lifetimes measured with weak dye laser pulses [1]. Defect accumulation causes the phonon lifetime to decrease. In Figure 4, both phonon lifetimes have decreased by about the same amount, 2x10$^9$/sec. Using the theory of phonon-defect scattering, we find that such a scattering rate can be obtained if the defect concentration is roughly 5%. Similar results

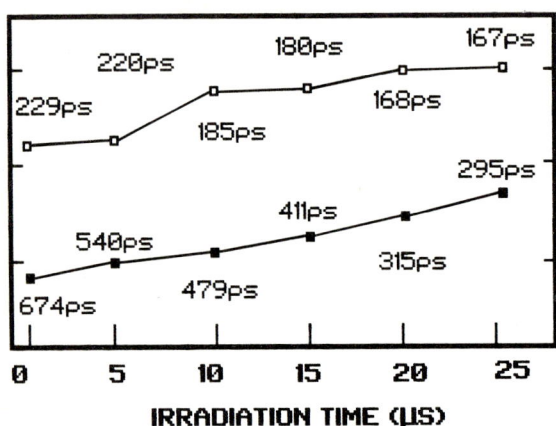

Figure 4. The destruction of the 10K crystal lattice during defect accumulation. The lifetimes decrease due to phonon-defect scattering. Open boxes = 38 cm$^{-1}$ Ag phonon. Solid boxes = 39 cm$^{-1}$ Bg phonon.

are obtained on phonons in crystals with several percent impurities. It is this large concentration of energetic defects which causes these crystals to "explode" [1].

This research was supported by the National Science Foundation, Solid-State Chemistry grant DMR 84-15070, and a fellowship from the Alfred P. Sloan Foundation to DDD.

References

* Current address: Hughes Aircraft Corporation, El Segundo, Ca. 90245.

1. J. R. Hill, T. J. Kosic, E. L. Chronister, and D. D. Dlott, Springer Proc. Phys. 4, 107 (1985); T. J. Kosic, J. R. Hill and D. D. Dlott, Chem. Phys. 104, 169 (1986).

2. R. E. Cline Jr., E. L. Chronister, T. J. Kosic, C. L. Schosser, and D. D. Dlott. Proceedings of the International Conference on Lasers, 1983, ed. R. C. Powell (STS Press, McLean, Va., 1985) p. 697.

3. T. J. Kosic, R. E. Cline Jr. and D. D. Dlott, J. Chem. Phys. 81, 4932 (1984); D. D. Dlott, Ann. Rev. Phys. Chem. 37, 157 (1986).

# Rotational Relaxation of Free and Solvated Rotors

*A.J. Bain, C. Han, P.L. Holt, P.J. McCarthy, A.B. Myers, M.A. Pereira, and R.M. Hochstrasser*

Department of Chemistry, University of Pennsylvania,
Philadelphia, PA 19104, USA

## 1. Introduction

In most experimental studies of molecular motion in liquids the signals are averaged over many collisional periods and the processes observed are diffusive. When studies are carried out with sufficiently short light pulses it should become possible to observe nondiffusive behavior. In the case of molecular reorientation dynamics the nondiffusive portion corresponds to nearly free rotation. It is of great interest to study these transient regimes because of the detailed information that can be obtained about collision dynamics in condensed phases. While these transients are well known through studies of Rayleigh scattering [1], infrared, and Raman lineshapes [2] and from time domain Kerr effect studies in neat liquids [3,4], we are not aware of studies of nondiffusive effects involving the electronically excited states of molecules in dilute solutions. In the present paper we discuss fluorescence and polarization spectroscopy as methods of approach and present preliminary results which bring us closer to realizing these goals.

For a short period after ultrashort pulse excitation a molecule in a solution can be considered to be freely rotating inasmuch as it will have definite values of the angular momentum, J, and its projections onto laboratory- and molecule-fixed axes. The collisions will cause the initial J and its projections to randomize [5]. The initial situation corresponds closely to that which would prevail in a gas at the same temperature. We therefore began our investigations with studies of gases to try to observe the free rotation transients. We expect that when free rotation is rapid compared with collisions the nondiffusive behavior should be dominant whereas in the other limit the diffusional regime should be reached before the molecule can rotate significantly and the nondiffusive effect will be vanishingly small.

## 2. Fluorescence and Polarization Methods

In a fluorescence experiment the sample is excited with a short pulse of linearly polarized light and the time evolution of the fluorescence anisotropy, $r(t) = (I_\parallel - I_\perp)/(I_\parallel + 2I_\perp)$, is measured. In time-resolved polarization spectroscopy the decay of the anisotropy induced by a polarized excitation pulse is measured by probing with a second pulse polarized at 45° to the first and detecting the intensity transmitted through crossed polarizers. In solution, if the reorientation can be described as rotational diffusion, $r(t)$ decays as a sum of exponentials and the polarization spectroscopy signal decays as $S(t) = \exp[-2t/\tau_{ex}][r(t)]^2$ where $\tau_{ex}$ is the excited state lifetime [6]. In a fluorescence experiment with collision-free rigid rotors, the intensity of $\alpha$-polarized emission following a z-polarized delta-function excitation pulse is given by [7]

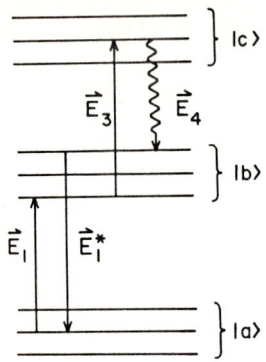

Fig. 1 Energy level diagram describing polarization spectroscopy and fluorescence anisotropy experiments. Levels b and b' may in general be the same or different states. In fluorescence, c is below b and b' in energy and the b → c and b' → c processes represent spontaneous emission.

$$I_{z\alpha}(t) \propto \sum_{a,c} P_a \left| \sum_b e^{-i\omega_b t} C_\alpha^{cb} C_z^{ba} \right|^2 , \tag{1}$$

where a, b, and c are the initial, intermediate, and final rotational states (see Fig. 1), $P_a$ is the initial population of state a, and $C_\alpha^{cb} = \langle c|C_\alpha|b\rangle$ is the direction cosine matrix element between the dipole axis of the molecule and space-fixed axis $\alpha$. The vibrational states factor out of the anisotropy if vibration and rotation are separable. The polarization spectroscopy signal is given by

$$S(T) \propto \int_T^\infty dt \, [\vec{P}^{(3)}(t) \cdot \hat{e}_4]^2 , \tag{2}$$

where T is the delay time between pump and probe pulses, and the macroscopic polarization $\vec{P}^{(3)}(t)$ is given for free rigid rotors by [7]

$$\vec{P}^{(3)}(t) \cdot \hat{e}_4 \propto i \, e^{i\omega_{CB}t} \sum_{a,c} P_a \{ \sum_b |C_1^{ba}|^2 C_3^{bc} C_4^{cb} e^{(i\omega_{cb}-\Gamma_{cb})(t-T)-\gamma_b T}$$

$$+ \sum_b \sum_{b' \neq b} C_1^{ba} C_1^{ab'} C_3^{b'c} C_4^{cb} e^{(i\omega_{cb}-\Gamma_{cb})(t-T)-(i\omega_{bb'}+\Gamma_{bb'})T} \} . \tag{3}$$

The pump, probe, and detected fields have polarizations $\hat{e}_1$, $\hat{e}_3$, and $\hat{e}_4$, $\Gamma_{cb}$ is the total dephasing constant between the final pair of levels, $\gamma_b$ and $\Gamma_{bb'}$ are the population decay and dephasing constants for the intermediate state, and $\omega_{CB}$ is the vibronic transition frequency.

For collision-free molecules r(t) does not in general decay to zero because the total angular momentum is conserved. For an ensemble of regular rotors at high J, there is a transient near time zero which decays to a constant value on a time scale of $\sim(I/kT)^{1/2}$ where I is the moment of inertia. This transient, which represents rigid-body rotational motion in a classical picture, arises in a quantum mechanical treatment from interferences between different intermediate rotational states that are coupled to the same pair of initial and final states [7] (Fig. 1). The anisotropy from a regular rotor in a particular initial J state would exhibit periodic recurrences, but the beating between different initial J's cancels all but one of these recurrences, leaving a constant long-time anisotropy for the isolated molecule that depends on the inertial ratios [8,9]. Collisions will cause the long-time anisotropy to approach zero, but even at high collision rates (i.e., in solution) the decay at very short times should resemble that for the free rotor.

The polarization spectroscopy signal for free rotors depends not only on the inertial ratios but also on the dephasing rate between the final pairs of coupled levels because this damps the macroscopic polarization in the medium. If dephasing is much faster than rotational periods, the polarization and fluorescence anisotropy decays are related by $S(t)=\exp[-2t/\tau_{ex}][r(t)]^2$ as for rotational diffusion. If dephasing is slow compared with rotation, a different polarization decay is expected which bears no simple relationship to the fluorescence anisotropy (see Fig. 2). In addition there are phase matching constraints to consider. In two-beam polarization spectroscopy, contributions from diagonal terms in the second-order density matrix (b'=b in Fig. 1) are exactly phase matched for all generated frequencies, while off-diagonal contributions (rotational CSRS processes) are not. The relative contributions to the signal from diagonal (non-oscillatory) and off-diagonal (oscillatory) terms thus depend on the spectral widths of the pulses relative to the spacing between P, Q, and R branches, the interaction length, and the angle between pump and probe beams [7]. Finally, even in the absence of vibration-rotation interaction, excitation of more than one intermediate <u>vibrational</u> state causes the nonlinear signal to decay at a rate determined by the energy spread of the coupled vibrational levels.

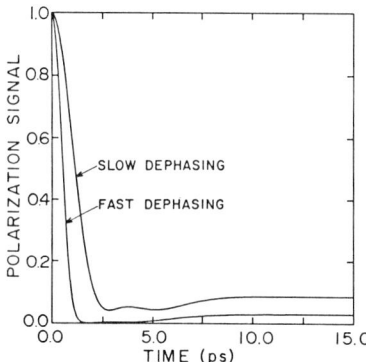

Fig. 2 Calculated polarization spectroscopy decays for fluorene modeled as a symmetric top in the fast and slow dephasing limits (see text). Rotational constants were $A=0.074$ cm$^{-1}$ and $B=0.0175$ cm$^{-1}$, at $T=443$ K. Delta-function pulses were assumed. Signal strengths are not to scale; the signal is much stronger in the slow dephasing limit.

## 3. Results and Discussion

We have recently observed a free rotation transient in the fluorescence anisotropy of stilbene vapor (Fig. 3) [10]. An overall time resolution of 4-5 ps was achieved by upconverting the uv fluorescence in potassium pentaborate (KB5). The 302 nm excitation pulse was obtained by amplifying and doubling the output of a hybrid mode-locked dye laser, while the remaining undoubled light at 604 nm was used to gate the fluorescence. $I_{\parallel}$ and $I_{\perp}$ were obtained on successive scans by rotating the polarization of the excitation pulse between horizontal and vertical while upconverting only vertically polarized fluorescence. Both the observed zero-time anisotropy of 0.16 and the general shape of the anisotropy decay are reproduced quite well by convoluting the theoretical regular rotor decay with a 5 ps instrument function. The observed long-time anisotropy of 0.069±0.003 lies between the value of 0.074 expected for a regular rotor with stilbene's inertial ratios [9] and the value of 0.056 expected for fully statistical rotation in which extensive vibration-rotation coupling generates a microcanonical distribution of K levels for

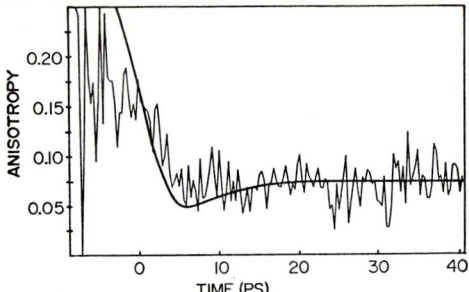

Fig. 3  Fluorescence anisotropy decay of <u>trans</u>-stilbene vapor at 463 K, ~1 torr, with 302 nm excitation. The smooth curve is the theoretical decay for a symmetric top with $A=0.0908$ cm$^{-1}$ and $B=0.00865$ cm$^{-1}$. A Gaussian instrument function of 5 ps FWHM was used.

each J state [8]. This indicates that the hot (463 K) stilbene molecule excited 650 cm$^{-1}$ above the origin undergoes partial but incomplete vibration-rotation energy transfer within approximately one rotational period. Lower long-time anisotropies were previously observed at higher excess energies [11]; the limiting anisotropy of 0.05 observed with excitation 5000 cm$^{-1}$ above the origin corresponds to essentially statistical rotation.

Figure 4 shows the polarization spectroscopy response of fluorene vapor. Excitation and probe pulses (296 nm, 4 ps and 592 nm, 6 ps) were derived from an excimer amplified synchronously pumped dye laser system [12]. The constant signal level observed from a few ps to 250 ps indicates the absence of any vibration-rotation coupling that could destroy the anisotropy on this time scale. Fluorescence experiments [13] yielded a value of 0.066 for the anisotropy of fluorene excited at its 0-0 transition, indicating that it behaves as a regular rotor. We have determined that the small spike at zero time in the polarization experiment originates from the cell windows. No free rotation transient due to the vapor can be discerned. However, this is consistent with calculations in which the appropriate pulse widths and phase matching conditions are included. This indicates the need for pulse widths that are closer to the 1.7 ps free rotation time of fluorene.

In addition to the vapor phase work, we have obtained fluorescence anisotropy decays for stilbene in low and high viscosity solvents [10]. The decays are fit well to single exponential rotational diffusion times of 11 and 78 ps

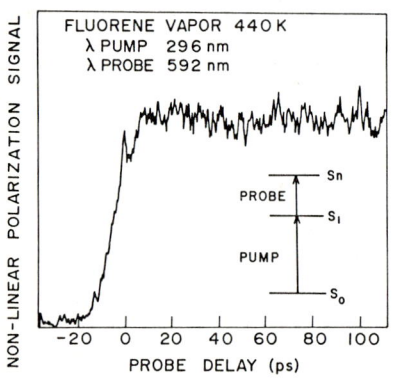

Fig. 4  The time-resolved polarization spectroscopy signal from collision-free fluorene. The spike at zero probe delay originates from window adsorbed species.

in isopentane and hexadecane, respectively, with corresponding zero-time anisotropies of 0.39 and 0.36. No nondiffusive behavior, which would be manifested as a rapid decrease in the anisotropy near t=0, is apparent.

Motions such as the in-plane spinning of benzene and substituted benzenes require little solvent displacement and appear more likely to exhibit reorientation by free rotation. Previous investigations using NMR and light scattering concluded that the rotational relaxation times for spinning are much shorter than predicted by standard hydrodynamic theory and are nearly independent of viscosity [1]. We have recently obtained fluorescence anisotropy decays for a substituted benzene, aniline, to directly measure these orientational correlation functions (Fig. 5). Fitting the data to single exponential decays yields rotational relaxation times that differ by only a factor of two between isopentane ($\eta$ =0.22 cp) and hexadecane ($\eta$ =3.3). The electronic transition is b-axis polarized (in the plane of the ring perpendicular to the CN bond) so the fluorescence can depolarize by rotation about both the a and c axes. "Slip" boundary conditions [14] predict rotational diffusion times of $\tau_a$=0.7 ps, $\tau_c$=0.1 ps in isopentane and $\tau_a$=10 ps, $\tau_c$=1.6 ps in hexadecane. The $\tau_c$ values are small because aniline is a near-oblate ellipsoid and rotation about the symmetry axis of a symmetric top experiences no friction in the slip limit. Presumably the above rotational diffusion times should be added to the free rotation times (the expected zero-viscosity intercepts) of $\tau_a$=0.4 ps, $\tau_c$=0.7 ps. It is difficult to determine how well our data follow slip hydrodynamics because we cannot reliably fit multiple exponentials to the anisotropy decays and because $\tau_a$, in particular, is sensitive to the axial ratios used [14]. However, even in hexadecane the overall reorientation time is within a factor of five of the free rotation time.

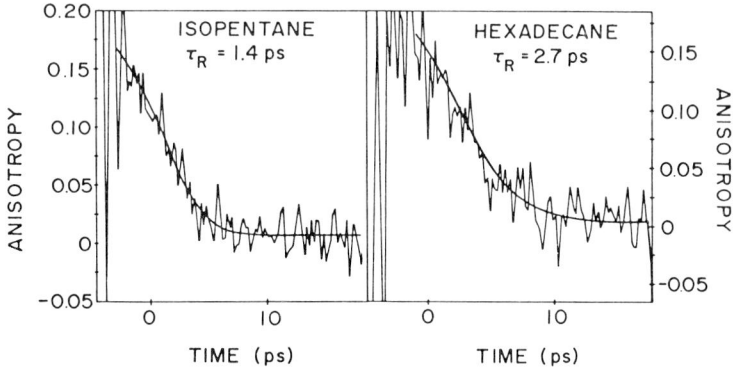

Fig. 5  Fluorescence anisotropy decays of aniline with 302 nm excitation. The smooth curves are the best fits to single exponential decays.

## 4. Conclusions

We have observed a transient in the fluorescence anisotropy of stilbene vapor that can clearly be attributed to free inertial rotation. In contrast, polarization spectroscopy did not resolve any such transient in fluorene vapor. As discussed above, several factors other than purely rotational dynamics contribute to the four-wave mixing decays and could render the free rotation transient less apparent. While these factors complicate the analysis of four-wave mixing data, they can also provide additional information. For example, by changing the interaction length or the angle between pump and probe beams it may be possible to experimentally separate the diagonal and off-diagonal contributions to the polarization signal. Comparison of polarization spectroscopy and fluorescence anisotropy decays may also allow determination of the dephasing constant $\Gamma_{bc}$ in (3). It should now be feasible to apply fluorescence methods to examine the extent of free rotation of medium-sized molecules in liquids, dense gases, and supercritical fluids. Preliminary results suggest that free rotation may contribute considerably to reorientation of aniline in solution. Such studies will provide an important link between previous work on rotational diffusion in larger molecules performed with lower time resolution and lineshape studies carried out on smaller molecules.

## 5. Acknowledgments

This work was supported by grants from NSF and NIH. A.B.M. is an NIH postdoctoral fellow.

## 6. References

1. B. J. Berne and R. Pecora: *Dynamic Light Scattering* (Wiley, New York 1976), chapter 7
2. J. H. R. Clarke: In *Advances in Infrared and Raman Spectroscopy*, ed. by R. J. H. Clark and R. E. Hester, vol. 4 (Heyden, London 1978)
3. J. M. Halbout and C. L. Tang: Appl. Phys. Lett. $\underline{40}$, 765 (1982)
4. B. I. Greene and R. C. Farrow: In *Picosecond Phenomena III*, ed. by K. B. Eisenthal, R. M. Hochstrasser, W. Kaiser, and A. Laubereau (Springer-Verlag, Berlin, Heidelberg 1982)
5. R. G. Gordon: J. Chem. Phys. $\underline{44}$, 1830 (1966)
6. A. B. Myers and R. M. Hochstrasser: IEEE J. Quantum Electron., in press
7. A. B. Myers and R. M. Hochstrasser: J. Chem. Phys., submitted
8. G. M. Nathanson and G. M. McClelland: J. Chem. Phys. $\underline{81}$, 629 (1984)
9. A. P. Blokhin and V. A. Tolkachev: Opt. Spectrosc. $\underline{51}$, 152 (1981)
10. A. B. Myers, P. L. Holt, M. A. Pereira, and R. M. Hochstrasser: Chem. Phys. Lett., submitted
11. D. K. Negus, D. S. Green, and R. M. Hochstrasser: Chem. Phys. Lett. $\underline{117}$, 409 (1985)
12. A. J. Bain, P. J. McCarthy, and R. M. Hochstrasser: Chem. Phys. Lett. $\underline{125}$, 307 (1986)
13. A. B. Myers and R. M. Hochstrasser: Unpublished results
14. G. K. Youngren and A. Acrivos: J. Chem. Phys. $\underline{63}$, 3846 (1975)

# Ultrafast Dynamics at the Interface: Probing the Transition from Solution to Surface Interactions in Charged Micelles

*E.F. Templeton, K. Brinker, S. Paone, and G.A. Kenney-Wallace*

Lash Miller Laboratories, University of Toronto, Toronto, Canada M5S 1A1

The charged and structurally complex nature of the interfacial region, in systems ranging from aqueous micelles to colloidal semiconductor particles or electrodes in solution, raises important questions concerning the energetics and dynamics of molecules residing in this quasi-ordered, microscopic layer and subject to strong field gradients. The dominant themes in the experiments reported here are the changing dynamics of a charged probe molecule as it undergoes a transition from free orientational motion in the bulk liquid to more confined motion in the interface region, the molecular mechanisms responsible for those changes, and the dynamics during the phase transition through the critical micelle concentration (CMC), which is a thermodynamically driven aggregation of individual surfactant molecules. Figure 1 illustrates a cross-section of a reverse (AOT) micelle in which possible probe locations are identified.

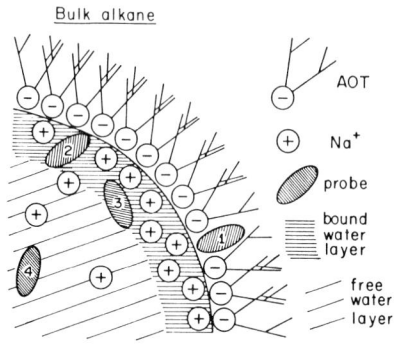

Fig. 1

We have carried out these experiments using difference frequency multiple modulation picosecond spectroscopy (DFMS) which monitors the relaxation of the time-dependent anisotropy in the orientational distribution of the ground state of a charged probe molecule to give its reorientation time $\tau_{rot}(1)$. A synchronously pumped, CW mode-locked argon ion-dye laser system was used in a pump-probe configuration for absorption spectroscopy, with typically 2ps pulses of 100 mW average power and a 580 nm $< \lambda <$ 620nm tuning range.

In the DFMS technique, both the polarized pump ($\omega_1$) and probe ($\omega_2$) pulses are individually modulated at different MHz frequencies with acousto-optic modulators. Only the transient depletion and recovery of the absorption $S_0 \rightarrow S_1$ of the probe molecule at the difference frequency (4.768 MHz) is detected, using a MHz lock-in-amplifier. The enhanced signal: noise, linearity of detection over $10^3$ dynamic range of light intensity, and the need for very small modulation depths (0.01%), make DFMS a very versatile spectroscopic approach for the study of ground state reorientational dynamics where $\omega_1 = \omega_2$ (1). Transmission of the probe molecules $\Delta T$ at $0°$, $90°$ and $54.7°$ for $\omega_2$, polarized with respect to $\omega_1$, permits direct measurement of the polarization anisotropy $R(t)$. Thus $\tau_{rot}$ and $\tau_{gsr}$ (which reflects all ground state recovery mechanisms for $S_1 \xrightarrow{gsr} S_0$)

$$R(t) \propto \sum_i A_i \exp(-t/\tau_{gsr}(i)) \cdot \exp(-t/\tau_{rot}(i)) \quad (1)$$

can be deduced for the population i of independently rotating probe molecules. In the case of identical molecular environments and hence dynamics, $R(t)$ assumes a single exponential decay.

Systems studied included the two probe dye molecules resorufin ($R^-$) and cresyl violet ($CV^+$) at $2 \times 10^{-5}$ M in (a) pure solvents, and (b) aqueous and alcohol electrolytes up to high ionic strengths (0.5 M ions), (c) aqueous micelles of sodium dodecyl sulphate (SDS, anion) and cetyl trimethyl ammonium bromide (CTAB, cation) with different surface charge and charge densities, and (d) reverse micelles of AOT (sodium diethylhexylsulfosuccinate) in alkane solvents. Typical results are tabulated below and will appear in detail elsewhere (2). In binary protic liquids such as water/alcohols, substantial and curvilinear deviations from hydrodynamic predictions of $\tau_{rot}$ are observed for probe ions of positive or negative charge, and yet are rationalized with concepts of dielectric friction to a surprisingly successful degree (2).

Table I ps reorientational times for $R^-$ in SDS, CTAB and AOT at 298 K.

| System | $\tau_{rot}$ (ps) | $\tau_{rot}$ (ps) | $\eta$ (cp) | $\lambda_{max}$ (nm) |
|---|---|---|---|---|
| $H_2O$ | 55 | | 0.95 | 575 |
| $10^{-5}$ M CTAB | 51 | | 0.95 | 575 |
| $10^{-4}$ M | 63 | | 0.94 | 575 |
| $10^{-3}$ M (CMC) | 173 | | 0.95 | 590 |
| $10^{-2}$ M | 650 | | 0.96 | 595 |
| $10^{-5}$ M SDS | 55 | | | 575 |
| $5 \cdot 10^{-2}$ M (>CMC) | 100 | | | 575 |
| AOT(w=5) | 30 | $2 \times 10^3$ | | 579 |
| w = 10 | 72 | 600 | | 576 |
| w = 15 | 82 | $\geq 10^3$ | | 574 |
| w = 20 | 71 | $\geq 10^3$ | | 575 |
| w = 25 | 77 | $\geq 10^3$ | | |
| w = 30 | 64 | $\geq 10^3$ | | 573 |
| w = 35 | 60 | $\geq 10^3$ | | |
| w = 40 | 58 | $\geq 10^3$ | | 573 |

In SDS solution, $R^-$ shows only minor changes in $\tau_{rot}$, consistent with constant $\eta$ and enhanced ion concentrations, even above the CMC. In CTAB, behaviour consistent with free reorientation times of 55ps in the bulk aqueous phase is seen at concentrations below the CMC, over a 3-fold increase at the CMC, and 10-fold above the CMC. Note the measured bulk viscosity $\eta$ is invariant within experimental error. The presence of $10^{-2}$ M CTAB, whose typical aggregation number is about 60, does not influence the bulk viscosity of 55M water. However, the miscroscopic location of $R^-$ has changed, and the question is whether or not the $R^-$ is located close to the interface double layer, where there is a surface gradient of $80mVA^{-1}$, or embedded in the exterior of the micelle, attracted by the positively charged ($N^+$) surfactant head group. The red spectral shift in the absorption maximum of $R^-$ which occurs at the CMC probably reflects the charge stabilisation of $R^-$ in the high polarity interface. However, $\tau_{rot}$ would be ms in duration if $R^-$ were deeply embedded and moving with micellar dynamics. Thus we conclude that the $R^-$ probe is only loosely associated with the micelle; for example out-of-plane librational motion of $R^-$ lying flat within the interface layer would lead to depolarization. In summary, the differences observed between $R^-$ in SDS and CTAB micelles indicate that the probe remains in the aqueous, electrolyte-like phase in the presence of repulsive coulombic interactions, but is significantly drawn into the interface by attractive interactions from the positive head groups, which are not effectively screened by their counter-ions.

For AOT system, $R^-$ unexpectedly shows 2 relaxation times ($\tau_1$, $\tau_2$) reflecting two coexisting locations for the probe. Reverse micelles encapsulate pools of water within which the density of counter ions and charged head groups can lead to very strong electrolyte conditions. If w = ratio of number density of water molecules to those of AOT, then at w = 5 the diameter of the water pool is $\sim 22\text{Å}$, and most $H_2O$ is hydrating the sulphonate and the counter ions. While the fast relaxation times for $R^-$ usually fell within $\tau_1 = 70 \pm 20$ps (the error arising from 2-exponential decay analysis) there was a consistent trend reflecting high ion density (location 3) to essentially pure water rotation values (location 4) over the sizes of micelles studied, namely $5 \leq w \leq 40$. At w = 5 a faster time of $\tau_1 = 30$ ps may reflect freer rotation of $R^-$ in a non-H-bonded water milieu.

The second population has highly inhibited ($\tau_2 \geq 1$ns) reorientational motion, indicating it is either bound to the interface or localized in the highly viscous hydration layer at the interface (location 2/3). Since the spectrum of $R^-$ is not markedly charged from that in free water, specific binding is unlikely, as is penetration into the monolayer region (location 1). Judging from the relative intensities of the two components as calculated from the ratio of the pre-exponentials, roughly 75% of the molecules are giving rise to $\tau_2$ even at w = 40, constant down to w = 10. Thus the change in the spectrum for w = 5 is probably due to the conformational changes in the reversed micelle, which result from incomplete hydration of the sulfonate head groups and the sodium ions (2). Since the spectrum of $R^-$ is not substantially affected in highly concentrated ionic solutions, this is consistent with the probe molecule localized in site 3 in the presence of high concentration of electrolyte but not bound to the interface. Very slow interchange between the interfacial and the central water pool was concluded previously on the basis of spin probe measurements (3) and it seems likely that this is the explanation for the large percentage of probe molecules present in the interfacial region, even when a very large free water pool is available. At w = 40, its diameter is estimated as $\sim$ 240 Å (4). It is difficult otherwise to explain probe localization in region 3 on the basis of solely themodynamic arguments, since binding is unlikely due to H-bonding: no spectral shifts are seen, H-bonding sites on $R^-$ are unfavourable, and coulombic repulsion due to negative charges on the $R^-$ and sulfonate head groups is present. In summary, these measurements are the first direct indication that ps-fast relaxation is possible and that a negative probe molecule such as $R^-$ is present in both regions 3 and 4 throughout the range of water pools studied. This population distribution will affect the reaction dynamics expected for species entrapped in reversed micelles, since far fewer molecules are seeing a bulk environment than expected from previous studies and models of AOT.

The principal overall conclusions are : (1) The major probe-micelle interaction is a screened coulombic repulsive or attractive interaction, depending on the effective ionic strength experienced by the probe in the microscopic boundary level comprising charged head groups and counter ions. (2) When there is significant attractive interaction as with $R^-$/CTAB, the dynamics and spectrum of the probe ion change dramatically and a large increase in $\tau_{rot}$ can be observed. (3) In reversed micelles, the data indicate two possible sites for the probe $R^-$ which permit fast (ps) and slow (ns) reorientation, where the ps time reflects free rotation within even the very small pools of water molecules and the ns time the restricted rotation of probes adjacent to the inner micelle surface. The majority of the $R^-$ molecules are located in this interfacial region even at higher water context.

1. E. Templeton and G.A. Kenney-Wallace J. Phys. Chem. **89**, 3238 (1985); ibid, in press July 1986.

2. E. Templeton and G.A. Kenney-Wallace, submitted for publication, J. Phys. Chem.

3. M. Wong, J.K. Thomas and T. Novak, J.A.C.S. **99**, 4730 (1977).

4. M. Zulauf and H. Eike, J. Phys. Chem. **83**, 480 (1979).

# Shock Moderated Photophysics and Photochemistry at Multi-kilobar Pressures

*B.L. Justus, A.L. Huston, and A.J. Campillo*

Molecular Optics Section, Code 6546, Naval Research Laboratory, Washington, DC 20375, USA

We recently demonstrated [1] the feasibility of using a picosecond laser to drive multi-kbar compressional shock waves into condensed phase samples. Using this generation technique, the properties of several condensed phase samples under conditions of shock compression were studied using added fluorescent molecules [2-5] as "molecular sensors" of the surrounding environment. We describe here the photophysical and photochemical properties of several fluorescent molecules under shock compression. Shock waves were generated with single, 10ps, 1054nm pulses selected from the output of a modelocked Nd:glass laser and amplified to energies of up to 30mJ. Details of the experimental apparatus are described elsewhere. [1-5]

The absorption and emission spectra of aromatic compounds such as anthracene often exhibit red shifts of 2000-4000 wavenumbers for applied pressures of up to 100 kbar. We studied the fluorescence spectra of anthracene in benzene solution at various shock pressures (see Fig. 1) following excitation with picosecond 353 nm laser pulses [2]. The shift in the large 0-1 vibronic peak versus pressure was found to be nearly linear and equal to 65 wavenumbers/kbar over the range 0-30 kbar.

The shock temperature of water was measured [3] in a previously inaccessible pressure range (1-10 kbar) with superior sensitivity and time response by studying the fluorescent properties of fluorescein in solution. Fluorescein shows a red shift in the absorption peak, as well as band broadening with increasing temperature causing an increase in the wing absorption of a green laser probe pulse. This results in enhanced laser-

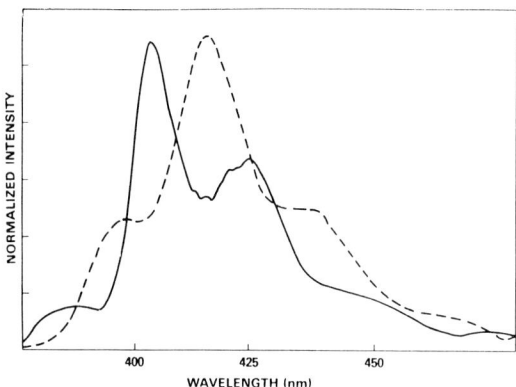

Fig. 1. Fluorescence spectra of anthracene in benzene solution at ambient pressure (solid line) and of shocked solution at 10 kbar (dashed line).

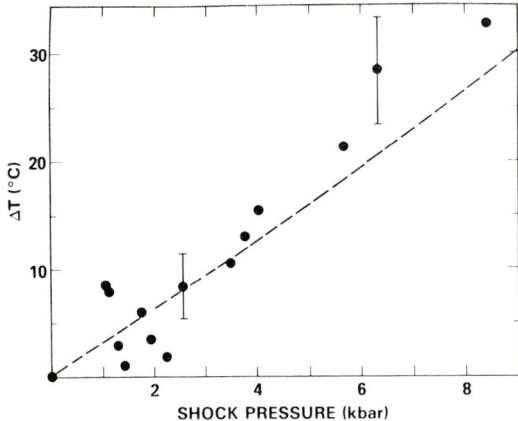

Fig. 2. Shock induced temperature rise in water vs shock pressure (circles). Dashed line follows calculated values.

induced fluorescence from the probe dye following singlet excitation. The magnitude of the fluorescence intensity provides a direct measure of the temperature increase. This can often be quite dramatic. In our work, a 30 degree temperature rise at 10 kbar shock loading resulted in a fourfold increase in the observed fluorescence. The rise in temperature vs shock pressure is shown in Fig. 2.

The influence of shock pressure on nonradiative decay processes within the molecule crystal violet was also studied [4]. The viscosity of the solvent, glycerol, was estimated from the streak camera measured fluorescence lifetime of crystal violet following picosecond green laser excitation. Structurally, the crystal violet molecule has phenyl groups which are free to rotate in low viscosity solvents. Torsional motions of these side groups are known to greatly affect the fluorescent lifetime of the excited state and are strongly affected by viscous drag which can be directly correlated with the macroscopic viscosity of the solution. Using this approach it was found that the fluorescence lifetime varied between

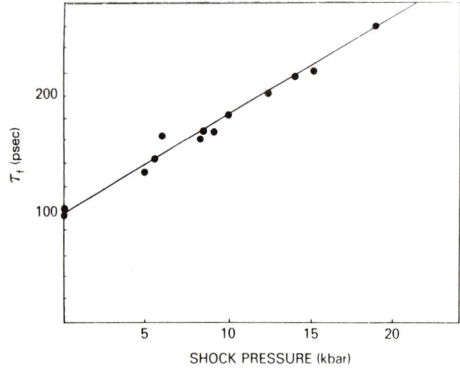

Fig. 3. Fluorescence lifetime of crystal violet in glycerol as a function of shock pressure.

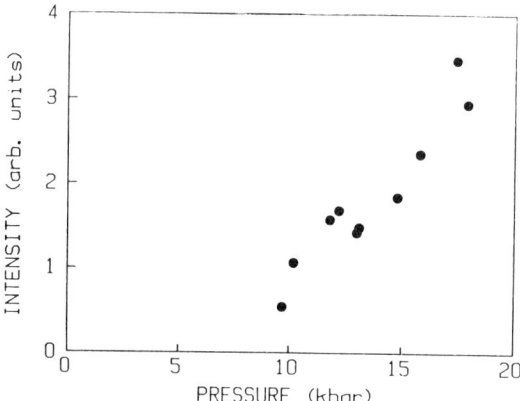

Fig. 4. Fluorescence enhancement of 9-anthraldehyde in cyclohexane vs shock pressure.

100 psec at ambient pressure to 260 psec at 19 kbar shock loading (see Fig. 3), corresponding to a viscosity change from 12 to 57 poise.

The observation of enhanced fluorescence efficiency of 9-Anthraldehyde under shock loading was also made. The fluorescence exhibits an approximately linear increase in the fluorescence efficiency vs pressure above 8 kbar (see Fig. 4). The shock data supports the hypothesis that the singlet-triplet energy gap increases with pressure, thereby increasing the activation energy for the thermally assisted intersystem crossing process.

Finally, we have studied the effect of shock compression on a simple diffusion-controlled chemical reaction. [5] The quenching of the fluorescence of Rhodamine 590 in isopropanol by iodide ions was studied by measuring the quenched fluorescence lifetime as a function of shock pressure. The shock wave causes both temperature and viscosity changes

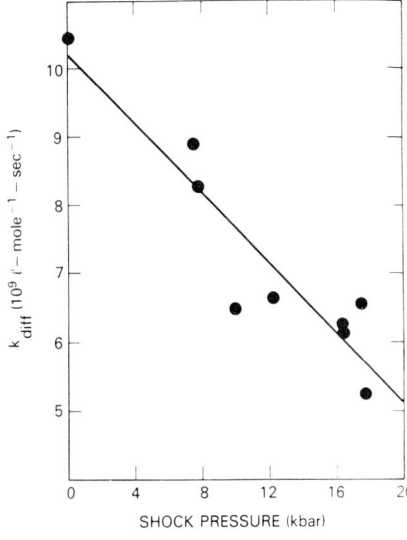

Fig. 5. Measured diffusion rate constant determined for Rhodamine 590 quenching by iodide vs. shock pressure.

which affect the diffusion-controlled bimolecular quenching rate constant (see Fig. 5). Future work will involve the study of shock-induced chemical reactions in which the shock wave initiates the chemical reactivity.


1. P.E. Schoen and A.J. Campillo, Appl. Phys. Lett. 45 (1984) 1049.
2. A.L. Huston, B.L. Justus and A.J. Campillo, Chem. Phys. Lett. 118(3) (1985) 267.
3. B.L. Justus, A.L. Huston and A.J. Campillo, Appl. Phys. Lett. 47 (1985) 1159.
4. A.L. Huston, B.L. Justus and A.J. Campillo, Chem. Phys. Lett. 122(6) (1985) 617.
5. B.L. Justus, A.L. Huston, and A.J. Campillo, Chem. Phys. Lett. (in press).

# Part IX

# Coherent Spectroscopic Techniques

# Phase Grating Approach to Susceptibility Tensors: Determination in Isotropic Media

*J. Etchepare, G. Grillon, I. Thomazeau, G. Hamoniaux, and A. Orszag*

Laboratoire d'Optique Appliquée, ENSTA, Ecole Polytechnique,
F-91128 Palaiseau Cedex, France

The nonlinearities of lowest order in transparent media arise from changes in the real part of the complex refractive index. They have been probed by a large variety of techniques; among them, the optical Kerr effect has remained one of the most useful, due in part to a high temporal resolution consecutive to the development of high peak power ultrashort light pulses. However, a severe limitation of this technique is inherent in the fact that only one inducing field is present; as a direct consequence of this geometrically degenerate configuration, one has access only to the knowledge of a linear combination of two (identical) elements of the third-order susceptibility tensor. We demonstrate here that with a temporal transient grating technique, extended to explore polarization properties of the diffracted signal, one can extensively describe the non-null elements of this nonlinear tensor. These experiments obviously have their counterpart in the frequency domain [1].

## 1. THEORETICAL BACKGROUND

In a typical grating experiment, conducted in the time domain, two intense laser pulses alter the refractive index of the material. The induced transient grating characteristics, which will be probed by a third pulse, are connected to the polarization states of the two exciting pulses; thus, a study of the polarization-dependent grating efficiency leads to information which may strengthen that deduced from temporal behavior alone.

In the general framework of the induced polarization P third order in the electric strength E, one recalls that

$$P_i(w_t) = X_{ijkl}(-w_t, w_t, w_p, -w_p) E_j(w_t) E_k(w_{p1}) E_l^*(w_{p2}) \quad (1)$$

when using two different pump fields at frequency $w_p$, and with polarizations along k and l, and a test field at frequency $w_t$ and polarized along j.

In order to study the whole tensor characteristics, three different pump pulse polarization directions are usually adopted: parallel, perpendicular or at 45° to each other. In the first two cases, one has direct access to $\delta_{ij} \delta_{kl}$ tensor elements, in the last case to either element, depending on the relative polarization states of the test and diffracted beams.

Our approach to the grating technique takes advantage of an exhaustive study of the polarization properties of the diffracted beam, in all cases where the polarization of one the three beams is at 45° to the others. Considering the $E_{p1}$ beam to be polarized along x, the nonlinear polarization at frequency $w_t$ analyzed along X, which makes an angle a with the x axis, is given by

$$P_X = (X_{xxxx} \cos a * \cos b + X_{yyxx} \sin a * \sin b ) \qquad (2)$$

when $E_{p2}$ polarization is along x and $E_t$ polarization at an angle b. Equivalent equations can be built up for other polarization arrangements. By performing measurements only as a function of the a angle value, one can study, with an optimum accuracy, the whole tensor elements, recalling that the isotropic symmetry condition: $X_{xxxx} = 2*X_{xyxy} + X_{yyxx}$, is fulfilled (the relation $X_{xyxy} = X_{xyyx}$ being always valid as one cannot differentiate $E_{p1}$ from $E_{p2}$ and vice versa). $P_X$ is in fact described by a sinusoidal function whose characteristic points depend on the relative values of $X_{ijkl}$ coefficients. Therefore, when the induced polarization is consecutive to various processes, this fact could be used to discriminate between phenomena inasmuch as their tensor elements are not connected through the same relationships. This is effectively the case for electronic and molecular processes, for which the third-order susceptibility coefficients obey the generally adopted relations [2]:

| model | $X_{1111}$ | $X_{1212}$ | $X_{1221}$ | $X_{2211}$ |
|---|---|---|---|---|
| electronic | e | 1/3 e | 1/3 e | 1/3 e |
| molecular | m | 3/4 m | 3/4 m | -1/2 m |

## 2. EXPERIMENTAL ARRANGEMENT and TEMPORAL CALIBRATION

We have used 100-fs duration optical pulses, centered at 620 nm frequency and issued from a CPM dye laser coupled to a four-stage amplifier [3], to create the transient grating. In order to avoid any coherent interaction via the test pulses, we chose to use for them a different wavelength, i.e. 650 nm, obtained by filtering, through an interference filter, in a continuum of light generated in a 1-mm water cell.

Two geometrical arrangements may be used, generally labelled as Bragg and Raman-Nath diffraction regimes. The second one has the advantage of giving rise to two symmetrically positioned diffracted beams with equal efficiencies, but depending (or not, as the case may be) on the same nonlinear parameters. For the specific application we describe here, we worked at Bragg

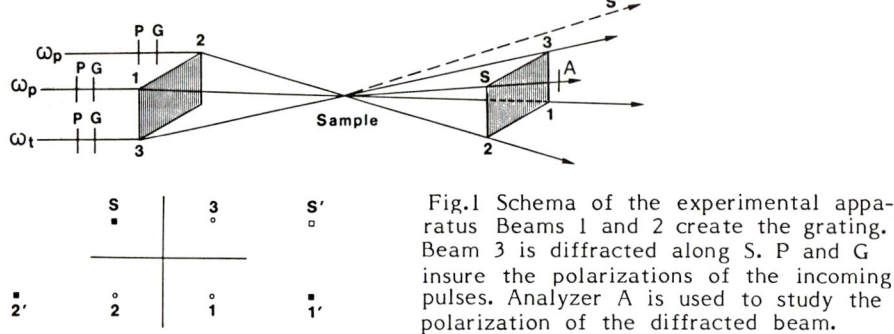

Fig.1 Schema of the experimental apparatus Beams 1 and 2 create the grating. Beam 3 is diffracted along S. P and G insure the polarizations of the incoming pulses. Analyzer A is used to study the polarization of the diffracted beam.

angle incidence in order to collect a maximum intensity of light in one direction and thereby maximize the signal-to-noise ratio (Fig. 1).

Figure 2 represents the correlation and diffracted signals obtained on a glass slab of 1 mm thickness. One can see the differences between the pump pulses' coherent temporal figure and the specific response of an instantaneous process; they come partly from the bandwidth of the interference filter, partly from dispersion effects which occur mainly in the liquid cell. The fit of these curves leads to the determination of the pulses' temporal parameters.

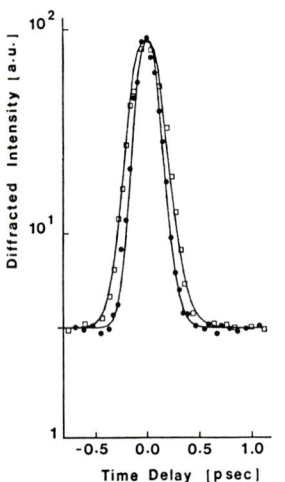

Fig.2 Coherence ● and temporal □ correlation functions of the pulses obtained with WG360 Schott glass ; the diffracted intensity is on a logarithmic scale.

## 3. RESULTS and DISCUSSION

### 3.1. Polarization analysis of the diffracted intensity of a transparent glass

Figure 3 represents measured values, at zero time delay, of the diffracted intensity of a 1-mm thick WG360 Schott glass with respect to the angle value of the analyzer. We had shown previously [4] that the relatively high value

506

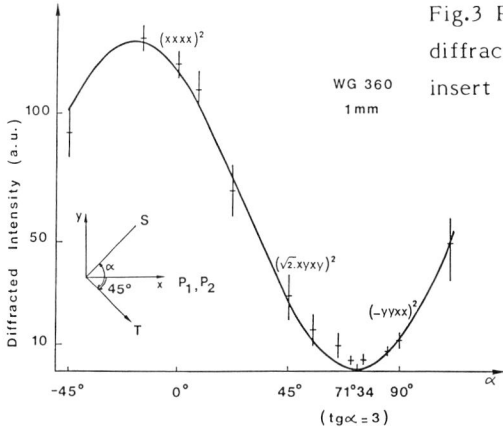

Fig.3 Polarization characteristics of the diffracted intensity of WG360 glass. Lower insert defines the polarization directions.

of its nonlinearity was due to the presence of PbO. The experimental points are well fitted by a sinusoidal function and the calculated elements of the susceptibility tensor have values connected through the Kleinman symmetry rule. Residual signal close to $\alpha = \tan^{-1}(3)$ may be attributed to static birefringence of the focusing lens and the glass slab.

## 3.2. The case of carbon disulfide

Liquid $CS_2$ is a good candidate for testing the foregoing as its nonlinearity is known to be made up of three processes, one of electronic origin, the two others being stated to be specific molecular reorientational motions with relaxation times of 0.20 and 1.50 ps.

On one hand, the $X_{1111}$ element of electronic origin has been measured recently [5] by non-phase-matched third harmonic generation of 1.06 μm pulses; its value being $30 \times 10^{-15}$ esu. On the other hand, time-resolved experiments succeeded in the measurement of the two non-instantaneous processes but failed in the determination of the electronic one, even when using 70-fs pulses [6,7]. Nevertheless the relative contribution to the third-order nonlinearity coming from the electronic process has been deduced from the comparison of the d.c. Kerr constant and Rayleigh scattering experiments; this leads to a value of $11 \pm 1\%$ [2].

In order to measure independently the magnitude of the electronic process in our frequency range, we performed a frequency mixing experiment, knowing that results of [5] can be enhanced due to multiphoton resonance [8]. Detection of the signal emitted at frequency $w_s = 2 w_p - w_t$ led to a $X_{1111}$ value of $18 \pm 2 \times 10^{-15}$ esu; furthermore, a polarization analysis of the process confirmed that out of resonance, $X_{1111} = 3 X_{2211}$.

507

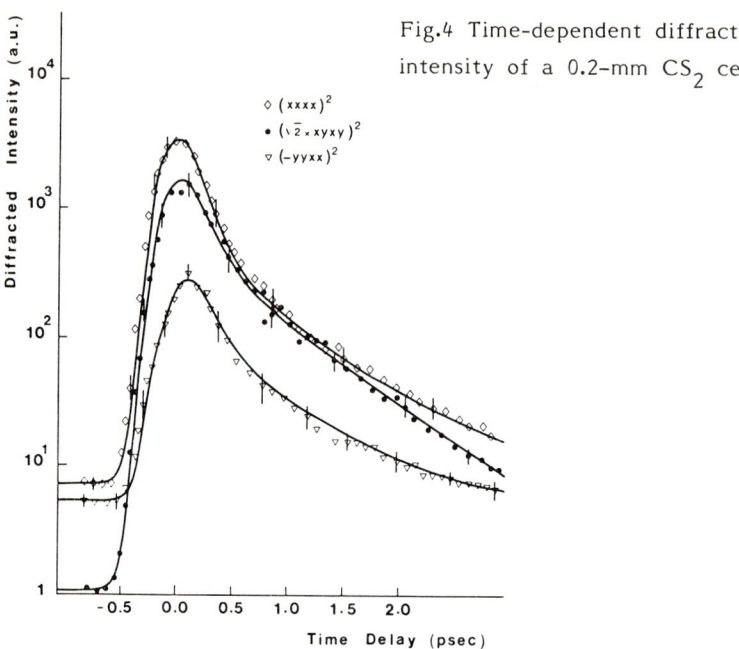

Fig.4 Time-dependent diffracted intensity of a 0.2-mm $CS_2$ cell.

Figure 4 presents three curves, for which the relative polarization directions lead to the measurement of $X_{1111}$, $X_{1212}$ and $X_{2211}$ elements. We have extensively studied the polarization properties of the transient grating efficiency by tuning the orientation of the analyzer. We outline here several of the general characteristic features of the emitted signal. Time-resolved diffracted intensities do not present bumps or dips in any case. This leads to the conclusion that the rapid and molecular reorientation processes are characterized by nonlinear coefficients having the same sign; on the other hand, such a special behavior, which does exist near zero time delay, cannot be detected because the temporal resolution of the apparatus is not good enough. As no optical element located before the sample cell has to be moved between different experiments, examination of the rising part and the maximum locations is used as a confirmation of the predicted variational behavior of the electronic process. Fitting of the curves gives, for the 1.50-ps molecular reorientation process, values which verify precisely the molecular model. Residual signal observed at a angle value (116°34) corresponding to $\tan^{-1}(-2)$ is attributed partly to noncompensated external birefringences (lens and cell) and partly to the fact that a non-null phase lag, evaluated to reach values of the order of several degrees, exist between the x and y components of the induced polarization. The behavior of the rapid process is still suspicious. The calculated parameters are roughly described by a sinusoidal function with characteristic points identical to those of the 1.50-ps process; nevertheless, a signal still remains around 116°34. Complementary experiments using different polarization arrangements are in progress to elucidate this point.

1. R. Trebino, A. E. Siegman: J. Chem. Phys. 79,3621(1983)
2. R. W. Hellwarth: Prog. Quantum Electron. 5,1(1977)
3. A. Migus, A. Antonetti, J. Etchepare, D. Hulin, A. Orszag: J. Opt. Soc. Am. B. 2,584(1985)
4. I. Thomazeau, J. Etchepare, G. Grillon, A. Migus: Opt. Lett. 10.223(1985)
5. M. Thalhammer, A. Penzkofer: Appl. Phys. Lett. 32,137(1983)
6. J. M. Halbout, C. V. Tang: Appl. Phys. Lett. 40,765(1982)
7. B. I. Greene, R. C. Farrow: J. Chem. Phys. 70,4779(1982)
8. R. T. Lynch, H. Lotem: J. Chem. Phys. 66,1905(1977)

# Nonlinear Response Function for Four-Wave Mixing: Application to Coherent Raman Lineshapes in Polyatomics and to the Optical Anderson Transition

S. Mukamel, Z. Deng, and R.F. Loring

Department of Chemistry, University of Rochester,
Rochester, NY 14627, USA

In this article, we summarize recent theoretical developments for calculating nonlinear susceptibilities and four-wave mixing (4WM) signals from complex molecular systems[1-5]. In particular, we make two applications to CARS lineshapes of polyatomic molecules in condensed phases and to the usage of transient grating techniques to probe the optical Anderson transition in molecular crystals.

## I. Nonlinear Susceptibilities and CARS Lineshapes of Polyatomic Molecules in Condensed Phases

We consider a polyatomic harmonic molecule with two electronic states, $|g\rangle$ and $|e\rangle$ and with N vibrational modes. Its Hamiltonian is given by

$$H = |g\rangle H_g \langle g| + |e\rangle [H_e + \omega_{eg}(t)] \langle e|, \tag{1a}$$

$$H_g = \frac{1}{2} \sum_{j=1}^{N} \omega_j'' (p_j''^2 + q_j''^2 - 1), \tag{1b}$$

$$H_e = \frac{1}{2} \sum_{j=1}^{N} \omega_j' (p_j'^2 + q_j'^2 - 1) - (i/2)\gamma. \tag{1c}$$

$p_j''$, $q_j''$, and $\omega_j''$ are the dimensionless momentum, coordinate, and frequency of mode j in the ground state and $p_j'$, $q_j'$, and $\omega_j'$ are the corresponding excited state quantities. We further have:

$$q_j' = (\omega_j'/\omega_j'')^{1/2} q_j'' + D_j. \tag{2}$$

$D_j$ is the dimensionless displacement between the equilibrium configuration of the two electronic states. $\omega_{eg}(t) = \bar{\omega}_{eg} + \delta\omega_{eg}(t)$, $\bar{\omega}_{eg}$ is the fundamental (0-0) transition frequency, and $\delta\omega_{eg}(t)$ is a stochastic modulation of the electronic energy gap by interactions with a thermal bath. $\delta\omega_{eg}(t)$ is assumed to be a Gaussian Markov process, with $\langle\delta\omega_{eg}(t)\rangle=0$ and

$$\langle\delta\omega_{eg}(t_1)\delta\omega_{eg}(t_2)\rangle = \Delta^2 \exp[-\Lambda|t_1-t_2|]. \tag{3}$$

The angular brackets in (3) denote an average over the stochastic process. $\Delta$ is the root-mean-squared amplitude, and $\Lambda^{-1}$ is the correlation time of the bath fluctuations. The choice of the stochastic Hamiltonian (Eq.(1)) is based on the assumption that the bath couples mainly to the electronic degrees of freedom, so that the ground state and the excited state manifolds are being stochastically modulated with respect to each other, but no modulation occurs for frequencies of levels belonging to the same electronic manifold. We have developed a general theory for any 4WM process in large polyatomic molecules characterized by the Hamiltonian (1) in condensed phases (e.g., solution, solid matrices, and glasses). The key quantity in the present formulation is the nonlinear response function $R(t_3,t_2,t_1)$, which contains all the microscopic information relevant for any type of 4WM. $R(t_3,t_2,t_1)$ is expressed in terms of a four-point correlation function of the dipole operator, which can be evaluated efficiently for this model in a closed form using Green function techniques. This eliminates the necessity of performing the multiple summations over vibronic states, which make such calculations tedious and impractical for large polyatomic molecules. The computational effort involved in the present Fourier transform method does not increase substantially as the molecular size increases. The nonlinear susceptibility $\chi^{(3)}$ is obtained by an appropriate Fourier transform of the response function R. This result is extremely helpful and allows a convenient calculation of the response functions, both in the time domain and in the frequency domain. As an illustration, we have calculated CARS lineshapes and compared them with ordinary spontaneous Raman lineshapes. The former are given by the absolute square of $\chi^{(3)}$, whereas the latter are related to the imaginary part of $\chi^{(3)}$. In Fig. 1 we show the result of a seven mode (N=7) calculation for the $S_0$-$S_1$ transition in

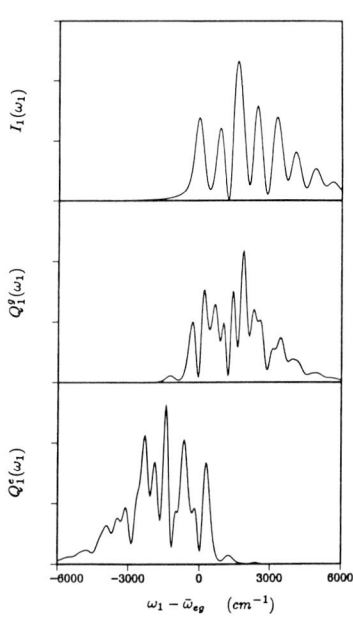

Fig. 1

The spontaneous Raman profile $I_1(\omega_1)$, the ground-state CARS $Q_1^g(\omega_1)$ and the excited state CARS $Q_1^e(\omega_1)$ for the fundamental (0-1) transition of the 1260 cm$^{-1}$ mode of azulene in $CS_2$ at 300°K. The broadening parameters are $\Delta$=180 cm$^{-1}$ and $\Lambda$=18 cm$^{-1}$ [3].

azulene[3]. Shown are Raman fundamentals for one of the modes with $\omega''=1260$ cm$^{-1}$, $\omega'=1193$ cm$^{-1}$, and D=0.77. The top panel shows the spontaneous Raman excitation profile $I_1(\omega_1)$ (intensity of the Raman line $\omega_1-\omega_2=1260$ cm$^{-1}$ vs. $\omega_1$). $\omega_1$ and $\omega_2$ being the excitation and the emitted frequency, respectively. In CARS, we can see both the ground-state profile $\omega_1-\omega_2=1260$ cm$^{-1}$ and the excited-state profile $\omega_1-\omega_2=1193$ cm$^{-1}$. These are denoted $Q_1^g(\omega_1)$ and $Q_1^e(\omega_1)$ respectively and are shown in the middle and in the bottom panels of Fig. 1. The broadening parameters in Fig. 1 are $\Delta=180$ cm$^{-1}$ and $\Lambda=18$ cm$^{-1}$.

## II. The Optical Anderson Transition as Probed by 4WM

A novel theory of quantum mechanical transport in disordered systems has been developed[4]. The theory is based on the Effective Dephasing Approximation (EDA), in which the ensemble-averaged Liouville space propagator is mapped into the propagator of an ordered lattice with an effective frequency-dependent dephasing rate. This generalized dephasing rate is determined self-consistently. This approach is applicable to strongly disordered systems and yields results that interpolate between the limits of coherent and incoherent excitation transport and that predict the optical analog of a metal-insulator phase transition (Anderson localization). Our results agree with the predictions of the scaling theory of the Anderson transition. We have applied the EDA to the calculation of the transient grating signal from a crystal with an inhomogeneously broadened absorption spectrum (static, site energy disorder). The transient grating experiment is shown to be a sensitive probe of the optical Anderson transition. We consider the tight-binding Hamiltonian,

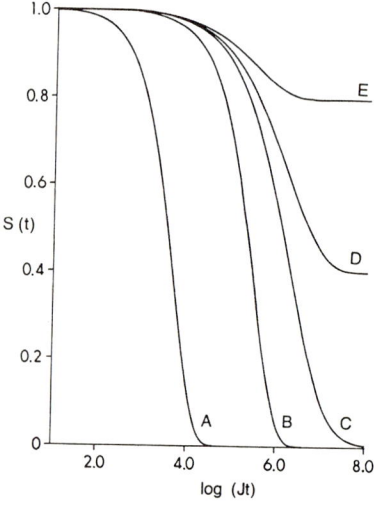

Fig. 2
The transient grating signal for a simple cubic lattice with varying degrees of site energy disorder (inhomogeneous line broadening). Curves A through E were calculated respectively for $\chi=0, 3.900, 3.957, 3.970,$ and $3.980$. The optical Anderson transition occurs at $\chi^*=3.957$. For $\chi < \chi^*$, the excitations can sample the entire system, and the signal decays exponentially to zero. For $\chi > \chi^*$, the excitations are localized, and the signal decays to a finite value at long times. The excited state lifetime is not included[4].

$$H = \sum_i E_i |i\rangle\langle i| + \sum_{i<j} J_{ij} (|i\rangle\langle j| + |j\rangle\langle i|). \tag{4}$$

$|j\rangle$ denotes a state in which the excitation resides on site j, $J_{ij}$ is the transfer matrix element between molecules located at sites i and j, and $E_j$ is the excited-state energy of the molecule at site j. We consider the case in which $J_{ij}$ has the value J, if sites i and j are nearest neighbors on the lattice, and is zero otherwise. The site energies $E_i$ are randomly distributed according to a given distribution P(E), whose second moment is denoted $\Delta^2$, i.e.

$$\Delta^2 = \int_{-\infty}^{\infty} dE\ E^2\ P(E)\ . \tag{5}$$

The site energy disorder of the lattice enters into the EDA expression for the TG signal through $\chi \equiv \Delta^2/J^2$. We predict the Anderson transition to occur at $\chi^* = 3.957$. The calculated transient grating signal for various values of $\chi$ both above and below the transition is displayed in Fig. 2.

Acknowledgments

The support of the National Science Foundation, The Office of Naval Research, the U.S. Army Research Office, the Camille and Henry Dreyfus Foundation, and of the donors of the Petroleum Research Fund, administered by the American Chemical Society, is gratefully acknowledged.

References

1. S. Mukamel and R. F. Loring , J. Opt. Soc. Am. B **3**, 595 (1986).
2. S. Mukamel, Phys. Rev. A. **28**, 3480 (1983); J. Chem. Phys. **82**, 5398 (1985).
3. Z. Deng and S. Mukamel, "Nonlinear Susceptibilities and Coherent and Incoherent Raman Spectroscopy of Polyatomic Molecules in Condensed Phases", J. Chem. Phys., **85**, (1986), (in press).
4. R. F. Loring and S. Mukamel, "Effective Dephasing Theory of the Optical Anderson Transition as Probed by Four-Wave Mixing Spectroscopy", J. Chem. Phys., **85**, (1986), (in press).
5. J. Sue, Y. J. Yan, and S. Mukamel, "Raman Excitation Profiles of Polyatomic Molecules in Condensed Phases - A Stochastic Theory", J. Chem. Phys., **85**, (1986), (in press).

# Picosecond Laser Pulse Shaping and Phase Shifting for Molecular Spectroscopy

M. Haner, F. Spano, and W.S. Warren

Department of Chemistry, Princeton University, Princeton, NJ 08544, USA

## 1. Introduction

Generating optical pulse sequences with better control of pulse shapes, phases, and delays is a necessary prerequisite to extending the applicability of coherent transient spectroscopy to complex molecules. The first coherent transient sequences, such as two pulse photon echoes, did not require specific pulse shapes (even an "incoherent pulse" gives an echo) [1] and were unaffected by phase shifting any individual pulse in the sequence. More recently, three pulse echoes [2,3], composite pulses [4], and photon locking [5,6] sequences have been demonstrated on a nanosecond timescale. Such sequences inherently require a specific phase relationship between individual pulses, and have generally used rectangular shapes. We have demonstrated that laser pulses can be shaped on a similar timescale (several nanoseconds resolution) [7], and used such shaped pulses to improve velocity selection in monitoring collisional dynamics[8]. Thus, in this time domain current technological capabilities are essentially complete, and essentially any conceivable sequence can be generated. However, relaxation times in condensed media or at high pressures limit the generality of nanosecond spectroscopy. At the opposite extreme, simple pulse shaping has been demonstrated with subpicosecond pulses (for example, making a pulse more rectangular)[9]. This pulse shaping is done in the frequency domain, and thus cannot be readily extended to tens or hundreds of picoseconds.

In this paper we demonstrate a technology which fills the four order of magnitude "gap" between the pulse shaping resolution of references [7] and [9]. We use GaAs FETs and our own design of microwave electro-optic modulator to give extremely accurate approximations to any shape we wish, with a theoretical minimum resolution of a few picoseconds. We have experimentally demonstrated (see Figure 1) complex pulse shaping with a few tens of picoseconds resolution, limited by the risetimes of our microwave sources and sampling oscilloscope. Changing pulse shapes does not require hardware changes, which is a considerable advantage. In addition, we show for the first time that phase shifts can be generated between high power, nonoverlapping picosecond or nanosecond pulses, using an injection locked laser system.

## 2. Picosecond Pulse Shaping

The extension of optical pulse shaping into the picosecond time regime presents several formidable technological problems. Arbitrary shaping of picosecond optical pulses has been recently reported by using a sophisticated spectral filter in conjunction with an optical fiber-grating pair pulse compressor [9]. This technique requires construction of precisely fabricated phase and amplitude masks to tailor the different Fourier components of the pulse. However, some of the shapes we commonly use (such as $(sech(at))^{1+5i}$) have highly structured Fourier transforms [7]. In addition, this approach would be difficult to implement in our experimental set-up where a wide range of pulse shapes are needed. We have previously generated laser pulse shapes by storing a digitized version of the desired envelope in memory, clocking it out into a video D/A convertor, using this voltage pulse to drive an r.f. modulator, and using the r.f. pulse to drive an acousto-optic modulator [7]. This gives software controllable pulse shapes, but the fastest risetimes we can obtain are typically on the order of 5 ns. The risetime limitation comes from a velocity mismatch between RF and optical waves in the substrate of approximately 5 orders of magnitude, as well as the speed limitation of IC logic families.

We surmount these limitations by developing high speed discrete circuitry capable of producing an arbitrary shaped analog waveform, and coupling this with a traveling wave electro-optic modulator. The electronic circuit consists of a linear cascade array of transistor switches using either GaAs FETs or high electron mobility transistors (HEMT). [10] Each transistor switch is biased with a different adjustable DC voltage. Triggering with a short pulse produces a replica pulse from each switch. Sequential triggering of the FETs with adjustable delays between transistor gates produces a summed envelope which is determined by the individual source biases and replicates the experimental pulse shape. The analog pulse shape is completely adjustable and has been used to produce arbitrarily shaped analog pulses of 100ps duration. The prototype circuit employs six of these high speed gates. Traveling wave electro-optic devices capable of phase and amplitude modulation have been designed with bandwidths of 8-12Ghz in the range of 500-850nm and approximately 18-20Ghz at 1.5μm.[11] Our directional coupler modulator has two 8μm diameter Ti diffused optical waveguides in the (001) plane of a c-cut $BaTiO_3$ crystal. Barium titanium oxide materials have been shown to have good electro-optic and optical damage characteristics in the visible region,[12] and excellent low-loss microwave properties.[13] Figure 1 shows the experimental apparatus and some shaped optical pulses. The optical pulse is resolution limited by the detection risetime (40 ps) and is the square of the voltage waveform. Our modulator is designed for a carrier frequency in the range of 12-20 Ghz, so the analog waveform modulates a c.w. microwave source. The coplanar strip electrode design[14] was fabricated using a photolithographic lift-off technique and vapor deposition of gold onto the modulator substrate. The electrodes have a characteristic impedance of 42Ω and the optical bandwidth of the modulator is 9.2Ghz measured at 514 nm.

3.Picosecond Phase Shifting

At the last Ultrafast Phenomena Conference, we reported an injection-locked pulsed dye laser system with sufficient power, narrow bandwidth, and the short pulse lengths required to observe fast dephasing times in molecular crystals.[15] Our current version, illustrated in Figure 2, consists of one or more atmospheric nitrogen lasers (PRA LN103), which produce 200psec uv pulses at a peak power

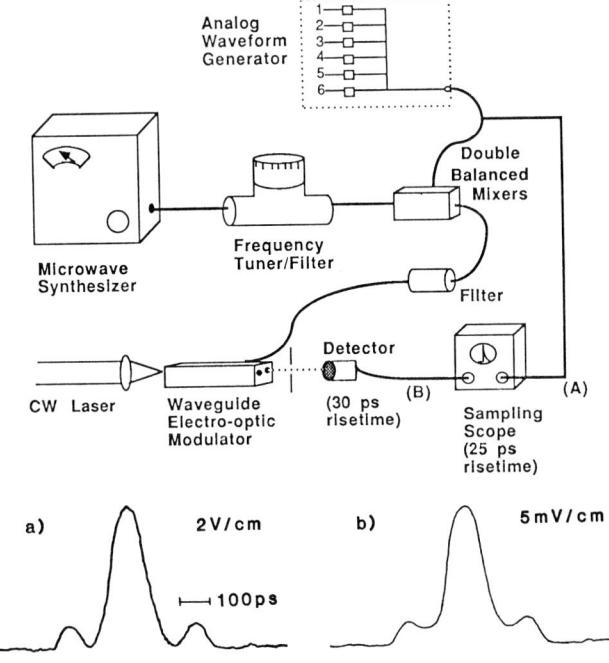

Figure 1. Experimental pulse shaping apparatus, a) analog waveform generated by the transistor array, b) optical pulse produced from the EOM.

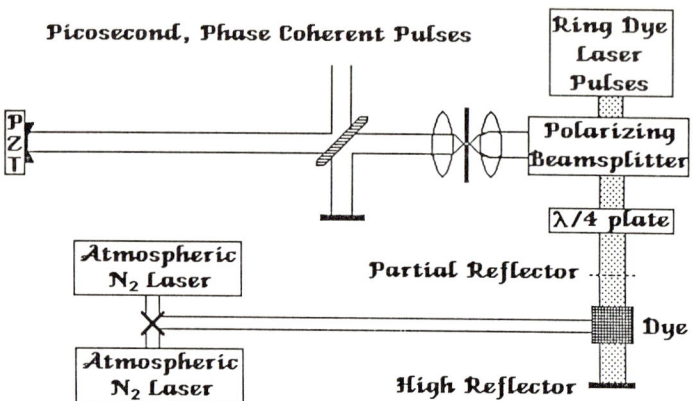

Figure 2. Experimental apparatus for subnanosecond, high power pulses with complete phase control

of 250kW, being used as pump sources for a specially designed dye laser. The dye laser was designed in house and has an adjustable cavity length (typically around 25-30 mm) which is optically stable to within $\lambda/10$. This was accomplished by using composite graphite materials for construction and incorporation of a piezoelectrically mounted rear mirror into the cavity. The threshhold for injection is remarkably low (25 milliwatts for R6G). The pulses produced from our laser are nearly transform limited, having a bandwidth of 4.25GHz. The temporal quality of the laser pulses was determined with a streak camera; in both the injected and uninjected cases the FWHM pulse length is approximately 150psec, however, in the injected pulses the spectral brightness is increased by roughly a factor of 20000. The pulse energy is $10\mu J$, resulting in a peak power of 50kW (a gain of $10^6$).

We have demonstrated that we can pump the dye cell with two independently triggered lasers, and efficiently lock both output pulses. The output should replicate the phase of the injecting electric field, so phase shifting the ring laser between the arrival times of the two pumping pulses generates a high power phase shifted sequence. This overcomes the peak power limitation of an acousto-optically modulated ring laser [7], and gives short pulses; however, phase shifts still required several nanoseconds.

We have overcome this problem by recognizing that the injection-locked pulse has a CW beam from the ring dye laser collinearly superimposed on it, which produces phase fringes on an interferometer. We therefore built an actively stabilized Michaelson interferometer as an optical delay line. A single injection-locked pulse is divided and reflected down the two arms of the interferometer, so the interpulse separation is determined by adjusting the difference in path lengths. Phase shifting is accomplished by modulating a piezoelectrically mounted mirror in the variable arm of the interferometer. The fringe pattern gives a clear indication of the phase relation between the two paths; for example, a dark fringe corresponds to a 180° phase shift and a bright fringe to no phase shift.

Originally we had great difficulty in getting reproducible results from this configuration. Even when the actively stabilized interferometer was clearly positioned to give a particular phase shift, spectroscopic demonstrations on iodine (such as looking for minimum fluorescence from two closely spaced pulses 180° out of phase) gave erratic results. We recently completed a detailed theoretical analysis of injection locking in this picosecond time domain, which correctly accounted for such complications as spatial holeburning and cavity detuning between the ring and the injected laser [16]. The output frequency is not in general identical to the injected frequency; even though the finesse of the injected cavity is extremely low, the output turns out to be concentrated on the nearest cavity mode to the injecting field. Our exact laser system has not been used by other workers, but this result would not be expected on the basis of some previously published work, [17] and has important implications for the interferometric approach to generating phase shifts; if the ring laser frequency differs from the pulse center frequency by, say, 1 Ghz, and the pulse delay is 2 nsec, a path length difference which is exactly an integral number of wavelengths for the ring (in phase) would be half integral for the pulsed laser. We have now generated and spectroscopically verified phase shifted picosecond pulse sequences with interpulse delays of 0.25-1nsec[18].

We note in passing that similar phase control could not be achieved by a normal pulsed laser with an interferometric delay line, stabilized for example by a helium-neon laser. A 1 nsec delay is 500,000 wavelengths of red light, so phase accuracy of $10^0$ (1/36 of a cycle) would require measurement of the center frequency of the laser pulse to 1 part in $36 \times 500{,}000 = 1.8 \times 10^7$. Similarly, the He-Ne and pulsed lasers would have to be collinear to better than $10^{-7}$ radians. However, for shorter (subpicosecond) delays this approach can work.

This work was principally supported by the National Science Foundation under grant CHE-8405944, and by the Petroleum Research Fund. W.S.W. is an Alfred P. Sloan Fellow, and F.S. was supported by a fellowship from the W. R. Grace Corporation. We wish to thank Harris Corp. for contributing GaAs electronic components.

1. R. Beach and S.R. Hartmann, Phys. Rev. Lett.,53, 663 (1984)
2. T. E. Orlowski and A. H. Zewail, J. Chem. Phys.,70, 1390 (1979)
3. W.S. Warren and A.H. Zewail, J. Chem. Phys.,75, 5956 (1981)
4. W.S. Warren and A. H. Zewail, J. Chem. Phys. 78, 2279 (1983)
5. E. T. Sleva, I. M. Xavier and A. H. Zewail, J. Opt. Soc. Am.,B3, 483 (1986)
6. Y. S. Bai, A. G. Yodh, and T.W. Mossberg, Phys. Rev. Lett.,55, 1277(1985)
7. W.S. Warren, J. Bates, M. McCoy, M. Navratil and L. Mueller, J. Opt. Soc. Am. B3, 488 (1986)
8. M. A. Banash and W. S. Warren, Laser Chemistry 6, 47 (1986)
9. a) A. M. Weiner, J. P. Heritage, and R. N. Thurston, Opt. Lett., 11, 153 (1986); b) R. N. Thurston, J. P. Heritage, A. M. Weiner, and W. J. Tomlinson, IEEE J. Quantum Electron.,QE-22, 682 (1986)
10. A. Barna and C. A. Liechti, IEEE J. Solid-State Circuits, SC-14, 708 (1979)
11. a) E. A. J. Marcatili, Appl. Opt., 19, 1468 (1980); b) K. Kubota, J. Noda, and O. Mikami, IEEE J. Quantum Electron., QE-16, 754 (1980); c) Ibid.,Appl. Opt., 19, 591 (1980)
12. J. Feinberg, D. Heiman, A. Tanguay, and R. Hellwarth, J. Appl. Phys., 51,1297 (1980)
13. H. M. O'Bryan, J. Thomson, and J. K. Plourde, J.Amer. Ceram. Soc., 57, 450 (1974)
14. a) R. C. Alferness, C. H. Joyner, L. L. Buhl, and S. K. Korotky, IEEE J. Quantum Electron., QE-19, 1339 (1983); b) S. K. Korotky and R. C. Alferness, J. Lightwave Technol., LT-1, 244 (1983)
15. F. Spano, F. Loaiza, M. Haner and W. S. Warren, Ultrafast Phenomena IV ,99(D. Auston and K. Eisenthal, editors;Springer, New York,1984)
16. F. Spano and W. S. Warren, Proc. O-E Lase '86 (in press)
17. A. J. Gibson and L. Thomas, J. Phys. D.,11, L59 (1978)
18. F. Spano, M. Haner and W. S. Warren, Phys. Rev. Lett. (submitted)

# Third-Order Nonlinear Optical Ineractions in Thin Films of Organic Polymers Investigated by Picosecond and Subpicosecond Four-Wave Mixing

P.N. Prasad, D. Narayana Rao, J. Swiatkiewicz, P. Chopra, and S.K. Ghoshal

Department of Chemistry, State University of New York at Buffalo, Buffalo, NY 14214, USA

This paper presents experimental measurements and theoretical analysis of the third-order optical susceptibility measurements by degenerate four-wave mixing in several classes of polymeric films which have conjugated π-electron structure. A high $\chi^{(3)}$ value derived from the π-electron conjugation effect is observed and a subpicosecond response is confirmed. The large $\chi^{(3)}$ value made it possible for us to investigate vibrational dephasing in polymer films of thickness in micron range by using time-resolved CARS and CSRS experiments at different temperatures.

Our experimental arrangement uses a CW mode-locked Nd-Yag laser (Spectra Physics model 3000), the beam from which after frequency doubling is split in two equal parts to sync pump two dye lasers (Spectra Physics model 375). The pulses are subsequently amplified in two separate amplifiers (Quanta Ray, model PDA) both of which are pumped by a pulsed Nd-Yag laser (Quanta Ray, model DCR-2A) to produce ~8ps pulses. For degenerate four-wave mixing, a backward interaction geometry was used in which two pump beams ($I_1$ and $I_2$) counter propagate and the probe beam ($I_3$) is incident at a small angle (4°). All three beams were obtained by splitting the beam from only one amplifier. For subpicosecond degenerate four-wave mixing, the pulses were shortened by adding a saturable absorber (DQOCI) to the dye in the sync pumped dye laser which produced ~500fs pulses at 605nm. To keep the pulses close to the transform limit, a saturable absorber jet was used in place of the final spatial filter in the amplifier. Even then, a two-fold broadening of the subpicosecond pulses in the amplifier was found.

For picosecond CARS and CSRS experiments, beams from both amplifiers were used where the best cross-correlation width found was 11ps. To obtain the vibrational dephasing, the pulse from one amplifier was split in two, one portion undergoing a variable delay. The coherent signal was observed at a phase-matched angle and detected by a spectrograph vidicon assembly system.

The first class of polymer is polydiacetylenes. Here we have investigated a specific polydiacetylene, poly-4-BCMU, which undergoes a red-to-yellow conformational transition at high temperature [1,2]. The conformational transition changes the π-electron conjugation which is considerably reduced in the yellow form. Also, poly-4-BCMU is a soluble polydiacetylene, therefore, can easily be cast as a film. A 4μm thick film of poly-4-BCMU was cast on a microscopic slide from a chloroform solution. The film at room temperature is in the red form. With the shortened pulse, the DFWM signal ($I_4$) obtained from the red form exhibits a response of ~700fs which is shorter than the autocorrelation width. A similar result is obtained for the yellow form. Therefore, the subpicosecond response of the electronic $\chi^{(3)}$ in conjugated polymeric systems is established. In order to obtain the value of $\chi^{(3)}$, a comparative study with $CS_2$ was conducted. The values obtained are $\chi^{(3)}_{1111} = 4 \times 10^{-10}$ esu for the red form and $\chi^{(3)}_{1111} = 2 \cdot 5 \times 10^{-11}$ esu for the yellow form and they are comparable for wavelengths 585 and 605 nm. No significant change in the value of $\chi^{(3)}$ is found as the film is rotated, indicating that it behaves isotropically. A 16-fold change in the value of $\chi^{(3)}$ in going from the red form to the yellow form with reduced conjugation demonstrates the dependence of $\chi^{(3)}$ on π-electron conjugation. Using a one-dimensional pseudopotential model [3] for conjugated π-electron system, the π-electron contributions to $\chi^{(3)}$ are estimated to be $9 \times 10^{-12}$ and $1 \cdot 5 \times 10^{-12}$ esu for the red and the yellow forms respectively.

The second class of polymer is a conjugated aromatic hetrocyclic polymer, poly-p-phenylenebenzobisthiazole (PBT) which has a very high mechanical strength due to its rigid rod conformation as well as environmental stability and a high laser damage threshold. The measurement in a 33μm thick as-spun biaxial film of PBT yields $\chi^{(3)}_{1111} = 9 \times 10^{-12}$ esu. Again, a subpicosecond response for the $\chi^{(3)}$ is found. The measurements at two wavelengths suggests that they are non-resonant $\chi^{(3)}$ values. The measured anisotropy of $\chi^{(3)}$ as a function of angular orientation of the film at two different sets of laser polarization is explained by using the fourth rank tensor properties of $\chi^{(3)}$ and assuming an orthorhombic symmetry for the PBT biaxial film.

The third class of organic polymer is a photoconductor, specifically polyphenylacetylene in the doped and undoped states. This class was investigated to examine the role of photoinduced charge carriers. The doped polyphenylactylene is a much better photoconductor [4,5]. We have measured $\chi^{(3)}$ values for films of polyphenylacetylene undoped and that doped with 2,3-dichloro-5,6-dicyano-p-benzoquinone (DDQ). No significant change in the DFWM signal was found. Therefore, we conclude that the third-order non-linearity is dominated by the π-electron contribution.

Another class of polymers investigated are those prepared electrochemically. Our initial result indicates a large $\chi^{(3)}$ value for polyazulene in the oxidized form.

The picosecond time-resolved CARS and CSRS studies have been performed on the poly-4-BCMU films of thickness ~10μm. Three different kinds of samples were investigated: (i) the crystalline blue form of poly-4-BCMU obtained by the solid state polymerization of the monomer film, (ii) the red form of poly-4-BCMU obtained by dissolving the polymer and casting the film. This form is amorphous and has a distribution of polymer chain lengths, (iii) the low molecular weight red form of poly-4-BCMU obtained by fractionating the smaller chain length polymer. The vibrational dephasing was studied for the -c=c- stretching vibration of the polymer backbone which in the blue form (lower band gap and, therefore, increased π-electron conjugation) occurs at ~1475cm$^{-1}$ but in the red form is at ~1530cm$^{-1}$. The vibrational dephasing by both CARS ($\omega_1 = 17076$cm$^{-1}$, $\omega_2 = 15544$cm$^{-1}$, $2\omega_1 - \omega_2 = 18608$cm$^{-1}$) and CSRS ($\omega_1 = 17102$ $\omega_2 = 15570$cm$^{-1}$, $2\omega_2 - \omega_1 = 14038$), in each case, was found to be within the resolution provided by the cross-correlation of pulses (~11ps), even at liquid helium temperature. In contrast, the ~1375cm$^{-1}$ ring mode of naphthalene at liquid helium temperature has a dephasing time of 100 ps [5]. The temperature dependence study does not indicate any phonon induced dephasing. Our initial interpretation is that the dephasing is strain induced. The polydiacetylene samples are known to exhibit strain effects on vibrational spectra [7].

1.  G. N. Patel, R. R. Chance, and J. D. Witt, J. Chem. Phys., 70, 4387 (1979).

2. K. C. Lim and A. J. Heager, J. Chem. Phys., <u>82</u>, 522 (1985).

3. C. Sauteret, J. P. Hermann, R. Frey, F. Pradere, J. Ducuing, R. H. Baughman, and R. R. Chance, Phys. Rev. Lett., <u>36</u>, 956 (1976).

4. E. T. Kang, A. P. Bhatt, P. Ehrlich and W. A. Anderson, Appl. Phys. Lett., <u>41</u>, 1136 (1982).

5. E. T. Kang, P. Ehrlich and W. A. Anderson, Mol. Cryst. Liq. Cryst., <u>106</u>, 305 (1984).

6. C. L. Schosser and D. D. Dlott, J. Chem. Phys., <u>80</u>, 1394 (1984).

7. D. Bloor, R. J. Kennedy, and D. N. Batchelder, J. Polym. Sci., Polym. Phys. Ed. <u>17</u>, 1355 (1979).

# Picosecond Raman-Induced Phase Conjugation Spectroscopy

*R. Dorsinville, P. Delfyett, and R.R. Alfano*

Institute for Ultrafast Spectroscopy and Lasers,
Physics and Electrical Engineering Departments,
The City College of New York, NY 10031, USA

## 1. Introduction

Non-linear optical phase conjugation (OPC) techniques have been used to study ultrafast dynamic phenomena in liquids and solids and have found numerous applications in optical image processing. Recently a new coherent Raman spectroscopy technique using OPC has been demonstrated by Saha and Hellwarth[1]. In this Raman induced phase conjugation technique (RIPC) two nanosecond single pulse laser beams at $\omega$ and $\omega-\Omega$ or $\omega+\Omega$ (where $\Omega$ corresponds to a vibrational frequency in a nonlinear medium) mix with a third laser beam to generate a fourth beam at $\omega-\Omega$ or $\omega+\Omega$, nearly phase conjugate to one of the beams at $\omega$. The resonance enhancement of the $\chi^3$ nonlinear coefficient generates the signals at the Stokes and anti-Stokes frequencies. The main advantages of this technique over other coherent Raman techniques are a wider frequency range (> 1000 cm$^{-1}$) and a broad acceptance angle for phase matching (> 30 mrad).[1]

In this paper we demonstrate for the first time the use of the RIPC technique in the picosecond regime. We have obtained the picosecond RIPC spectra of several liquids and solids and by delaying one of the interacting beams relative to the other pump and probe beams we were able to determine with picosecond resolution the intensities of the phase conjugate beams at the Stokes frequencies as a function of time.

## 2. Experimental

In the experiment, a frequency-doubled, 35 ps laser pulse was generated from a Quantel Nd:YAG laser system. The fundamental and second harmonic were separated using a harmonic beam splitter. The second harmonic was divided into two beams using a beam splitter and then recombined in the sample at a small angle. The fundamental was weakly focused into a 5 cm cell containing $D_2O$ to produce a broad band supercontinuum extending throughout the visible spectrum, providing a broad band of light at the Stokes frequencies. In order to increase the continuum intensity in the visible, a second harmonic KDP crystal was placed on the path of the fundamental laser radiation before the focusing lens. The continuum was then collimated and focused into the sample cell along with the other two beams, in a phase conjugate geometry. The signal beam, phase conjugate to the probe beam, was directed by a wedge glass plate toward the entrance slit of a spectrometer. The spectra were recorded and analysed with an optical multichannel analyzer. For the time-dependent measurements, the monochromatic pump or the probe beam was delayed relative to the other beams and the intensity of the Raman induced phase conjugate beam was measured for different delays. By delaying the probe beam either the coherence or the intensity correlation of the exciting beam could be measured.[2] Transient grating effects could be investigated with picosecond resolution by delaying the monochromatic pump beam.[3]

## 3. Results

Low resolution RIPC spectra were obtained for nitrobenzene, nitrotoluene, carbon disulfide, calcite and lithium niobate. The spectra span a 2000 cm$^{-1}$ range. The nitrobenzene spectrum shows three strong lines at 1000 cm$^{-1}$, 1345 cm$^{-1}$, and 1585 cm$^{-1}$. These bands correspond to the strongest Raman lines in the spontaneous Raman spectrum. By slightly varying the crossing angle one can select any one of these lines. The carbon disulfide spectrum had one strong band located at 656 cm$^{-1}$. For calcite when the electric vector E was parallel to the c axis, $E||c$, a strong conjugated signal was observed from a 1 cm crystal at 564 nm which corresponds to the 1086 cm$^{-1}$ optical phonon. For the same configuration ($E||c$) two phonon lines at 260 cm$^{-1}$ and 630 cm$^{-1}$ were observed in a 2 cm lithium niobate crystal. Typical spectra for nitrobenzene and calcite are shown in fig.1a and 1b.

For all samples the RIPC signal disappeared completely when any one of the three interacting beams were blocked and was nearly collinear and counter propagating to the probe beam. Careful phase matching alignment was not needed to generate a strong signal. The spectra shown are averages of at least 10 shots, but one shot spectra contained all the main features and were not substantially different. The shot to shot variations were due to continuum fluctuations.

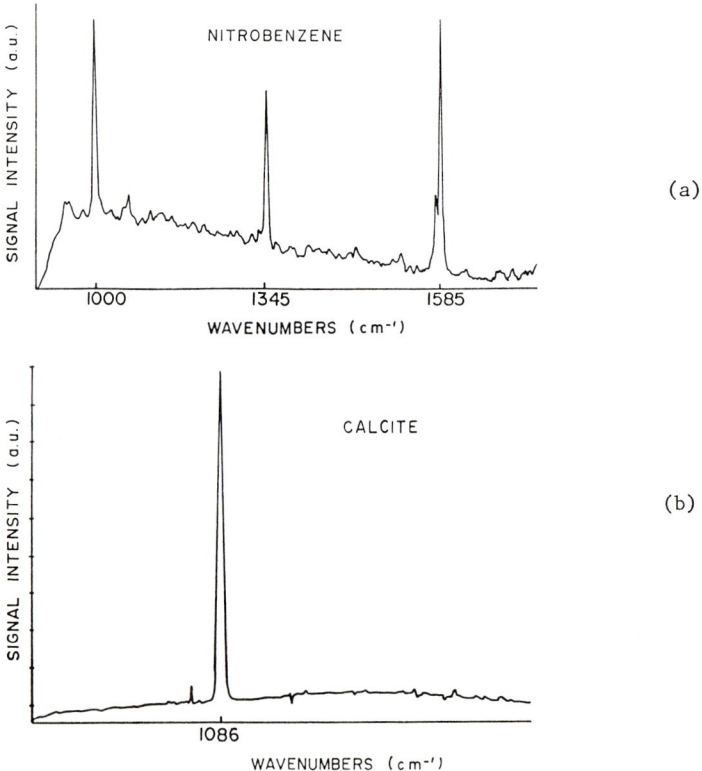

Fig. 1 a) RIPC spectrum of nitrobenzene b) RIPC spectrum of calcite

We were able to determine with picosecond resolution the intensities of the phase conjugate beams at the Stokes frequencies as a function of time. With our geometry, in a transparent medium, a measurement of the signal intensity at Stokes frequencies versus delay time between one of the pump beams and the continuum and probe should be a measure of the relaxation time of the laser-induced grating". The grating is produced by the periodic changes in optical properties of the material in the overlap region of the probe and continuum beams. In RIPC experiments the modulation of the medium is considerably enhanced by the presence of the resonance Raman lines. For a delta-function excitation the transient grating should disappear with the relaxation of the vibrational mode and the time resolved RIPC experiment provides a direct measurement of the relaxation time of the excited vibrational modes or phonons. A typical picosecond time resolved curve for the 1086 cm$^{-1}$ vibration in calcite is shown in fig. 2. For all the samples the vibrational decay times were not resolved, showing an upper limit of < 20 ps for the coherent lifetimes.

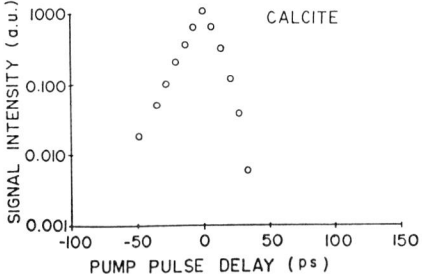

Fig. 2 Dependence of RIPC signal for 1086 cm$^{-1}$ optical phonon of calcite on pump pulse delay.

These measurements show that the time resolved RIPC technique is potentially a powerful tool and experiments with a laser system capable of delivering shorter pulses (<5ps) should allow the accurate measurement of vibrational relaxation times for different liquids and solids. We hope our work will stimulate others to use this technique for phonon lifetime measurements.

This research is supported by ASFOR.

## 4. References

1. S.K. Saha and R.W. Hellwarth, Phys.Rev. A,27, 919 (1983).
2. J. Buchert, R. Dorsinville, P. Delfyett, S. Krimchansky, and R.R. Alfano, Optics Comm.,52, 433 (1985).
3. H.J. Eichler, Optica Acta, 24, 631 (1977).
4. R.K. Jain and R.C. Lind, J. Opt. Soc. Am.,73, 647 (1983).

# Polarization Dependence of Time-Resolved CARS in Liquids

N. Kohles and A. Laubereau

Physikalisches Institut, Universität Bayreuth,
D-8580 Bayreuth, Fed. Rep. of Germany

Ultrafast excitation and subsequent probing of molecular vibrations in condensed matter and gases have received increasing interest in recent years. In numerous investigations time domain coherent Raman spectroscopy has provided information on vibrational dephasing times /1/ and on different parts of the third-order nonlinear susceptibility /2/; a Fourier transform technique has also been developed for high frequency resolution /3/. In previous work special polarization conditions with parallel polarization of the excitation pulses and of the probe field components were used /4/.

Very recently we have investigated the very different physical situation of time-resolved CARS under more general polarization conditions /5/. For liquids three factors were shown to contribute to the observed probe scattering signal:
(i) scattering off the isotropic part of the resonant material excitation (dephasing time $T_2$),
(ii) scattering off a second oriented component of the resonant material excitation (anisotropic decay time $\tau_{an}$),
(iii) four wave mixing via the non-resonant part $\chi_{NR}$ of the third-order susceptibility.

The observed time evolution of the scattering signal is a result of the relative role of the three mechanisms for the various polarization geometries. For example, the ratio $\chi_{NR}/\chi_{res}$ was determined using this polarization dependence, where $\chi_{res}$ represents the resonant part of the third-order susceptibility /5/. In this summary two new effects will be discussed which have been recently predicted /5/ and experimentally studied: observation of a coherence peak and of the anisotropic decay time $\tau_{an}$.

The coherence peak occurs for parallel polarization of all four light fields involved in CARS and equal frequencies of the exciting and probing laser pulses. It is generated for short delay times and related to the signal enhancement observed in degenerate four wave mixing. The effect is explained by pump-probe-coupling where the laser pump pulse ($E_L$) and probe pulse ($E_p$) interchange their relative role in the scattering process.

Some experimental data are shown in Fig. 1. The fundamental mode of liquid $N_2$ is investigated. The experimental system using a mode-locked glass laser and a stimulated Raman generator-amplifier set-up was described recently /5/. The open triangles in Fig. 1 represent the experimental data for a conventional

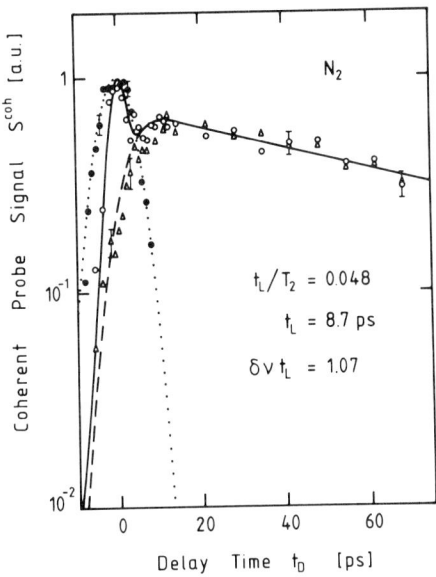

Fig. 1

Time-resolved CARS in liquid $N_2$ (2326 cm$^{-1}$) for two polarizations; experimental points, theoretical curves.
a) $E_L \| E_S \| E_P \| E_A$ (o, solid curve) observing a coherence peak;
b) $(E_L \| E_S) \perp (E_P \| E_A)$ without coherence peak ($\Delta$, broken line);
instrumental response function as measured by off-resonant CARS in $H_2O$ (●, dotted curve)

polarization geometry, where the polarization plane of the pump fields ($E_L \| E_S$) is rotated by 90 degrees with respect to the probe field components ($E_P \| E_A$). The signal curve rises to a delayed maximum with subsequent exponential decay with $T_2/2$. Strikingly different is the signal transient, when the four field components of the scattering process have same polarization (open circles, calculated solid line). The signal curve displays a more rapid increase and a sharp maximum at $t_D = 0$. After a short decay the signal curve finally reaches the same exponential asymptote. The coherence peak is readily seen from the Figure, which is reported here for the first time. Working with non-transform-limited pulses ($\delta\nu \times t_L = 1.07$) the temporal width of the signal enhancement is notably reduced as compared with the convolution of the intensity of the involved light pulses (dotted curve in Fig. 1). Taking into account the coherence peak is important for determining experimental decay times.

Perpendicular polarization planes of the exciting laser and Stokes field ($E_L \perp E_S$) and, correspondingly, of the field components of the probing process ($E_P \perp E_A$) lead to drastically different signal transients. For this situation, mechanism (i) mentioned above is absent and only factors (ii) and (iii) generate the probe scattering signal. Most important, the signal decreases asymptotically with half the time constant $\tau_{an}$, the decay time of the oriented part of the resonant material excitation. For statistically independent vibrational and rotational relaxation, $\tau_{an}$ is simply determined by the vibrational dephasing time $T_2$ and the rotational relaxation time $\tau_R$ /5/:

$$\tau_{an}^{-1} = \tau_R^{-1} + T_2^{-1} . \tag{1}$$

In recent work, values of $\tau_R$ have been derived by deconvolution techniques from spontaneous Raman spectra. The first direct observation of $\tau_{an}$ and $\tau_R$ is reported by the present paper.

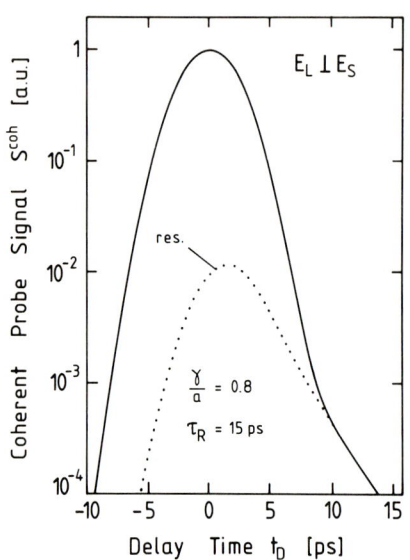

Fig. 2

Calculated CARS signal vs delay time $t_D$ for perpendicular polarization $(E_L \perp E_S) \parallel (E_P \perp E_A)$ generated by non-resonant four wave mixing $(\chi_{NR}/\chi_{res} = 0.037)$ and scattering off the oriented resonant material excitation (coupling coefficient $\gamma/a = 0.8$); $t_L = 4.5$ ps; $\tau_{an} = 5.0$ ps.

A numerical example for time-resolved CARS with perpendicular polarization is presented in Fig. 2. The values assumed for the coupling coefficients $\chi_{NR}$ (non-resonant excitation) and $\gamma$ (anisotropic part of the Raman polarizability $\partial\alpha/\partial q$) are listed in the figure caption. $\gamma$ is known from the depolarization factor $\rho_s$ of spontaneous Raman scattering. Of particular interest is the signal decay (solid line) with $\tau_{an}/2$ for long delay time. The

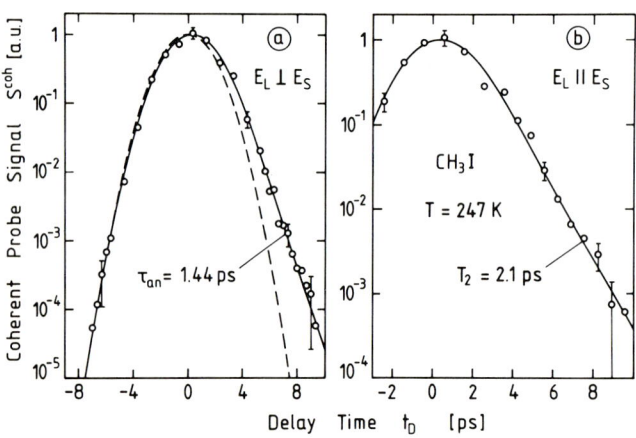

Fig. 3: Time-resolved CARS of the $\nu_3$-vibration of $CH_3I$ at $T = 247$ K; a) perpendicular polarization $(E_L \perp E_S) \parallel (E_P \perp E_A)$ observing for the first time the dephasing time $\tau_{an} = 1.44$ ps of the oriented, resonant material excitation; b) scattering off the isotropic resonant material excitation for effectively parallel polarization; calculated solid curves $(\chi_{NR}/\chi_{res} = 0.021;$ $\rho_s = 0.18$ /6/); instrumental response function (broken line)

dotted curve indicates the contribution of the oriented resonant material excitation.

An experimental verification is depicted in Fig. 3. The $\nu_3$-mode (525 cm$^{-1}$) of $CH_3I$ is investigated at T = 247 K. The coherent scattering signal for effectively parallel polarization is depicted in Fig. b) yielding the vibrational dephasing time $T_2$ = 2.10 $\pm$ 0.16 ps at this temperature. Of particular interest are the data of Fig. a) for perpendicular polarization. For large $t_D$ the signal decays much more rapidly with time constant $\tau_{an}/2$ = 0.72 $\pm$ 0.05 ps. Experimental time resolution of 0.35 ps is considerably shorter as illustrated by the instrumental response function (broken line). The accelerated signal decay of Fig. a) compared with b) originates from rotational relaxation. Using Eq. 1 we find from the data of Fig. 3 a) and b) $\tau_R$ = 4.6 $\pm$ 1.2 ps. Our results on $\tau_{an}$ and $\tau_R$ are fully consistent with estimates from spontaneous Raman spectroscopy /7/.

In conclusion we point out that novel features of time-resolved CARS have been predicted and observed yielding new information on molecular dynamics in liquids.

References

1. See, for example, Ultrafast Phenomena IV, eds. D.H. Auston, K.B. Eisenthal, Springer Ser. Chem. Phys. 38 (Springer, Berlin, 1984)
2. W. Zinth, A. Laubereau and W. Kaiser, Opt. Commun. 26, 457 (1978);
J. Kuhl and W.E. Bron, Sol. State Commun. 49, 935 (1984)
3. H. Graener and A. Laubereau, Opt. Commun. 54, 141 (1985)
4. A. Laubereau and W. Kaiser, Rev. Mod. Phys. 50, 3607 (1978)
5. N. Kohles and A. Laubereau, Appl. Phys. B 39, 141 (1986)
6. G. Döge, R. Arndt and A. Khuen, Chem. Phys. 21, 53 (1977)
7. R.B. Wright, M. Schwartz and C.H. Wang, J. Chem. Phys. 58, 5125 (1973)

# Direct Measurement of Wave-Vector-Dependent Polariton Energy Velocity and Dephasing in NH$_4$Cl

G.M. Gale, F. Vallée, and C. Flytzanis

Laboratoire d'Optique Quantique du C.N.R.S., Ecole Polytechnique,
F-91128 Palaiseau Cedex, France

The creation and evolution of short pulses of propagating collective excitation in a crystal is a new and important problem in solid state physics both intrinsically and for the information it yields on crystal behaviour in the infra-red and far infra-red. In recent experiments [1], femtosecond far infra-red pulses were generated and used to perform coherent time-domain spectroscopy of infra-red active modes in semiconductor crystals. The selectivity of this technique is, however, limited.

We show here that it is possible to create and follow the temporal and spatial evolution of a picosecond pulse of propagating Raman active phonon-polaritons, using a time and space-resolved CARS technique. This method gives, for the first time, direct access to the energy velocity and dephasing time of phonon-polaritons over a large portion of the polariton dispersion curve. Similar results, on the group velocity of exciton-polaritons, have been obtained from the propagation delays of picosecond pulses in thin slabs, [2,3] but this time of flight technique is difficult to extend to phonon-polaritons in the far infra-red and the dephasing time of the excitation is difficult to obtain [4].

We have performed the first demonstration of the time and space-resolved CARS technique on the phonon polariton associated with the $\nu_4$ polar vibration of ammonium chloride, whose dispersion curves [5,6] and Raman spectra [7] have been extensively studied. These spectra indicate strong coupling of the $\nu_{4t}$ polariton with many-phonon states. In addition, NH$_4$Cl undergoes an orientational order/disorder phase transition at 243 K. Thus one expects polariton dephasing times to be highly sensitive to polariton frequency and crystal temperature.

The principle of the experiment is outlined in Fig.1. Two space and time-coincident optical picosecond pulses of frequencies $\omega_L$ and $\omega_S$, with $\omega_L - \omega_S = \omega(k)$ the frequency of the polariton of wave vector k, create by coherent Raman scattering a polariton wave packet at time t=0 and position X=0 (see fig.1). This wave packet then propagates in the crystal in a direction determined by the phase-matching condition : $\underline{k}_{POL} = \underline{k}_L - \underline{k}_S$. The temporal and spatial evolution of the propagating excitation is followed by coherent anti-Stokes Raman scattering of a third pulse $\omega_p$ displaced with respect to the excitation in time (by $t_D$) and space (by $X_D$). The measured dependence of $X_D$ on $t_D$ is directly related to the energy propagation characteristics of the polariton wave-packet.

The experimental system is driven by a mode-locked Nd$^{3+}$/glass laser producing a single 5ps pulse at 1.054μm which is frequency converted to gene-

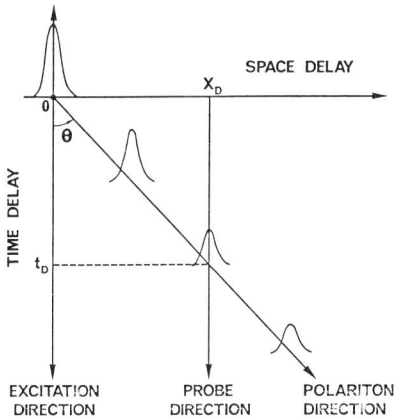
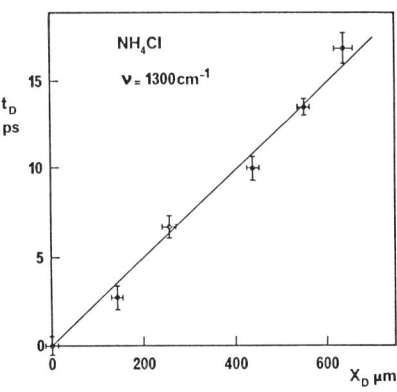

Fig.1. Space time-resolved picosecond CARS for a propagating polariton wave packet.

Fig.2. Peak coherent excitation position $X_D$ as a function of probe delay $t_D$.

rate $\omega_p$ (18 975 cm$^{-1}$), $\omega_S$ (16 935 cm$^{-1}$) and $\omega_L$ (tunable 17900 - 18400 cm$^{-1}$). These three beams are focused into a 3 or 10 mm long NH$_4$Cl sample in a non-collinear k-matching geometry and the anti-Stokes signal is detected by a photomultiplier. The results obtained at $\omega(k)=1300$ cm$^{-1}$ are shown in fig.2 where we plot the measured position $X_D$ of the peak coherent excitation as a function of probe delay $t_D$. The experimental points are well described by a straight line of slope P passing through the origin and the polariton group velocity may thus be calculated from

$$v_g = P/(\sin\theta + pn\cos\theta/c),$$

where n is the crystal refractive index at the probe wavelength [8], $\theta$ is the angle between the $\omega_L$ beam and the polariton propagation direction, and c is the speed of light in vacuum.

The energy velocity measured for various polariton frequencies is shown in Fig.3. The observed rapid decrease of polariton energy velocity (by more than a factor of 20) as polariton frequency increases reflects the change of polariton character (from photon-like to phonon like) in this frequency region. The polariton dispersion law in NH$_4$Cl has been extensively studied by Raman scattering [5,6] and may be reproduced (neglecting damping) by a Lyddane-Sachs-Teller expression

$$\frac{k}{2\pi\nu} = \left(\varepsilon_\infty \prod_j \frac{\nu_{jL}^2 - \nu^2}{\nu_{jT}^2 - \nu^2}\right)^{1/2}$$

containing the large-k frequencies $\nu_{jT}$, $\nu_{jL}$ of the four strongest polar states [6]. The full line in fig.3 shows the group velocity, $v_g = \partial\omega/\partial k$, calculated from the above expression. Agreement with the experimental points is excellent (there are no fitted parameters) and gives an indirect confirmation of our results.

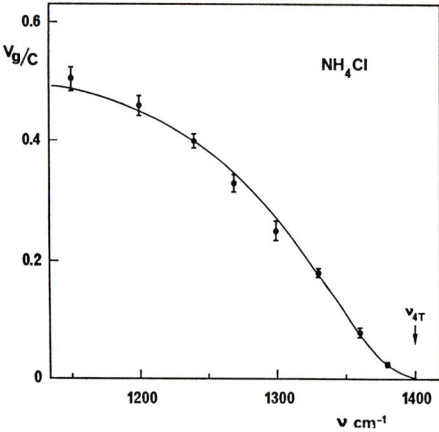

Fig.3. Measured polariton group velocity as a function of polariton frequency (points). The full line is calculated (see text)

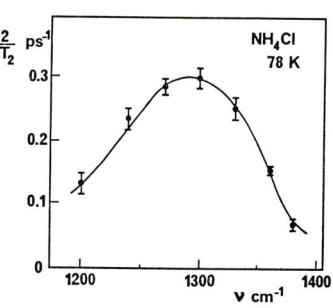

Fig.4. Polariton decay rate $2/T_2$ plotted versus polariton frequency at 78 K.

Spatial or temporal spreading (due to velocity dispersion) of the polariton wave-packet is small in our experiment and hence the intensity decrease of the peak ("followed polariton") coherent anti-Stokes signal with time (and space) delay of the probe pulse can be simply related to loss of coherence ($T_2$) of the polariton packet. Exponential time decay of polariton coherence is observed for all polaritons studied (between 1200 and 1380 cm$^{-1}$) except when the polariton interacts with the crystal exit surface. The results of these measurements at a temperature of 78 K are shown in fig.4 where we plot $\Gamma = T_2^{-1}$ ps$^{-1}$ as a function of polariton frequency ($\tilde{\nu}$ cm$^{-1}$). A resonance type dependence is observed with $T_2$ decreasing below 7 ps at intermediate frequency. We have also measured the temperature dependence of $T_2$ for various polariton frequencies. Above about 130 K a similar dependence is seen for all polaritons with $T_2$ decreasing very rapidly as one approaches the phase transition temperature ($T_c = 243°K$) [A similar temperature dependence is also observed for the longitudinal component of $\nu_4$ at 1418 cm$^{-1}$]. Experimentally, we may separate the observed inverse relaxation time into two components

$$\Gamma = \Gamma_s(\omega,T) + \Gamma_f(T) ,$$

where $\Gamma_s$ is frequency dependent and varies only slowly with temperature and $\Gamma_f$ is frequency independent and varies rapidly with temperature. The above observations are consistent with our hypothesis that dephasing is caused by two independent mechanisms, one of which is correlated to the phase transition and can be neglected at low temperature when the crystal is ordered.

In an ordered crystal the damping of polaritons is mainly due to the finite lifetime of their mechanical component, which is anharmonically coupled to many-phonon bands [9]. This mechanism is particularly important in low-temperature ammonium chloride and can explain the frequency variation of the polariton damping rate (Fig.4) which, (taking the phonon strength factor into account [10]) reflects the many phonon density of states. This process gives only a weak temperature dependence of $\Gamma$, via Bose-Einstein occupation numbers.

The second dephasing mechanism may be related to the order/disorder phase transition in $NH_4Cl$ at 243 K. Below this temperature residual disorder can produce polariton dephasing in several ways, including the effects of off-diagonal disorder and relaxation of symmetry selection rules. Thus anharmonic decay and scattering processes not allowed in the totally ordered phase may appear, leading to new relaxation channels. The above effects may be related to the ammonium ion orientational order parameter which is strongly temperature dependent in the vicinity of $T_\lambda$ [11] and gives, at least qualitatively, a temperature variation of $\Gamma_f$ in agreement with observation.

In conclusion, we have shown for the first time that the space and time evolution of a propagating Raman active polariton pulse may be directly followed using a space and time - resolved CARS technique. This method has allowed the study of polariton/crystal and polariton/surface interactions and should give direct access to other important problems such as polariton/polariton and parametric interactions in crystals.

References

1. D.H. Auston and K.P. Cheung, J.Opt.Soc.Am.B. 2, 606 (1985)
2. R.G. Ulbrich and G.W. Fehrenbach, Phys.Rev.Lett. 43, 963 (1979)
3. Y. Masumoto, Y. Unuma, Y. Tanaka, and S. Shionoya, J.Phys.Soc.Japan 47, (1844)(1979)
4. Y. Masumoto, S. Shionoya, and T. Takagahara, Phys.Rev.Lett. 51, 923 (1983)
5. V.S. Gorelik, O.P. Maximov, G.G. Mitin, and M.M. Sushchinskii, Solid St.Comm. 21, 615 (1977)
6. G.G. Mitin, V.S. Gorelik, L.A. Kulevskii, Yu.N. Polivanov and M.M. Sushchinskii, Sov.Phys.JETP 41, 882 (1976)
7. C.H. Wang and R.B. Wright, J.Chem.Phys. 58, 1411 (1973), and references therein
8. G. Poinsot and J.P. Mathieu, Ann.de Phys. 10, 481 (1955)
9. S. Ushioda, J.D. Mc Mullen, and M.J. Delaney, Phys.Rev. B8, 4634 (1973)
10. R. Loudon, J.Phys. A3, 233 (1970)
11. N.G. Parsonage and L.A.K. Staveley, Disorder in Crystals (1978)

# Impulsive Stimulated Rayleigh, Brillouin, and Raman Scattering: Experiments and Theory of Light Scattering Spectroscopy in the Time Domain

*M.R. Farrar, L.R. Williams, Yong-Xin Yan, Lap-Tak Cheng, and K.A. Nelson*

Department of Chemistry, Massachusetts Institute of Technology, Cambridge, MA 02139, USA

An "impulsive" stimulated scattering (ISS) technique has been developed through which ultrashort laser pulses can be used to excite coherent vibrational motion and to probe the vibrating material at various stages of vibrational distortion. [1-2]. The general importance of ISS in femtosecond optics has been discussed [3]. Here we report results of recent impulsive stimulated quasielastic, Brillouin and Raman scattering (ISQS, ISBS, ISRS) experiments in crystalline solids and discuss the connection between ISS (in the linear response or low vibrational amplitude regime) and conventional (frequency-domain) light scattering. The theory shows that, as expected, the time- and frequency-domain experiments are complementary and in principle the results from one can be used to predict the results of the other. However, in practice there are often decided advantages offered by one or the other approach. The utility of ISS for small-angle scattering and for examination of low-frequency or heavily damped modes has been demonstrated [1-4].

The ISS experiment is shown schematically in Figure 1. Two ultrashort pulses centered at the same frequency $(\omega_L, \vec{k}_1)$ and $(\omega_L, \vec{k}_2)$ are crossed inside a sample to "impulsively" excite coherent phonons of the difference wave vector, $\vec{k} = \pm(\vec{k}_1 - \vec{k}_2)$. The inherently broad excitation-pulse bandwidth must exceed the phonon frequency, $\omega(\vec{k})$, so that phonon excitation can occur via stimulated scattering involving the Fourier components of the pulses. Equivalently, the excitation pulse duration must be short compared to the time of a single vibrational cycle so that the crossed pulses exert a temporally impulsive, spatially periodic force on the vibrational mode. This results in coherent standing-wave oscillations which can be probed by coherent scattering of a third, variably delayed pulse.

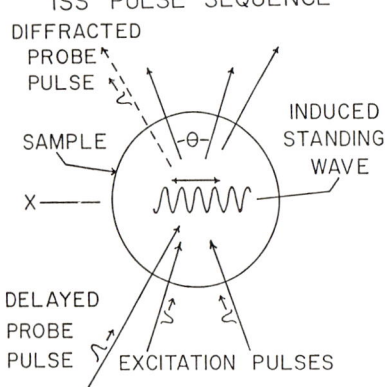

Fig. 1. Schematic illustration of the impulsive stimulated scattering experiment. The crossed excitation pulses, with wave vector difference $\vec{k}$, excite coherent phonons which form a standing wave of wave vector $\pm\vec{k}$. The phonon frequency is derived from mixing of frequency components contained within the excitation pulse bandwidth. Time-dependent vibrational oscillations are monitored by coherent scattering of variably delayed probe pulses.

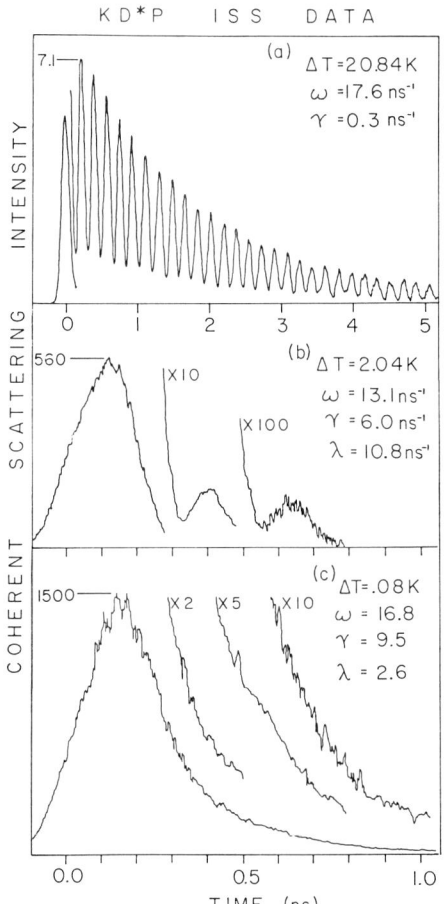

Fig. 2. ISS data from $KD_2PO_4$ crystals at various temperatures, with constant excitation and probe conditions. The two ~70-ps, 40-μJ, 532-nm excitation pulses were polarized perpendicularly to each other (vertically and horizontally V-H) and crossed at a 60° angle such that the phonon wave vector was aligned along one of the equivalent crystallographic ā axes. The ~65-ps, 20μJ, 590-nm probe pulse was V polarized (along the other ā axis) and the coherently scattered light was H polarized. (a) ISBS data from the transverse acoustic phonon with the crystal temperature above the phase transition temperature by $\Delta T$=20.84K. The data are well described by a simple time-dependent form $(e^{-\gamma t}\sin\omega t)^2$ with the parameters shown due to the damped acoustic phonon. (b) ISBS/ISQS data from mixed acoustic phonon and optic phonon/polarization modes at $\Delta T$=2.04K. The data are well-described by the time-dependent form $(Ae^{-\lambda t}+Be^{-\gamma t}\sin(\omega t))^2$ with the parameters shown due to the coupling of the two modes. (c) Same as (b) with $\Delta T$= 0.08K. The acoustic mode is now very heavily damped so that the material displacements, and therefore the signal, monotonically decrease after reaching a maximum.

Picosecond pulses have been used to excite and monitor coherent acoustic waves and other low-frequency modes through impulsive stimulated Rayleigh-Brillouin scattering. Figure 2a shows ISBS data from transverse acoustic waves in a $KD_2PO_4$ (D-KDP) crystal. The oscillations in the data correspond to time-dependent shear wave oscillations in the crystal. The acoustic phonon frequency and attenuation are determined from the time-dependent data. In Figs. 2b and 2c, the temperature of the (paraelectric) crystal is closer to the temperature, $T_c$=213K, of a structural phase transition into a ferroelectric low-temperature phase. At these temperatures the shear acoustic mode is strongly mixed with an overdamped optic phonon/polarization mode which drives the phase transition [5]. The coherent scattering observed in Figs. 2b and 2c is due to <u>both</u> modes, and the polarization relaxation time as well as the acoustic frequency and attenuation can be determined. These ISQS/ISBS experiments, carried out over a wide range of wave vectors and temperatures, permit quite thorough characterization of the coupled-mode dynamics of the phase transition [5]. The difficulties of frequency-domain light scattering studies of this system and others like it have been demonstrated and discussed at length [6].

Improved time resolution will permit direct observation of the polarization response at all temperatures and measurement of its <u>overdamped</u> <u>vibrational</u> (ratner than purely <u>relaxational</u>) character [7].

Using femtosecond pulses, higher frequency optic phonons have been excited through impulsive stimulated Raman scattering and their coherent oscillations and decay observed in real time. Temperature-dependent phonon dephasing times as short as 1 ps were measured with ISRS in the organic molecular crystal α-perylene [2]. Figure 4 shows ISRS data from a 61-$cm^{-1}$ mode in the cooperative Jahn-Teller crystal, $TbVO_4$. To our knowledge, this is the first report of this mode in $TbVO_4$. Our observation of a mode of identical frequency and symmetry in the isomorphic crystal $DyVO_4$ leads us to believe that we are observing coherent optic phonon oscillations in both crystals. It is unlikely that both rare-earth ions have Raman-active electronic transitions at this frequency.

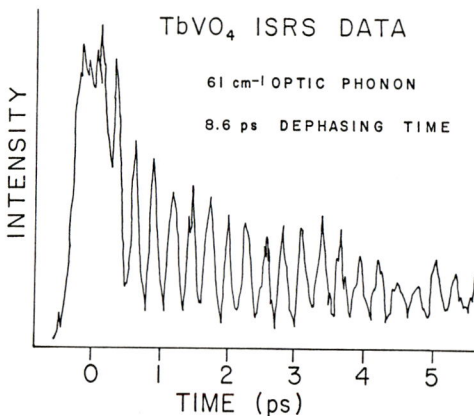

Fig.3. ISRS data from terbium vanadate crystal at room temperature. The 3.65-THz oscillations in the data correspond to the coherent optic phonon oscillations in the crystal. The 70-fs, 620-nm, 2-μJ excitation and probe pulses were polarized along the crystallographic ā axis and crossed at a small angle such that the phonon wave vector was aligned along the c̄ axis. (Data taken in collaboration with J.M. Huxley, W.Z. Lin, J.G. Fujimoto and E.P. Ippen.)

To relate ISS time-domain data to frequency-domain results, we use the familiar expression for the light-scattering spectrum

$$S(\omega) \propto \frac{kT}{\omega} G''(\omega) , \tag{1}$$

where $G''(\omega)$ is the imaginary part of the frequency-dependent Green's function for the dielectric constant. In the ISS experiment, coherent time-dependent variations in the dielectric constant give rise to scattered signal of the form

$$S(t) \propto [G(t)]^2 , \tag{2}$$

where $G(t)$ is the time-dependent Green's function which is the Fourier transform of $G(\omega)$.

Figure 4 shows several time- and frequency-domain simulations for scattering from a single vibrational mode with various degrees of damping. It is clear that for modes with strong damping, the time-domain approach can be advantageous. The frequency-domain data cannot adequately distinguish between overdamped vibrational motion and relaxational motion, even neglecting the practical problems of parasitic elastic scattering, which often masks the true low-frequency response.

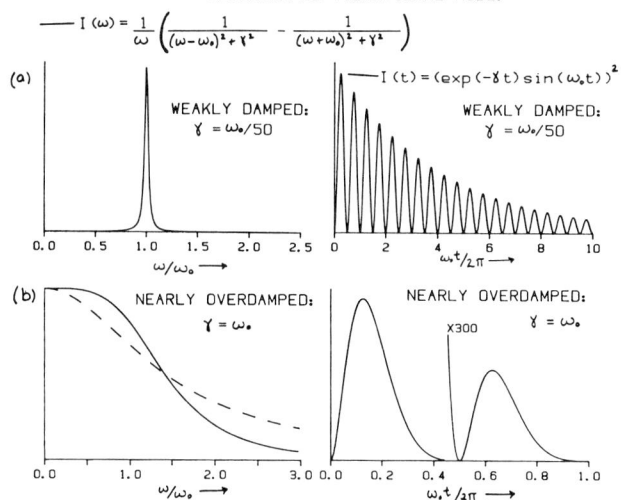
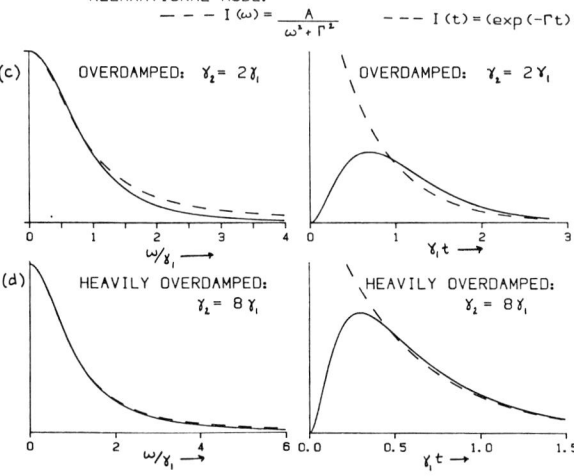

Figure 4. Simulated frequency- and time-domain data of scattering from vibrational modes. a) Weakly damped vibrational mode. b) Nearly overdamped vibrational mode. c) Overdamped vibrational mode. The dashed lines show time- and frequency-domain simulations for scattering from a purely <u>relaxational</u> mode. The frequency domain curves are normalized to the same amplitude and half-width; and the time-dependent simulations are calculated using the same parameters. d) Heavily overdamped mode. All simulations calculated as in (c). In cases (c) and (d) the overdamped vibrational vs. relaxational dynamics are distinguished clearly in the time domain but poorly in the frequency domain. In cases (a) and (b), either type of data permits adequate characterization of the dynamics.

This research was supported in part by NSF DMR-8306701 and grants from Research Corp. and PRF(ACS). The ISRS work was supported by US ARO DAAL03-86-K-0002. LRW acknowledges AT&T Bell Laboratories Scholarship.

1. M.M. Robinson et al., Chem. Phys. Lett. <u>112</u>, 491 (1984).
2. S. De Silvestri et al., Chem. Phys. Lett. <u>116</u>, 146 (1985).
3. Y.-X. Yan et al., J. Chem. Phys. <u>83</u>, 5391 (1985).
4. M.R. Farrar et al., IEEE J. Quantum Electron., in press.
5. M.R. Farrar et al., in preparation.
6. R.L. Reese et al., Phys. Rev. B7, 4165 (1973).
7. P.S. Peercy, Phys. Rev. B <u>12</u>, 2725 (1975).

# Ultrafast Transient Spectroscopy with Broadband Non-Transform-Limited Light Sources

T. Yajima and N. Morita

Instiute for Solid State Physics, University of Tokyo, Roppongi, Minato-ku, Tokyo 106, Japan

The progress of ultrafast transient spectroscopy in the picosecond and femtosecond regions has hitherto been promoted mainly by the continual effort of generating transform-limited ultrashort light pulses. There still remain, however, difficulties in the generation of practically usable pulses and their applications in the extremely short time region. Recently, another approach of studying ultrafast phenomena in the time domain without using ultrashort pulses has been developed with an aim to overcome these difficulties. It has been shown that a broadband and strongly non-transform-limited light source can play a similar role to a short pulse with the duration corresponding to the reciprocal bandwidth in a number of correlation-type optical processes. This means that in these processes the time-resolution is governed by the field correlation time $\tau_c$ (reciprocal bandwidth) rather than the light duration $\Delta t$, and therefore ultrashort resolution times far below $\Delta t$ can easily be achieved.

One of the relevant light sources for this purpose is the incoherent light (not of thermal origin but based on laser technique). Another example is the phase-modulated coherent pulse as produced by self-phase modulation and other mechanisms. In general any broadband light field with arbitrary coherence and modulation properties can be the light source of interest. The purpose of the present paper is to present the outline of three topics of our recent study concerning this subject in order to show various aspects of the problem. We will show how widely this new method can be applied, and also show advantages and disadvantages in the use of various kinds of broadband light sources in comparison with each other including ultrashort pulses.

## 1. Degenerate Four-Wave Mixing with Light Sources of Arbitrary Coherence and Regular Phase Modulation

The degenerate four-wave mixing (DFWM) with two or three incident light beams in a resonant material is a basic nonlinear optical process relevant to apply new kind of light sources. The process involves the photon echo as a limiting process and serves particularly for the determination of dephasing time $T_2$ of the material. Several theoretical and experimental studies for this process with incoherent light have already been reported.[1-10] We present here the result of a comprehensive computer simulation study of the process (two input beams) for a variety of light sources with different coherence and modulation characteristics.[11] The aim is to seek what kind of light is really suitable for the ultrafast transient spectroscopy.

In order to represent various light fields in a unified manner, we adopted a multi-mode model of light with a fixed power spectrum. By giving various regular and irregular phase relations among many modes, we can represent a variety of lights including transform-limited short pulse,

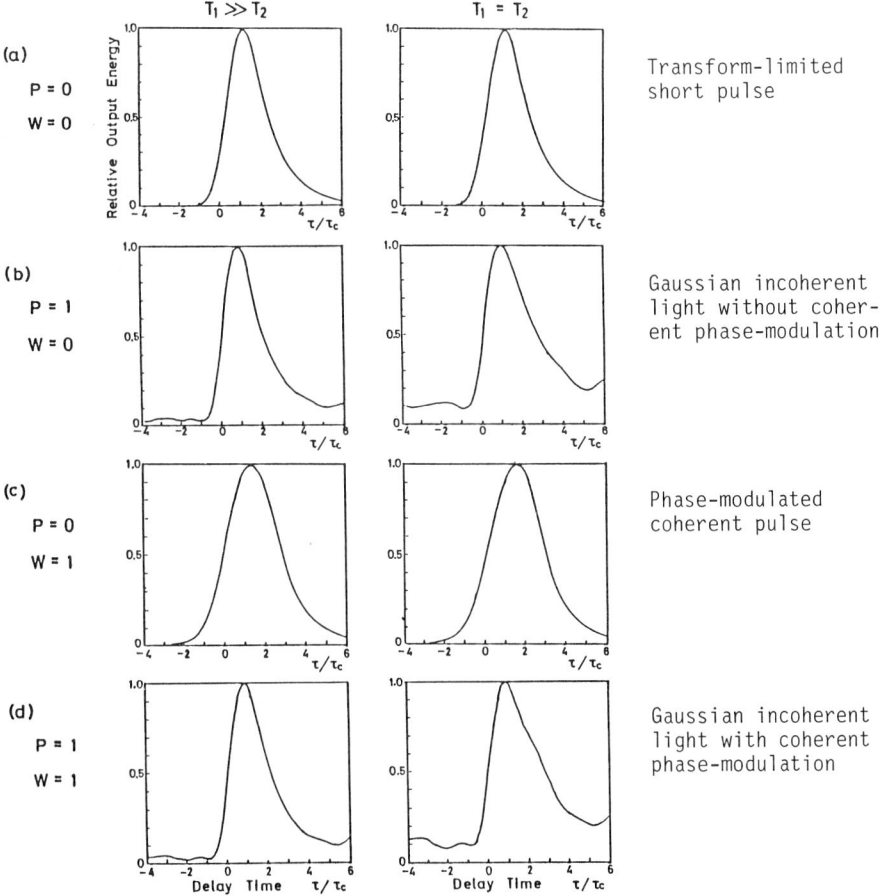

Fig. 1 Calculated correlation traces of the degenerate four-wave mixing with two incident light beams of various kinds for an inhomogeneously broadened transition under the condition of $T_2/\tau_c = 5$.[11] $T_1$: energy relaxation time, $T_2$: phase relaxation time, $\tau_c$: light correlation time (reciprocal bandwidth), P: coherence parameter (extent of random phase distribution), W: dispersion parameter ($\simeq 5$ times pulse-broadening for W = 1)

Gaussian incoherent light, partially coherent light, phase-modulated coherent pulse and more complicated ones with combined coherent and incoherent modulations. Thus the output characteristics of the DFWM process for different types of light field can be compared under the condition of a fixed correlation time $\tau_c$ resulting from fixed spectrum. As a model of resonant material, we first consider an inhomogeneously broadened two-level system with relaxation times $T_1$ (longitudinal) and $T_2$ (transverse). In the DFWM process, two noncollinear incident light beams with wave-vectors $\vec{k}_1$ and $\vec{k}_2$ produce output light beams with wave-vectors $\vec{k}_3 = 2\vec{k}_2 - \vec{k}_1$ and $\vec{k}_4 = 2\vec{k}_1 - \vec{k}_2$. The integrated or averaged power of one output beam versus the delay time $\tau$ between two incident beams (called the correlation trace) provides information on relaxation times and energy level structure of the material.

Examples of calculated correlation traces with $T_2/\tau_c = 5$ for an inhomogeneously broadened transition are shown in Fig. 1 for four typical kinds of light fields : (a) transform-limited short pulse, (b) Gaussian incoherent light without coherent phase-modulation, (c) phase-modulated coherent pulse and (d) Gaussian incoherent light with coherent phase-modulation. The parameter P represents the range of random mode-phase distribution, which is $2\pi$ for P = 1. The parameter W is a measure of the linear dispersion effect which brings about coherent phase-modulation and pulse broadening (about five times for W = 1). All traces show basically similar behaviors, but some specific features associated with different sources will be remarked. The traces for incoherent light (b) show clear relaxation decay with the same risetime (corresponding to the resolution time) as that for a transform-limited short pulse (a). This result is not affected by the presence of material dispersion (d), i.e., the time-resolution is not deteriorated. This is a very important feature in the application of incoherent light in the extremely short time region. For phase-modulated coherent pulse (c), the risetime also becomes much shorter than the pulse width, but considerably longer than those for (a) and (b), i.e., it does not ultimately reach the correlation time $\tau_c$. The phase-modulated pulse by linear dispersion exhibits a similar chirping behavior to that of self-phase-modulated pulse except that in the former the pulse profile is broadened under the fixed spectral profile, and vice versa in the latter.

The presence of cross relaxation in materials generally brings about a change in correlation traces such that the central coherence peak is enhanced relative to the component representing $T_2$ decay. It was revealed that this tendency becomes prominent in the case with incoherent light particularly for $T_1 \gg T_2$ and is appreciably dependent on the dulation of incoherent light. The result indicates that in materials with very fast cross-relaxation, the use of incoherent light for measuring $T_2$ becomes not appropriate. On the other hand, in the case with phase-modulated coherent pulse, the traces were found to be relatively insensitive to the cross relaxation effect, and to approach those with transform-limited short pulses at high cross-relaxation rate.

In summary, different types of light sources have their own merits and demerits, and will play complementary roles in ultrafast spectroscopy.

## 2. Energy Relaxation Time Measurement by Transient Four-Wave Mixing with Incoherent Light

Measurement of energy relaxation time ($T_1$) is one of the major subjects of ultrafast spectroscopy. Incoherent light has been shown to be useful for this purpose in the pump-probe method.[12] We present here an alternative method based on the degenerate four-wave mixing with three input incoherent light beams, which corresponds to the stimulated echo (S.E.) for short pulse input and has a merit of easy separation of input and output beams.[13] As in the case of S.E., the averaged power of the output beam with the wave-vector $\vec{k}_4 = \vec{k}_2 + \vec{k}_3 - \vec{k}_1$ ($\vec{k}_1$, $\vec{k}_2$, $\vec{k}_3$ being input ones) versus the delay time $\tau$ between the $\vec{k}_2$ and $\vec{k}_3$ beams (correlation trace) can give information on $T_1$. The delay time between the $\vec{k}_1$ and $\vec{k}_2$ beams is then kept within $T_2$.

A theory for an inhomogeneously broadened two-level system with a Gaussian incoherent light predicts that the correlation trace consists of three components : (i) exponential decay with a time constant $T_1$, (ii) central coherence peak, and (iii) flat background. In the stimulated echo, (ii) and (iii) components are absent, and the time constant is $T_1/2$ in (i).

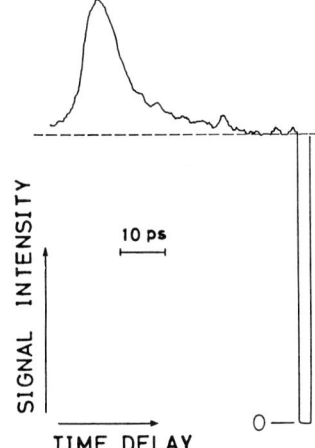

Fig. 2 Observed correlation trace of the degenerate four-wave mixing with three input incoherent light beams ($\Delta t \simeq 1$ ns, $\tau_c \simeq 6$ ps) for a dye solution (brilliant green in ethylene glycol), showing central coherence peak, exponential decay ($T_1 = 11$ ps), and flat background

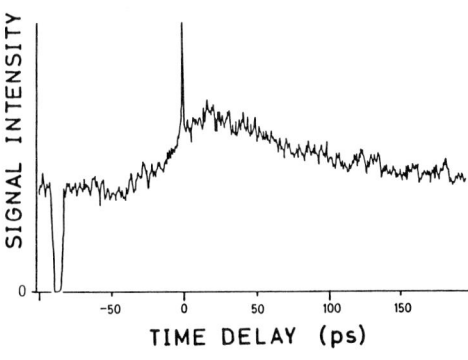

Fig. 3 Another type of observed correlation trace under similar conditions to those in Fig. 2 with different kind of incoherent light source and sample (DQOCI in ethylene glycol solution) with $T_1 = 85$ ps

An experimental demonstration is performed using a properly controlled Q-switched-YAG-laser-pumped dye laser ($\Delta t \simeq 1$ ns, $\tau_c \simeq 6$ ps) as an incoherent source and a dye solution (brilliant green in ethylene glycol) as a sample. Typical measured correlation trace as shown in Fig. 2 reproduced fairly well the theoretical predictions, and the value of $T_1 = 11$ ps deduced from this result is in good agreement with that using a 1 ps short pulse. Thus, this method is proved to be useful for the $T_1$ measurement, although the central peak and the background are generally obstructive.

We observed sometimes a different type of correlation trace as shown in Fig. 3, where the central peak is much narrower and smaller than that in Fig. 2. We could also explain theoretically this result by considering an incoherent light having two very different values of correlation times, the shorter determining the width of the central peak and the longer determining the resolution of the $T_1$ decay. Several physical models of the generation of such kind of light have also been examined. The result suggests that the reduction of the obstructive component is possible by the proper control of the coherence properties of the incident light.

## 3. Coherent Propagation Effect of Incoherent Light

When a weak ultrashort pulse is incident on a resonant medium with sufficiently long relaxation times, it approaches a $0\pi$ pulse during propagation, showing an oscillatory behavior which reflects the coherent excitation of the material. The same behavior is expected to be observed for incoherent light, if some correlation technique is utilized.

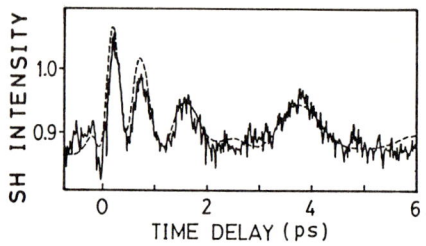

Fig. 4 Observed cross-correlation profile between the incoherent light pulses before ( $\Delta t \simeq 34$ ps, $\tau_c \simeq 0.35$ ps) and after passing through the Na vapor, showing the $0\pi$ pulse behavior. Dashed curve is the theoretical profile.[14]

This phenomenon has been demonstrated experimentally in the subpicosecond region, using the 3S - 3P transition of Na atoms and the light from an imperfectly mode-locked cw dye laser ( $\Delta t \simeq 34$ ps, $\tau_c \simeq 0.35$ ps ).[14] The expected result was obtained by measuring the cross-correlation between the propagated incoherent light and the initial light through the second harmonic generation outside the sample. A typical cross-correlation trace is shown in Fig. 4, where both the $0\pi$ pulse behavior and the additional structure arising from the splitting of the 3P state appear. The whole behavior observed is basically the same as seen in the propagation of a weak coherent pulse of 0.35 ps duration, which is equal to the correlation time of the incoherent light used here, except the appearance of a large background. The observed result is also in good agreement with that of the theoretical analysis based on a Gaussian stochastic field, except the appearance of a dip extending below the background level.

Summarizing all the results, we can say that the broadband non-transform-limited light is widely applicable to the investigation of a variety of ultrafast phenomena, as far as we recognize the specific features associated with each kind of light source and control its coherence and modulation properties.

References

1. S. Asaka, H. Nakatsuka, M. Fujiwara and M. Matsuoka: Phys. Rev. A29, 2286 (1984)
2. N. Morita, T. Yajima and Y. Ishida: In Ultrafast Phenomena IV, ed. by D.H. Auston and K.B. Eisenthal (Springer, Berlin, Heidelberg 1984) P. 239.
3. T. Yajima, N. Morita and Y. Ishida: J. Opt. Soc. Am. B1, 526 (1984)
4. R. Beach and S.R. Hartmann: Phys. Rev. Lett. 53, 663 (1984)
5. N. Morita and T. Yajima: Phys. Rev. A30, 2525 (1984)
6. H. Nakatsuka, M. Tomita, M. Fujiwara and S. Asaka: Opt. Commun. 52, 150 (1984)
7. M. Fujiwara, R. Kuroda and H. Nakatsuka: J. Opt. Soc. Am. B2, 1634 (1985)
8. R. Beach, D. DeBeer and S.R. Hartmann: Phys. Rev. A32, 3467 (1985)
9. M. Defour, J.C. Keller and J.L. LeGouët: J. Opt. Soc. Am. B3, 544 (1986)
10. J.E. Golub and T.W. Mossberg: J. Opt. Soc. Am. B3, 554 (1986)
11. T. Yajima and N. Morita: In Methods of Laser Spectroscopy, ed. by Y. Prior (Plenum, New York 1986)
12. M. Tomita and M. Matsuoka: J. Opt. Soc. Am. B3, 560 (1986)
13. N. Morita and T. Yajima, to be published
14. N. Morita, K. Torizuka and T. Yajima: J. Opt. Soc. Amer. B3, 548 (1986)

# Picosecond Dephasing Time Measurement by CSRS Using Temporally Incoherent Nanosecond Laser with Short Correlation Time

T. Kobayashi, T. Hattori, and A. Terasaki

Department of Physics, Faculty of Science, University of Tokyo, Hongo 7-3-1, Bunkyo, Tokyo 113, Japan

Dephasing dynamics of vibrational states in polyatomic molecules in liquid phase or in solution can be investigated by picosecond coherent Raman experiments [1]. In these experiments, coherent vibrations were excited and probed by picosecond laser pulses and the time resolution was considered to be limited by the pulse durations of them. Since the dephasing time of vibrations in liquid molecules are usually shorter than 10 ps at room temperature, available light sources of short and intense enough pulses are limited, and they are not handy.

However, it was suggested [2], and experimentally verified later [3], that the time resolution is determined not by the pulse durations but by the correlation time of the optical field for dephasing measurements by degenerate four-wave mixing.

We have applied this principle to time-resolved coherent Raman measurements (see Fig. 1), where two beams are originated from a single incoherent light source, with a delay time $\tau$, and the other beam is of different wavelength and assumed to be coherent. A pair of one of the incoherent light and the coherent light excite the vibration and the other incoherent light probes it. When the delay time $\tau$ is negative the roles of two incoherent lights are exchanged. The correlation time of the incoherent light field should be shorter than the dephasing time while pulse durations of all the beams may be much longer than the dephasing time.

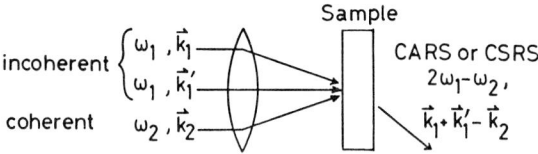

Fig. 1. Directions of the incident beams and the coherent Raman signal

The electric field is given as

$$E(\mathbf{r},t) = E_1(t)\exp[i(\mathbf{k}_1\mathbf{r}-\omega_1 t)] + E_1'(t)\exp[i(\mathbf{k}_1'\mathbf{r}-\omega_1 t)]$$
$$+ E_2(t)\exp[i(\mathbf{k}_2\mathbf{r}-\omega_2 t)] + c.c. ,$$

where $E_1$ and $E_1'$ are the envelopes of the electric field of the incoherent light, and $E_2$ is assumed to be constant. The third-order polarization with frequency $\omega_S = 2\omega_1 - \omega_2$ and wavenumber $\mathbf{k}_S = \mathbf{k}_1 + \mathbf{k}_1' - \mathbf{k}_2$ is given as

$$P^{(3)}(k_S, \omega_S) \propto \exp[i(k_S r - \omega_S t)] \int_{-\infty}^{t} dt' \exp[-(t-t')/T_2]$$
$$\times E_2^*[E(t)E(t'-\tau) + E(t')E(t-\tau)],$$

where

$$E(t) \equiv E_1(t) = E_1'(t+\tau),$$

and $T_2$ is the dephasing time between the ground state and the vibrationally excited state.

The output signal intensity is calculated as a function of the delay time as follows for the case when the correlation time is much shorter than the dephasing time:

$$I(\tau) \propto 1 + G(\tau) + (2\tau_c/T_2)\exp(-2|\tau|/T_2).$$

Here $\tau_c$ is the correlation time of the incoherent light field, and $G(\tau)$ is the autocorrelation function of the field normalized by its peak value. Since the time resolution of this method is limited only by $\tau_c$, picosecond or subpicosecond dephasing times can be measured with the use of nanosecond laser pulses.

We have performed experiments using a second harmonic (532 nm) of a Q-switched Nd:YAG laser as coherent light and a broad-band dye laser pumped by the second harmonic as an incoherent light source (see Fig. 2). Since the correlation time of the second harmonic of Nd:YAG laser is about 30 ps, it can be regarded as coherent light on the time scale of observation. The dye laser was constructed with no tuning elements and the wavelength (around 630 nm) was changed by varying the concentration of the laser dye (Rhodamine 640). The correlation time of the optical field was about 100 fs, which was obtained from the autocorrelation measurement by second harmonic generation. The durations and the energies of the Nd:YAG laser pulses were about 10 ns, 250 μJ respectively, and the dye laser 6 ns, 100 μJ, respectively, and the pulses of the two lasers were focused in the sample with a lens of focal length 30 cm. The spot size in the focal plane was approximately 50 μm. The relative delay time between two dye laser beams was varied by a translation stage which was driven by a stepping motor. The CSRS signal was detected by a photomultiplier and digitized with an A/D converter, and the data were stored in a microcomputer for further data processing.

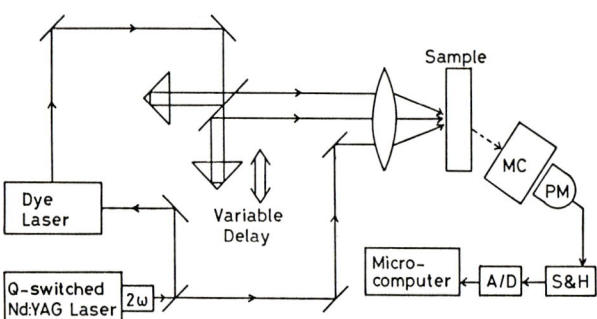

Fig. 2. Schematic diagram of the experiment. MC: monochromator, PM: photomultiplier, S&H: sample and hold circuit, A/D: A/D converter

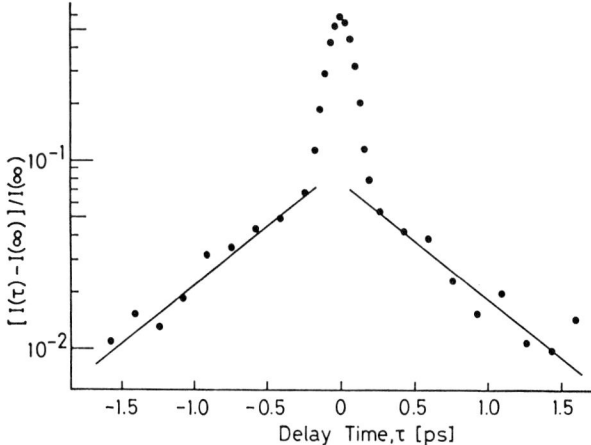

Fig. 3. CSRS signal intensities which are normalized by the background intensity $I(\infty)$ after it is subtracted

Figure 3 shows the result for a C-H stretching mode (2915 cm$^{-1}$) in dimethylsulfoxide. The obtained value of $T_2$ is in agreement with the value 1.4 ps obtained using short pulses [4], and the intensity ratio of the peak to the tail is seven, which is also in good agreement with the theoretical value.

1. A. Laubereau and W. Kaiser: Rev. Mod. Phys. 50, 607 (1978).
2. N. Morita and T. Yajima: Phys. Rev. A 30, 2525 (1984).
3. S. Asaka, H. Nakatsuka, M. Fujiwara, and M. Matsuoka: Phys. Rev. A 29, 2286 (1984).
4. S. M. George, H. Auwester, and C. B. Harris: J. Chem. Phys. 73, 5573 (1980).

# Anomalous Pulse Duration Dependence of the Quasicontinuum Absorption Spectrum

P. Mukherjee and H.S. Kwok

Department of Electrical and Computer Engineering,
State University of New York at Buffalo, Amherst, NY 14260, USA

## 1. Introduction

The study of energy transfer processes in the vibrationally dense quasicontinuum (QC) of states in a polyatomic molecule has been a field of considerable interest. Detailed understanding of the mechanisms which underlie the interaction between different vibrational modes at such high density of states has remained elusive. Considerable understanding, using the fluorescence technique, has been obtained, for example, in the work of Felker et al. [1-2].

In the present work, absorption studies in the QC provide dynamical information on the relatively featureless envelopes of the absorption spectra. It will be seen that the multi-tier classification of energy levels in the QC can physically explain the observed experimental spectra of $C_3F_7I$, both in terms of the pulse duration dependent and the collision-dependent spectra observed. Unlike fluorescence measurements, involving electronic excitation, our spectra probe absorption within the ground electronic manifold.

## 2. Experimental Technique

The experiments reported here are performed on $C_3F_7I$, which was obtained in liquid form from Morton-Thiokol Inc. The vapor pressure at room temperature was sufficient to enable experimental gas cell pressures, the highest of which was 30 Torr. The experimental system has been described in detail elsewhere [3]. As described there, the optical free induction decay (OFID), variable duration, picosecond $CO_2$ laser pulses were used for the experiments. Since $C_3F_7I$ is in its QC at room temperature, a dual-pulse pump-probe arrangement was not required. The transmission of the laser pulse was measured accurately for various pulse durations and wavelengths. The small intensity spectra were obtained by an extrapolation to zero intensity.

## 3. Experimental Results

The collisionless picosecond IR spectra obtained using the OFID picosecond pulses as well as the 80 ns TEA $CO_2$ pulses are shown in Fig. 1. There is no relative shift between the various peaks consistent with their collisionless characteristic. However, a narrowing of the spectrum for shorter pulses is clearly evident. To quantitatively show the dependence of the absorption spectrum on the laser pulse duration, the FWHM of the spectra in Fig. 1 are plotted as a function of the probing pulse duration in Fig. 2. The horizontal bar in Fig. 2 represents the 80 ns data. It can be seen that a very significant change in the spectral width has occurred.

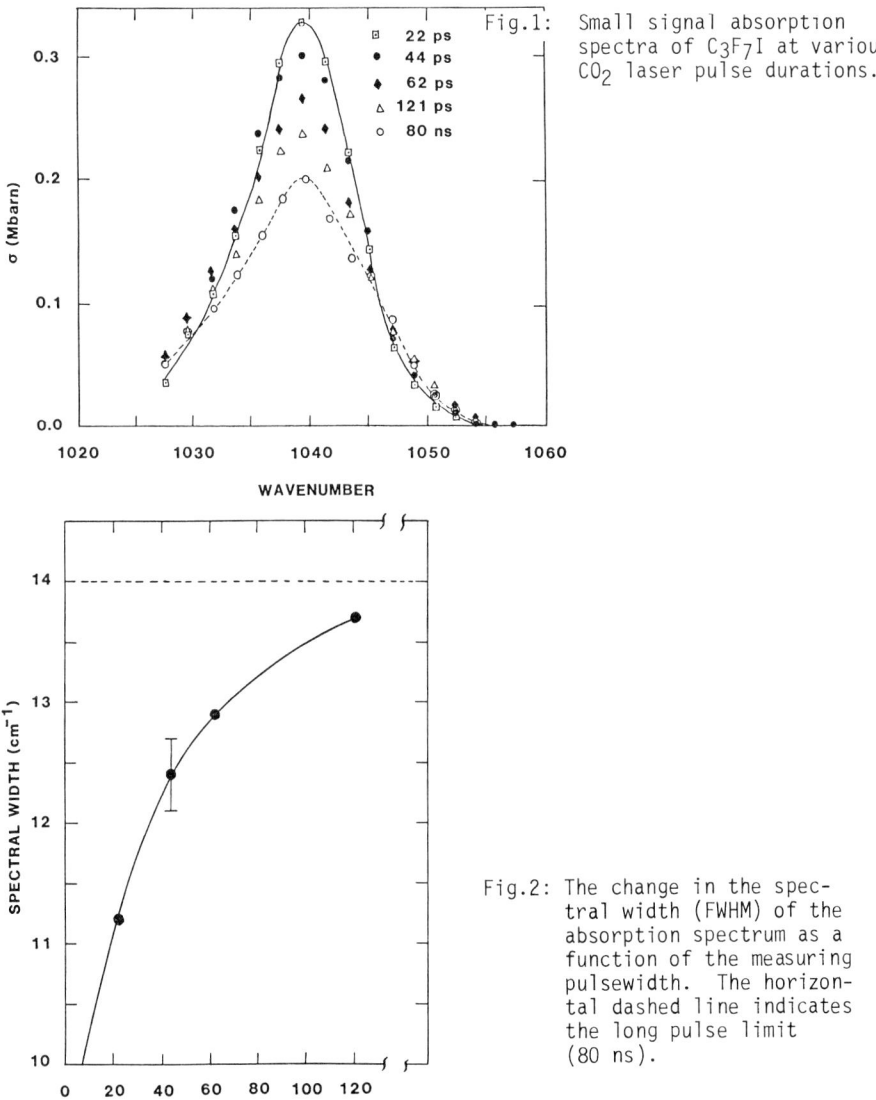

Fig.1: Small signal absorption spectra of $C_3F_7I$ at various $CO_2$ laser pulse durations.

Fig.2: The change in the spectral width (FWHM) of the absorption spectrum as a function of the measuring pulsewidth. The horizontal dashed line indicates the long pulse limit (80 ns).

## 4. Discussion

The spectral narrowing at smaller pulse durations is obviously unexplainable on the basis of the increasing linewidth of the shorter pulses, which would dictate the opposite trend. The low fluences used in the experiment (< 2.5 µJ/cm$^2$) rule out possible saturation effects on measured spectral widths. The explanation of this and the collision-dependent redshift and broadening [4] lies in the multi-tier energy level description discussed with regard to vibrationally hot $C_2F_5Cl$ [3].

545

In the multi-tier picture, the dense ensemble of eigenstates is divided
into many tiers. The first tier states interact with the photon field
directly. The second tier states are connected to the first tier states due
to anharmonicity or Coriolis coupling. The second tier is coupled to the
third and so on until the classification of the states is completed. As
seen in Fig. 3, the laser photon at $\nu_0$ excites the resonant pumped mode
$|\nu_R>$ which is coupled to subsequent tiers. With increasing time of interaction, via a longer pulse duration, more states are coupled. At an infi-
nitely long time all states sharing oscillator strength with $|\nu_R>$ (repre-
sented by $|\nu_R^{(\infty)}>$) will interact with the photon field at $\nu_0$. The observed
absorption spectrum reflects this increased dynamical coupling as a broadening of the spectrum with increasing picosecond pulse duration. In fact,
the experimentally observed increase is an ensemble average over all the
first tier states involved in the fundamental spectrum.

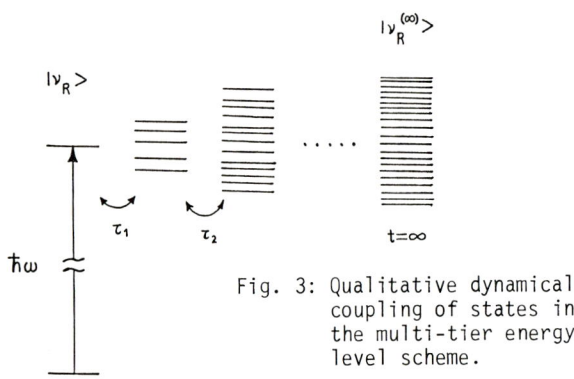

Fig. 3: Qualitative dynamical
coupling of states in
the multi-tier energy
level scheme.

The same multi-tier classification scheme explains the red-shift and
broadening of the collisional spectrum observed in a related experiment [4].
The quiescent state of interaction among the energy levels for the 80 ns
pulse under collisionless conditions changes in the presence of collisions
to allow different second tier ensembles, not coupled in the collisionless
case, to interact with each other. The forced mixing causes the associated
homogeneous vibrational ensembles to be larger as reflected in the broadening.
The predominantly anharmonic interaction explains the red-shift of the
broadening observed in reference [4].

## 5. Conclusions

Based on the experimental spectra for $C_3F_7I$ presented here, the multi-tier
energy level classification scheme emerges as a plausible physical choice
for explaining laser-molecule interaction in the dense QC of states in a
polyatomic molecule. The pulse duration dependent spectra demonstrate
that absorption spectrum measurements as a function of the coherent interaction time are potentially capable of yielding dynamical information vis
a vis intramolecular vibrational energy coupling. It is believed that the
multi-tier picture is a general one for all absorption spectra of polyatomic
molecules in the QC. In fact, recent experiments conducted by us on vibrationally hot $SF_6$ and $C_2F_5Cl$ clearly show similar trends and will be discussed elsewhere.

References

1. P.M. Felker, A.H. Zewail: J. Chem. Phys. 82, 2961 (1985)
2. P.M. Felker, W.R. Lambert, A.H. Zewail: J. Chem. Phys. 82, 3003 (1985)
3. P. Mukherjee, H.S. Kwok: J. Chem. Phys. 84, 1285 (1986)
4. P. Mukherjee, H.S. Kwok: Chem. Phys. Lett. 125, 101 (1986)

# Index of Contributors

Abeles, J.H. 127
Abul Haj, N.A. 330
Aechtner, P. 27
Alfano, R.R. 157,251,521
Amano, K. 134
Andersson, A. 86
Anfinrud, P.A. 442,447
Antonetti, A. 75,197,308, 393,419
Arakawa, Y. 231
Arjavalingam, G. 370
Arzumanyan, G. 166
Atkinson, G.H. 409
Auston, D.H. 284

Badan, J. 75
Baer, T. 51
Bain, A.J. 489
Balk, M.W. 341
Barbara, P.F. 299
Barber, J. 406
Beaud, P. 54
Becker, M. 374
Beddard, G.S. 322
Ben-Amotz, D. 350
Benoist d'Azy, O. 89
Berendzen, J.R. 433
Blanchard, D. 409
Blankenship, R.E. 374
Bloom, D.M. 98
Boggess, T.F. 5,203,207
Bohnert, K. 203,207
Bokor, J. 123
Bonkhofer, T. 17
Bouvier, M. 38
Boyer, K. 366
Bozler, C.O. 103
Brack, T.L. 409
Bréhéret, E. 89
Breton, J. 393
Brinker, K. 495
Brito Cruz, C.H. 179
Brown, J.K. 326
Brucker, G.A. 319

Brun, A. 20
Burrus, C.A. 114

Campbell, B.F. 423
Campillo, A.J. 499
Canonica, S. 341
Carruthers, T.F. 131
Casassa, M.P. 465
Castner, Jr., E.W. 303
Catherall, J.M. 24
Causgrove, T.P. 442
Cavanagh, R.R. 465
Chattopadhyay, S.K. 334
Chekalin, S.V. 402
Chen, H.H. 65
Cheng, Lap-Tak 532
Chesnoy, J. 14
Chi, C.-C. 110,120
Cho, Y. 169
Chopra, P. 518
Chronister, E.L. 482
Clemens, B.M. 257
Clifton, B.J. 103
Cooper, D.E. 117
Corcoran, T.C. 280
Corkum, P.B. 149
Cote, M.J. 433
Cotter, D. 274
Courtney, S.H. 341
Cowan, J.A. 322
Craig, B.B. 334

Danielzik, B. 182
Dawson, M.D. 5
Declémy, A. 312
Delfyett, P. 521
Deng, Z. 510
Diamond, S.K. 98
Diels, J.-C. 2,71,166
Dlott, D.D. 433,482,485
Dobler, J. 379
Dohlus, R. 458
Doland, C. 86,242
Dorsinville, R. 521

Downer, M.C. 238
Duling III, I.N. 110,120
Dykaar, D.R. 103

Eaton, W.A. 430
Eesley, G.L. 257,260
Eisenstein, G. 68,114
Eisenthal, K.B. 293
El-Sayed, M.A. 280
Elsayed-Ali, H. 264
Etchepare, J. 504

Fabricius, N. 182
Farrar, M.R. 532
Fauchet, P.M. 248
Fini, L. 14
Fischer, S.F. 454
Fleming, G.R. 303,341, 413
Fluegel, B.D. 270
Flytzanis, C. 528
Fork, R.L. 179,277
Fox, A.M. 210
Frank, H.A. 388
Frauenfelder, H. 433
Freeman, J.L. 98
French, P.M.W. 11
Friedman, J.M. 423
Fujimoto, J.G. 193,260
Fujita, T. 134

Gale, G.M. 528
Gallagher, W.J. 110
Garvey, D.W. 5
Gauduel, Y. 308
Ghoshal, S.K. 518
Gibbs, H.M. 197
Giorgi, L.B. 398
Glownia, J.H. 153,370
Goddi, A. 43
Göbel, E.O. 234,254
Goltsos, W.C. 218
Golub, J.E. 164
Gomes, L. 280

549

Gordon, J.P. 62
Gore, B.L. 398
Graener, H. 458
Grangier, P. 20
Greene, B.I. 472,478
Grillon, G. 504
Grischkowsky, D. 110,120
Gustafson, T.L. 245

Halas, N.J. 110
Halbout, J.-M. 120
Hall, K.L. 68
Hamoniaux, G. 248,504
Han, C. 489
Haner, M. 514
Harris, C.B. 326,350
Harrison, R.J. 322
Hart, D.E. 447
Hattori, T. 541
Head, D.F. 51
Heeger, A.J. 475
Hefetz, Y. 218
Heilweil, E.J. 465
Heinz, P. 27
Heinz, T.F. 370
Henry, E.R. 430
Heritage, J.P. 34
Hermes, P. 182
Herpers, U. 17
Hicks, J.M. 293
Hill, J.R. 433,482,485
Hirlimann, C. 43
Ho, P.-T. 30
Ho, P.P. 157
Hochstrasser, R.M. 344, 430,489
Hodel, W. 54
Hoff, A.J. 384
Hollis, M.A. 103
Holt, P.L. 489
Honold, A. 201
Houde, D. 419
Hsiang, T.Y. 103
Huang, Z. 92
Hulin, D. 75,197,248
Huppert, D. 315
Huston, A.L. 499
Hynes, J.T. 288

Ide, J.P. 406
Ippen, E.P. 193,213
Ishikawa, M. 8
Islam, M.N. 46

Jackson, W.B. 86,242
Jain, R.K. 107
Jamasbi, N. 2

Jang, Du-Jeon 280
Jara, H. 366
Jimbo, T. 157
Johann, U. 366
Johnson, A.M. 123,160
Jongeward, K.A. 427
Jopson, R.M. 68
Jun Cha, I. 169
Junnarkar, M.R. 251
Justus, B.L. 499

Kafka, J.D. 51
Kaiser, W. 267,379,454
Kalt, H. 207
Kash, J.A. 188
Kauffman, J.F. 433
Kelley, D.F. 319,330
Kemnitz, K. 438
Kenney-Wallace, G.A. 495
Kesler, M.P. 213
Ketchen, M.B. 110,120
Kim, S.K. 341
Klein, M.B. 203
Klein, M.V. 223
Kleinsasser, A.W. 110
Klem, J. 223
Klug, D.R. 406
Knox, W. 179
Knox, W.H. 277
Kobayashi, T. 134,231, 416,541
Koch, S.W. 270
Kohles, N. 524
Koishi, M. 169
Kolodzey, J. 248
Kosic, T.J. 485
Kottis, Ph. 312
Kühlke, D. 17
Kuhl, J. 201,234
Kuhlbrandt, W. 406
Kwok, H.S. 544

Laubereau, A. 27,458,524
Ledoux, I. 75
Lee, C.H. 30
Lee, D. 218
Lee, M. 344
Lee, W. 92
Lee, Y.C. 65
Li, G.P. 120
Li, Q.X. 157
Lin, P.S.D. 127
Lin, W.Z. 193,260
Linde, D., von der 17, 182
Ling, J.D. 30
Liu, S.N. 338

Logan, R.A. 193
Longworth, J.W. 413
Loo, R.Y. 207
Loring, R.F. 510
Luk, T.S. 366
Luty, F. 280
Lyon, S.A. 227

Magde, D. 427
Maine, P. 38
Manassah, J.T. 157
Manning, R.J. 210
Marcus, R.B. 127
Maroncelli, M. 303
Marsh, J.H. 210
Marsters, J.C. 427
Martin, J.L. 308,393,419
Martin, M.M. 89
Masselink, W.T. 197
Mataga, N. 449
Matveets, Yu.A. 402
May, P.G. 120
McCarthy, P.J. 489
McDonald, J.D. 433
McGuire, M. 353
McIntyre, I.A. 366
McLendon, G. 353
McPherson, A. 366
Meech, S.R. 384
Menyuk, C.R. 65
Meyer, Y.H. 89
Mialocq, J.C. 334,362
Middendorf, D. 374
Migus, A. 75,197,308,393
Millard, R.R. 472,478
Miller, A. 210
Misewich, J. 153
Mitschke, F.M. 58,62
Miwa, M. 169
Mollenauer, L.F. 46,58,62, 277
Morhange, J.-F. 43
Morimoto, A. 134
Morita, N. 536
Morkoç, H. 197,223
Moses, D. 475
Moss, S.C. 117
Mossberg, T.W. 164
Mounet, R. 43
Mourou, G. 38,264
Mourou, G.A. 103
Mukamel, S. 510
Mukherjee, A. 166
Mukherjee, N. 166
Mukherjee, P. 544
Murphy, R.A. 103
Myers, A.B. 489
Mysyrowicz, A. 197

Nagarajan, V. 299
Nakashima, N. 438
Narayana Rao, D. 518
Nelson, K.A. 532
New, G.H.C. 24
Nichols, K.B. 103
Ning, C. 92
Nordlund, T.M. 353
Norris, T. 264
Norwood, D.P. 207
Nurmikko, A.V. 218
Nuss, M.C. 267,284

Oberli, D.Y. 223
Ohtani, H. 416
Olbright, G.R. 270
Orenstein, J. 472
Orszag, A. 248,393,504

Paddock, C.A. 257
Paige, M.E. 326
Paone, S. 495
Parson, W.W. 374
Penzkofer, A. 469
Pereira, M.A. 489
Pessot, M. 264
Petrich, J.W. 413,419
Peyghambarian, N. 197, 270
Pier, T.J. 107
Pines, E. 315
Piskarskas, A. 142
Podenas, D. 142
Polland, H.-J. 234
Porter, G. 398,406
Postlewaite, J.C. 482
Poyart, C. 419
Prasad, P.N. 518

Raybon, G. 68
Ressl, M.G. 107
Rhodes, C.K. 366
Roberts, D.M. 245
Rodwell, M.J.W. 98
Roger, G. 20
Rojas, O.L 419
Rolland, C. 149
Rosker, M.J. 461
Rothenberg, J.E. 78
Rudolph, W. 71
Rullière, C. 312
Rumbles, G. 409
Russell, D.J. 326

Saito, H. 254
Sakaki, H. 231

Sanders, J.K.M. 322
Sarger, L. 2
Sato, T. 8
Scheuermann, M. 120
Schoenlein, R.W. 260
Schultheis, L. 201,234
Schwarzenbach, A.P. 366
Scott, T.W. 338
Seddiki, O. 43
Seilmeier, A. 454
Shank, C.V. 179,238
Siemankowski, L. 409
Simpson, W.M. 123,160
Sinclair, M. 475
Sitzmann, E.V. 293
Smirl, A.L. 5,203,207
Smith, D.E. 326
Smyth, M. 120
Sobolewski, R. 103
Sommer, J.H. 353
Sorokin, P.P. 153
Spaepen, F. 174
Spano, F. 514
Sperber, P. 469
Srinivasan-Rao, T. 149
Stabinis, A. 142
Steinbach, P.J. 433
Stephenson, J.C. 362,465
Stix, M.S. 213
Stolen, R.H. 46,160
Storz, R.H. 123
Strickland, D. 38
Struve, W.S. 442,447
Sueta, T. 134
Sukowski, U. 454
Swiatkiewicz, J. 518

Tamai, N. 444,449
Tang, C.L. 461
Taylor, A.J. 114,461
Taylor, J.R. 11
Templeton, E.F. 495
Terasaki, A. 541
Thomazeau, I. 504
Thurston, R.N. 34
Tiede, D.M. 388
Torizuka, K. 8
Tsai, C.C. 242
Tsang, J.C. 188
Tsuda, M. 416
Tu, C.W. 201,234
Tucker, R.S. 114

Uchiki, H. 231
Uemura, T. 134

Umbrasas, A. 142
Umeda, T. 169

Valdmanis, J.A. 82
Vallée, F. 528
Valley, G.C. 203
Vandersall, M.T. 293
Varanavichius, A. 142

Wagner, S. 248
Wai, P.K.A. 65
Wake, D.R. 223
Waldeck, D.H. 347
Wang, W. 92
Warren, W.S. 514
Wasielewski, M.R. 388
Webb, S.P. 303
Weber, H.P. 54
Weidner, M. 469
Weiner, A.M. 34,127
Weingarten, K.J. 98
Weller, J.F. 131
Whalen, M.S. 68
Whitaker, J.F. 103
Wiersma, D.A. 384
Wiesenfeld, J.M. 114
Williams, L.R. 472,532
Williamson, S. 38
Wise, F.W. 461
Woodbury, N.W. 374

Yajima, T. 536
Yamada, N. 308
Yamashita, M. 8
Yamazaki, I. 444,449
Yamazaki, T. 444,449
Yan, L. 30
Yan, Yong-Xin 532
Yang, C.H. 227
Yankauskas, A. 142
Yartsev, A.P. 402
Yonushauskas, G. 142
Yoshihara, K. 438

Zeglinski, D.M. 347
Zewail, A.H. 356
Zhang, Huanwen 137
Zhang, X.-C. 107
Zinth, W. 267,379
Zyss, J. 75
Zysset, B. 54

# Ultrafast Phenomena

## Ultrafast Phenomena IV
Proceedings of the Fourth International Conference Monterey, California, June 11-15, 1984

**Editors: D. H. Auston, K. B. Eisenthal**

1984. 370 figures. XVI, 509 pages. (Springer Series in Chemical Physics, Volume 38). ISBN 3-540-13834-X

**Contents:** Part I: Generation and Measurement Techniques. – Part II: Solid State Physics and Nonlinear Optics. – Part III: Coherent Pulse Propagation. – Part IV: Stimulated Scattering. – Part V: Transient Laser Photochemistry. – Part VI: Molecular Energy Redistribution, Transfer, and Relaxation. – Part VII: Electronics and Opto-Electronics. – Part VIII: Photochemistry and Photophysics of Proteins, Chlorophyll, Visual Pigments, and Other Biological Systems. – Index of Contributors.

## Picosecond Phenomena III
Proceedings of the Third International Conference on Picosecond Phenomena, Garmisch-Partenkirchen, Federal Republic of Germany, June 16-18, 1982

**Editors: K.B. Eisenthal, R. M. Hochstrasser, W. Kaiser, A. Laubereau**

1982. 288 figures. XIII, 401 pages. (Springer Series in Chemical Physics, Volume 23). ISBN 3-540-11912-4

**Contents:** Advances in the Generation of Ultrashort Light Pulses. – Ultrashort Measuring Techniques. – Advances in Optoelectronics. – Relaxation Phenomena in Molecular Physics. – Picosecond Chemical Processes. – Ultrashort Processes in Biology. – Applications in Solid-State Physics. – Index of Contributors.

## Picosecond Phenomena II
Proceedings of the Second International Conference on Picosecond Phenomena, Cape Cod, Massachusetts, USA, June 18-20, 1980

**Editors: R. Hochstrasser, W. Kaiser, C. V. Shank**

1980. 252 figures, 17 tables. XII, 382 pages. (Springer Series in Chemical Physics, Volume 14). ISBN 3-540-10403-8

**Contents:** Advances in the Generation of Picosecond Pulses. – Advances in Optoelectronics. – Picosecond Studies of Molecular Motion. – Picosecond Relaxation Phenomena. – Picosecond Chemical Processes. – Applications in Solid State Physics. – Ultrashort Processes/Biology. – Spectroscopic Techniques. – Index of Contributors.

## Picosecond Phenomena I
Proceedings of the First International Conference on Picosecond Phenomena, Hilton Head, South Carolina, USA, May 24-26, 1978

**Editors: C. V. Shank, E. P. Ippen, S. L. Shapiro**

1978. *Out of print*

Springer-Verlag
Berlin Heidelberg New York
London Paris Tokyo

RAYMOND H. FOGLER LIBRARY
**DATE DUE**

BOOKS ARE SUBJECT TO
RECALL